STRUCTURE AND FUNCTION OF A CHIHUAHUAN DESERT ECOSYSTEM

LONG-TERM ECOLOGICAL RESEARCH NETWORK SERIES
LTER Publications Committee

Grassland Dynamics: Long-Term Ecological Research in Tallgrass Prairie
Edited by
Alan K. Knapp, John M. Briggs, David C. Hartnett, and Scott L. Collins

Standard Soil Methods for Long-Term Ecological Research
Edited by
G. Philip Robertson, David C. Coleman, Caroline S. Bledsoe, and Phillip Sollins

Structure and Function of an Alpine Ecosystem:
Niwot Ridge, Colorado
Edited by
William D. Bowman and Timothy R. Seastedt

Climate Variability and Ecosystem Response at Long-Term Ecological Sites
Edited by
David Greenland, Douglas G. Goodin, and Raymond C. Smith

Biodiversity in Drylands: Toward a Unified Framework
Edited by
Moshe Shachak, James R. Gosz, Steward T.A. Pickett, and Avi Perevolotsky

Long-Term Dynamics of Lakes in the Landscape:
Long-Term Ecological Research on North Temperate Lakes
Edited by
John J. Magnuson, Timothy K. Kratz, and Barbara J. Benson

Alaska's Changing Boreal Forest
Edited by
F. Stuart Chapin III, Mark W. Oswood, Keith Van Cleve, Leslie A. Viereck and David L. Verbyla

Structure and Function of a Chihuahuan Desert Ecosystem: The Jornada Basin Long-Term Ecological Research Site
Edited by
Kris M. Havstad, Laura F. Huenneke, and William H. Schlesinger

STRUCTURE AND FUNCTION OF A CHIHUAHUAN DESERT ECOSYSTEM: THE JORNADA BASIN LONG-TERM ECOLOGICAL RESEARCH SITE

Edited by

Kris M. Havstad

Laura F. Huenneke

William H. Schlesinger

LTER

OXFORD
UNIVERSITY PRESS

2006

OXFORD
UNIVERSITY PRESS

Oxford University Press, Inc., publishes works that further
Oxford University's objective of excellence
in research, scholarship, and education.

Oxford New York
Auckland Cape Town Dar es Salaam Hong Kong Karachi
Kuala Lumpur Madrid Melbourne Mexico City Nairobi
New Delhi Shanghai Taipei Toronto

With offices in
Argentina Austria Brazil Chile Czech Republic France Greece
Guatemala Hungary Italy Japan Poland Portugal Singapore
South Korea Switzerland Thailand Turkey Ukraine Vietnam

Published by Oxford University Press, Inc.
198 Madison Avenue, New York, New York 10016

www.oup.com

Library of Congress Cataloging-in-Publication Data
Structure and function of a Chihuahuan Desert ecosystem: the Jornada
Basin long-term ecological research site / edited by Kris M. Havstad,
Laura F. Huenneke, William H. Schlesinger.
p. cm.
Includes bibliographical references and index.
ISBN-13 978-0-19-511776-9
ISBN 0-19-511776-X
1. Desert ecology—Chihuahuan Desert. 2. Jornada Experimental Range.
I. Havstad, Kris M. II. Huenneke, L. F. III. Schlesinger, William H.
QH107.S77 2006
557.54'0972'16—dc22 2005020889

9 8 7 6 5 4 3 2 1

Printed in the United States of America
on acid-free paper

We dedicate this volume to the hundreds of students and staff since 1912 who have worked so diligently and professionally in pursuit of science in and about this desert environment. Their efforts were and are the foundation of this synthesis and any clarity of thinking that results from reading the enclosed chapters.

Preface

The Jornada Basin in southern New Mexico is truly a long-term research site, and this history is reflected in the databases described in many of the chapters in this book. Research in this basin formally began in 1912 with the creation by Presidential Order within the U.S. Department of Agriculture of the 77,000-ha Jornada Range Reserve. Just months after New Mexico was granted statehood, these lands were withdrawn from the public domain due to the efforts of Elmer Otis Wooton, a professor at the New Mexico College of Agriculture and Mechanical Arts, and Charles T. Turney, a prominent rancher and farmer in the Las Cruces area. Wooton's published assessments of rangeland conditions throughout the Southwest at the turn of the twentieth century painted a bleak picture of resource deterioration due to a combination of drought and mismanagement. As a consequence of this assessment, he proposed creation of the Jornada Range Reserve (later named the Jornada Experimental Range) to determine proper management practices through research, then demonstrate those practices to stock producers in the region. By 1915, under the direction of scientists with the U.S. Forest Service, long-term research objectives were established and implemented with C. T. Turney as the cooperating rancher who stocked the reserve. Early publications, primarily USDA bulletins or journal articles in *Ecology* and the *Botanical Gazette*, appeared in print beginning in 1917. Researchers such as C. L. Forsling, J. T. Jardine, R. S. Campbell, and R. H. Canfield authored many of these early works, which are both classics in rangeland management and part of the foundation of our LTER studies decades later. The creation in 1927 of the Chihuahuan Desert Rangeland Research Center (CDRRC; initially known as the College Ranch) under the jurisdiction of New Mexico State University on 24,000 ha adjacent to the Jornada Range established the 101,000-ha area for long-term research in the basin that operates today.

The research during the early twentieth century had numerous themes. Much of the emphasis was on management related to agricultural production. Many of the principles that have been developed about livestock grazing management and rangeland improvement practices in the Southwestern United States and for arid lands in general can be traced to studies in the Jornada Basin. Yet during these earlier decades there were also classic papers published on vegetation dynamics, resource redistribution, and small-mammal ecology. Over 40 years ago the first interdisciplinary studies were initiated; these studies served as a forerunner for the collaborative studies of the International Biological Programme (IBP) of the 1970s. They also prompted the involvement of USDA scientists with the Agricultural Research Service which assumed control of the Jornada Experimental Range in 1954. Intensive studies of grassland and shrubland sites within the Jornada Basin formed important elements in the comparative IBP investigations of ecosystem productivity and processes. Those IBP projects set the stage for the next phase of the research program.

The establishment of a National Science Foundation Long-Term Ecological Research (LTER) site in the Jornada Basin of south-central New Mexico in 1981, as part of the original cohort of LTER sites, followed this long record of scientific investigation of ecosystem processes and management. The LTER program catalyzed and coalesced efforts of scientists across a multitude of disciplines and affiliations, and facilitated the understanding of Chihuahuan desert ecosystems synthesized in this volume. Much of that research effort, evident in the chapters of this book, has been creating an abiotic-based understanding of this ecosystem given the overwhelming importance of these bottom-up processes in the Jornada Basin. That understanding has now matured into a more complex set of studies on biotic and abiotic interactions across multiple spatial and temporal scales. The chapters of this book reflect this long history of research, report on many of the key intensive studies that have characterized this program for decades, but also include the more integrative, across-scale investigations that have developed in recent years.

The vision for our research program in the twenty-first century is easily described—innovative, carefully executed, and insightful research relevant to the public in the problems being addressed, effective in providing solutions to those problems, collaborative across disciplines, global in its application to arid lands, and ethical in its conduct and its explanation. Although socioeconomic studies have not been the explicit focus of our programs, increasingly we see our studies expanding to include comparisons of human impact in the cross-border region of the United States and Mexico. Within this region, it is increasingly important that we develop the knowledge and technologies for mitigating the impacts to desert environments and for improving environmental conditions for the increasing human population that lives here. An understanding of basic processes of desert ecosystems must form the basis of management and remediation techniques. Predicting how arid land ecosystems here and elsewhere around the globe will respond to human impacts, and ameliorating these impacts are critical goals for our future research.

Our progress is built on the efforts of numerous scientists, students, technical staff, and collaborators that have been involved over the past 100 years. Several deserve mention and our thanks because of their length of service working in this desert and the importance of their contributions to our understanding of this system today. Robert Gibbens, Fred Ares, Carlton Herbel, Rex Pieper, Reldon Beck, Gary Donart, Gary Cunningham, John Ludwig, John Tromble, and Walt Whitford are a few of these names. We are fortunate that some of these people are still working with us in retirement today and have made significant contributions to this synthesis.

The cover photograph presents several important elements of this desert environment intended to convey a few useful perspectives. In the foreground are plant species representing different classes of vegetation that are characteristic of the northern Chihuahuan Desert. Of significance is the soap-tree yucca (*Yucca elata*) on the right side of the front cover. This endemic species is common within desert grasslands and areas that formerly were grasslands, and it serves as an indicator of this area's highly dynamic nature. The windmill in the background reminds us of the importance of historical legacies, in this case livestock use, in shaping the environment we see today. The backdrop of mountains illustrates that the Jornada Basin is within the Mexican Highlands region of the basin and Range Physiographic Province.

The preparation of this volume could not have been completed without the assistance of many people. We would like to express our thanks to Peter Prescott of Oxford University Press for guiding the project to publication; Eddie Garcia and the many staff at the USDA-ARS Jornada Experimental Range; Rob McNeely, Calvin Bailey, and the staff at the NMSU CDRRC; Ken Ramsey, Jim Lenz, and Valerie LaPlante for networking and data support; Brenda Irish, Bernice Gamboa, Durga Pola, Monica Silva, and Tanya Beeman of the USDA-ARS Jornada Experimental Range for their assistance with all phases of the preparation of this text; Barbara Nolen for her excellent work in preparing numerous figures and graphics; Dara Parker and her associates for many hours in the field collecting data that are central to this book; and, in particular, John Anderson for his years of dedicated and professional work as site manager for the Jornada Basin LTER. John has been the only site manager in the history of this LTER location, and we have all benefited greatly from his contributions, presence, and good humor. We acknowledge with gratitude the sustained funding from the National Science Foundation and the USDA that allowed the progressive development of the understanding summarized here. We are quite grateful to the many colleagues that provided reviews of chapters, including Leland Gile, Mitch McClaran, Sue Ann Monger, and Arlene Tugel. Finally, we wish to sincerely thank Scott Collins for his careful, thorough, and professional review of the initially assembled volume.

The following two Web addresses provide the full spectrum of information on our research activities, including archived and active data sets, current projects, research protocols, access authorization request forms, program descriptions, and an interactive bibliography of the history of publications from the Jornada Basin: http://usda-ars.nmsu.edu and http://jornada-www.nmsu.edu.

Contents

Contributors

Athol D. Abrahams
Department of Geography
University at Buffalo, State University of New York
Buffalo, NY 14261

Brandon T. Bestelmeyer
USDA-ARS Jornada Experimental Range
P.O. Box 30003, MSC 3JER
New Mexico State University
Las Cruces, NM 88003-8003

Joel R. Brown
USDA-ARS Jornada Experimental Range
P.O. Box 30003, MSC 3JER
New Mexico State University
Las Cruces, NM 88003-8003

Mark J. Chopping
Earth and Environmental Studies
1 Normal Avenue
Montclair State University
Upper Montclair, NJ 07043

Roberto Fernández
Department of Botany
Duke University
Durham, NC 27708

Ed Fredrickson
USDA-ARS Jornada Experimental Range
P.O. Box 30003, MSC 3JER
New Mexico State University
Las Cruces, NM 88003-8003

Qiong Gao
Institute of Resources Science
Beijing Normal University
Beijing 100875, P.R. China

Robert P. Gibbens
USDA-ARS Jornada Experimental Range
P.O. Box 30003, MSC 3JER
New Mexico State University
Las Cruces, NM 88003-8003

Leland H. Gile
USDA-NRCS (retired)
2600 Desert Drive
Las Cruces, NM 88001

Dale Gillette
USEPA, NOAA, Air Resources Laboratory
Mail Drop #81
RTP, NC 27711

Vincent P. Gutschick
Department of Biology
New Mexico State University
P.O. Box 30003, Department 3Q
Las Cruces, NM 88003

Kris M. Havstad
USDA-ARS Jornada Experimental Range
P.O. Box 30003, MSC 3JER
New Mexico State University
Las Cruces, NM 88003-8003

Jeffrey E. Herrick
USDA-ARS Jornada Experimental Range
P.O. Box 30003, MSC 3JER
New Mexico State University
Las Cruces, NM 88003-8003

David A. Howes
Compliance Services International
1112 Alexander Avenue
Tacoma, WA 98421

Laura F. Huenneke
Deans Office, College of Arts and Sciences
Northern Arizona University
P.O. Box 5621
Flagstaff, AZ 86011-5621

Paul R. Kemp
Department of Biology
University of San Diego
5998 Alcala Park
San Diego, CA 92110

William Kustas
USDA-ARS Hydrology and Remote Sensing Laboratory
BARC-West Bldg. 007, Rm. 104
Beltsville, MD 20705-2350

Greg H. Mack
Department of Geological Sciences
New Mexico State University
Box 30003, Department 3Q
Las Cruces, NM 88003

Katherine A. Mitchell
Southwestern Indian Polytechnic Institute
P.O. Box 10146
Albuquerque, NM 87196

H. Curtis Monger
Department of Agronomy and Horticulture
New Mexico State University
Box 30003, Department 3Q
Las Cruces, NM 88003

Melissa Neave
Department of Geography
University of Sydney
New South Wales 2006, Australia

Barbara A. Nolen
USDA-ARS, Jornada Experimental Range
P.O. Box 30003, MSC 3JER
New Mexico State University
Las Cruces, NM 88003-8003

Kiona Ogle
Department of Ecology and Evolutionary Biology
Princeton Environmental Institute
Guyot Hall, Room 102
Princeton, NJ 08544

Anthony J. Parsons
Department of Geography
University of Leicester
Leicester LE1 7RH, UK

Debra P. C. Peters
USDA-ARS, Jornada Experimental Range
P.O. Box 30003, MSC 3JER
New Mexico State University
Las Cruces, NM 88003-8003

Albert Rango
USDA-ARS, Jornada Experimental Range
P.O. Box 30003, MSC 3JER
New Mexico State University
Las Cruces, NM 88003-8003

James F. Reynolds
Department of Biology
Nicholas School of the Environment and Earth Sciences, Phytotron Bldg.

Duke University
Durham, NC 27708-0340

Jerry Ritchie
BARC-West Building 007
10300 Baltimore Avenue
Beltsville, MD 20705 USA

William H. Schlesinger
Nicholas School of the Environment and Earth Sciences
Duke University
Durham, NC 27708

Sebastian M. Schmidt
Nicholas School of the Environment and Earth Sciences
Duke University
Durham, NC 27708

Tom Schmugge
Gerald Thomas Chair for Sustainable Agriculture
College of Agriculture and Home Economics
New Mexico State University
Las Cruces, NM 88003

Keirith A. Snyder
USDA-ARS
920 Valley Road
Reno, NV 89512

Sandy L. Tartowski
Las Cruces, NM 88003-8003

John Wainright
Department of Geography
Sheffield Centre for International Drylands Research
University of Sheffield
Sheffield S10 2TN, UK

Walter G. Whitford
USDA-ARS Jornada Experimental Range
P.O. Box 30003, MSC 3JER
New Mexico State University
Las Cruces, NM 88003-8003

Jianguo Wu
Department of Life Sciences (2352)
Arizona State University West
4701 Thunderbird Road
Glendale, AZ 85306

STRUCTURE AND FUNCTION OF A CHIHUAHUAN DESERT ECOSYSTEM

1

Introduction

Kris M. Havstad
William H. Schlesinger

Arid lands throughout the world are often degraded or increasingly at risk of degradation. These lands, including those at the border of arid regions, commonly exhibit accelerated soil erosion, losses of productivity, and impaired economic potential to support human populations. Human history is replete with the collapse of great civilizations of the hot and dry subtropics that suffered severe soil resource depletions in their midst or at their margins. Given that over 1 billion people currently inhabit the arid lands of the world, it is critical that we have the knowledge and resulting technologies to mitigate our impacts and improve environmental conditions of these lands and their resources. This book describes our understanding of basic processes of arid ecosystems resulting from nearly a century of research in one desert locale, the Jornada Basin of southern New Mexico. Much of our understanding comes from both extensive and intensive studies in a landscape that has drastically changed over that time.

The loss of ecological, economic, and social capital is called "desertification" (Dregne et al. 1991). The 1992 United Nations Desertification Convention defined *desertification* as "land degradation in arid, semiarid and dry subhumid areas resulting from various factors, including climatic variations and human activities." In the future, we can expect that the shifting border between arid and semiarid lands will be one of the most sensitive indicators of global change.

Desertification involves human and environmental drivers of change but is a regional symptom that emerges from degradation at finer spatial scales (Reynolds and Stafford Smith 2002). Desertification does not describe cyclic phenomena, as when decadal variations of precipitation lead to periods of drought and to losses of vegetation cover that are fully restored when rains return (Tucker et al. 1994). An updated and revised desertification paradigm has been developed by Reynolds et al. (2003; table 1-1). An important feature of this conceptual model is that both biophysical and socioeconomic factors are jointly involved in desertification. This paradigm clearly recognizes critical points, or thresholds, in system dynamics, yet these points may be manageable for increasing system resilience. Central to the model is the recognition that managing degradation is possible if drivers of change for a particular region are properly understood and actions are geared to those drivers.

This book is about our understanding of a particular arid ecosystem and its recent changes. We draw on a long history of research by U.S. Department of Agriculture (USDA) scientists working in the Jornada Basin since 1912. In 1981, a group of scientists based in Las Cruces, New Mexico, and associated with New

Table 1-1. The nine assertions of the Dahlem Desertification Paradigm and some of their implications (Reynolds et al. 2003). These assertions are not all-encompassing but provide the framework for a new paradigm.

Assertion 1. Desertification always involves human and environmental drivers	Always expect to include both socioeconomic and biophysical variables in any monitoring or intervention scheme
Assertion 2. "Slow" variables are critical determinants of system dynamics	Identify and manage forth small set of "slow" variables that drive the "fast" ecological goods and services that matter at any given scale
Assertion 3. Thresholds are crucial and may change over time	Identify thresholds in the change variables at which there are significant increases in the costs of recovery and quantify these costs, seeking ways to manage the thresholds to increase resilience
Assertion 4. The costs of intervention rise nonlinearly with increasing degradation	Intervene early where possible, and invest to reduce the transaction costs of increasing scales of intervention
Assertion 5. Desertification is a regionally emergent property of local degradation	Take care to define precisely the spatial and temporal extent of and processes resulting in any given measure of local degradation. But don't try to probe desertification beyond a measure of generalized impact at higher scales
Assertion 6. Coupled human–environment system change over time	Understand and manage the circumstances in which the human and environmental subsystems become decoupled
Assertion 7. The development of appropriate local environmental knowledge (LEK) must be accelerated	Create better partnerships between LEK and conventional scientific research, employing good experimental design, effective adaptive feedback and monitoring
Assertion 8. Systems are hierarchically nested (manage the hierarchy!)	Recognize and manage the fact that changes at one level affect others; create flexible but linked institutions across the hierarchical levels, and ensure processes are managed through scale-matched institutions
Assertion 9. Limited suite of processes and variables at any scale makes the problem tractable	Analyze the types of syndromes at different scales, and seek the investment levers that will best control their effects—Awareness and regulation where the drivers are natural, changed policy and institutions where the drivers are social

Mexico State University proposed an expanded program of long-term ecological research in the Jornada Basin to gain a better understanding of processes that determine the structure and function of arid land ecosystems. Given its existing history of research, the Jornada Basin was a natural candidate as an initial site for the Long-Term Ecological Research (LTER) program being organized by the National Science Foundation. Today, it is one site in a 26-site network located throughout the United States and elsewhere (figure 1-1).

The Jornada Basin group hoped that this expansion would translate into more ecologically-based knowledge, strategies, and practices for effectively managing these increasingly human-affected environments and restoring degraded areas. To an extent, this book represents both a synthesis of that effort and a benchmark of our progress since the early twentieth century. The history behind this long-term presence of scientists in this basin is an important part of our synthesis.

Early Ranching in the Jornada Basin

Though livestock were introduced from Mexico into southern New Mexico during the early part of the sixteenth century (Hastings and Turner 1965; see also chapter 13), for over 250 years grazing was limited to the Rio Grande Valley and adjacent slopes because of lack of surface water in the surrounding basins, including the Jornada Basin (see figure 2-1 in chapter 2 for general features of the Jornada Basin). Some water could be found in springs and seeps in the mountains away from the Rio Grande Valley, but supplies were ephemeral and livestock use was sporadic.

The Jornada Plain began to be settled following passage of the Homestead Act of 1862 and the end of the American Civil War in 1865. The first well on the plain was dug in 1867 at the Aleman ranch along the southern portion of the Santa Fe Trail north of the Doña Ana Mountains. Yet it was not until 1888 that the Detroit and Rio Grande Livestock Company pumped water from the river to a tank on the mesa and piped it to troughs 10 km inside the Jornada Basin so that cattle *(Bos taurus)* could graze these upland grasslands. Originally owned by former U.S. Army cavalry officers from Michigan, the Detroit Company began to assemble grazing rights across the Jornada Plain during this period, and grazing use increased. In the 1880s, it was estimated that the Bar Cross brand (and the 20 or so other purchased brands) of the Detroit Company could be found on 20,000 cattle, including 1,000 bulls on the Jornada Plain. The number of other stock, especially horses *(Equus caballus)*, is unknown but assumed to have been substantial.

This level of stocking (approximately 50 head of cattle per section of land) may have resulted in an annual forage consumption of 30–60 g/m^2 (see chapter 13). With the tremendous spatial and temporal variability of aboveground net primary productivity (see chapter 11) in this environment, this amount of forage consumption during drought years would have greatly exceeded any reasonable concept of carrying capacity. In addition, this amount of forage consumption would have been spatially heterogeneous, and even in productive years some areas

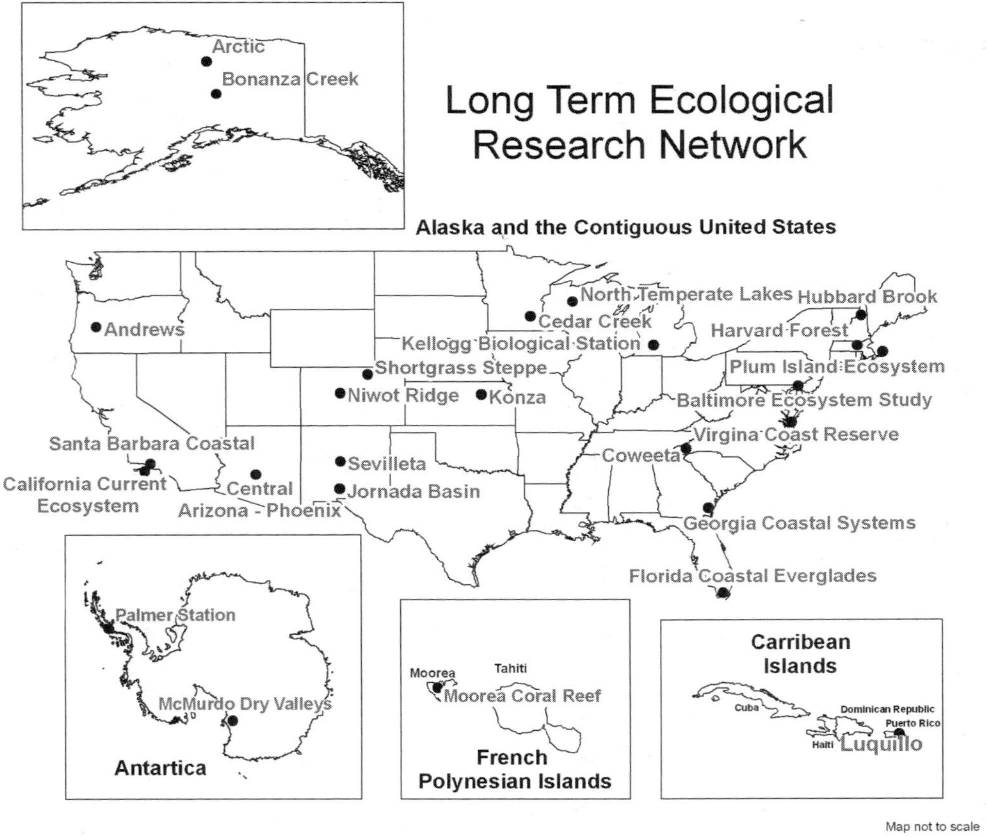

Figure 1-1. The network of the National Science Foundation's LTER sites in North America, and the two additional sites in Antarctica.

would have been heavily grazed. The impacts of this level of use over several years would have been substantial and long lasting (see chapters 13 and 17).

Grazing was more limited on the intermittently distributed homesteads located at springs on the far eastern side of the Jornada Basin. At Goldenburg Springs on the east side of the basin, the three Goldenburg brothers reportedly watered 1,800 cattle in the very early 1900s. Henry Summerford was reported to water 1,400 cattle at a well at the base of the northern side of the Doña Ana Mountains (Mount Summerford) in 1905 (Lohmiller 1963). Lack of developed watering systems in the central regions of the plain limited livestock distribution in the area. The first wells at the current site of the Jornada Experimental Range were drilled in 1903 by Harvey Ringer. Ringer had begun to purchase portions of the Bar Cross ranch from the Detroit Company as it was dissolved, following the severe drought of the 1890–93 period.

One of the key people involved in expanding cattle grazing onto the Jornada Basin was Charles Travis Turney. Born in 1857 in Sutton County, Texas, and on his own from the age of eight, Turney spent his life working as a cowboy. Like some of his fellow Texans, he had decided that land had become too expensive in Texas at the end of the nineteenth century and viewed the New Mexico territory as a land of opportunity. In 1902, he began moving a herd of cattle from Texas to southern New Mexico. He was delayed in the Pecos area for a year due to animal quarantine restrictions but eventually arrived in Doña Ana County in 1904. In January, Turney purchased three 16-ha water lots in the Jornada Basin and $4,000 worth of cattle from Harvey Ringer. During the following eight years, Turney purchased additional wells from other homesteaders, including Joe Taylor and Hugo Seaburg, and shipped or trailed to Jornada an additional 3,000 cattle. By 1912, Turney held deed to 120 ha and nine wells that provided grazing rights to over 80,000 ha on the Jornada Plain for his 4,000–5,000 head of cattle. In fact, by 1912, he had already constructed fence around a portion of his ranch.

Concurrent with expansion of livestock grazing in the late 1800s throughout the Southwest region was a noticeable decline in rangeland conditions (Smith 1899). In 1908, E. O. Wooton documented deteriorated rangeland conditions in New Mexico. Wooton, a professor at New Mexico Agriculture and Mechanical Arts College in Las Cruces, spent years evaluating and characterizing rangeland conditions throughout the state (Allred 1990). In a thorough report on the subject, Wooton (1908) included the results of a survey questioning Southwestern cattlemen on the condition of regional rangelands. Of the 118 responses, 16 stockmen felt that rangeland conditions had improved in recent years, and 102 responded there had been significant declines in grazing capacities. Of these 102 responses, 69 attributed the declines to overgrazing and 33 blamed drought conditions. Earlier observations with similar results led to the first organized research efforts in the Texas Panhandle seeking to develop range management principles and improvement practices (Smith 1899). Wooton initiated his own experiments in cooperation with ranchers such as Turney by 1904, but he was frustrated by the lack of a suitably large area for research that would have application to the large ranches typical in the Southwest. Wooton, Turney, and L. B. Foster, president of the New Mexico Agricultural and Mechanical College, developed a plan to withdraw from the public domain Turney's water lot holdings and the surrounding 72,000 ha as a land base for rangeland research (Ares 1974). In May 1912, five months after New Mexico was granted statehood, the Jornada Range Reserve (later named the Jornada Experimental Range) was created by presidential Executive Order with Turney as the first livestock cooperator.

In 1919, Turney expanded his personal holdings outside the Jornada when he purchased several ranches that gave him control over all of the rangeland in northeastern Doña Ana County. These purchases included the Henry Summerford ranch. Later, financial problems forced Turney to sell some of these holdings to his children. His daughter, Maude, and her husband, Max Vander Stucken, bought the Summerford ranch in 1919, but unfortunately, this ranch went into foreclosure by 1926. In 1927, the New Mexico College of Agriculture and Mechanic Arts purchased 70 ha associated with water rights at three locations and the rights to about 2,430 ha of New Mexico state land grazing leases that comprised the holdings of the Vander Stucken ranch. Also in 1927, Congress deeded approximately 23,490 ha of public land associated with these water rights to New Mexico A&M for the purpose of conducting range livestock research. This holding provided a 25,920-ha facility adjacent to but separate from the federally operated Jornada Experimental Range (JER) for the university to pursue its research objectives. Using cattle purchased from area ranchers, including Turney, the Chihuahuan Desert Rangeland Research Center (CDRRC) was immediately used for livestock nutrition research, particularly studies on supplemental feeds (New Mexico Agricultural Experiment Station 1949, 1950). By the early 1930s rangeland research was being actively conducted by university and federal scientists on over 100,000 ha of desert rangeland within the Jornada Basin (see figure 2-1, chapter 2).

Our Setting

The Chihuahuan Desert is the largest desert in North America (figure 1-2). The habitats of the Jornada Basin are representative of the northern Chihuahuan Desert (the Trans-Pecos region) and the Mexican Highland region of the Basin and Range Physiographic Province. The only North American desert located east of the Continental Divide, this region is a transition between the short-grass prairies of the central United States and the shrub dominated ecosystems of the Sonoran and Mojave Deserts to the west.

The Chihuahuan Desert dates to about 9,000 years ago. It has been hypothesized that during the past 3,000 years there have been three transitions from grasslands to shrublands, each followed by a return of grasslands in southern New Mexico (Van Devender 1995). Today, perennial grasses, such as black grama, may represent a relict community from more mesic climatic conditions in the mid-Holocene. Some modern plant species endemic to the desert were certainly present when large herbivores, such as ground sloths *(Megalonyx)* and mammoths *(Mammuthus)* grazed the Chihuahuan Desert, but apparently this region has not been subject to high levels of herbivory by bison *(Bison)* and other native

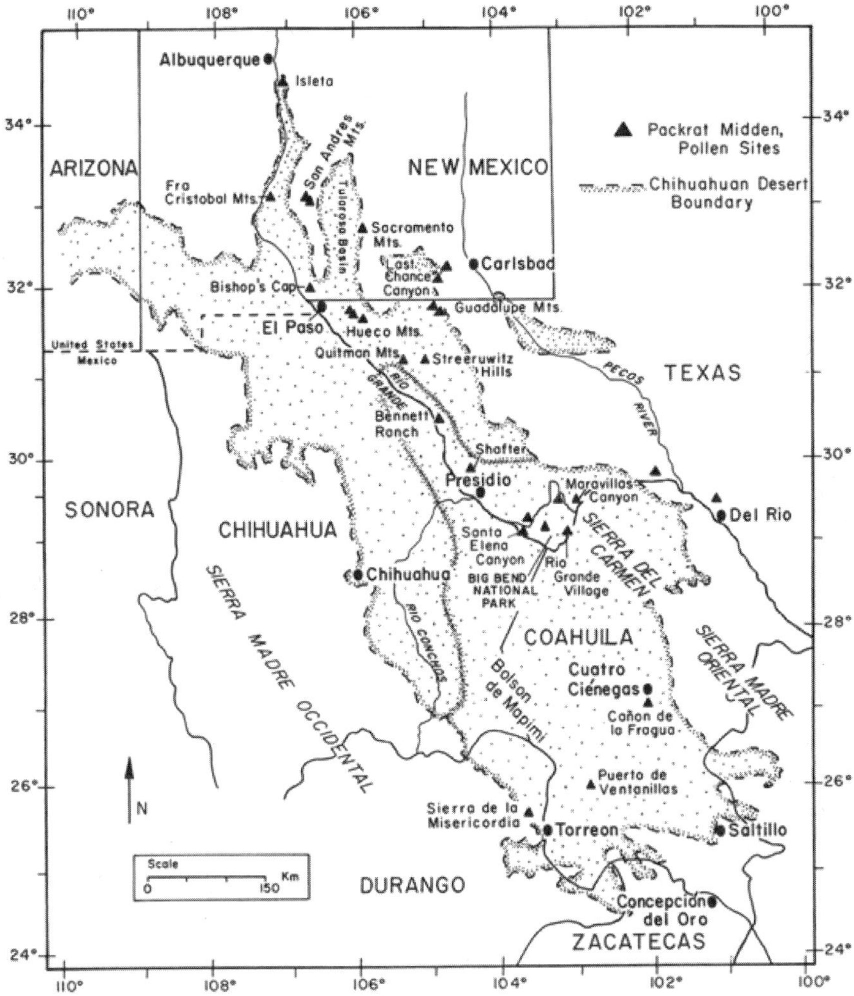

Figure 1-2. Map of the full extent of the Chihuahuan Desert. From VanDevender (1990).

ungulates for the past 10,000 years (Mack and Thompson 1982; Bock and Bock 1993).

The Jornada Basin receives an average of 245 mm/year of precipitation, about half in monsoonal storms that derive from the Gulf of Mexico during the late summer, and the remainder in synoptic weather systems stemming from the Pacific Ocean during the winter months. Rainfall shows large interannual variability that controls the relative growth of C_3 shrubs during the spring and C_4 grasses in summer. Measured evaporation is about 220 cm/year (chapter 3), so the Bowen ratio—the dissipation of sensible versus latent heat—is very high (chapter 8). During the summer, the mean maximum temperature is 36°C, and often little or no precipitation is recorded in the months of May and June.

The Jornada Basin is typical of the closed-basin topography found in many arid regions of the Southwestern United States (chapter 2). Parallel, north–south, block-faulted mountains separate individual valleys, which have a predominance of internal drainage ways terminating in intermittently flooded lakes, known as playas. In the Jornada Basin, soils are largely derived from alluvial deposits from the mountains, as well as from floodplain deposits laid down by an ancient watercourse of the Rio Grande through the Jornada Valley (chapter 4). The entire surface is subject to wind erosion and eolian redistribution of soil materials (chapter 9).

Recent changes in ecosystem structure and function in southern New Mexico may represent a degradation process that is driven by both environmental and human impacts in combination with climatic stress, particularly prolonged drought. Anthropogenetic factors have contributed to local degradation (Fredrickson et al. 1998; see also chapter 13). Grassland sites of low resistance to grazing disturbance have shifted to alternate, stable states in which shrubs dominate (Bestelmeyer et al. 2003a). Shifts are likely to have been accelerated by a decline in the proportion of summer precipitation (which favors C_4 perennial grasses), and an increase in the proportion of precipitation during the winter, which favors C_3 shrubs (Neilson 1986). This shift may reflect an increasing frequency of Pacific El Niño events, which enhance wintertime synoptic rainfall in the Southwestern United States.

High rates of human population growth, low per capita income, and land-use changes driven by an increasing globalization of the world's economy impact natural ecosystems in most desert environments throughout the world. The ecosystems of the Chihuahuan Desert are no exception. In southern New Mexico, for example, the population of Doña Ana County (which includes much of the Jornada Basin) grew by nearly 81% in the decades of the 1980s and 1990s as population density went from 65 people per km^2 to nearly 120 people per km^2 (U.S. Census Bureau 2000). These population densities may pale in comparison to the urban centers of the world, but their relative impacts are substantial in these harsh water- and nutrient-limited environments. Similar high rates of population growth are found throughout much of the Chihuahuan Desert of Mexico and across the Southwestern United States. In addition to the direct space needed for human occupancy, the infrastructure needed to support this population is rapidly expanding. The region is increasingly traversed by roads, power lines, and aqueducts, and construction activities leave barren soils subject to wind erosion, reroute and linearize natural drainage ways, and replace native arid land vegetation with more profligate users of water. This desert area, like others, is in flux as the various pressures to increase economic production to support a growing human population are applied in a region of sparse and unevenly distributed resources.

Prior Research Themes

The history of research in the Jornada Basin consisted of several main themes over the first five to six decades: community ecology, rangeland management, animal husbandry, rangeland improvement, ecosystem sciences, and interdisciplinary studies (Havstad and Schlesinger 1996). An initial motivation for many studies was the expansive vegetation change within the Jornada Basin. These changes, now well recorded (Gibbens et al. 2005), were dramatic, progressive losses of desert grasslands dominated by black grama (*Bouteloua eriopoda*) and an invasion of desert shrubland species, predominately creosotebush (*Larrea tridentata*) and mesquite (*Prosopis glandulosa*) (Buffington and Herbel 1965; see figures 10-1 and 10-2, chapter 10).

Though the impacts of the nineteenth-century cattle industry and associated developments in southern New Mexico were severe, in reality, there was little formal, mechanistic understanding of what might have caused the complete reconfiguration of vegetation and soil resources on the Jornada landscape. It was possible that numerous factors, including fire suppression, rising concentrations of atmospheric carbon dioxide, and changes in the seasonal distribution of rainfall, had contributed to large changes in ecosystem structure and function. It is also quite possible that shrub encroachment was a result of a series of events occurring over centuries, and not a response to recent livestock overgrazing (Fredrickson et al. 2005). Current vegetation patterns may more broadly reflect historical legacies, dynamic patterns of ecological variables, various transport processes (fluvial, eolian, animal), resource redistributions across landscapes, and different cross-scale interactions (Peters and Havstad 2005).

Irrespective of our theories on causes and effects, an unequivocal answer would likely not derive from traditional short-term research studies. The simple facts that many of the plants in question require well over 10 years to be established in new areas and changes in the seasonal distribution of precipitation (which might lead to changes in vegetation) occur on time scales of decades and longer would require a research perspective with a more expansive temporal scale.

As in all LTER sites, the LTER studies in the Jornada Basin were initially organized into the following five core areas (Callahan 1984):

1. Pattern and control of primary production.
2. Spatial and temporal distribution of populations selected to represent trophic structure.
3. Pattern and control of organic matter accumulation in surface layers and sediments.
4. Patterns of inorganic input and movements through soils, ground water, and surface water.
5. Pattern and frequency of disturbance to the research site.

Over recent decades, we have developed a long-term monitoring program and archival data sets in each of these areas to provide a baseline of information regarding the response of this Chihuahuan Desert ecosystem to climatic fluctuations and regional changes in climate (chapter 3). Research on disturbance is of particular interest: Any insight gained into roles of environmental and human drivers would surely aid the ongoing national effort for the management of rangelands, public and private, throughout the United States (chapter 13). In addition, development of remediation technologies for degraded landscapes requires a thorough understanding of the processes associated with disturbance (chapter 14).

Prior studies of plant growth had shown little difference in the annual aboveground net primary production (ANPP) between black grama grasslands and the various shrubland habitats of the Jornada Basin. Rather, plant production seemed determined by landscape position, with greater plant growth in areas where runoff water accumulated and lower plant production in areas of limited soil moisture (Noy-Meir 1985; Ludwig 1986). The similarity of ANPP between grassland and shrubland habitats suggested that plant production per se was not a good index of desertification, to the extent that this term is appropriate to the historical loss of productive desert grassland in southern New Mexico (chapters 10 and 11).

This research group has also focused on changes in the distribution of essential soil resources during the transition from grassland to shrubland habitats. For example, sampling shrublands at a scale of 10- to 100-cm intervals, we found enormous variation in the content of nitrogen among soil samples. When we sampled grasslands at the same spatial scale, the soil samples seemed rather homogeneous in basic soil characteristics (Schlesinger et al. 1990, 1996). Of course, ecologists have long recognized patches, or "islands," of fertility from under shrub vegetation, which leads to a heterogeneous distribution of soil resources in deserts. What was new to our work was that we hypothesized that desertification of semiarid grasslands may not be so much associated with a change in vegetation production as with an increase in the spatial heterogeneity of soil resources (Schlesinger et al. 1990) (see figure 1-3). The heterogeneity of soil resources created by invading

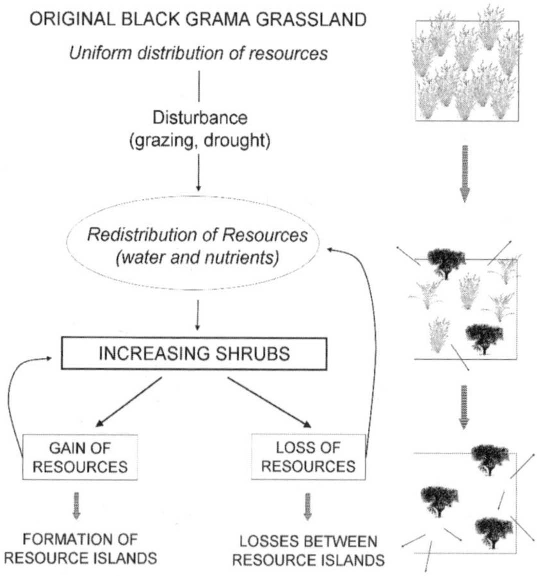

Figure 1-3. A model of autogenic factors that lead to the development of soil resource heterogeneity and degradation of desert grasslands.

shrubs is followed by a further localization of soil resources under shrub canopies promoting further invasion, increase, and persistence of shrubs. In this teeter-totter model, the patchy distribution of soil resources leads to heterogeneity in the distribution of soil microbial biomass (Herman et al. 1995), nematodes (Freckman and Mankau 1986), and microarthropods (Santos et al. 1978). The patches of soil fertility created by plants and animals are preferred sites for the establishment of shrubs (Chew and Whitford 1992; Silvertown and Wilson 1994; see also chapter 12). Barren areas between shrubs are increasingly subject to the physical removal of soil resources by water and wind erosion and to the potential for long-term depletion or loss of soil nutrients (Ritchie et al. 2003).

Beyond our studies of the causes of degradation in southern New Mexico, we were also interested in its regional effects. Using the Jornada as an example, we have postulated the role of deserts in determining regional to global characteristics of the Earth's climate and biogeochemistry (Schlesinger et al. 1990; see also chapter 7). Loss of vegetation raises regional albedo as well as regional air temperatures (Bryant et al. 1990). Barren soils are a source of windborne dust, which can affect the radiative balance of the planet, depending on the mineralogy of dust and its persistence in the atmosphere (Tegen and Fung 1995; Sokolik and Toon 1996; Okin 2002; see also chapter 9). Loss of vegetation lowers the infiltration of rainfall, leading to higher runoff losses of rain water, greater losses of soil nutrients, and the persistence of regional desertification (Abrahams et al. 1995; see also chapters 5, 6, and 7). Thus, studies in the Jornada Basin treated both cause and effect relationships associated with the shrub invasion of desert grassland habitats. Now in the twenty-first century, our research group and its collective missions have continued to evolve.

Our general research objectives now are (1) to understand and explain historic landscape scale dynamics characteristic of arid lands, (2) to understand current landscape structure and function, and (3) to predict future dynamics, including those of managed landscapes (Peters and Havstad 2005). All of these objectives still provide a base from which to develop strategies to manage arid lands and restore degraded areas, and our work has been providing these strategies and practices (Bestelmeyer et al. 2005; Herrick et al. 2005).

Conclusions

This book describes many of the basic processes operating over time and within and among the communities of the Jornada Basin. These descriptions and understandings should translate to other arid ecosystems around the world. Yet, these desert systems actually are intricate, interactive, and connected sets of patches within communities, within sites, within landscapes, and within regions that are dynamic over time (chapters 16, 17, and 18). Our research group understands that we now need to focus on describing the linkages among these units and identifying important influences on emerging ecosystem dynamics and biotic patterns. We are concentrating on questions of temporal scale and spatial patterns. (chapter 18). Over recent years, we have effectively linked the research efforts of the LTER and the USDA Agricultural Research Service (which has operated the JER since 1954 as part of the USDA research network) programs into a synergistic set of scientific objectives. This maturation and melding is a logical outcome because these groups share similar goals. This book, with its many coauthored chapters, is evidence of this collaboration. In reading this volume it often may not be evident what research is USDA-, university-, or LTER-based, because there is no longer a distinction in our minds. Given the enormity of arid landscapes not only in this region but around the world, working together to understand how resources are distributed and redistributed across these landscapes and among landscape units is crucial to developing a process-based understanding of the causes and consequences of degradation. Perhaps more important, this understanding will create a basis for both predictions of landscape dynamics and technologies for remediating or restoring degraded conditions.

2

Regional Setting of the Jornada Basin

H. Curtis Monger
Greg H. Mack
Barbara A. Nolen
Leland H. Gile

Within the area around Las Cruces, New Mexico, is a network of studies both at the Jornada Experimental Range (JER) and the Chihuahuan Desert Rangeland Research Center (CDRRC), which includes the Desert Soil-Geomorphology Project (figure 2-1). All of these research entities are in the Chihuahuan Desert; this chapter describes the geologic history and development of landscapes that are important elements to our understanding of this ecosystem and its dynamic nature.

Southern New Mexico consists of C_3 shrubs and C_4 grasses in the lower elevations surrounded by C_3 woodlands and juniper savannas in the higher elevations (Dick-Peddie 1993; see also chapter 10). Although the boundary of the Chihuahuan Desert has differed slightly depending on whether it was based on vegetative or climatic criteria, the boundary most widely used is based on a de Martonne aridity index of 10 (Schmidt 1979). Like many deserts of the world, the Chihuahuan Desert is in a zone of dry, high-pressure cells near 30° latitude; is relatively far from marine moisture sources; and occupies a position in which mountains scavenge moisture from weather fronts (Strahler and Strahler 1987). The orographic influence is especially important in Mexico, which contains about three-fourths of the Chihuahuan Desert, because the desert is bounded to the west by the Sierra Madre Occidental and to the east by the Sierra Madre Oriental Mountains.

The basin and range province consists of north–south trending mountain ranges and broad intervening desert basins (figure 2-2) that at the Jornada Basin site is the product of the Rio Grande rift tectonic system, which has been active since middle tertiary time (Seager 1975; Seager et al. 1984). Erosion of the steeply tilted mountain ranges has been the main source of sediment for the filling of the

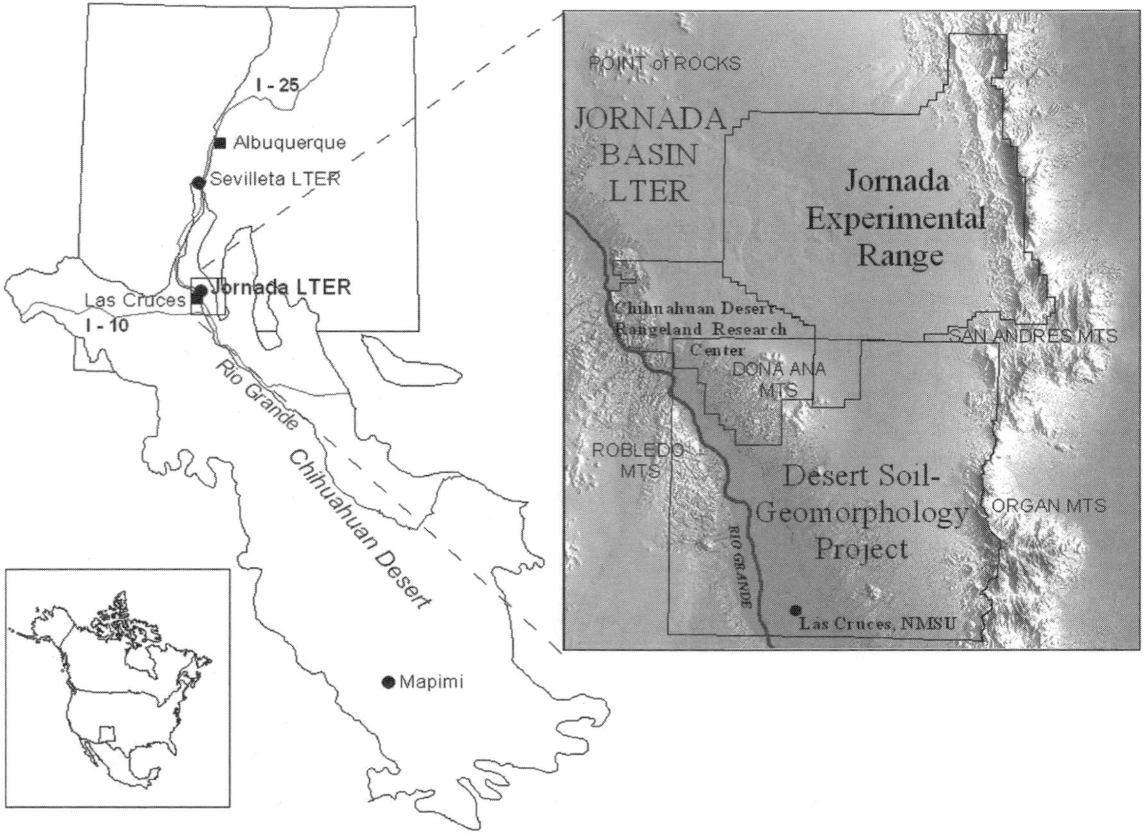

Figure 2-1. Location of the Jornada Basin LTER site in relation to the Chihuahuan Desert as defined by Schmidt (1979). The two research entities where LTER research is conducted are the Jornada Experimental Range (JER), and the Chihuahuan Desert Rangeland Research Center (CDRRC). The Desert Soil-Geomorphology Project was covered by a portion of these two facilities.

intermontane basins. Vertically, sedimentary deposits are separated by paleosols that indicate quiescent times following periods of erosion and sedimentation. As the basins filled with sediment, pore spaces between sediment particles filled with water from rain, making these structural basins large reservoirs of ground water (Hawley and Lozinsky 1992; Hawley and Kennedy 2004). The water table in the Jornada and neighboring basins dropped when the Rio Grande entrenched below the basin floors in the middle Pleistocene (Hawley and Kottlowski 1969). Today, the water table is 76 m (250 feet) or more deep in some places in the Jornada Basin but as shallow as 1 m near the river on the modern floodplain (King and Hawley 1975).

The major landscape components (or physiographic parts) of the Jornada Basin LTER site are: (1) mountains and hills, (2) piedmont slopes (bajadas), (3) basin floors, and (4) the Rio Grande Valley (figure 2-2). Elevations range from 2,747 m (9,012 feet) in the mountains to 1,180 m (3,870 feet) in the Rio Grande Valley. Details of local climate are presented in chapter 3.

Compared to the geologic setting of other deserts, being located in a tectonically active zone with great topographic diversity is a characteristic that the Chihuahuan Desert shares with other North American deserts (Mojave, Sonoran, and

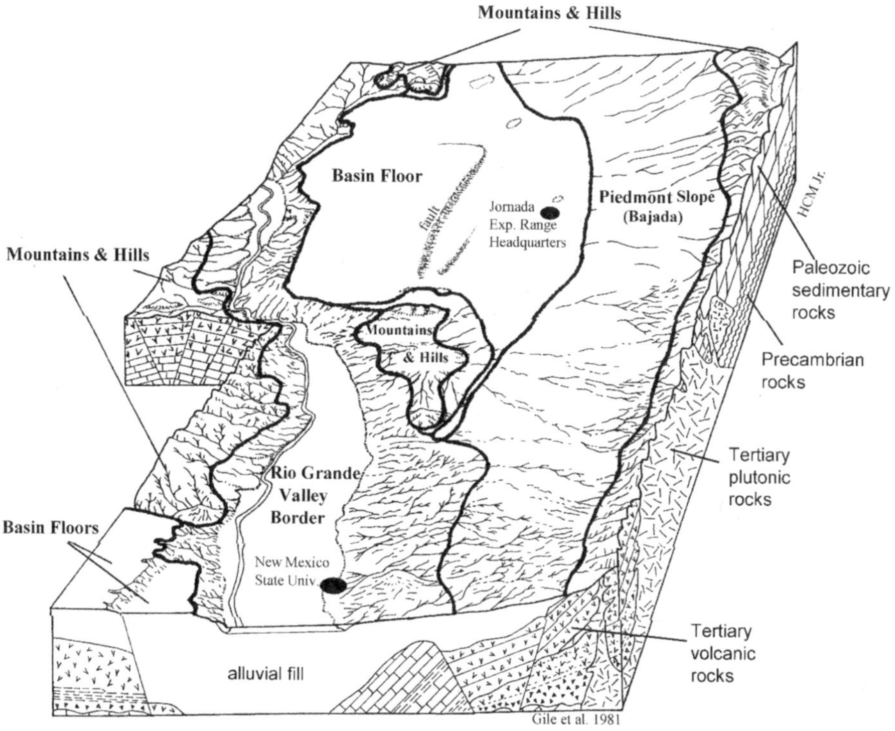

Figure 2-2. Block diagram of the Jornada Basin and Desert Project areas illustrating the major landscape components and subsurface composition.

Great Basin), the Monte Desert of South America, and deserts of the Middle East. These deserts contrast with the tectonically stable Thar Desert of India/Pakistan, the Sahara Desert, and Australian deserts (Cooke et al. 1993). Although the Chihuahuan, the other North American deserts, the Monte, and the Middle Eastern deserts have similar geologic settings, these deserts exhibit their own unique appearance in proportion to their degree and duration of aridity (Oberlander 1994). In the American Southwest alone, the Chihuahuan and Mojave Deserts, both of which are in the basin and range province, display different characteristics due to different degrees of aridity.

Geologic History

Precambrian Rocks

Precambrian rocks constitute the crystalline basement of the Jornada Basin site and are exposed near the base of the larger fault-block mountain ranges, such as the San Andres and Caballo Mountains, which border the Jornada Basin (Seager 1981; Mack et al. 1998a, 1998b). The primary rock types are schist, gneiss, amphibolite, phyllite, and metaquartzite locally intruded by granitic rocks (Condie and Budding 1979). Precambrian rocks in New Mexico range in age from 1.0 to 1.8 billion years and occupy three east-trending chronologic provinces that become younger southward (Condie 1982). These metamorphic provinces represent the deeply eroded roots of collisional mountain ranges and record the accretion of continental crust onto the southern margin of North America (Condie 1982).

Paleozoic Rocks

Paleozoic sedimentary rocks are major components of most mountain ranges in the Jornada area, including the San Andres, Caballo, Robledo, Doña Ana, Franklin, and Sacramento Mountains. Based on rock types, unconformities, and sediment accumulation rates, Paleozoic rocks in the Jornada area can be divided into two sequences: a lower and middle Paleozoic sequence and an upper Paleozoic sequence associated with the Ancestral Rocky Mountains.

The lower sequence ranges in age from the late Cambrian through the Mississippian and consists of approximately 1 km of sedimentary rocks. The majority of these rocks originated as fosilliferous limestone deposited in shallow, warm tropical or subtropical seas that periodically occupied the stable interior of North America (Kottlowski et al. 1956; Kottlowski 1965). Some of the limestone was subsequently altered to dolostone (dolomite) during burial. Also present are shallow marine and shoreline sandstone in the basal formation (Bliss sandstone) and an upper Devonian formation enriched in dark shales (Percha shale) deposited on a poorly oxygenated sea floor. The lower sequence of Paleozoic rocks is characterized by numerous unconformities that represent erosion during periods of sea-level fall.

The upper Paleozoic sequence includes the Pennsylvanian and Permian. During these periods, New Mexico and adjacent areas experienced a mountain-building event, referred to as the Ancestral Rocky Mountains, which developed hundreds of kilometers inland of the collision between North America and Gondwanaland (Kluth and Coney 1981; Kluth 1986). The Ancestral Rocky Mountains were characterized by north- and northwest-trending, fault-bound mountain ranges separated by sedimentary basins in which were deposited approximately 2 km of mixed marine and nonmarine sedimentary rocks (Kottlowski 1960b, 1965). In south-central New Mexico, the main tectonic features were the Pedernal Mountains, whose axis was located just east of the present-day Sacramento Mountains, and the Orogrande Basin, whose western margin includes the Jornada area. Rocks of the sedimentary basin are a mix of sandstone and shales. Among the nonmarine depositional systems are fluvial and alluvial-fan red beds and well-sorted eolian sandstone (Mack and Suguio 1991; Mack et al. 1995). Meter-scale interbedding of marine and nonmarine rocks was probably a response to global sea-level changes driven by growth and shrinkage of continental ice sheets in Gondwanaland (Mack and James 1986). The presence of calcic paleosols, widespread gypsum precipitation, and a paleoflora dominated by gymnosperms suggest that Permian paleoclimate in southern New Mexico was relatively dry (Mack et al. 1995).

Mesozoic Rocks

Triassic and Jurassic rocks are absent in the Jornada area and throughout most of southern New Mexico. Only about 50 m of marine shales and sandstone of early Cretaceous age are exposed east of the Jornada area in the southern San Andres Mountains (Seager 1981), although lower Cretaceous sedimentary rocks up to 3.5 km thick were well exposed in the extreme southwestern part of New Mexico and were deposited within the northern edge of the extensional Brisbee Basin (Mack et al. 1986). The Jornada area has a thin lower Cretaceous section because it occupied the rift shoulder of the Brisbee Basin and was a site of erosion.

Upper Cretaceous sedimentary rocks were originally deposited throughout the state of New Mexico, but exposures of these rocks in the Jornada area are restricted today to the southern San Andres Mountains and the northeastern flank of the Caballo Mountains (Seager 1981; Mack et al. 1998a, 1998b). Approximately 1 km thick, upper Cretaceous rocks are composed of interbedded marine and nonmarine sandstones and shales deposited within and adjacent to the Western Interior Seaway. Deposition of dark, organic-rich shales suggests the floor of the seaway was poorly oxygenated. A subhumid to humid paleoclimate is suggested by the presence of in situ petrified angiosperm and conifer stumps, whose spacing is indicative of an open canopy forest, as well as by paleosols characterized by well-developed A, E, and argillic B horizons (Mack 1992).

Laramide Orogeny (Latest Cretaceous-Ecocene) and Post-Laramide Volcanism (Late Eocene)

Compressional mountain building, referred to throughout the Western United States as the Laramide orogeny, affected southern New Mexico in latest Cretaceous (late Campanian and Maastrichtan ages) and early tertiary (Paleocene and Eocene epochs) time. In the Jornada area, reconstructed Laramide tectonic features include the northwest-trending Rio Grande uplift and the complimentary Love Ranch sedimentary basin to the northeast (Seager and Mack 1986; Seager et al. 1986, 1997; Seager and Mayer 1988). The Rio Grande uplift was bordered on the northeast by southwest-dipping thrust faults and associated folds, which are exposed today in the southern San Andres Mountains, San Diego Mountains, and the Caballo Mountains.

The Love Ranch Basin contains approximately 1.4 km of nonmarine sedimentary rocks separated into two formations. The older McRae formation consists of fluvial conglomerates, sandstone, and shales derived from carapace of Cretaceous andesitic volcanic rocks that were eroded from the crest of the Rio Grande uplift. With the exception of a few isolated exposures of volcanic vents, the Cretaceous volcanic rocks exist only as clasts in the Love Ranch Basin. The McRae formation is also noted for the presence of a late Maastrichtan dinosaur fauna, including *Alamosaurus, Tyrannosaurus, Torosaurus*, and an ankylosaurid (Lozinsky et al. 1984; Gillette et al. 1986; Wolberg et al. 1986; Lucas et al. 1998). Plant fossils in the lower member of the McRae formation consist of silicified tree stumps, many of which are in situ, and leaf impressions from about 40 different species of ferns, cycads, conifers, monocots (including palms), and dicots (Upchurch and Mack 1998). The paleobotanical evidence suggests that early in the history of deposition of the McRae formation, the paleoclimate was warm and subhumid with little seasonal variation in precipitation and supported a subtropical to paratropical open-canopy forest. By the end of McRae deposition, however, the paleoclimate became drier and perhaps more seasonal, as indicated by the presence of calcic paleosols with vertic features (Buck and Mack 1995).

The upper part of the sedimentary sequence in the Love Ranch Basin, the Love Ranch formation, consists of alluvial fan, fluvial and lacustrine conglomerates, sandstones, and shales that record a major pulse of uplift and eventual post tectonic onlap of the Rio Grande uplift. Calcic paleosols and gypsum in the Love Ranch formation suggest a relatively dry climate in early tertiary (Seager et al. 1997).

Following Laramide mountain building, volcanism of intermediate composition spread into southern New Mexico in late Ecocene time. Ranging from 100 to 600 m thick, these rocks consist of bouldery lahars, volcanistic river deposits, lava flows, and ash flows derived from a series of stratovolcanoes, the remains of only one of which, in the Doña Ana Mountains, has been identified in southern New Mexico (Seager et al. 1976).

Rio Grande Rift

Crustal extension associated with the Rio Grande rift began in southern New Mexico in the Oligocene epoch and continues to the present day. Traditionally, the history of the Rio Grande rift has been divided into two phases, an early phase and a late phase that began in the latest Miocene or Pliocene and continues to the present (Seager 1975; Seager et al. 1984).

Oligocene Bimodal Volcanism and Miocene Uplift and Deposition in Closed Basins

The initial phase of extension of the Rio Grande rift in southern New Mexico began in the Oligocene epoch with bimodal volcanism and minor uplift and basin development. The rhyolitic end of the volcanic spectrum involved explosive volcanism and caldera development. The remnants of three calderas can be identified in southern New Mexico, including the Organ Mountains' and Doña Ana Mountains' calderas, which are partially exposed in mountain ranges of the same name, and the Emory caldera, the eastern half of which is exposed in the Black Range. The exposed portion of the Organ and Doña Ana calderas consists of 600 m to 3.3 km of ash flow tuffs that erupted between 36 and 35 million years ago and were trapped within the calderas as they collapsed, as well as exposures of plutonic igneous rocks that crystallized in the magma chambers beneath the calderas (Seager et al. 1976; Seager 1981; McIntosh et al. 1991). The preserved portion of the Emory caldera contains caldera-fill, ash flow tuffs that erupted about 35 million years ago, as well as tuffs, lava flows, and lava domes that were emplaced during a resurgence event that followed the main eruptive cycle (Elson et al. 1975). Also related to eruption of the Emory caldera is the Kneeling Nun tuff, which spilled out of the erupting caldera and spread for tens to hundreds of kilometers across the surrounding countryside. Outflow sheets of ash flow tuff from other calderas in southern New Mexico were preserved, along with sedimentary rocks, in the Goodsight–Cedar Hills half graben located west of the Jornada area (Mack et al. 1994a). Simultaneous with development of the Oligocene calderas were eruptions of basalt and basaltic andesite lava flows. The thickest, the Uvas basaltic andesite of late Oligocene age, is 250 m thick and consists of as many as 14 distinct lava flows (Clemons and Seager 1973). Oligocene volcanic rocks erupted during the initial phase of extension of the Rio Grande rift are interpreted to have been derived from partial melting of the lithospheric mantle (McMillan 1998).

The initial phase of the Rio Grande rift in southern New Mexico culminated in the Miocene epoch with block uplifts and complementary basin subsidence, but little or no volcanism. Basin-fill sedimentary rocks of the Miocene age have a maximum thickness of almost 2 km and consist of conglomerate, sandstone, shales, and gypsum deposited on alluvial fans and in ephemeral lakes (Mack et al. 1994b). There is no evidence at this time of a through-going river analogous to the modern Rio Grande.

Latest Miocene to Recent Uplift, Basin Subsidence, and Basalt Volcanism

The more recent phase of crustal extension of the southern Rio Grande rift began in latest Miocene or Pliocene time and continues to the present day and is responsible for producing the modern topography. Many of the fault blocks that initially arose in the Oligocene or Miocene continued to be active, although to a lesser extent. However, new uplifts, such as the Robledo Mountains, San Diego Mountains, and Rincon Hills, arose through former rift basins resulting in a series of more closely spaced, north-trending basins and uplifts that existed in the Miocene. The sedimentary record of this phase of extension is the Pliocene–early Pleistocene Camp Rice formation, which consists of a maximum of about 150 m of conglomerates, sandstones, and shales deposited on alluvial fans and by the Ancestral Rio Grande (Strain 1966; Hawley et al. 1969). Volcanism resumed during the recent phase of extension in the form of basaltic lava flows that erupted from fissures and individual cinder cones (Seager et al. 1984). These young basalts differ from those of the early phase because they were derived from partial melting of the asthenospheric mantle (McMillan 1998). Extensional tectonism remains an important geologic process in the Jornada area today. Two major Quaternary faults cut through the Jornada Basin (Seager 1975; Mack et al. 1994c; Seager and Mack 1995). Moreover, the Organ and Caballo Mountains are bordered by normal faults with evidence of Holocene movement (Seager 1981; Machette 1987). Uplifts up to 5 m have taken place within the last 1,000 years along the Artillery Range Fault on the east side of the Organ Mountains (Gile 1986, 1994).

The Ancestral Rio Grande

During the more recent phase of extension of the Rio Grande rift, the Ancestral Rio Grande arrived in southern New Mexico. The fluvial facies of the Camp Rice formation consist of interbedded, pebbly sands and sandstones deposited in channels and floodplain mudstones and siltstones, some of which contain calcic paleosols (Mack and James 1992). The channel sands contain distinctive "mixed-rounded gravels" of quartz, chert, quartzite, and other siliceous rocks transported from northern and central New Mexico (Strain 1966; Hawley et al. 1969). Maps of the fluvial sediments reveal that the Ancestral Rio Grande alternately spilled into and filled several basins in the region (Kottlowski 1960a; Ruhe 1962; Gile et al. 1981; Seager et al. 1987; Seager 1995; Mack et al. 1997). Reversal megnetostraitography and $^{40}Ar/^{39}Ar$ dates of pumice in the fluvial sediments indicate that the Ancestral Rio Grande reached southern New Mexico about 5 million years ago (Mack et al. 1993, 1996, 1998a, 1998b; Leeder et al. 1996). Pumice dated from a trench 7 km west of the Jornada–Agricultural Research Service (ARS) headquarters reveals that the Ancestral Rio Grande was in the Jornada Basin about 1.6 million years ago (Mack et al. 1996).

The geomorphic surface that caps the fluvial facies of the Camp Rice formation in the Jornada and neighboring basins is the La Mesa surface (Kottlowski 1953; Ruhe 1967; Gile et al. 1981; Seager et al. 1987). The antiquity of the La Mesa surface and its soils is indicated by substantial accumulations of pedogenic carbonate, ranging from stage III to stage V petrocalcic horizons (Gile 1967; Monger 1993; Herbel et al. 1994). The La Mesa surface has member surfaces of three ages (table 2-1): lower and upper La Mesa (Gile 1967) and JER La Mesa (Gile 2002). Lower La Mesa is highly significant because its soils illustrate the morphological transition from the plugged stage III horizon to initial development of the stage IV horizon in nongravelly materials (Gile et al. 1981). In contrast, most

Table 2-1. Geomorphic surfaces and stages of carbonate accumulation in soils of the Desert Project, JER, and CDRRC.

Geomorphic Surface[2]			Carbonate Stage		Estimated Soil Age
Rio Grande Valley Border	Piedmont Slope	Basin Floor	Nongravelly Materials	Gravelly Materials	(years B.P. or epoch)
Coppice dunes	Coppice dunes	Whitebottom			Historical (since 1850 A.D.)
		Lake tank			Present to 150,000
					Middle and Late Holocene
Fillmore	Organ		0,I	I	100–7,000
	III		I	I	100(?)–1,000
	II		I	I	1,100–2,100
	I		I	I	2,200–7,000
Leasburg	Isaacs' ranch		II	II,III	Latest Pleistocene (10,000–15,000)
Butterfield	Baylor		III	III	Late Pleistocene (15,000–100,000)
Picacho	Jornada II	Petts tank	III	III, IV	Late to Middle Pleistocene (100,000–250,000)
Tortugas			III	IV	Late to middle Pleistocene (250,000–500,000)
Jornada I	Jornada I	Jornada I	III	IV	Late to middle Pleistocene (500,000–700,000)
	Doña Ana			IV	(> 700,000)
Lower La Mesa		Lower La Mesa	III,IV		Middle Pleistocene (780,000)
JER La Mesa		JER La Mesa	IV,V		Middle Pleistocene to late Pliocene (780,000–2,000,000)
Upper La Mesa		Upper La Mesa	V		Late Pliocene (2,000,000–2,500,000)

[2]Geomorphic surfaces after Ruhe (1967), Hawley and Kottlowski (1969), Gile et al. (1981, 1995), and Gile (2002). Lower, JER, and upper La Mesa are included with the Rio Grande Valley because they form part of a stepped sequence with the valley border surfaces. Coppice dunes have not been formally designated a geomorphic surface but are considered separately here because of their extent and significance to soils of the area. The Butterfield and Baylor surfaces were formerly designated as late phases of the Picacho and Jornada II surfaces, respectively (Gile et al. 1995).

of the next-older soils of the JER La Mesa have prominent stage V horizons (Gile 2002). These soils have been remarkably well dated by pumice at about 1.6 million years (Mack et al. 1996; Gile 2002). Parts of JER La Mesa younger than this occur in places (such as along scarps and some areas bordering the Rio Grande Valley), and the soils of the JER La Mesa are estimated to range in age from about 0.78 to 2.0 million years (table 2-1). Paleomegnetic dates indicate that the soils of upper La Mesa are probably about 2.0–2.5 million years old (Mack et al. 1993; table 2-1).

Quaternary Geology

Entrenchment of Ancestral Rio Grande

Various models have explained the entrenchment of the Rio Grande and the establishment of a through-flowing drainage from southern Colorado to the Gulf of Mexico. In one model, entrenchment is thought to have occurred in the middle Pleistocene when large lakes in northern Mexico, southern New Mexico, and west Texas, which were fed by the Ancestral Rio Grande, spilled over a topographically low area near El Paso (Kottlowski 1953) or at the southeast end of the Hueco bolson (Seager et al. 1984). In this model, the entrenchment coincided with the integration of the Upper Rio Grande, which existed north of El Paso, and the Lower Rio Grande, which existed south of El Paso (Kottlowski 1958; Hawley and Kottlowski 1969). In another model, integration of the Upper and Lower Rio Grande is thought to have occurred during deposition of Camp Rice sediments, with entrenchment being the result of sediment supply climatically driven by glacial-interglacial cycles (Gustavson 1991a, 1991b).

Several forms of evidence shed light on the timing of the Rio Grande's entrenchment. These include analysis of vertebrate fossils (Strain 1966; Hawley et al. 1969; Tedford 1981), tephrachronology (Seager and Hawley 1973; Seager et al. 1975a, 1975b; Izett and Wilcox 1982; Izett et al. 1988), K-Ar dates of basalt flows (Seager et al. 1984), the amount of pedogenic carbonate in La Mesa soils (Machette 1985; Gile 2002), and radiometrically dated pumice deposits and magnetostratigraphy (Mack et al. 1993, 1996). Two ash layers have been important for determining the time of entrenchment: Lava Creek B ash (0.62 million years, Izett and Wilcox 1982) and Bishop ash (0.76 million years, Sarna-Wojcicki and Pringle 1992). Earlier investigations suggested that the Lava Creek and Bishop ashes were stratigraphically beneath the La Mesa surface in southern New Mexico (Hawley et al. 1969; Seager and Hawley 1973; Seager et al. 1975a, 1975b; Hawley 1981). This stratigraphic position indicated that valley cutting by the Rio Grande did not begin until after 0.62 million years ago. However, downstream from El Paso, near Candelaria, Texas, Lava Creek B ash was found in an inset terrace (Hawley 1975b, 1981). In the Rincon Arroyo in the northern Jornada Basin, the Bishop ash was also found in an inset terrace (Mack et al. 1993). Therefore, it now appears that entrenchment began before the ashes were deposited.

Magnetostratigraphy of one area of lower La Mesa west of Las Cruces is Matayuma (Mack et al. 1998a, 1998b), whereas another area of lower La Mesa to the south is Brunhes (Vanderhill 1986). Thus the true age of lower La Mesa could be close to 0.78 million years, the boundary between Matayuma and Brunhes. This age agrees with the start of valley downcutting at Rincon Arroyo (very near or at the Matayuma–Brunhes boundary, Mack et al. 1998a, 1998b).

Geoarchaeology

The Jornada Basin area has been occupied by humans since the latest Pleistocene, perhaps before the Clovis dates of 12,000 years ago (MacNeish and Libby 2004). With the minor exceptions of manipulating plant species and building rock terraces (Sandor et al. 1990), prehistoric humans appear to have had little impact as a geomorphic agent in southern New Mexico. Whether Paleo-Indian and Archaic hunters and gatherers or Mogollon agriculturalists, these prehistoric peoples adapted to long-term changes (e.g., climate change) and short-term changes (e.g., drought) by migration and by modifying their dependence on plants and animals for food, water for drinking or for crops, and other natural resources for shelter, medicine, and tools (Kirkpatrick and Duran 1998).

The cultural history of south-central New Mexico can be divided into four general categories (Kirkpatrick and Duran 1998): (1) Paleo-Indian (12,000 to 7,500 years B.P.). This group, which is subdivided into Clovis and Folsom people, were hunters who subsisted on bison, mammoth, and limited foraging. Clovis people made large spear points and hunted mammoth. Folsom people made large spear points with flutes and hunted bison. (2) Chihuahua Archaic (8,000 to 1,750 years B.P.). These people were hunters and gatherers who migrated through vegetation zones using available resources. They rarely built houses and made no pottery. They used spears with atlatls, spear throwers. (3) Jornada Mogollon (AD 400 to 1400 or 1450). These people were pit house–dwelling agriculturalists who made brown-ware ceramics. In the later stages, the people lived in aboveground adobe structures of contiguous rooms and made brown-ware ceramics decorated with red and black paint. (4) Masons (1450 to present). These people were possibly the descendants of the Jornada Mogollon. They intermarried with refugees from northern pueblos after the Pueblo Revolt in 1680.

Geomorphic environments where the archaeological stratigraphy is best preserved occur on the piedmont slopes in Holocene alluvial sediments and on basin floors in Holocene eolian sediments (Monger 1995). In the deflated areas of the basin floors, wind erosion has impacted archaeological stratigraphy by preferentially removing finer particles, causing artifacts to collapse vertically onto an erosional surface (Davis and Nials 1988). During this process, wind erosion can modify the surface distribution of artifacts and make large, single archaeological sites appear to be several smaller sites (Leach et al. 1998). Such sites are common in the basin floor of the Jornada Basin. Other prominent archaeological remains at the Jornada include remnants of a pueblo (Buffington and Herbel 1965) and petroglyphs on boulders.

Late Quaternary Climate

Information about late Quaternary climate in the Jornada Basin region (see also chapters 3 and 4) has come from several sources. The most prominent sources are studies of pluvial lakes (Hawley 1993), alpine glaciers (Richmond 1986), rock glaciers (Blagbrough 1991), packrat middens (Van Devender 1990), fossil pollen (Freeman 1972), landscape stability (Gile and Hawley 1966), polygenetic soils (Gile et al. 1981), depth of carbonate (Marion et al. 1985), and carbon isotopes in soils (Connin et al. 1997a; Monger et al. 1998; Buck and Monger 1999).

Climate during the late Pleistocene glacial maximum in New Mexico, approximately 20,000 years ago, is interpreted as being cooler than today by 5–7°C (Phillips et al. 1986). Cooler temperatures are also indicated by rock glaciers (which require interstitial ice to form) that existed at elevations as low as 2,360 m (Blagbrough 1991) and by glacial ice on Sierra Blanca Peak (Richmond 1986). It has been a matter of debate whether the climate was wetter than today (Galloway 1970, 1983; Brackenridge 1978; Van Devender 1990). Nevertheless, there is ample evidence for more "effective" moisture at the Jornada Basin site during the last full glacial based on soil, geomorphic, and botanical evidence (Gile 1966a; Wells 1979; Hawley 1993).

Compared to the last glacial maximum climate, the last 10,000 years that constitute the Holocene have been warmer and drier with short intervals of greater moisture. Packrat midden records from limestone cliffs of the Hueco Mountains near El Paso indicate a unidirectional change in that environment from (1) oak-juniper woodland in the early Holocene to (2) desert grassland in the middle Holocene to (3) desert scrub in the late Holocene (Van Devender 1990). In the lower elevations of the piedmont slopes, soil-geomorphic and fossil pollen records indicate (1) grassland in the early Holocene, (2) desert scrub in the middle Holocene, and (3) a return of grassland with intermittent periods of desert scrub in the late Holocene (Freeman 1972; Gile et al. 1981; Buck and Monger 1999; Monger 2003).

Landforms of the Jornada Basin

There are four major landscape components of the Jornada Basin site: (1) the mountains and hills, (2) piedmont slopes (bajadas), (3) basin floors, and (4) Rio Grande Valley (figure 2-2). Subdivisions of these landscape components into landforms are illustrated in figure 2-3. Landforms at the Jornada Basin were collected into a single map by combining existing maps of the northern Desert Project area (Gile et al. 1981) with a new landform map made for the Jornada Basin. The new map was constructed by delineating landforms on true-color, stereo-pair aerial photographs (scale 1:32,000), Landsat images (bands 1, 2, 3), and digital elevation models. The landform map is presented as figure 2-4 and is also available on the Jornada Basin Web site (http://jornada-www.nmsu.edu/maps/jer_map/viewer.htm). General descriptions of the landforms are given in the following discussion; more detailed definitions are given in table 2-2.

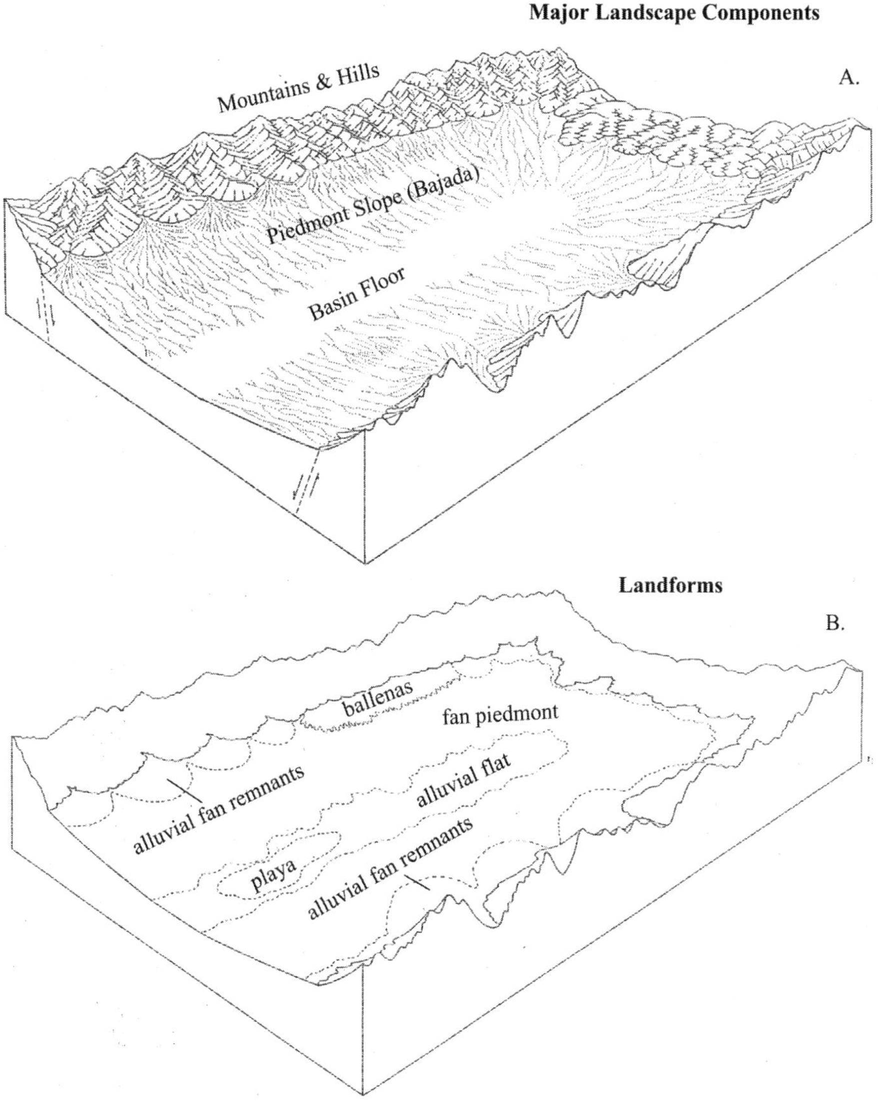

Figure 2-3. Illustrations of terms used to describe landforms. (a) Major landscape components, or physiographic parts, of a basin and range landscape. (b) Subdivisions of physiographic parts into major landforms. Modified from Peterson (1981).

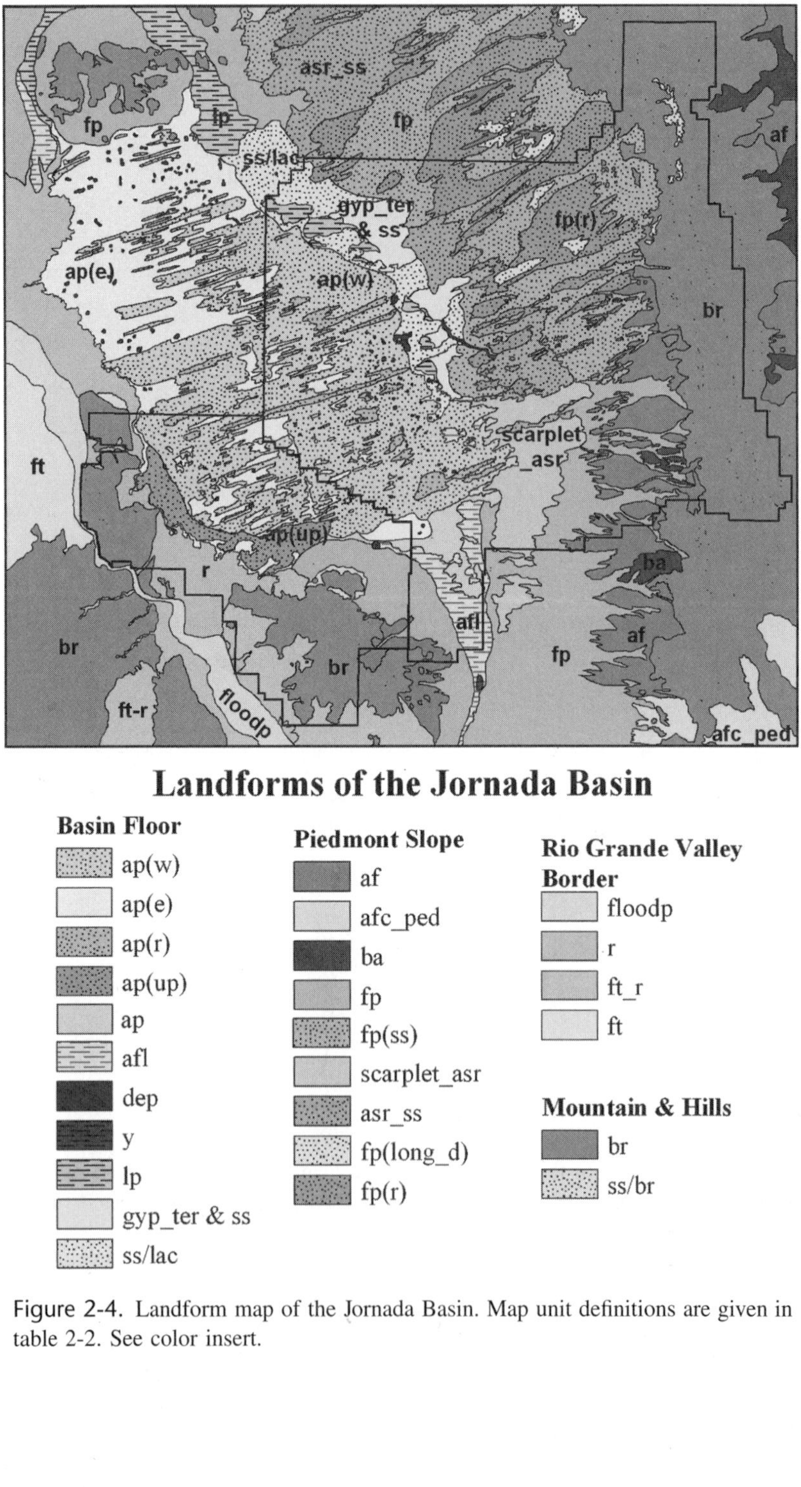

Figure 2-4. Landform map of the Jornada Basin. Map unit definitions are given in table 2-2. See color insert.

Table 2-2. Definitions of landform map units (shown in figure 2-4) of the Jornada Basin.

Symbols	Landform Definitions
Basin Floor: Nearly level surface of an intermontane basin composed of various alluvial, eolian, and lacustrine landforms (P, DP).[a] The Jornada basin floor typically has a gradient less than 1%, although steeper short slopes occur along fault scarps and dune faces.	
ap(w)	***Alluvial Plain Wind Worked.*** Low-relief surface of relict fluvial fans and floodplains of the ancestral Rio Grande. The fluvial sediments are the Camp Rice formation fluvial facies characterized by rounded pebbles of mixed lithology derived from upstream sources (DP, M). Surficial sediments have been reworked by strong winds that have produced deflational and depositional features with an east–northeastern orientation. The alluvial plain corresponds to the La Mesa geomorphic surface with prominent stage IV and V petrocalcic horizons in all but the southeastern portion, where younger surfaces are present (G).
ap(e)	***Alluvial Plain Eroded.*** Wind- and water-eroded areas of the alluvial plain characterized by exhumed or shallow petrocalcic horizons of the La Mesa geomorphic surface.
ap(r)	***Alluvial Plain Reddish Brown Sand Sheets.*** Deposits of reddish brown sand sheets and coppice dune fields linearly oriented with prevailing wind direction.
ap(up)	***Alluvial Plain Uplifted.*** Tectonically uplifted area of the alluvial plain (S). Southern portion occupied by a coppice dune field.
ap	***Alluvial Plain.*** Surface of relict fluvial fans and floodplains of the ancestral Rio Grande (i.e., Camp Rice formation fluvial facies) with geomorphic surfaces younger than the La Mesa surface. This map unit differs from the wind-worked alluvial plain to the north because (1) it lacks prominent deflation-depositional features oriented to the east–northeast, (2) it has soils with stage III calcic rather than stage IV and V petrocalcic horizons, or (3) it has brown and pinkish gray rather than reddish brown surface colors.
afl	***Alluvial Flat.*** Landform characterized by its slightly lower topographic position than the juxtaposed alluvial plain and by its alluvial sediments predominantly carried by sheet floods from neighboring piedmont slopes. These locally derived sediments overlie ancestral Rio Grande (Camp Rice) deposits (DP).
dep	***Depressions (playettes).*** Karst-like concavities of the alluvial plain. Differentiated from playas by their smaller size, coarser texture, and presence of vegetation. The linear depression north–northeast of JER Headquarters is tectonic in origin, whereas the two large depressions northwest of headquarters associated with the map unit *ss/lac* are lacustrine and eolian in origin.
y	***Playas.*** Ephemerally flooded, nongypsiferous, clayey floors of depressions that are more barren of vegetation than surrounding terrain (DP, P).
lp	***Lake-Plain Playas.*** Relict clayey to loamy bottoms of late Pleistocene pluvial lakes. Differentiated from playas by their less frequent flooding, greater vegetative cover, and gypsiferous lithology.
gyp_ter & ss	***Gypsiferous Lake-Plain Terraces and Sand Sheets.*** Gypsiferous lake plain terraces slightly higher than neighboring lake plain playas. These benches are composed of two or more levels separated by low scarps. Such levels are typically formed by recessional stands of a lake (P). Contains complex patterns of deflation and eolian deposits of reworked gypsum, as well as thin deposits of quartzose sand from the alluvial plain.

(continued)

Table 2-2. (*continued*)

Symbols	Landform Definitions
ss/lac	***Sand Sheets over Gypsiferous Lacustrine Landforms.*** Quartzose sand sheets and coppice dune fields blown from the alluvial plain that overlie and are mixed with underlying gypsiferous landforms of lacustrine origin.
Piedmont Slope (Bajada): Sloping land rising from the boundary of the basin floor to the boundary of mountains and hills.	
af	***Alluvial Fan Remnants.*** Older, higher geomorphic surfaces with petrocalcic horizons that border mountain fronts (DP). Where these alluvial deposits have debouched from mountain valleys they display the characteristic alluvial fan shape.
afc_ped	***Alluvial Fan Collar and Pediments.*** Holocene sediments washed from adjacent mountain slopes. The map unit also contains areas of pediments consisting of thin alluvial veneer on erosional surfaces cut across bedrock (DP, P).
ba	***Ballena.*** Eroded remnants of highest and oldest alluvial fans characterized by their high rounded ridges (P).
fp	***Fan Piedmont.*** Landform comprising most of the piedmont slope. It descends from the alluvial fan remnants and mountain fronts to the basin floor. In the middle and lower portions, it consists of coalescent alluvial mantles. In the higher portion, it comprises the interfan valleys (DP, P). Progressively greater amounts of eolian reddish quartzose sand are mixed with fan piedmont sediments northeast of JER Headquarters.
fp(ss)	***Sand Sheets and Coppice Dunes on Fan Piedmont.*** Sand blown from the alluvial plain overlies and is interbedded with fan-piedmont alluvium derived from sedimentary bedrock upslope. Coppice dunes are common in areas containing mesquite. In upper regions, sand directly overlies sedimentary bedrock.
scarplet_asr	**Erosional Scarplets and Arcuate Sand Ridges.** Portion of fan piedmont characterized by low winding ridges of reddish brown quartzose sand blown from the basin floor. Sand ridges occur above arcuate erosional scarplets cut into underlying silty alluvium washed from sedimentary bedrock upslope (R, DP).
asr_ss	***Arcuate Sand Ridges and Sand Sheets.*** Portion of fan piedmont consisting of broad arcuate reddish brown sand ridges and intervening sand sheets that overlie and are interbedded with fan-piedmont alluvium derived from sedimentary bedrock upslope.
fp(long_d)	***Longitudinal Dunes on Fan Piedmont.*** Small longitudinal dunes and large coppice dunes of reddish brown quartzose sand blown from the basin floor.
fp(r)	***Reddish Brown Sand Sheets and Dunes on Fan Piedmont.*** Reddish brown sand sheets and coppice dunes. Sand overlies and is interbedded with fan-piedmont alluvium derived from sedimentary bedrock upslope.
Rio Grande Valley: Incised valley cut into basin fill and piedmont slope sediments.	
floodp	***Valley Floor.*** Modern Rio Grande floodplain. Valley floor consisting of meandering river system inset beneath older alluvium through which the river and its tributaries downcut. The longitudinal slope of the valley floor is less than 0.1% (DP).
r	***Ridges and Interridge Valleys.*** Erosional ridges and structural benches of the Camp Rice fluvial facies. Includes scarps, colluvial wedges, and arroyo channels (DP).

Symbols	Landform Definitions
ft_r	***Remnant Fan Terraces, Ridges, and Interridge Valleys.*** Primarily erosional surfaces of rounded fan terraces graded to former base levels of the Rio Grande. Map unit also contains interridge valleys, scarps, and arroyo channels (DP).
ft	***Fan Terraces.*** Fan terraces graded to former levels of the Rio Grande base level. Similar to map unit *ft_r* but differentiated by having more broad, flat, and stable geomorphic surfaces.
Mountains and Hills: Landscape masses with bedrock cores that rise steeply from surrounding piedmont slopes. Hills rise less than 1,000 ft (305 m) above piedmont slopes, whereas mountains are higher. Both typically have slopes steeper than 15% (P).	
br	***Bedrock Outcrop.*** Exposed bedrock, pediments, small areas with thin alluvial and colluvial deposits, and narrow mountain valley channels, floodplains, and terraces.
ss/br	***Eolian Sand Sheets on Bedrock.*** Reddish quartzose sand sheets blown from the basin floor overlying bedrock.

[a]Letters in parentheses refer to the following references: (P) Peterson 1981; (DP) Gile et al. 1981; (M) Mack et al. 1993, 1997, 1998a; (G) Gile 1999; (S) Seager et al. 1987; (R) Ruhe 1967; (H) Hawley and Kottlowski 1969; Hawley 1975a.

Mountains and hills include the San Andres, Organ, Doña Ana, and Robledo Mountains and the Point of Rocks Hills (figures 2-1, 2-2, and 2-5). This map unit consists of elevated landscapes with bedrock cores that rise steeply from surrounding piedmont slopes (figure 2-5). Hills rise less than 305 m (1,000 feet) above piedmont slopes, whereas mountains are higher. For the basin and range in general, both hills and mountains have slopes steeper than 15% (Peterson 1981), which holds true at the Jornada Basin site (Gile and Grossman 1979). Most mountains and hills consist of bedrock outcrop with intervening areas of mountain valley channels, floodplains, terraces, and slopes covered with thin alluvial and colluvial deposits.

Piedmont slopes (bajadas) are composed of three dominant landforms, ballena, alluvial fan remnants, and fan piedmont (figure 2-3b). Ballenas, which are prominent on the eastern side of the San Andres Mountains, are the remains of the highest and oldest alluvial fans that have been eroded to linear ridges (figure 2-6). Alluvial fan remnants are the high geomorphic surfaces that border mountain fronts, including alluvial fans that debouched from mountain valleys. There is also an alluvial fan collar of Holocene sediments that borders mountains composed of intrusive rock, such as Mount Summerford (Wondzell et al. 1987, 1996). The fan piedmont consists mainly of coalescent alluvial mantles that descend to the basin floor but also includes interfan valleys in the higher zones (figure 2-6a). The fan piedmont at the Jornada Basin site is subdivided into landforms formed by alluvium, erosion (i.e., scarplets), and sand deposition (i.e., sand sheets, dunes, and arcuate ridges) blown from the basin floor (figure 2-7a).

Basin floor at the Jornada site is dominated by the alluvial plain landform and its subcategories (figure 2-4; table 2-2). The alluvial plain consists of fluvial fan

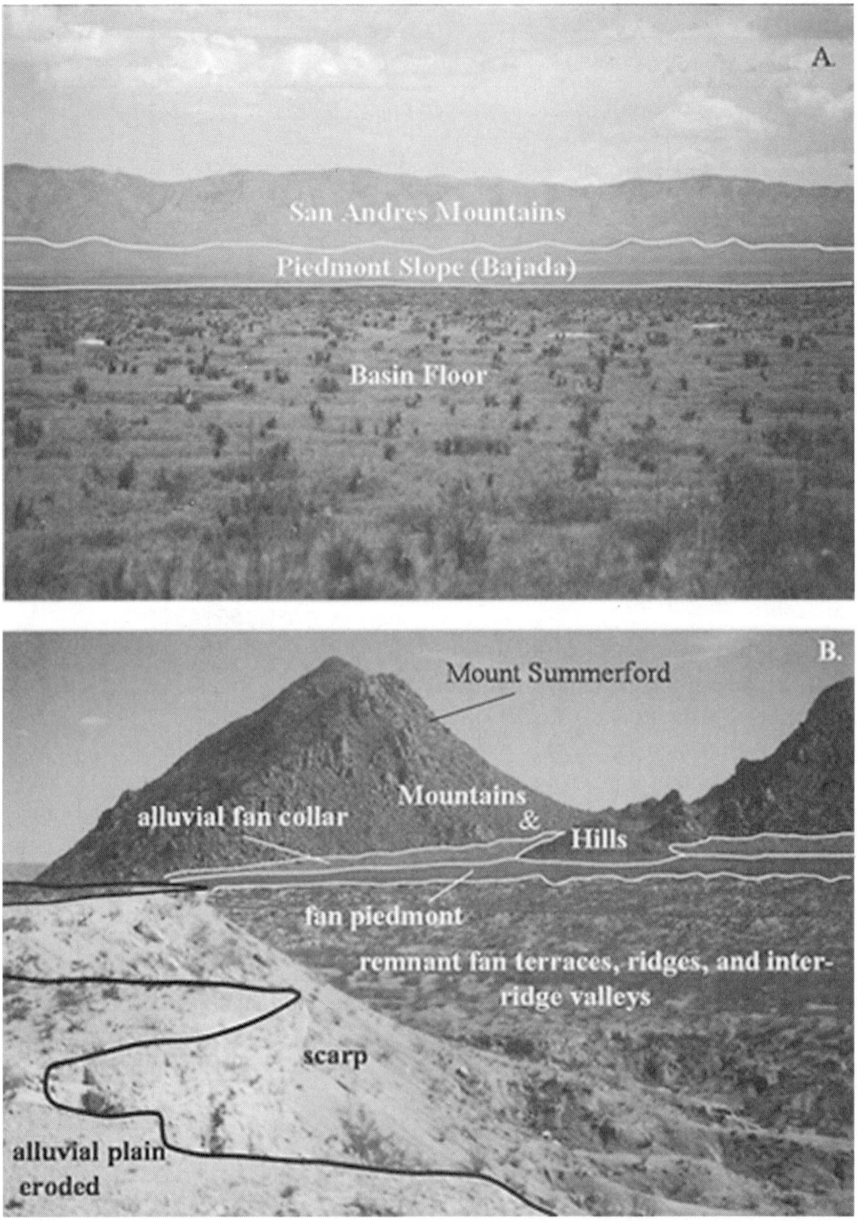

Figure 2-5. Landforms at the Jornada Basin LTER site. (A) Example of three of the four physiographic parts: the basin floor, piedmont slope, and mountains and hills (San Andres Mountains). View toward east; photograph taken July 2000. (B) Example of five landforms: mountains and hills, alluvial fan collar, fan piedmont, alluvial plain (eroded), and remnant fan terraces, ridges, and interridge valleys. View toward east; photograph taken December 2003.

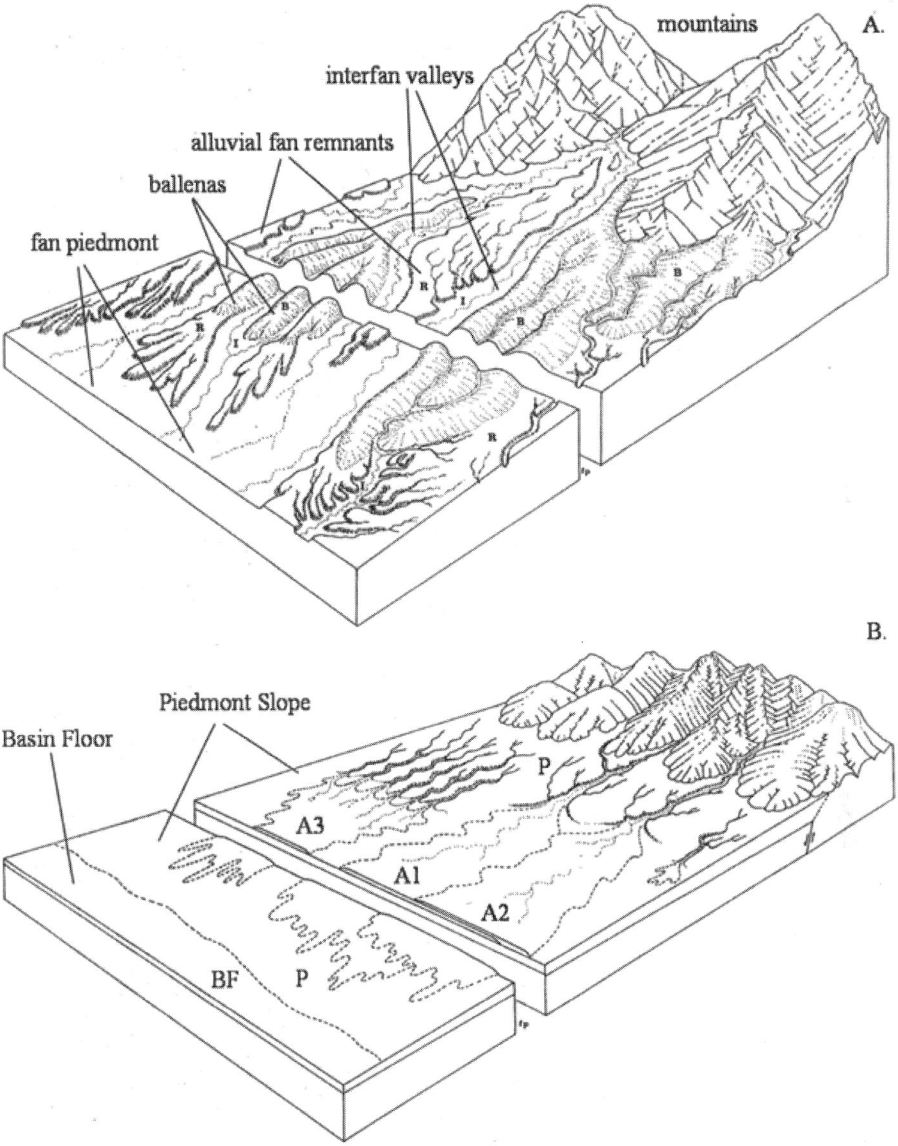

Figure 2-6. Block diagrams of piedmont slope landforms and stratigraphy. (A) Illustration of interfan valleys that are inset below and are younger than alluvial fan remnants, which, in turn, are inset below and are younger than ballenas. The interfan valleys coalesce downslope to form the fan piedmont, which descends to the basin floor. (B) Deposit A2 and its geomorphic surface, based on superposition relations, is younger than A1, which might or might not be younger than A3. Each surface (A1, A2, A3) is younger than surface P because they are cut into and are inset below P upslope, and overlie P downslope. Modified from Peterson (1981).

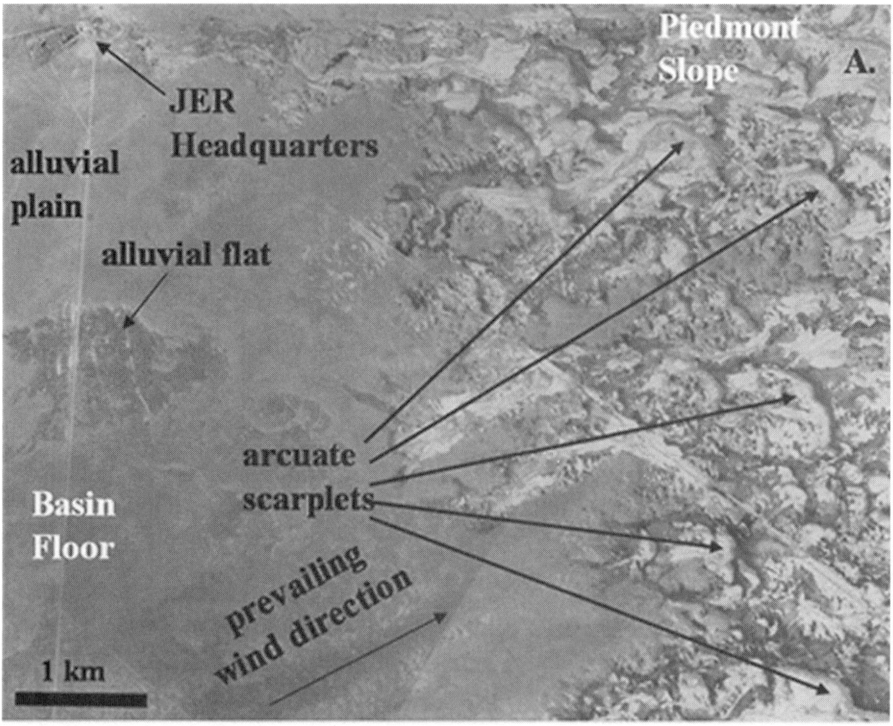
Piedmont
Slope
A.
JER
Headquarters
alluvial
plain
alluvial flat
arcuate
scarplets
Basin
Floor
prevailing
wind direction
1 km

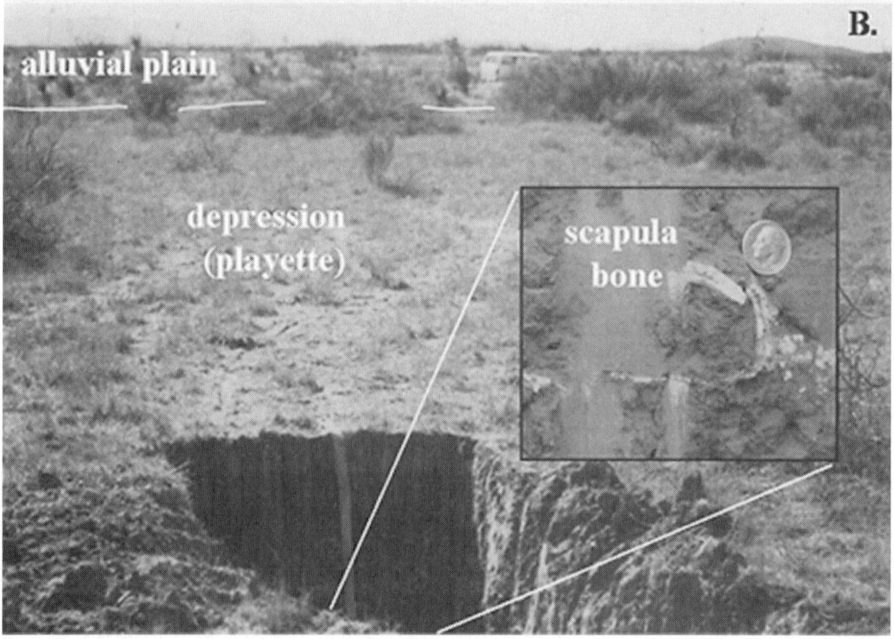
B.
alluvial plain
depression
(playette)
scapula
bone

and floodplain deposits of the Ancestral Rio Grande (Gile et al. 1981; Seager et al. 1987; Mack et al. 1997). Other basin floor landforms include the alluvial flat, playas, depressions (playettes), and gypsiferous landforms related to Pleistocene pluvial lakes and their associated eolian deposits. Depressions (playettes) are the smallest landforms in figures 2-4. These karst-like features, which penetrate through petrocalcic horizons, contain paleosols (figure 2-7b).

The Rio Grande Valley comprises the landforms formed by the episodic down-cutting and partial backfilling that occurred during the incision of the Rio Grande through basin fill and piedmont slope sediments (Hawley and Kottlowski 1969; Hawley 1975a; Gile et al. 1981). The youngest landform is the valley floor, which is the floodplain of the modern Rio Grande. Adjacent to the valley floor is the valley border, which contains a stepped sequence of fan terraces that graded to progressively higher base levels of the Rio Grande during the late quaternary. Other Rio Grande Valley landforms are ridges, inter-ridge valleys, structural benches, scarps, colluvial wedges, and arroyo channels (Gile et al. 1981).

Geomorphic Surfaces

Geomorphic surfaces are based on age. They are mappable surfaces of landforms that developed when climatic and base-level controls were relatively constant (see Daniels et al. 1971; Hawley 1972; Ruhe 1975; Gile et al. 1981). Many geomorphic surfaces at the Jornada Basin site have both an erosional and constructional (depositional) component. Examples of geomorphic surfaces and their age relations are shown in figure 2-6b. In this illustration a fan piedmont surface (P), which descends from a mountain front to a basin floor, is inset by and overlain by younger surfaces A1, A2, A3, inset higher on the slope and overlain lower on the slope. Geomorphic surface A1 is older than surface A2 but may or may not be older than A3. Soil morphology evidence would be needed to determine if A3 had greater profile development and was therefore older than either A1 or A2.

Geomorphic surfaces at the Jornada Basin site have been mapped for the Desert Project study area (Gile et al. 1981) where, for example, nine geomorphic surfaces occur in the Rio Grande Valley border as a stepped sequence (table 2-1). The ages of the geomorphic surfaces increase with increasing elevation of the steps, each step representing a temporary halt in river entrenchment and partial valley

Figure 2-7 (opposite page). Photograph of scarplets and depression soil. (A) Aerial photograph of basin floor and piedmont slope landforms (BLM 1980 color series). It shows reddish brown sands, blown from the alluvial plain to the west, encroaching on the piedmont slope, which in this area, is dominated by low winding ridges of quartzose sand paralleling arcuate erosional scarplets cut into silty alluvium from sedimentary rock (mainly limestone) of the San Andres Mountains. (B) Photograph of depression (playette) at Mayfield Well showing soil pit and scapula bone encountered at 60 cm on the surface of a paleosol. The excavated bone was identified by Dr. Art Harris (University of Texas, El Paso, 1996) as potentially being from a bison. Radiocarbon age of the organic material in the bone was 3,890 ± 59 years BP (Beta-109920). View to west; photographed March 1996.

backfilling (Hawley 1975b). Surfaces of similar age occur on the piedmont slopes (table 2-1; Gile et al. 1981). Downslope these descending surfaces overlap in places and occur as a sequence of buried paleosols (figure 4-7).

The age of a geomorphic surface and the age of its soils are the same (Gile et al. 1981). On a constructional (depositional) surface, for example, both the surface and the soil would date from the approximate time that sedimentation stopped and soil development began. Similarly, on an erosional surface, the age of the surface and soil would date from the time erosion stopped and pedogenesis began.

More than one soil type generally occurs on a single geomorphic surface. At the Jornada Basin multiple soil types are most commonly the result of lateral changes in surficial sediments in which soils form (Gile and Grossman 1979). Lateral changes in climate and vegetation on a single geomorphic surface also give rise to different soil types.

Soil Parent Materials of the Jornada Basin

Parent materials at the Jornada Basin site are highly contrasting in age, ranging from Precambrian granite to historical alluvium, and for unconsolidated parent materials, highly contrasting in textures, ranging from sand to clay (figures 2-8; table 2-3). The effect of parent material on soil texture, mineralogy, and chemical properties is important for vegetation in several ways. These include water-holding capacity, infiltration, permeability, aeration, erodibility, organomineral complexes, nutrient status, pH, toxicity, clay illuviation, and shrink-swell characteristics. For example, extrusive igneous rocks, such as rhyolite, disintegrate more slowly than intrusive igneous rocks, like monzonite (Gile et al. 1981). Consequently, soils derived from rhyolite contain more gravel and, therefore, have less plant available water than soils derived from monzonite (see Herbel et al. 1994). Another example is the effect that limestone parent material has on argillic horizon formation. With other soil-forming factors held constant, high-carbonate parent material derived from limestone will suppress the formation of argillic horizons, in contrast to similar soils derived from igneous rocks, in which argillic horizons are common (Gile et al. 1981).

Parent materials of the Jornada Basin, as with landforms, were compiled into a new Geographic Information System (GIS) layer for ecological studies. This map was made by combining new mapping with existing maps by Ruhe (1967), Gile et al. (1981), and Seager et al. (1987), as well as the 1918 and 1963 soil maps (chapter 4). In the new map, parent materials were grouped into six categories: igneous bedrock, sedimentary bedrock, alluvium from these two bedrock types, sediments derived from the Ancestral Rio Grande, and gypsiferous sediments (figure 2-8; table 2-3). Igneous bedrock consists of mostly tertiary felsic intrusive and extrusive rocks, though basaltic andesite and Precambrian granite are also in the area (Seager et al. 1987). Sedimentary bedrock is mostly Paleozoic marine carbonates with interbedded terrigenous clastic rocks. These range from Cambrian sandstone to interbedded limestone, gypsum, and sandstone of Permian age, with minor areas of Paleocene to Miocene conglomerate, sandstone, and mudstone from local sources (Seager et al. 1987).

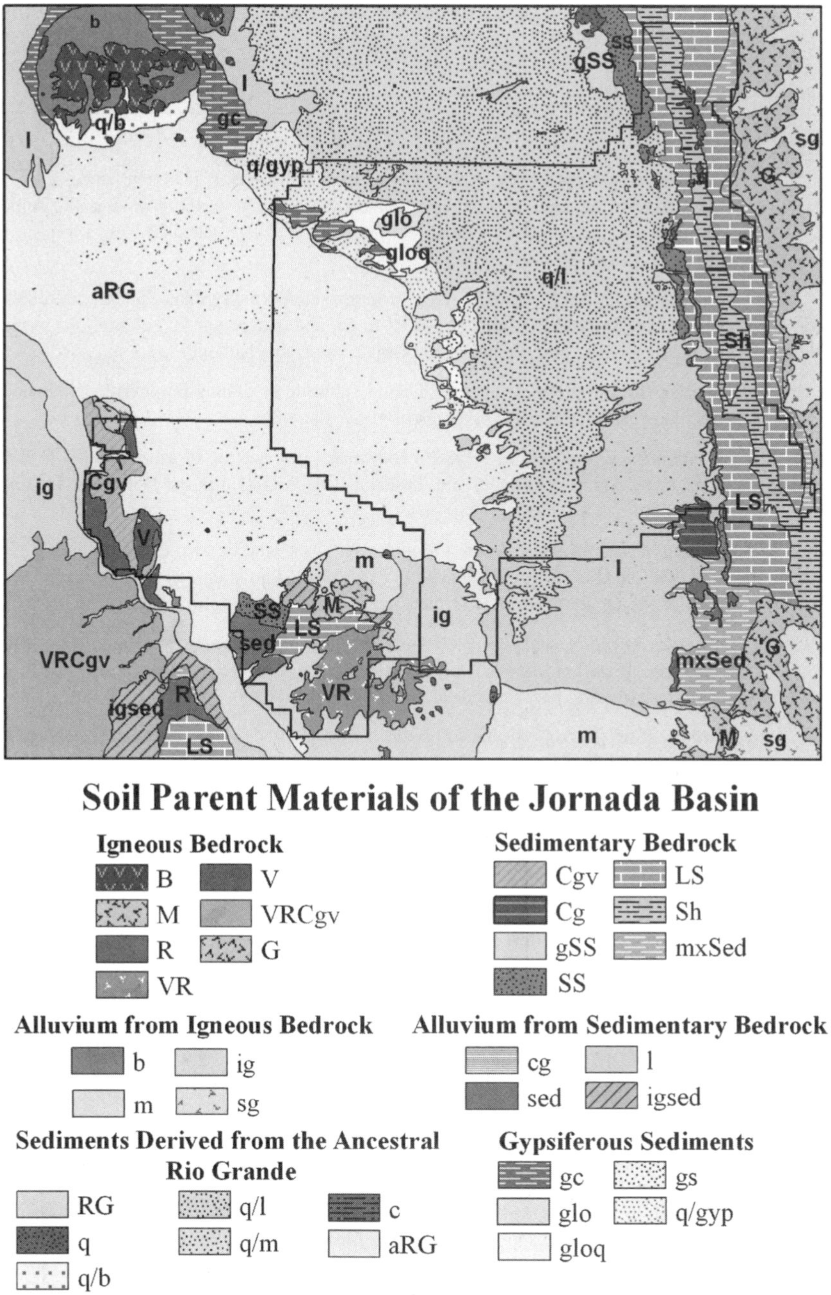

Figure 2-8. Soil parent material map of the Jornada Basin LTER site. Map unit definitions are given in table 2-3. See color insert.

Table 2-3. Definitions of soil parent material map units (shown in figure 2-8) of the Jornada Basin.

Symbols	Soil Parent Material Definitions
Igneous Bedrock	
B	***Vesicular Basaltic-Andesite.*** Rock outcrop and small areas of piedmont slope sediments of black to medium gray, vesicular basaltic-andesite of the Uvas Basaltic Andesite. Bedrock unit is a few hundred feet thick. Tertiary age, dated 25.9 ± 1.5 m.y., 26.1 ± 1.4 m.y., 27.4 ± 1.2 m.y. (S).[a]
M	***Silicic Plutonic Rocks.*** Rock outcrop of monzonite-porphyry dikes, stocks, and a laccolith of the Doña Ana Mountains (33.7 m.y.), and quartz-monzonite porphyry, granite, quartz syenite, and syenite of the Organ Mountains batholith (32.8 m.y.) (S).
R	***Rhyolite Intrusives.*** Rock outcrop of nonporphyritic to slightly porphyritic, siliceous sills, dikes, and small rhyolitic plugs of tertiary age (S).
VR	***Volcanic Rocks and Rhyolite Undifferentiated.*** Rock outcrop of ash-flow tuffs of the Doña Ana rhyolite (33.0 m.y.) and the Palm Park formation in the Doña Ana Mountains (S).
V	***Volcanic Rocks of Intermediate Composition.*** Rock outcrop of purplish epiclastic strata with beds of laharic breccia of the Palm Park formation and other volcaniclastic rocks of the Orejon Andesite. Tertiary age, dated 51 m.y., 43 m.y., and 42 m.y. (S).
VRCgv	***Volcanic, Rhyolitic, and Conglomeratic Rocks Undifferentiated.*** Volcanic rocks of the Palm Park formation, flow-banded rhyolite, Rincon Valley conglomerate, and small areas of basalt flows, basaltic andesite, and fanglomerate (S).
G	***Granite.*** Rock outcrop of pink to brown, coarse-grained granite of Precambrian age and associated amphibolites and quartzites (S).
Sedimentary Bedrock	
Cgv	***Conglomerates and Mudstones from Local Volcanic Rocks.*** Fanglomerate, conglomerate, conglomeratic sandstone, mudstone, and gypsiferous playa sediments of the Rincon Valley and Hayner Ranch formations (Miocene) (S).
Cg	***Conglomerate, Sandstone, Mudstone from Local Sedimentary Rocks.*** Pebble to boulder conglomerate, red mudstone, and sandstone of the Love Ranch Formation (Paleocene-Eocene). Sediments derived from local sedimentary rocks and Precambrian granite (S).
gSS	***Interbedded Sandstone, Gypsum, Limestone.*** Light-colored sandstones, gypsum, and fetid limestone of the Yeso formation (Permian)(S).
SS	***Siltstone, Fine Sandstone, Shale.*** Reddish-brown siltstone, fine sandstone, arkosic sandstone, and shale of the Abo formation (Permian) (S).
LS	***Limestones.*** Limestones and associated sedimentary rocks of the Hueco formation (Permian) in the Doña Ana Mountains, and the Hueco, Lead Camp formation (Pennsylvanian), and Mississippian and Devonian limestones and shales in the San Andres Mountains (S).
Sh	***Interbedded Shale, Sandstone, Siltstone, Gypsum.*** Brown to gray shale, sandstone, siltstone, gypsum, and fine-grained laminated limestone of the Panther Seep formation (Pennsylvanian) (S).
mxSed	***Mixed Sedimentary Rocks.*** Dolomites, limestones, and terrigenous clastic sedimentary rocks of the Fusselman dolomite (Silurian), Montoya group (Ordovician), El Paso group (Ordovician), and Bliss sandstone (Cambrian-Ordovician) in the eastern San

Symbols	Soil Parent Material Definitions
	Andres Mountains; and limestones mainly of the Hueco and Lead Camp formations and interbedded terrigenous clastics and gypsum of the Panther Seep formation in the southern San Andres Mountains (S).
Alluvium from Igneous Bedrock	
b	***Alluvium from Basaltic Andesite.*** Gravelly[b], arkosic[c] piedmont slope alluvium derived from the Uvas Basaltic Andesite in the Point-of-Rocks hills.
m	***Alluvium from Silicic Plutonic Ingneous Rocks.*** Arkosic piedmont slope alluvium derived from coarse-grained felsic and intermediate intrusive rocks, including quartz monzonite, porphyry, granite, quartz syenite, and syenite. Alluvium is gravelly near bedrock source but breaks down into finer sediments downslope.
ig	***Alluvium from Felsic and Intermediate Extrusive Igneous Rocks.*** Arkosic piedmont slope alluvium derived from the Doña Ana Rhyolite (felsic) and Palm Park formation (intermediate) in the Doña Ana Mountains; from felsic granites and quartz monzonites in the San Agustin Mountains (between the San Andres and Organ Mountains); and from mixed volcanic rock sources west of the Rio Grande. Alluvium is gravelly on the upper and middle piedmont slope and loamy in the lower zones.
sg	***Alluvium from Granite and Sedimentary Rocks.*** Gravelly, arkosic piedmont slope alluvium derived from Precambrian granite and Paleozoic sedimentary rocks on the eastern slopes of the San Andres Mountains. Map unit includes a small area at San Diego Mountain west of the Jornada basin floor, and an area containing map unit *ig* east of the San Agustin Mountains.
Alluvium from Sedimentary Bedrock	
cg	***Alluvium from Conglomerates.*** Gravelly, high-carbonate[d] piedmont slope alluvium derived from the Love Ranch formation of the San Andres Mountains.
sed	***Alluvium from Limestone and Siltstone.*** Gravelly, high-carbonate piedmont slope alluvium derived from limestones of the Hueco formation and siltstones of the Abo formation in the Doña Ana Mountains. Map unit also contains small areas of ancestral Rio Grande alluvium.
l	***Alluvium from Sedimentary Rocks, Mainly Limestone.*** High-carbonate piedmont slope alluvium derived from Paleozoic limestones and associated terrigenous clastic rocks. Alluvium is gravelly near bedrock sources but becomes loamy, then silty farther downslope.
igsed	***Alluvium from Mixed Igneous and Sedimentary Sources.*** Loamy, arkosic piedmont slope alluvium in the Doña Ana and Robledo Mountains derived from igneous rocks, with minor amounts of limestone detritus.
Sediments Derived from the Ancestral Rio Grande	
RG	***Modern Rio Grande Alluvium.*** Floodplain alluvium of the Rio Grande consisting of stratified fluvial deposits ranging from coarse-grained bedload sediments in the current and former meandering river channels to clay deposits formed in small oxbow lakes. Flooding and sediment deposition has been inactive since the completion of Elephant Butte Dam in 1917.
q	***Eolian Quartzose[e] Sand over Bedrock.*** Deposits of reddish brown quartzose sand blown from the basin floor by strong winds from the west–southwest.

(continued)

Table 2-3. *(continued)*

Symbols	Soil Parent Material Definitions
q/b	***Eolian Quartzose Sand over Basaltic Andesite Alluvium.*** Deposits of reddish brown quartzose sand from the basin floor that overlie and are mixed with piedmont slope alluvium derived from the Uvas basaltic andesite in the Point-of-Rocks hills.
q/l	***Eolian Quartzose Sand over Limestone Alluvium.*** Deposits of reddish brown quartzose sand blown from the basin floor that overlie and are mixed with piedmont slope alluvium derived from Paleozoic limestones and associated terrigenous clastic rocks.
q/m	***Eolian Quartzose Sand over Alluvium from Silicic Plutonic Rocks.*** Deposits of reddish brown quartzose sand blown from the basin floor that overlie and are mixed with piedmont slope alluvium derived from coarse-grained plutonic rocks of the northern Doña Ana Mountains.
c	***Clayey Playa Sediments.*** Clay-rich, nongypsiferous alluvium with shrink-swell properties. Vertisols occur in the larger playas (DP).
aRG	***Ancestral Rio Grande Alluvium (Camp Rice Formation, Fluvial Facies).*** Fluvial sediments from upstream sources deposited as fluvial fans and floodplains by the ancestral Rio Grande (H, DP). Sediments consist of arkosic sand, rounded pebbles of mixed lithology, floodplain mudstones and siltstones, occasional clay rip-up clasts less than 1 long, sporadic groundwater calcretes, and infrequent pumice lenses. Deposition spanned a period from about 5 million years ago when the river system reached the Jornada region to about 0.8 million years ago when the river began downcutting through its previously deposited sediments (M). The constructional top of the Camp Rice formation fluvial facies is the La Mesa surface, though younger surfaces are present in the southeastern most portion of the Jornada Basin (G). Strongly developed soils with thick petrocalcic horizons are typically components of the La Mesa surface. Wind erosion of those soils above the petrocalcic horizon has been the main source of quartzose sand accumulated on downwind piedmont slopes and bedrock areas.
Gypsiferous Sediments	
gc	***Gypsiferous Lake Bed Clay.*** Reddish brown calcareous clay to silty clay loam lake sediments containing rounded, gravel-size gypsum crystals. These fine grained sediments overlie gypsum beds.
glo	***Gypsiferous Loamy Sediments.*** Loamy, calcareous, gypsiferous sediments of lake plain terraces and playas. In some areas, gypsum beds have been exposed by erosion.
gloq	***Gypsiferous Loamy Sediments and Quartzose Sand.*** Similar to map unit *glo* except this unit contains additions of reddish quartzose sand blown from the adjacent alluvial plain.
gs	***Grayish Gypsiferous Sands.*** Sandy, grayish, calcareous deposits high in gypsum blown from adjacent lake beds by strong winds. Unit also contains significant amounts of grayish quartzose sand.
q/gyp	***Eolian Quartzose Sand over Gypsiferous Sediments.*** Sandy deposits of reddish quartzose sand blown from the adjacent alluvial plain. These deposits overlie and are mixed with gypsum-rich lacustrine and eolian sediments.

[a]Letters in parentheses refer to references listed in table 2-2.

[b]Gravelly, sandy, loamy, silty, and clayey are used qualitatively but are estimated equivalents of skeletal, sandy, loamy, and fine as used in Soil Taxonomy (SS).

[c]Arkosic designates notable quantities of feldspar (≥25%) in addition to quartz, the dominant mineral.

[d]High-carbonate designates parent material with greater than about 15% $CaCO_3$ (Gile et al. 1981).

[e]Quartzose designates deposits that are almost totally quartz.

Alluvium from igneous and sedimentary bedrock occurs downslope from bedrock outcrops. Some general characteristics are that alluvium derived from limestones tends to be highest in silt, whereas alluvium derived from rhyolites tends to be highest in gravel. Alluvium from limestone has high carbonate content, and alluvium from igneous bedrock has notable quantities of feldspar ($\approx$25%) in addition to quartz, which is the dominant mineral. Alluvium from limestone contains more mixed-layer clay minerals than alluvium from igneous rocks, which contains more kaolinite and smectite (Monger and Lynn 1996). Illite occurs in similar quantities in both sedimentary and igneous alluvial types.

Sediments derived from the Ancestral Rio Grande are fluvial sediments from upstream sources (table 2-3). Also included in this unit are clayey playa sediments, modern Rio Grande alluvium, and eolian quartzose deposits. The eolian quartzose deposits have blown from eroded soils of the alluvial plain and accumulated over a variety of bedrock and alluvial units (figure 2-9a). The red color, which is readily detectable by satellite images, is the result of coatings of clay and admixed iron oxide on quartz and other sand grains (figure 2-9b).

Gypsiferous sediments, which extend from the JER headquarters to the northwest, originated in pluvial lakes of the late quaternary. These gypsiferous clayey sediments of lake-plain playas, loamy sediments of lake-plain terraces, and sandy sediments of eolian deposits associated with the lakes are similar but in miniature to those of White Sands National Monument on the eastern side of the San Andres Mountains. Unlike White Sands, however, the Jornada gypsiferous sediments have received large influxes of quartzose sand blown from the adjacent alluvial plain.

Conclusions

The coevolution of ecosystems and landscapes in the area that is now the Jornada Basin site extends back into the Precambrian. However, it is the late phase of the Rio Grande rift system that is mainly responsible for the modern Jornada landscape. This phase began in the latest Miocene or Pliocene and consisted of renewed extensional tectonics, the associated tilting of strata, progressive uplift of north–south trending mountain chains, development of intervening structural basins, filling of those basins with sediments and rain water, the arrival of the Ancestral Rio Grande and fluvial fan deposits, followed by the entrenchment of the Rio Grande below the level of the basin floors and the formation of the Rio Grande Valley. During most of this time, the climate was dry enough that carbonate occurred in paleosols (Mack et al. 1994c; chapter 4). Yet the climatic swings were high enough that during glacial maxima lakes formed in southern New Mexico in general (Hawley 1993) and the Jornada Basin in particular (Gile 2002).

With C_3 woodlands in the mountains of the Jornada Basin site, C_3 shrublands in the lower and drier areas and C_4 grasslands in between, the Jornada Basin is and has been a site of dynamic competition between these plant communities as climatic swings intermittently tilts conditions in favor of one group or another. Based on modern measurements (Wainwright et al. 2000) and the stratigraphic

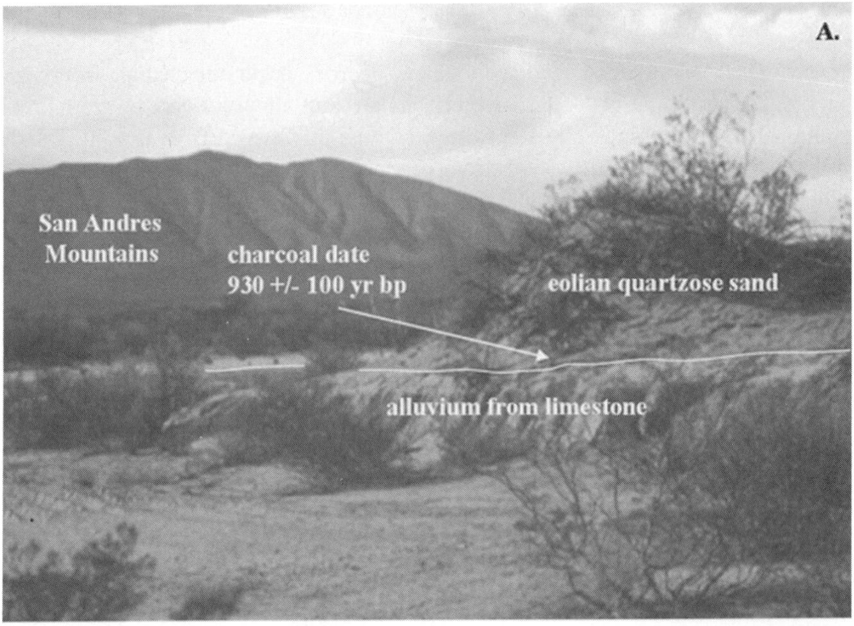

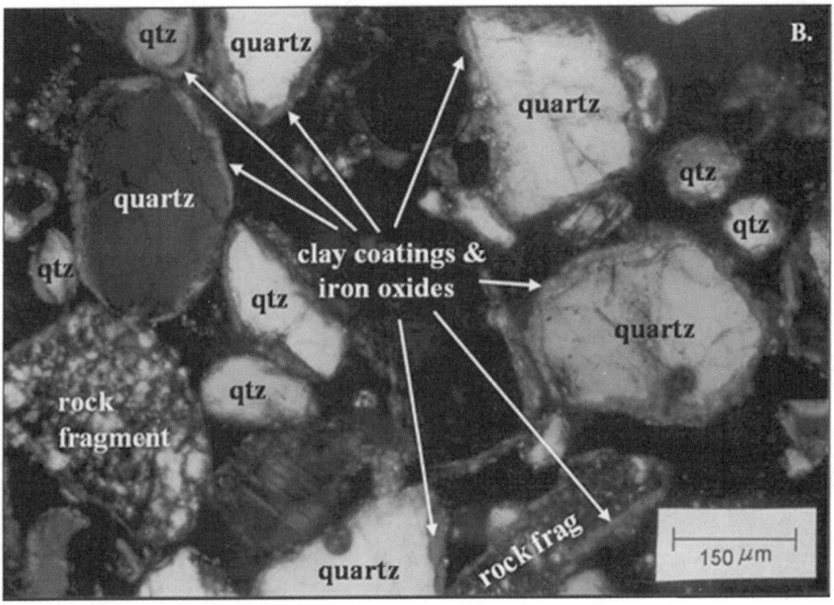

Figure 2-9. Photograph of coppice dunes and petrographic thin section of quartzose sand. (A) Photograph on fan piedmont 4.5 km northeast of JER Headquarters showing eolian quartzose sand disconformably overlying silty alluvium from sedimentary rock (mainly limestone) of the San Andres Mountains. A charcoal layer exposed at the sedimentary contact provided a radiocarbon date of 930 ± 100 years BP (Beta-109919) indicating the dune formed sometime after that time. View taken to southeast; photographed 1996. (B) Photomicrograph of a thin section of reddish brown sand collected from the land surface near Red Lake in the northwest region of the JER. It shows that clay and admixed iron oxides coat sand grains, which are mainly quartz, and impart the prominent reddish brown color to the sandy landscapes of the Jornada basin floor and leeward piedmont slope.

record of paleosols (chapter 4), erosion paralleled the decline of grasslands as progressive amounts of bare ground associated with increasing shrublands left soil unprotected and vulnerable to detachment and transport by running water and strong wind.

The four major landscape components at the Jornada Basin site are (1) mountains and hills, (2) piedmont slopes (bajadas), (3) basin floors, and (4) the Rio Grande Valley. These major landscape components are subdivided into landforms, such as ballenas, playettes, and the valley floor (table 2-2). Landforms are discrete pieces of the landscape that impact vegetation by their configuration, which influences redistribution of runoff water (Wondzell et al. 1996) and microclimate (Dick-Peddie 1993). Landforms also impact vegetation by their soil-related physical and chemical properties, such as water-holding capacity and soluble salt content. Landforms are subdivided, based on age, into geomorphic surfaces that in turn are subdivided, based on lateral variability, into soil types.

Parent materials at the Jornada Basin occur as colluvial, alluvial, fluvial, lacustrine, eolian, and residual bodies derived from assorted intrusive to extrusive, mafic to felsic igneous rocks, as well as various marine to terrigenous clastic sedimentary rocks (table 2-3). Like landforms, parent materials have a controlling influence on soil physical and chemical properties important for vegetation. For example, intrusive igneous rocks give rise to gravelly soils with lower plant-available water because they break down less readily than extrusive igneous rocks.

Owing to its assorted parent materials, multifaceted landforms, and bioclimatic setting, the Jornada Basin site is diverse and dynamic. In modern times, there are increasing human impacts superimposed on the natural diversity, which has resulted in desertification (Schlesinger et al. 1990; see also chapter 1). Quantifying human-induced desertification in the context of natural cycles of desertification will be a challenge but will provide important insights about the effect humans can have on the arid–semiarid environment.

3

Climate and Climatological Variations in the Jornada Basin

John Wainwright

The purpose of this chapter is to review the climatic data for the Jornada Basin over the period for which instrumental records exist. Over this time period, up to 83 years in the case of the Jornada Experimental Range (JER), we can deduce both the long-term mean characteristics and variability on a range of different spatial and temporal scales. Short-term variability is seen in individual rainstorms. Longer-term patterns are controlled spatially by factors such as large-scale circulation patterns and basin and regional orography and temporally by the large-scale fluctuations in atmospheric and oceanic circulation patterns. Variability can have significant impacts on the biogeography of a region (Neilson 1986) or its geomorphic processes (Cooke and Reeves 1976), which may set in motion a series of feedbacks, most important those referring to desertification (Schlesinger et al. 1990; Conley et al. 1992). Understanding the frequency and magnitude of such variability is therefore fundamental in explaining the observed landscape changes in areas such as the Jornada Basin.

The Instrumental Climate Record of the Jornada Basin

The patterns observed for different climatic variables within the available instrumental records for the Jornada Basin are defined in a hierarchical series of temporal scales, starting with the patterns that emerge from long-term average conditions and moving to seasonal and monthly, daily, and subdaily time scales. Two further analyses are made because of their potential importance to the hydrological and ecological characteristics of the basin, namely, the occurrence of extreme rainfall events and of longer-term changes. The effects of El Niño events in controlling the rainfall over decadal time scales will be addressed in particular. Spatial variability is an additional important concern, especially when characterizing dryland areas such as the Jornada Basin, where spatial variability tends to be high.

The overall climate of the basin can be defined according to the Köppen classification as being cool and arid, belonging to the midlatitude desert zone (BWk). However, interannual variability is important, and occasionally, the annual conditions are more characteristic of the semiarid steppe (BSk) zone. The higher rainfall rates in the higher altitudes of the basin are also more characteristic of semiarid conditions. Using the Thornthwaite approach, the basin is defined as arid (zone E) and only rarely crosses the threshold into semiarid in the higher altitudes. This spatial and temporal variability in classification probably explains the fact that previous overviews have placed the Jornada in both arid and semiarid zones.

Description of Data Available for the Jornada Basin

Instrumental climatic data for the Jornada Basin are available from these four sources: the Long-Term Ecological Research (LTER) project itself, the USDA JER, the New Mexico State University Chihuahuan Desert Rangeland Research Center (CDRRC), and the U.S. Geological Survey (USGS) (figure 3-1). The

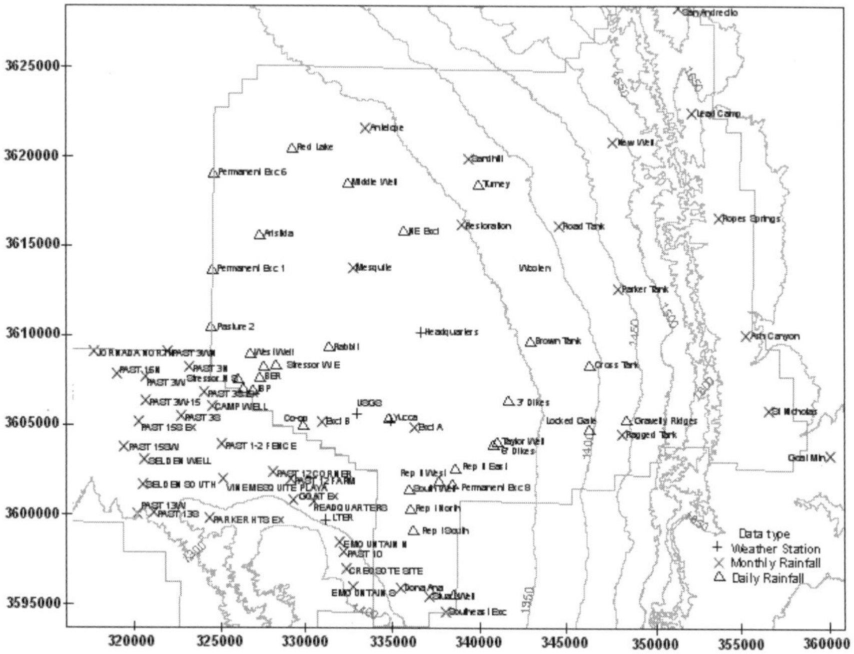

Figure 3-1. Map of locations in the Jornada Basin at which climatic variables have been recorded. Locations are displayed within a typical UTM grid.

LTER data principally derive from the recording area weather station and the rain gauge at the base of Summerford Mountain (lower left of figure 3-1), which started in 1983 and 1981, respectively. The USDA data set contains the largest series of records, and those extend back over the longest period of time. Rainfall and temperatures have been recorded at the JER headquarters continuously since June 1914. Monthly pan evaporation was also recorded between January 1953 and August 1979. Rainfall data have also been recorded as monthly totals collected at 33 rain gauges, as well as at the headquarters, of which 23 have a record longer than 50 years. Daily rainfall data have been recorded at 58 locations since 1976, although most of these data do not cover the entire period since 1976. The CDRRC data are made up of monthly rainfall totals for 26 locations, the earliest being the CDRRC headquarters rain gauge, which was first operated in 1930. A number of these stations are discontinuous or have short records. The USGS data relate to the wind erosion project (described in chapter 9) in the central part of the basin and include daily records of temperature, precipitation, and wind speed since January 1990. Maximum rainfall intensities are recorded at a monthly level for durations of 15 minutes to 3 hours.

This impressive quantity of data makes it possible to look at temporal changes for most of the 20th century with a good temporal resolution for at least 10 years. Spatial patterns are well represented over a 50 year period in terms of monthly rainfall and over a 20 year period at a daily resolution. Comparisons of temperature and wind speed are much more restricted.

Annual Patterns

The long-term average rainfall for the Jornada headquarters between 1915 and 1995 is 245.1 mm with a standard deviation of 85.0 mm (table 3-1). The coefficient of variation is 34.7%. The minimum recorded value for a complete year is 77.0 mm, which occurred in 1953, with the maximum of 507.2 mm falling in 1984. The average annual temperature between 1915 and 1993 is 14.70°C with a standard deviation of 0.58°C. The minimum average annual temperature was 13.54°C in 1987, and the maximum average annual temperature of 16.25°C occurred in 1954. As already noted, measurements of evaporation (standard class "A" pan) were made over a shorter time period, between 1953 and 1979, with values ranging from 1,565.2 mm (1976) to 2,832.6 mm (1971) and with a mean value of 2,204.1 mm and a standard deviation of 279.0 mm. There is a high interannual variability, particularly of rainfall and potential evaporation. On the annual time scale, there is always a large moisture deficit (evaporation minus precipitation) averaging 1,960.3 mm.

Comparisons with other sites in the basin are only possible with rainfall data. Because of the temporal variability, it is not possible to use all the available rain gauges to carry out this analysis. Thus, long series of data are derived from 37 sites for which there exist more than 30 years of data from the period since 1947. For those sites within this series that have longer sequences, there appear to be no significant differences between the total sequence and the sequence after 1947.

Table 3-1. Long-term (1915–95) averages, minimums, and maximums for precipitation (P), ambient air temperature, potential evaporation (PE), and moisture balance (PE-P) as recorded at the instrumented station at the JER headquarters.

	Jan	Feb	Mar	Apr	May	Jun	Jul	Aug	Sep	Oct	Nov	Dec	Annual
Precipitation mm													
Average	12.80	9.76	7.66	5.78	10.94	16.01	46.66	50.05	35.61	23.79	11.85	17.65	245.09
SD	11.52	9.30	9.17	9.14	15.26	22.24	29.54	33.24	28.76	20.73	14.47	18.72	85.58
SE	1.26	1.02	1.01	1.00	1.69	2.43	3.24	3.65	3.16	2.28	1.59	2.05	9.57
Minimum	0.00	0.00	0.00	0.00	0.00	0.00	6.35	1.52	0.00	0.00	0.00	0.00	76.96
Median	10.67	7.11	4.32	2.03	5.08	7.01	38.10	42.42	30.99	18.50	6.86	13.46	227.84
Maximum	49.53	39.88	47.50	51.31	78.49	135.10	147.32	167.13	114.05	86.61	80.26	100.33	507.24
Average temperature °C													
Average	3.78	6.43	9.57	14.00	18.59	23.88	26.03	24.81	21.51	15.15	8.19	3.97	14.70
SD	1.86	1.65	1.44	1.57	1.35	1.47	1.01	1.10	1.37	1.20	1.42	1.35	0.58
SE	0.21	0.19	0.17	0.18	0.16	0.17	0.12	0.13	0.16	0.14	0.17	0.16	0.07
Minimum	−0.61	1.83	6.65	8.43	12.72	18.89	23.89	21.88	19.37	12.08	5.17	0.71	13.54
Median	3.79	6.16	9.56	13.94	18.52	24.01	26.13	24.68	21.27	15.32	8.07	3.87	14.68
Maximum	8.33	11.18	12.96	17.82	22.03	26.83	28.60	27.37	26.94	18.14	11.10	7.43	16.25
Minimum temperature °C													
Average	−5.99	−3.71	−0.94	3.24	7.67	13.10	17.11	16.08	12.17	4.98	−2.29	−5.46	4.70
SD	2.56	2.35	1.84	2.03	1.75	2.12	1.47	1.68	1.93	2.10	2.02	2.88	1.03
SE	0.29	0.27	0.21	0.23	0.20	0.24	0.17	0.19	0.22	0.24	0.24	0.34	0.13
Minimum	−13.17	−11.17	−6.77	−3.33	2.67	6.63	12.39	10.57	6.44	−1.34	−7.78	−10.11	1.51
Median	−5.78	−3.56	−0.57	3.43	7.48	13.46	17.42	16.38	12.34	5.48	−2.37	−5.68	4.79
Maximum	0.08	2.22	1.87	7.24	11.88	17.28	20.27	19.30	16.02	8.69	3.11	11.48	6.76

(continued)

Table 3-1. (*continued*)

	Jan	Feb	Mar	Apr	May	Jun	Jul	Aug	Sep	Oct	Nov	Dec	Annual
Maximum temperature °C													
Average	13.50	16.55	20.16	24.76	29.46	34.69	34.96	33.52	30.84	25.41	18.65	13.73	24.74
SD	2.22	1.89	2.08	2.14	1.73	1.51	1.46	1.38	1.77	1.72	2.06	2.40	0.83
SE	0.25	0.22	0.24	0.24	0.20	0.17	0.17	0.16	0.21	0.20	0.24	0.28	0.10
Minimum	6.61	12.18	13.79	15.98	20.76	30.11	31.59	30.46	27.35	21.04	13.28	8.53	22.17
Median	13.36	16.30	20.31	24.91	29.63	34.69	35.00	33.49	30.63	25.56	18.49	13.79	24.76
Maximum	17.94	20.99	24.83	28.41	32.58	38.06	38.15	36.31	36.67	28.96	22.32	23.52	26.45
Evaporation mm													
Average	58.65	96.60	181.99	255.47	299.26	323.82	270.65	236.14	191.94	142.93	88.18	54.37	2204.05
SD	25.69	39.72	41.97	38.90	38.14	34.56	38.51	35.45	34.94	25.43	27.92	29.64	279.04
SE	4.94	7.64	8.08	7.49	7.34	6.65	7.41	6.82	6.99	4.99	5.48	5.81	54.72
Minimum	0.00	0.00	106.93	182.88	223.27	257.81	178.31	126.75	136.40	91.95	37.59	0.00	1565.15
Median	57.15	91.19	174.75	246.63	301.50	321.56	280.92	245.62	185.17	140.21	86.61	56.77	2215.39
Maximum	108.97	213.11	306.32	330.96	395.22	387.86	361.44	303.53	255.02	184.15	168.15	141.73	2832.61
Moisture balance mm													
Average	−47.5	88.5	−175.1	−251.9	−293.3	−312.2	−220.5	−185.1	−157.3	−116.9	−76.7	−40.9	−1977.6
SD	29.8	43.1	46.6	40.6	40.3	36.9	57.1	56.6	62.4	40.2	37.9	35.4	313.5
SE	5.7	8.3	9.0	7.8	7.8	7.1	11.0	10.9	12.5	7.9	7.4	6.9	62.7
Minimum	−109.0	−213.1	−306.3	−329.9	−395.2	−384.6	−340.1	−280.2	−255.0	−184.2	−168.1	−127.3	−2637.0
Median	−48.5	−85.9	−170.2	−242.3	−300.5	−316.7	−222.5	−193.8	−159.3	−125.5	−82.2	−43.7	−1992.4
Maximum	3.0	17.3	−59.4	−172.7	−210.6	−224.8	−57.2	−31.0	−40.9	−39.4	26.2	40.9	−1289.1

Shorter sequences commencing after 1947 show an increasing divergence from the long-term values, and thus the sample cannot be assumed to be comparable. The average annual rainfall for these sites is 242.3 mm, which is close to the Jornada headquarters value of 247.1 mm since 1947 (compared to 245.1 mm for 1915–95). However, this average is made up of values ranging from 212.8 mm to 348.9 mm. Similarly, the values of the coefficient of variation range from 24.8% to 43.1%, with a median value of 36.5% (compare with 34.7% for Jornada headquarters). The variability is largely due to the altitude of the rain gauge (figure 3-2a–d). For example, the highest rainfall (348.9 mm) is recorded from the Rope Springs gauge (see figure 3-1 for gauge name and location) at an altitude of 1,725 m in the San Andres Mountains. Linear regression analysis shows that 77% of the variance in annual average rainfall is controlled by altitude. However, there is a great deal of scatter due to two locational factors: (1) a general gradient of higher rainfall in the south of the basin, and (2) an increase away from the central north–south axis of the basin. Incorporating these two variables gives a general model for basin annual average rainfall since 1947 of

$$r = 3{,}140.3 + 0.342z - 9.297 \times 10^{-4}y - 1.475 \times 10^{-3}d_m, \qquad (3-1)$$

where r is mean annual rainfall (mm), z is altitude (m), y is the UTM northing (m), and d_m is the absolute distance from the medial N–S axis of the basin (m) in which 4.5% of the variance is explained by y and 2.7% by d_m. Similarly, there is a tendency for the variability of the annual average rainfall to increase with altitude, although this is countered by a decrease with distance from the north–south axis of the basin. This relationship is described by

$$sd_r = -61.62 + 0.114z - 9.454 \times 10^{-4}d_m \qquad (3\text{-}2)$$

in which sd_r is the standard deviation of the annual average rainfall (mm). The total variance explained by this relationship is 47.3%, of which 39.4% is explained by variations in altitude. It is likely that these relationships reflect the orographic control of the San Andres Mountains (right side of figure 3-1) on rainfall and the distance from the source of the rainfall (i.e., from the Gulf of Mexico). It is possible that the decrease in rainfall and its variability reflects a vegetation feedback on precipitation with the more bare central areas of the basin leading to more atmospheric uplift. Associated convective rainfall activity might then be reflected in the decreasing variability away from the center of the basin.

Monthly and Seasonal Patterns

Within the annual patterns described, there are important variations at the monthly and seasonal level. The Jornada headquarters data show a peak in rainfall between July and October, the maximum being in August, with a much smaller secondary peak in November to February (table 3-1). This seasonality is related to the different source of rainfall. In the summer months, rainfall is monsoonal in origin with moisture derived from the Gulf of Mexico. These months are characterized by thunderstorm activity, which frequently leads to relatively large precipitation

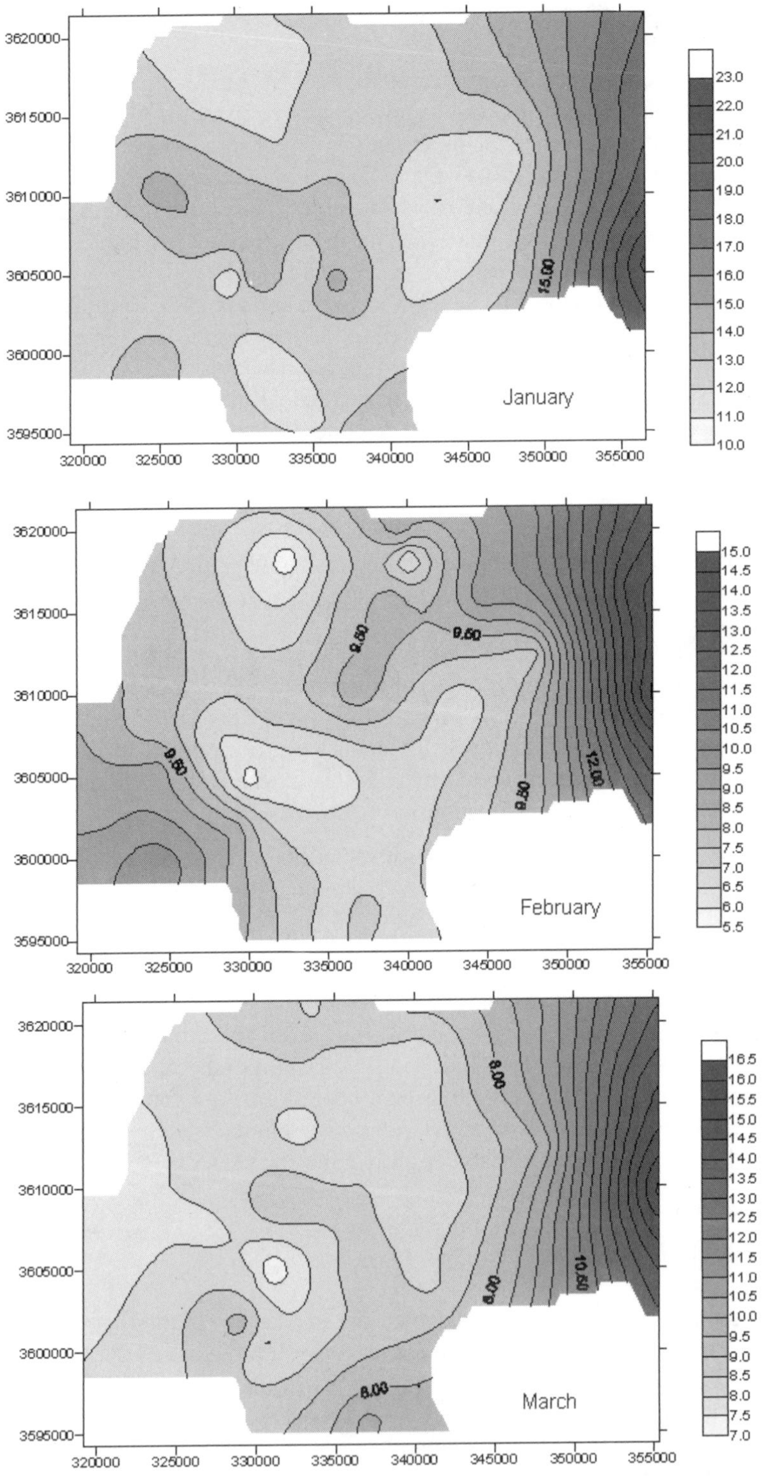

Figure 3-2. Monthly distributions of precipitation in the Jornada Basin: (a) January–March; (b) April–June; (c) July–September; (d) October–December. Distributions are displayed within a UTM grid.

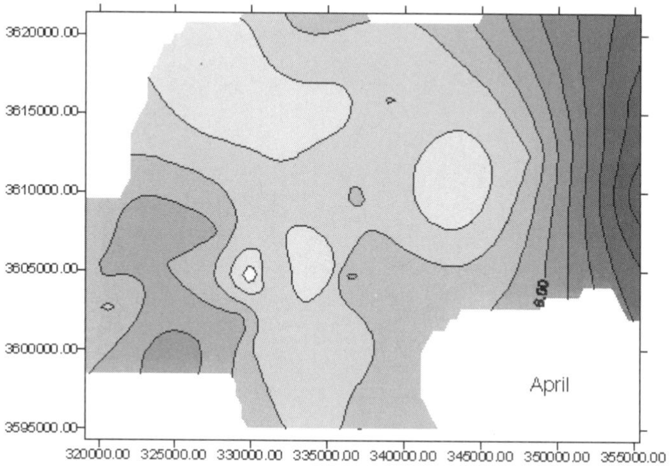
3620000.00
3615000.00
3610000.00
3605000.00
3600000.00
3595000.00
320000.00 325000.00 330000.00 335000.00 340000.00 345000.00 350000.00 355000.00
6.00
April
9.0
8.5
8.0
7.5
7.0
6.5
6.0
5.5
5.0
4.5
4.0
3.5

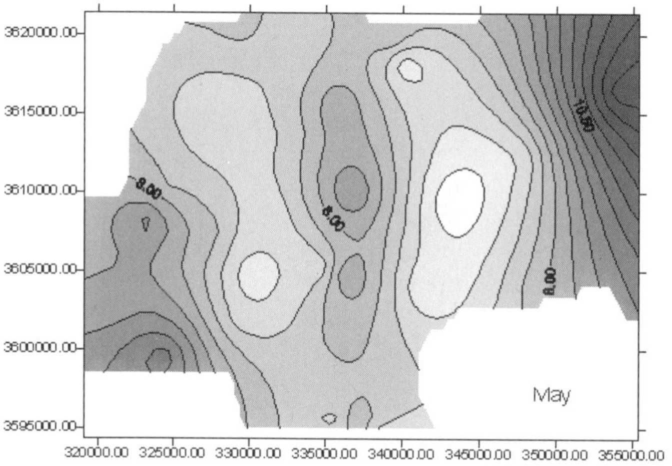
3620000.00
3615000.00
3610000.00
3605000.00
3600000.00
3595000.00
320000.00 325000.00 330000.00 335000.00 340000.00 345000.00 350000.00 355000.00
8.00
8.00
8.00
10.50
May
22.0
21.0
20.0
19.0
18.0
17.0
16.0
15.0
14.0
13.0
12.0
11.0
10.0
9.0
8.0

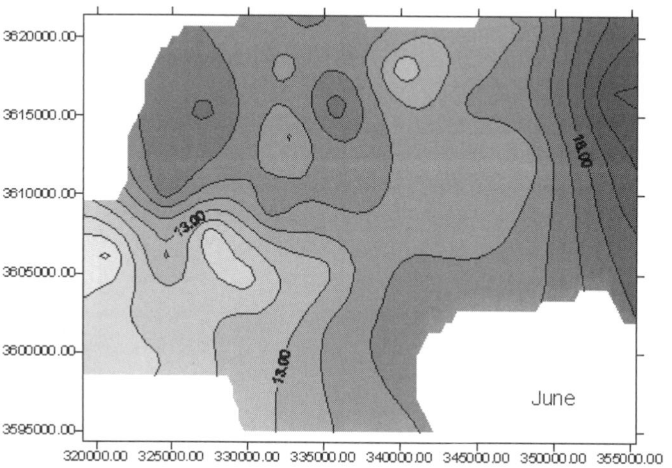
3620000.00
3615000.00
3610000.00
3605000.00
3600000.00
3595000.00
320000.00 325000.00 330000.00 335000.00 340000.00 345000.00 350000.00 355000.00
13.00
13.00
18.00
June
13.0
12.5
12.0
11.5
11.0
10.5
10.0
9.5
9.0
8.5
8.0
7.5
7.0
6.5
6.0
5.5

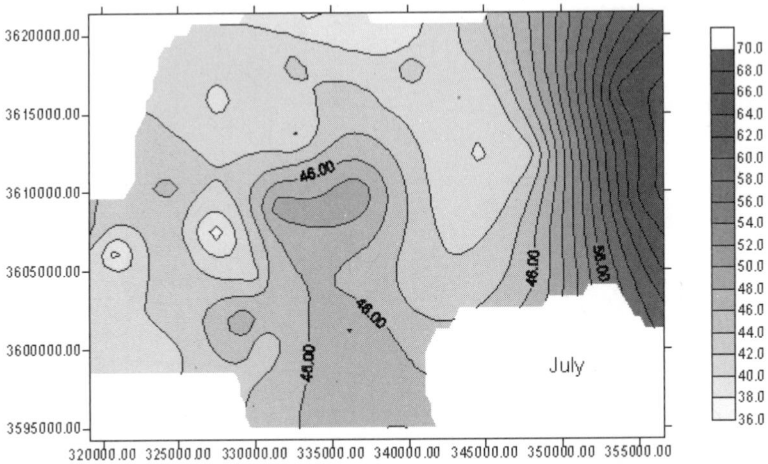
3620000.00
3615000.00
3610000.00
3605000.00
3600000.00
3595000.00
320000.00
325000.00
330000.00
335000.00
340000.00
345000.00
350000.00
355000.00
46.00
46.00
46.00
46.00
56.00
July
70.0
68.0
66.0
64.0
62.0
60.0
58.0
56.0
54.0
52.0
50.0
48.0
46.0
44.0
42.0
40.0
38.0
36.0

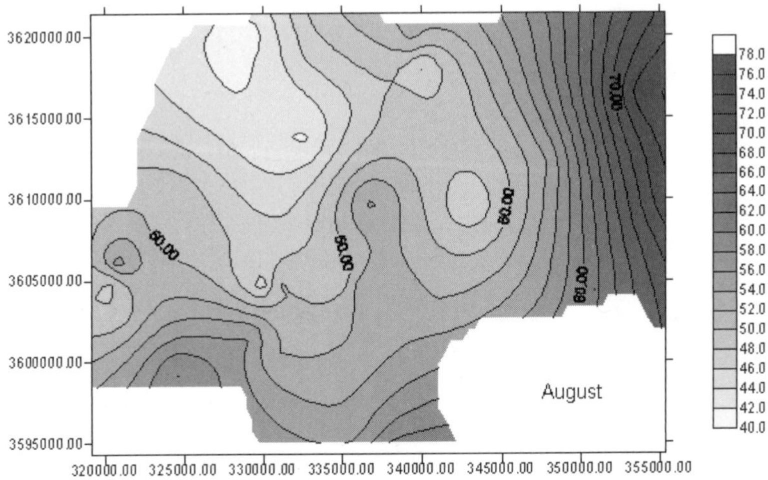
3620000.00
3615000.00
3610000.00
3605000.00
3600000.00
3595000.00
320000.00
325000.00
330000.00
335000.00
340000.00
345000.00
350000.00
355000.00
70.00
60.00
50.00
50.00
60.00
August
78.0
76.0
74.0
72.0
70.0
68.0
66.0
64.0
62.0
60.0
58.0
56.0
54.0
52.0
50.0
48.0
46.0
44.0
42.0
40.0

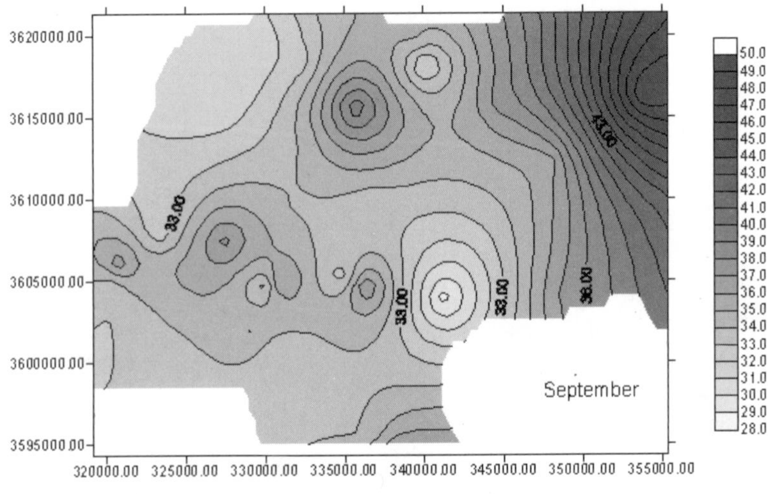
3620000.00
3615000.00
3610000.00
3605000.00
3600000.00
3595000.00
320000.00
325000.00
330000.00
335000.00
340000.00
345000.00
350000.00
355000.00
43.00
33.00
33.00
33.00
36.00
September
50.0
49.0
48.0
47.0
46.0
45.0
44.0
43.0
42.0
41.0
40.0
39.0
38.0
37.0
36.0
35.0
34.0
33.0
32.0
31.0
30.0
29.0
28.0

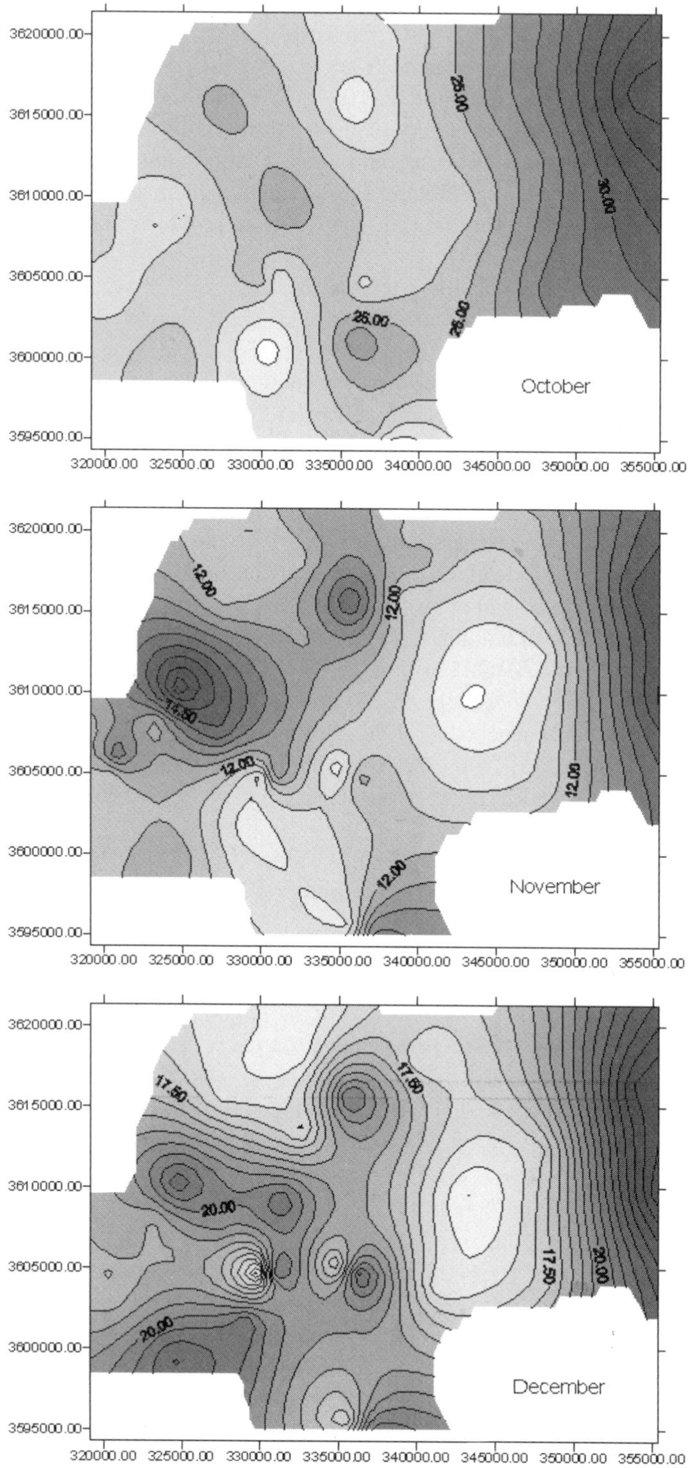
October
3620000.00
3615000.00
3610000.00
3605000.00
3600000.00
3595000.00
320000.00 325000.00 330000.00 335000.00 340000.00 345000.00 350000.00 355000.00
25.00
30.00
26.00
26.00
34.0 33.0 32.0 31.0 30.0 29.0 28.0 27.0 26.0 25.0 24.0 23.0 22.0 21.0 20.0
November
12.00
14.50
12.00
16.5 16.0 15.5 15.0 14.5 14.0 13.5 13.0 12.5 12.0 11.5 11.0 10.5 10.0 9.5
December
17.50
20.00
24.5 24.0 23.5 23.0 22.5 22.0 21.5 21.0 20.5 20.0 19.5 19.0 18.5 18.0 17.5 17.0 16.5 16.0 15.5 15.0

events in the basin. The rainfall in the winter months tends to be more frontal in character, and moisture largely comes from the Pacific coast. Therefore, both because of the lower intensity rainfall and the greater rain shadow effect, rainfall tends to be lower in total. There is important variability in the monthly totals with minimum values of zero in all but July and August. Maximum monthly rainfalls exceeding 100 mm have been recorded in July, August, September, and December. The maximum recorded monthly rainfall was 167.1 mm, which occurred in August 1984.

Average temperatures are at their lowest in January with a mean value of 3.78°C. The lowest recorded value of −0.61°C occurred in January 1919, whereas the following January the average temperature reached its maximum of 8.33°C. Apart from the month of March, the first six months of the year tend to show greater variability, which is seen not in the standard deviation but in the recorded extremes. Peak average monthly temperatures occur in July when the average is 26.03°C, with a minimum recorded value of 23.89°C (1962) and a maximum of 28.60°C (1951). Minimum and maximum temperatures follow the same cycle. Average January minimum temperatures are −5.99°C, the lowest recorded value being −13.17°C in 1963. Indeed, on only one occasion since 1915 has the January minimum temperature averaged above freezing. That was in 1916 when the value of 0.08°C was recorded. Maximum daytime temperatures average 13.5°C in January with a range from 6.61°C (1919) to 17.94°C (1920) and rise to an average of 34.96°C in July. The coldest July on record was in 1916, when the maximum daytime temperature only reached 31.59°C, and the warmest July saw mean daytime temperatures of 38.15°C in 1980. Minimum temperatures peak at 17.11°C in July, although the observed range of values is from 12.39°C (1963) to 20.27°C (1935).

The onset of frosts starts on average on October 22, although dates as early as September 14 (1959) and as late as December 1 (1932) have been recorded. There is a similar variability in the last frost of the year. The average date is April 29, the earliest date is March 20 (1990), and the latest is June 10 (1963). The average period with frosts (i.e., the length of time between the first frost in one year and the last frost the next) is 188 days, although the range of values recorded varies from 145 (1922–23) to 244 (1962–63). The actual number of days with frost per year is again somewhat shorter, averaging 128 days, but again with a wide range from 97 (1940–41) to 186 days (1929–30). There is a weak linear correlation between the length of the period with frosts and the actual number of days with frost ($r = 0.49$).

Measured evaporation rates range from a minimum average of 54.37 mm in December (with a range from zero in 1975 and 1977) to a maximum of 141.73 mm in 1970. Evaporation rises rapidly in the first half of the year, peaking at average figure of 323.82 mm in June, before falling more slowly in the second half of the year. Evaporation thus peaks before either temperature or rainfall. June values range from a minimum of 257.81 mm recorded in 1976 to a maximum of 387.86 mm two years previously in 1974. However the maximum recorded value is 395.22 mm, which occurred in May 1971. Patterns of moisture balance generally follow the trend in evaporation because this variable dominates over precipitation. On the other hand, the decrease in the deficit in July to August is more marked than the decrease in evaporation because of the high summer rainfall. Only the months of November to February have recorded positive moisture balances, the maximum being 40.9 mm in December 1960. A positive water balance only occurred for seven months in the recording period from January 1953 to August 1979.

Wind patterns have only been recorded over much shorter periods. At the USGS site (5.9 km southwest of Jornada headquarters), daily records have been analyzed for the period from January 1990 to August 1996 to give monthly summaries. Average monthly wind speeds are highest in April with a value of 12.4 km/h, declining to a value of 7.9 km/h in August (figure 3-3). There is a second maximum of 8.8 km/h in November. The least windy month is December, when the average velocity is 7.6 km/h. In terms of the average peak gust, there is a less well-defined annual pattern, although peaks occur in April (78.6 km/h) and July (78.8 km/h). Thus the summer months have lower average wind speeds, but they have important gusts, usually relating to local atmospheric convection in the afternoons, leading to the characteristic dust devils, which can commonly be seen tracking along the center of the basin. The dominant direction of the peaks gusts is from the WSW, which occurs in seven months from the end of October through May. January and April have similar directions with peaks arriving from the west and southwest, respectively. In the summer months the direction shifts distinctly so there is no dominant wind direction throughout the month of July. These shifts are due to the onset of monsoonal activity derived from the Gulf of Mexico.

For the shorter time period between 1983 and 1997, the LTER weather station on the Summerford Mountain bajada presented a more detailed record of climate (see data sets available online at http://jornada-www.nmsu.edu). Temperature patterns illustrate that dew temperatures are rarely reached, and the soil temperature at a depth of 5 cm lies midway between the average and maximum air temperature. Wind velocities show a major peak between March and June and a smaller peak in November. Relative humidity decreases during the early part of the year to a minimum in June, when the average minimum is 13.5% and the average maximum 45.6%. Relative humidity then increases rapidly during the onset of the monsoonal season, reaching a peak in August. The minimum values stay relatively constant until reaching a peak with the winter rainfall season in December (average 35.9%). The maximum values decline over the same period but again rise in December (average 75.7%).

Solar radiation is asymmetrically distributed through the year, rising rapidly to a peak value in May (671.1 MJ/m^2) and then declining more slowly to a minimum in December (244.5 MJ/m^2). The average annual solar radiation received at the LTER weather station is 6,250.1 MJ/m^2.

Spatial Patterns

Spatial patterns of monthly rainfall can be reconstructed for the Jornada Basin using 37 gauges. At the monthly time scale, most months show spatial patterns similar to those described for the annual time scale. However, there are some notable variations (figure 3-2a–d). There are greater altitudinal gradients in the months of July and August compared to the other months. Higher rainfall occurs locally in the west and central parts of the basin in November and December, whereas in May, June, and October there are central bands of higher rainfall trending either north–south or northwest–southeast. There is a localized peak in rainfall to the north of Summerford Mountain in February, May, and December. These different patterns probably reflect the different orographic effects of the different mountain ranges surrounding the basin, as they affect moisture arriving in the basin from difference sources. Semivariograms of the monthly and seasonal patterns follow a Gaussian pattern with semivariance generally increasing rapidly beyond a lag distance of around 20 km.

The spatial variability of temperatures can be seen with reference to the LTER, Jornada headquarters, and USGS weather station data (figure 3-4). There is a consistent ordering of the average monthly temperatures, with the highest values at the LTER weather station, followed by the USGS, and the lowest values at Jornada headquarters. Temperatures at the LTER site are between 0.4°C and 2.8°C warmer than at the USGS site, which is in turn between 0.3°C and 1.5°C warmer than at Jornada headquarters. These patterns relate to the fact that minimum monthly temperatures are much higher at the LTER site than the other two (2.7–5.0°C warmer than USGS and 3.6–5.9°C warmer than JER headquarters). The LTER weather station has the lowest maximum temperatures of the three locations in every month except May (0.9–2.7°C cooler than USGS and 1.1°C cooler to

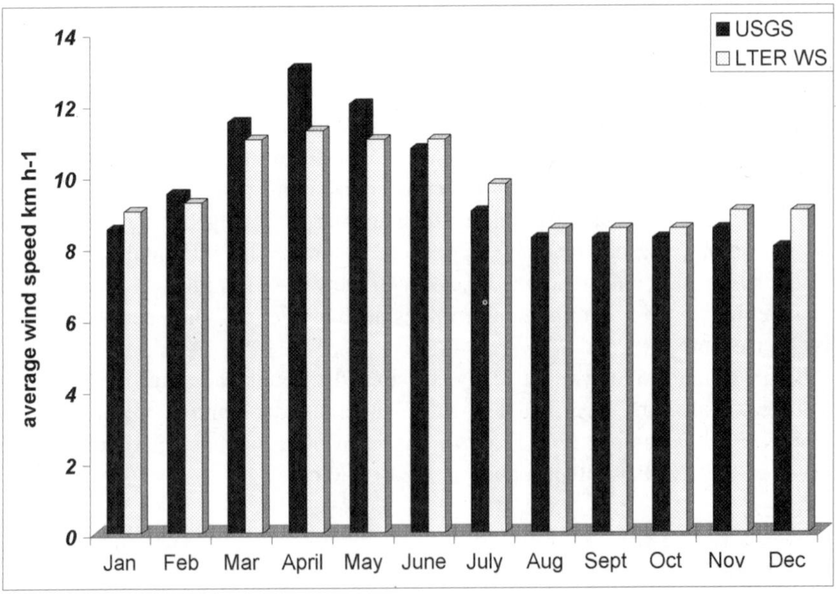

Figure 3-3. Comparison of average wind speed as recorded at the LTER weather station (1983–97) and the U.S. Geological Service (USGS) weather station (1990–96).

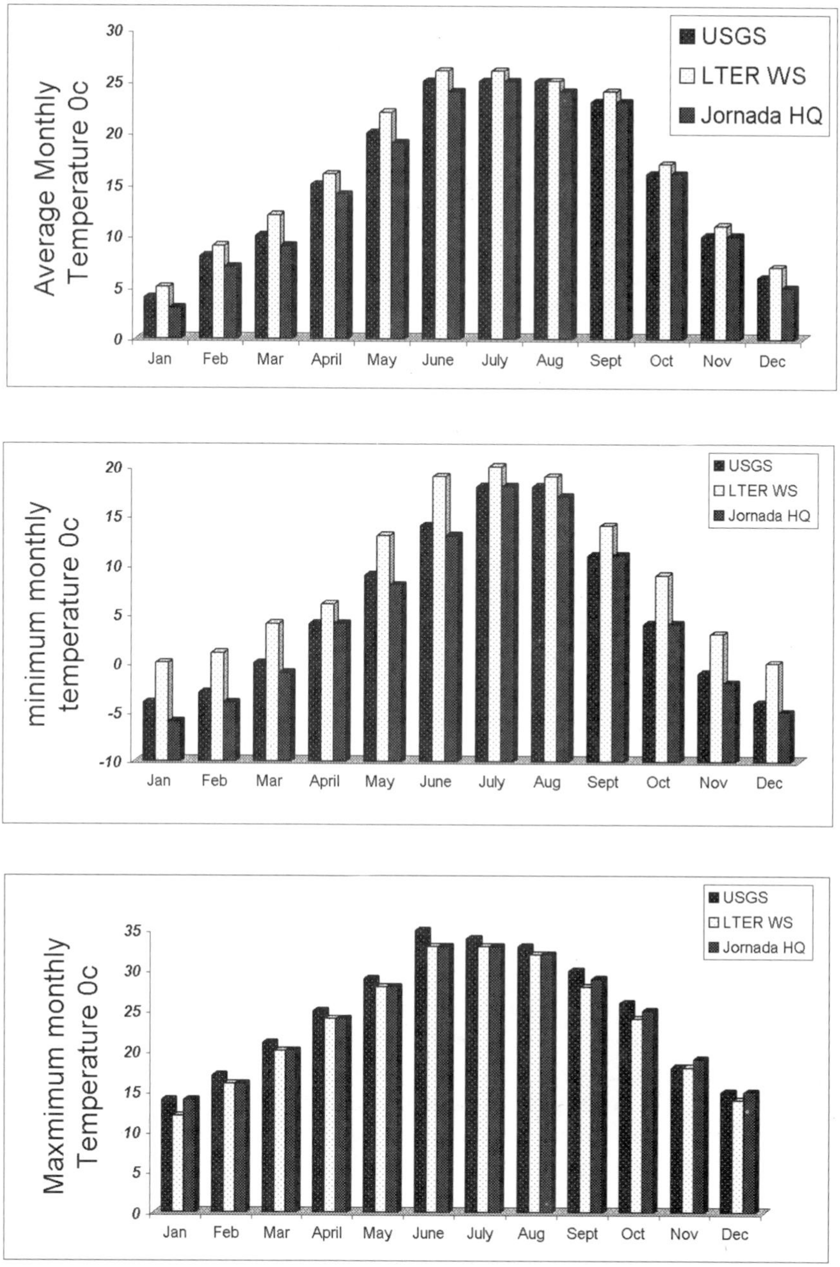

Figure 3-4. Comparisons of temperature records at the LTER, USGS, and JER headquarters weather stations: (top) average monthly temperature; (middle) minimum monthly temperature; and (bottom) maximum monthly temperature.

0.1°C warmer than Jornada headquarters), and there tends to be less diurnal variability at the LTER site. Wind data can be compared between the LTER and USGS weather stations, although due to differences in the way these data are recorded, the comparison is limited to average wind speeds. The average wind speed tends to be higher at the LTER weather station between the months of June and January (figure 3-3). The change in June coincides with the change from the modal direction at the LTER weather station from SSW to S, although the modal direction reverts to SSW in October.

In a number of climate variables, but most dominantly precipitation, there is a major shift in the climatic character of the basin for June through September, associated with the onset and development of the summer monsoonal system. For example, at the JER headquarters, on average, 61% of the annual rainfall occurs in these summer months. Precipitation increases and a change in the dominant wind direction are associated with changes in the other variables observed. Thus, a bipartite division of the Jornada climate has been made, with summer conditions relating to the June through September period and winter conditions referring to the October through May period. The difference in rainfall between these periods is clearly marked in terms of quantity of rainfall and, to a lesser extent, in terms of rainfall variability (figures 3-5a and 3-5b). This division contrasts with that of Conley et al. (1992) who used a tripartite division of winter (November–March), spring (April–June), and summer (July–October). Although this allows more detail for the winter period, it is thought that this is of little significance in terms of application to understanding plant growth patterns (Reynolds et al. 1999b).

Daily Patterns

Specific daily climatic patterns of interest include distributions of rainfall events, diurnal ranges of temperature, and relative humidity. The occurrence of extreme rainfall events and their spatial and temporal distributions will be considered separately in the next section.

The mean number of rain days recorded at Jornada Experimental Range headquarters is 42.6, and at the LTER weather station the value is 50.7. These values are predominantly composed of rain days in July and August during the peak of the monsoon. The winter peak in rainfall is dominated by the three to five rain days in December and January, with a steady decline until the minimum in April. Across the basin, the number of rain days ranges from a minimum of 41.6 to a maximum of 51.1 (median 45.2). The spatial pattern of the number of rain days is different from that of the annual average rainfall, with higher values occurring on the higher ground to the north, northeast, and southwest and the lowest values in the north-central and southeastern parts of the basin. The seasonal number of rain days does not differ substantially between summer—with a minimum if 21.3, median of 23.4, and maximum of 29.1 across the basin—and winter—minimum 19.7, median 21.2, maximum 24.9. This pattern reflects the typically much lower intensity event in the winter months.

An important parameter of rainfall is its persistence (table 3-2). Where runoff

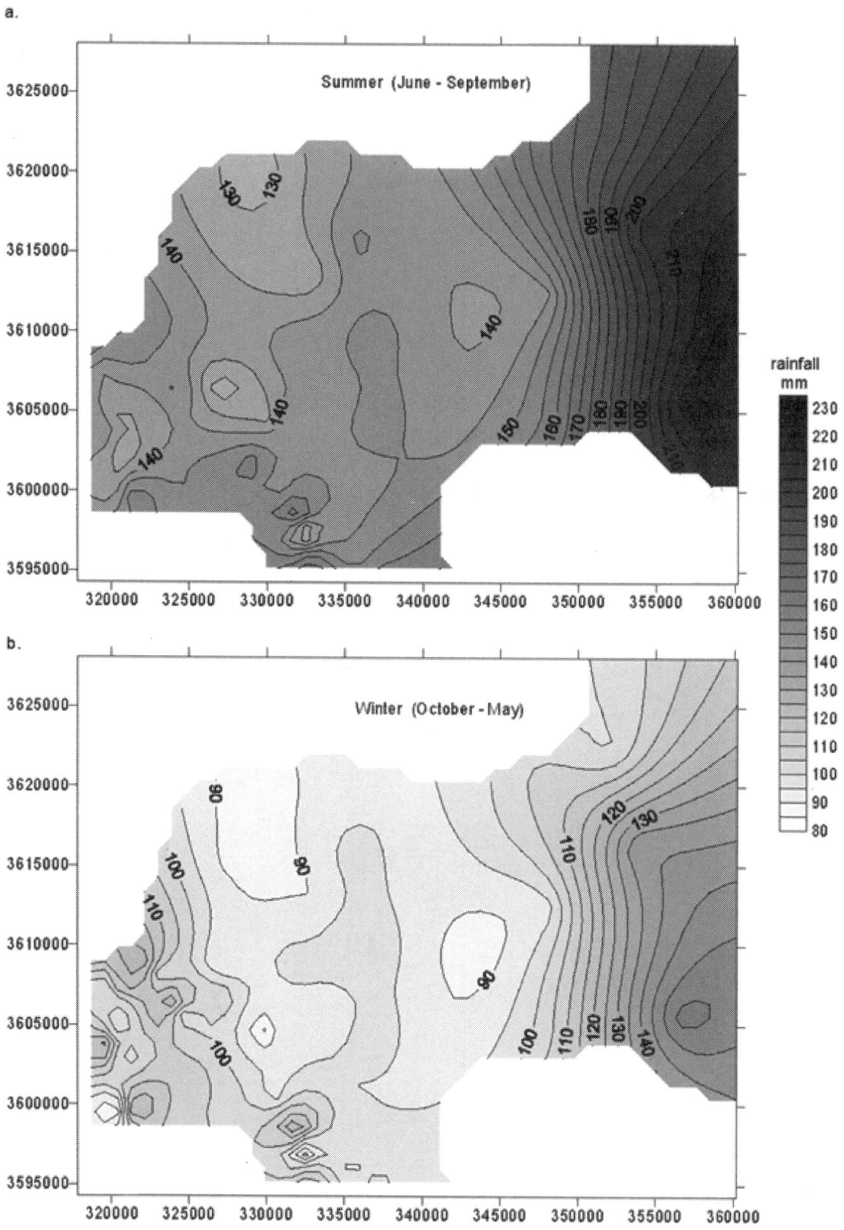

Figure 3-5. Seasonal patterns of precipitation in the Jornada Basin: (a) summer (June–September); and (b) winter (October–May). Patterns are displayed within an UTM grid.

Table 3-2. Measured and predicted probabilities for the JER headquarters and the Jornada Basin LTER weather station locations for persistence (successive days) of 2, 3, or 4 successive dry days (dd, ddd, or dddd, respectively) or 2, 3, or 4 successive days with precipitation (ww, www, or wwww, respectively).

	Measured Probabilities						Predicted Probabilities					
	Annual		Summer		Winter		Annual		Summer		Winter	
Location	*Pdd*	*Pww*	*Pdd*	*Pww*	*Pdd*	*Pww*						
JER HQ	0.76	0.04	0.31	0.01	0.96	0.05						
LTER	0.69	0.05	0.29	0.02	0.83	0.06						
	Pddd	*Pwww*	*Pddd*	*Pwww*	*Pddd*	*Pwww*	*Pddd*	*Pwww*	*Pddd*	*Pwww*	*Pddd*	*Pwww*
JER HQ	0.668	0.011	0.280	0.0044	0.831	0.0139	0.578	0.0016	0.0961	0.0001	0.9216	0.0025
LTER	0.598	0.015	0.260	0.0033	0.719	0.0200	0.476	0.0025	0.0841	0.0004	0.6889	0.0036
	Pdddd	*Pwwww*	*Pdddd*	*Pwwww*	*Pdddd*	*Pwwww*	*Pdddd*	*Pwwww*	*Pdddd*	*Pwwww*	*Pdddd*	*Pwwww*
JER HQ	0.585	0.0040	0.256	5.5×10^{-4}	0.724	0.00932	0.439	6.4×10^{-5}	0.0298	1.0×10^{-6}	0.885	1.3×10^{-4}
LTER	0.521	0.0066	0.236	0.00210	0.622	0.00472	0.329	1.3×10^{-5}	0.0244	8.0×10^{-6}	0.572	2.2×10^{-4}

production, sediment transport, or plant growth are involved, the occurrence of repeated rainfall events can be significant. Conversely, the persistence of days without rain can signal drought conditions. One way of observing such persistence is to calculate Markov transition probabilities for the occurrence of days with rain followed by a second day of rain, and similarly for days without rain followed by a second day without rain. Higher orders of persistence can be observed by repeating the calculation for three successive days, and so on. The probability for two successive wet days at the annual level is 0.04 at the Jornada headquarters and 0.05 at the LTER weather station, whereas the respective probabilities for two successive dry days are 0.76 and 0.69. At the seasonal level, the wet-wet probability increases in the summer months to 0.06 at the Jornada headquarters and 0.08 at the LTER weather station. In other words, as might be expected, there is much greater persistence of dry than wet periods. The annual probability of three successive wet days is 0.011 at the Jornada headquarters and 0.015 at the LTER weather station. Based on the probabilities for two successive dry days, we can calculate the probability of having three successive events by chance, assuming independence of events. This probability is simply the square of the first probability. The calculated probabilities are 0.0016 for the Jornada headquarters and 0.0025 at the LTER weather station. The fact that these are lower than the measured transition probabilities suggests much stronger persistence of wet episodes. Extending this analysis to four successive wet days gives predicted probabilities of 0.000064 for the Jornada headquarters and 0.000125 at the LTER weather station compared to measured transition probabilities of 0.004 and 0.00657, respectively. Repeating this analysis for sequences of dry periods and for both wet and dry periods according to season shows a similar result, except for the sequences of dry periods in winter. Thus persistence is a feature of both dry and wet periods. This conclusion contrasts with the results of Chin (1977) who suggested that rainfall at El Paso, Texas, 80 km to the south, was only first-order dependent in summer and second-order dependent in winter. An additional difference is seen in comparing persistence at different temporal scales. Whereas persistence seems to be an important feature at the daily scale, it has already been seen to be far less significant at an interannual scale.

The intensity of rainfall is significant in producing runoff and therefore in its role in distributing water through the basin catchment. Relative to soil infiltration rates, the rainfall intensity determines the extent of runoff during rainfall events of the same magnitude. Because most rainfall events at Jornada last for much less than one day due to their convective nature, daily rainfall data do not give good estimates of rainfall intensities. Data that can provide a better indication occur at the LTER weather station, where hourly intensities have been recorded since 1992, and at several tipping-bucket rain gauges employed within the LTER project. Peak intensities for time periods between 5 and 180 minutes are also included in the USGS summaries on a monthly basis.

The majority of events (55%) recorded at the LTER weather station last for an hour or less, with an exponential decline in the length of event. The mean event is 2 hours long. The longest event took place in 1994, when 19.8 mm of rain fell in 19 hours between 1000 on December 5 and 0500 on December 6. The median hourly rainfall intensity is 0.76 mm/h, with an average of 1.7 mm/h. The maximum value of 25.2 mm/h fell at 1900 on July 24, 1992, although this may be an underestimate of the true peak intensity as this whole event took place within a single hour.

Data from three tipping-bucket rain gauges also show an exponential decline in intensities (figure 3-6). Maximum intensities recording by the tipping-bucket gauges are 137.3 mm/h, although this intensity was only maintained for 1 minute (on two occasions, June 26 and September 14, 1996). During the June 26 event, the peak 5-minute intensity was 100.6 mm/h; the peak 10-minute intensity, 75.3 mm/h; the peak 30-minute intensity, 48.2 mm/h; and the peak hourly intensity, 36.3 mm/h. The September 14 event had corresponding 5-, 10-, 30-minute, and hourly values of 82.3 mm/h, 50.6 mm/h, 40.3 mm/h, and 38.6 mm/h, respectively. These values compare with the maximum recorded hourly rainfall at LTER weather station, which is 25.2 mm/h.

Diurnal temperature ranges have been calculated for the long sequence at the Jornada headquarters (1914–95), as well as for the shorter LTER weather station (1983–97) and USGS (1990–96) sites. On average, the LTER weather station diurnal range is 14.6°C, which is significantly lower than either the Jornada headquarters (20.0°C) or the USGS (20.1°C) sites. A similar difference is also noted for monthly averages. The most likely explanation for this difference is the more

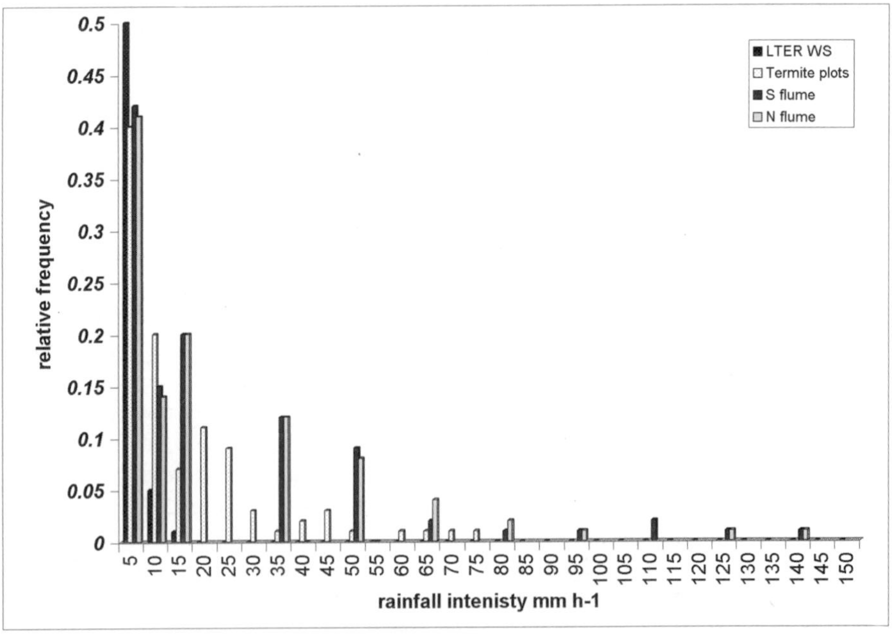

Figure 3-6. Rainfall intensities recorded at three tipping-bucket raingauges (Termite plots, S flume, and N flume) on LTER sites on the Summerford Mountain bajada, compared with hourly intensities as recorded at the LTER Weather Station (LTER WS).

exposed location of the Jornada and USGS sites toward the central part of the basin. In all cases, there is a seasonal pattern with low values in December and January, an increase to May or June, followed by a decrease to the annual minimum in August, with a second cycle peaking in October.

The diurnal range of relative humidity at the LTER weather station shows a single oscillation through the year, with a rapid rise to a peak corresponding to the onset of the monsoonal rains, followed by a more gradual decrease. Low ranges are typically recorded for the months of April to June after the end of the main winter rains, although in certain years, such as 1992, high values have been recorded in May.

Occurrence of Extreme Events

Maximum daily rainfall shows significant variation through the basin (figure 3-7). For those sites with data for 10 years or more, the maximum recorded values range from 47 mm to 105 mm. There is no significant relationship between the maximum daily rainfall and the mean average rainfall ($r = 0.19$, $p = 0.29$), or with the altitude of the rain gauge ($r = 0.17$, $p = 0.35$). The larger recorded events seem to cluster around the central, northwestern, and southeastern part of the basin.

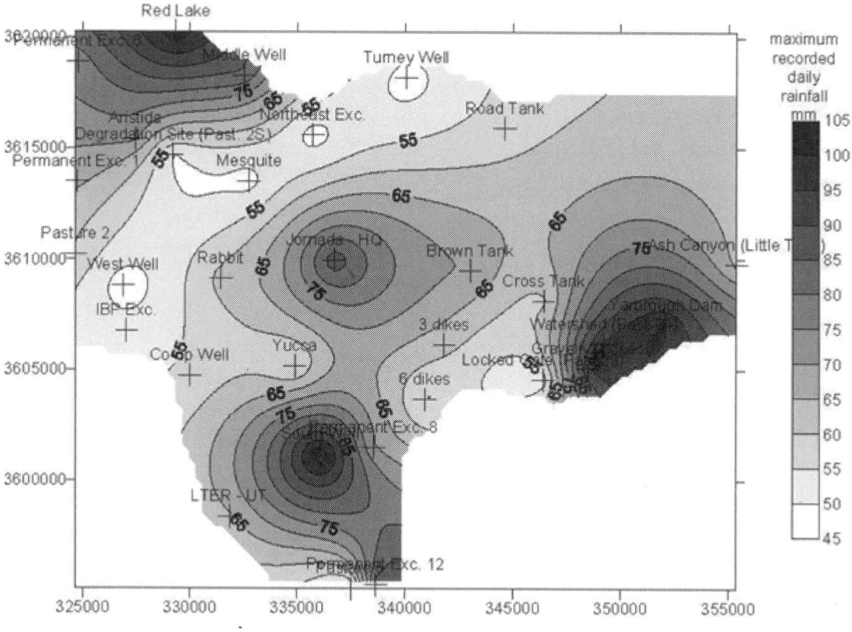

Figure 3-7. Patterns of maximum daily rainfall throughout the Jornada Basin. Patterns are displayed within an UTM grid.

Intensity-duration-frequency values demonstrate the importance of relatively frequent high-intensity rainfalls, albeit lasting for relatively short periods (table 3-3). Because of the convective character of the rainfall, these high intensities are rarely sustained for more than 30 minutes. Because only hourly data were available for short time periods from the LTER weather station, it was not possible to carry out a more detailed analysis for the Jornada Basin itself, beyond that proposed by the National Oceanographic and Atmospheric Administration (NOAA) (Miller et al. 1973). A comparison for short return periods using the method of Chen (1983) showed a correspondence with the figures given in Miller et al. (1973) for return periods of less than 10 years, so in these cases the LTER figures are presented. However, because of extrapolation from a short data set, the intensity of the lower frequency events was lower, and other means were used to estimate these values.

Return periods for daily rainfall have also been calculated, using standard Gumbel analysis, for each of the daily rain gauges in the basin to explore spatial patterns of large events. Daily rainfall within the basin varies between 27.5 and 38.5 mm for a 2-year return period, between 35.8 and 57.6 mm for a 5-year return period, between 41.3 and 70.2 mm for a 10-year return period, between 47.2 and 87.4 mm for a 25-year return period, between 51.5 and 100.3 mm for a 50-year return period, and between 55.8 and 113.1 mm for a 100-year return period. Three peaks of generally higher rainfall tend to appear in these calculations in the northwest, southwest, and southeast of the basin area studied. There is a relatively complex pattern in the central part of the basin, which varies according to the length of the return period, although this may, in part, be due to the generally shorter record lengths (albeit > 10 years) of the gauges in this part of the basin.

A sample of extreme events was selected by examining the rainfall across the basin for the largest recorded event at each rain gauge. The spatial pattern of these events can be divided into three broad groups. First, there are a small number of

Table 3-3. Intensity (mm/h) of precipitation events observed for different storm durations over different time intervals (years).

Storm Duration	Rainfall Intensity in mm/h for Given Return Interval: 2 year	5 year	10 year	25 year	50 year	100 year
5 min	86.6‡	97.5	109.7	135.7	161.2	189.6
10 min	68.7‡	80.6	90.8	106.9	131.8	156.4
15 min	57.2‡	67.6	79.5	95.5	110.4	130.5
30 min	38.6‡	48.1	55.7	69.0	80.4	96.0
60 min	24.0‡	27.9†	31.7†	35.6†	40.6†	45.9†
6 hr	5.67‡	6.77†	7.62†	9.52†	10.58†	12.28†
24 hr	1.72‡	1.96†	2.38†	2.91†	3.39†	3.81†

†Values based directly on Miller et al. (1973) and should be considered as underestimates.

‡Values derived from Jornada Basin LTER weather station data using Chen (1983) method, but may still be underestimates due to the length of available record.

events that have a single center but may be more or less continuous across the basin. The maximum rainfall at the center of these events may still be large, for example the event of July 4, 1988, when 103.6 mm of rain was recorded at the Middle Well and Red Lake gauges. Only 3 out of 32 of the sampled events fell into this category. Second, there are 11 events that have multiple centers but that have a discontinuous cover over the basin. The third type contains multiple cells that cover the entire basin and makes up the majority of the sampled storm events (17 out of 32). There appears to be no correlation with the type of extreme event and the time within the monsoonal season at which it occurred in the chosen sample. There are, on the other hand, a significant number of events where the third pattern occurred outside the main monsoonal season (e.g., May 3, 1980; April 29, 1981; December 9, 1982; November 2, 1983; November 2, 1986; and, May 22, 1992), producing significant rainfalls.

Longer Term Changes and the Effects of El Niño and Other Large-Scale Atmospheric Phenomena

The analysis of the longer term records from the Jornada Basin to observe any significant trends through time is composed of two parts. First, the records are assessed for any statistically significant patterns that may relate to climatic variability or change at moderate to long time scales. Second, the records are subdivided to allow the impact of El Niño events to be assessed. For the most part, these analyses have been carried out at the annual and seasonal level.

Conley et al. (1992) used regression and autocorrelation analysis to examine trends in the rainfall and temperature at the Jornada headquarters station as well as nearby sites in Las Cruces. They found no significant linear or quadratic trends for the annual precipitation data (1914–84) but a significant quadratic relationship for mean temperature. Analyses for trends have been carried out in the same way here, except that the sequence of data are now extended to cover the period until 1996; the minimum, maximum, and average annual temperatures have been analyzed separately; and the evaporation data from 1953–78 have also been analyzed (table 3-4). Outliers have not been ignored, as in the analysis of Conley et al. (1992), but years for which data are incomplete (which is often the cause of major outliers) have been excluded from the analysis.

The average temperature data at the Jornada headquarters station show significant linear, quadratic, and quartic trends; however, the implications of the different models are important. The linear model describes a general decrease of the average temperature of 0.0092°C per year, whereas the quadratic model shows a slight increase of temperature from 1914 to around 1940, when a more rapid decrease takes place. On the other hand, the third-order polynomial model shows a similar but more rapid increase in the earlier time period followed by a more gradual decrease until the mid-1980s when the trend starts to reverse. A similar general pattern is observable with the average minimum temperature, albeit with a much stronger upturn since the mid-1970s. In the case of average maximum temperatures, only the quadratic and third-order models are significant, both of

Table 3-4. Results of regression analyses of long-term records from the JER Range headquarters for temperature, precipitation, evaporation, and frost days. Results are graphed in figures 3-8, 3-9, and 3-10.

Variable	Constant	Year	Year²	Year³	Year⁴	r^2	p
Average	32.68	−0.0092	−0.00024	1.183×10^{-5}		0.124	0.005
annual	−914.083	0.9590	−0.0696			0.161	0.006
temperature	−89,327.9	136.61				0.195	0.006
Average	36.49	−0.0162	0.00022	3.593×10^{-5}		0.122	0.006
minimum	895.71	−0.8949	−0.21059			0.131	0.017
temperature	−26,7769.5	411.325				0.229	0.002
Average	28.91	−0.0021	−0.00079	-1.128×10^{-5}		0.003	0.661
maximum	−3003.03	3.0982	0.06537			0.190	0.002
temperature	81,326.70	−126.293				0.205	0.003
Average	−1378.65	0.8303	0.02740	0.00031		0.051	0.044
annual	103402.5	−106.350	−1.79283			0.074	0.051
precipitation	-2.216×10^{6}	3452.82				0.076	0.111
Average	−994.411	0.58379	0.00954	6.534×10^{-5}		0.059	0.027
summer	35471.71	−36.7273	−0.37365			0.066	0.065
precipitation	-4.526×10^{5}	712.361				0.066	0.142
Average	−643.444	0.38030	0.02524	0.00065	-3.319×10^{-7}	0.036	0.089
winter	95,811.9	−98.3069	−3.75863	0.00324		0.107	0.012
precipitation	-4.724×10^{6}	7298.935	−11.354			0.126	0.015
	-9.551×10^{6}	17,186.77				0.127	0.034
Average	−14,774.96	8.63852	−0.28449	−0.25513		0.056	0.244
annual	-1.114×10^{6}	1126.99	1504.089			0.059	0.498
evaporation	1.936×10^{9}	-2.956×10^{6}				0.336	0.027

First frost day	467.234	−0.08780				0.025	0.159
Last frost day	−149.518	0.13750	−0.00606	−0.00021	-7.635×10^{-7}	0.050	0.047
	−23,322.2	23.847	1.25005	0.00576		0.091	0.025
	1.5767×10^{6}	−2431.71	−16.2533			0.113	0.027
	-9.567×10^{6}	20,375.0				0.116	0.053
Number of frost days	−770.78	0.45485	−0.01259	5.383×10^{-5}		0.138	7.5×10^{-4}
	−48,878.7	49.6773	−0.32832			0.185	4.2×10^{-4}
	-4.510×10^{5}	666.89				0.186	1.4×10^{-3}

which suggest a general increase in the first half of the century followed by a subsequent marked decrease. Annual average precipitation seems to increase slightly over the period (figure 3-8a). Subdivision into summer and winter precipitation suggests that both summer and, more recently, winter precipitation have contributed to this increase (figure 3-8b and 3-8c). The results of these analyses suggest that the period of drought experienced in the 1930s was a result of higher temperatures, but the drought of the 1950s was in response to lower precipitation levels. Evaporation shows a strong oscillation and thus only shows significant trends with third- and fourth-order polynomials (figure 3-9). The monthly trends in evaporation show the oscillation occurred in all months, albeit more markedly in April to June and September to December. It is likely that the oscillation is a function of two factors, the generally lower minimum temperatures in the years just before and after 1960 and followed by an increase around 1970 and, to a lesser extent, the slight increase in maximum temperatures during the latter period. There is no significant trend in the day of the occurrence of the first frost, but both the date of the last frost and the total number of frost days show similar patterns. Linear models suggest continuous increases in frost occurrence (i.e., longer periods with frost with more frost days), but higher order models suggest an increase until the mid-1960s followed by a subsequent decrease (figure 3-10a–c).

Removing the linear trend from each of these sequences allows analysis of higher frequency changes using Fourier analysis. The average temperature shows spectral peaks at periods greater than 40 years (supporting the existence of the longer term oscillations shown by the regression analysis), as well as between 11 and 20 years. There are a number of strong peaks at periods of between two and five years. The minimum and maximum temperature signals are similar, except that both have a relatively stronger long-term cyclicity compared to the 11–20-year peaks. The minimum temperature signal shows no two-year cyclicity but has an extra peak at seven years, whereas the maximum temperature has an extra peak at around nine years. The annual precipitation has a strong spectral response through its range, although decomposition into the summer and winter rainfall suggests this is due to the interplay of seasonal factors that occur at difference frequencies. The summer rainfall shows stronger peaks at frequencies of around 3, 6, 12, and 30 years. As well as the 3- and 6-year signals, the winter rainfall has a stronger response at 7, 9, and 18 years, as well as over periods of more than 40 years (again supporting the regression analysis). The 3- and 6-year cycles may relate to the El Niño phenomenon (see further analysis to follow). Persistence in these sequences was also tested by means of runs tests, but only the strong oscillation in the evaporation data showed any significant persistence at the 95% confidence level.

It is also interesting to compare long-term trends in the precipitation at other sites in the basin for which there are long data records. A total of 28 gauges, including the Jornada headquarters, were selected for this analysis on the basis of having data series starting before 1940, with more than 40 years of total records. Of these sites, 14 had significant linear trends, 19 significant quadratic trends, and 21 significant third-order polynomial trends at a 95% confidence level (table 3-

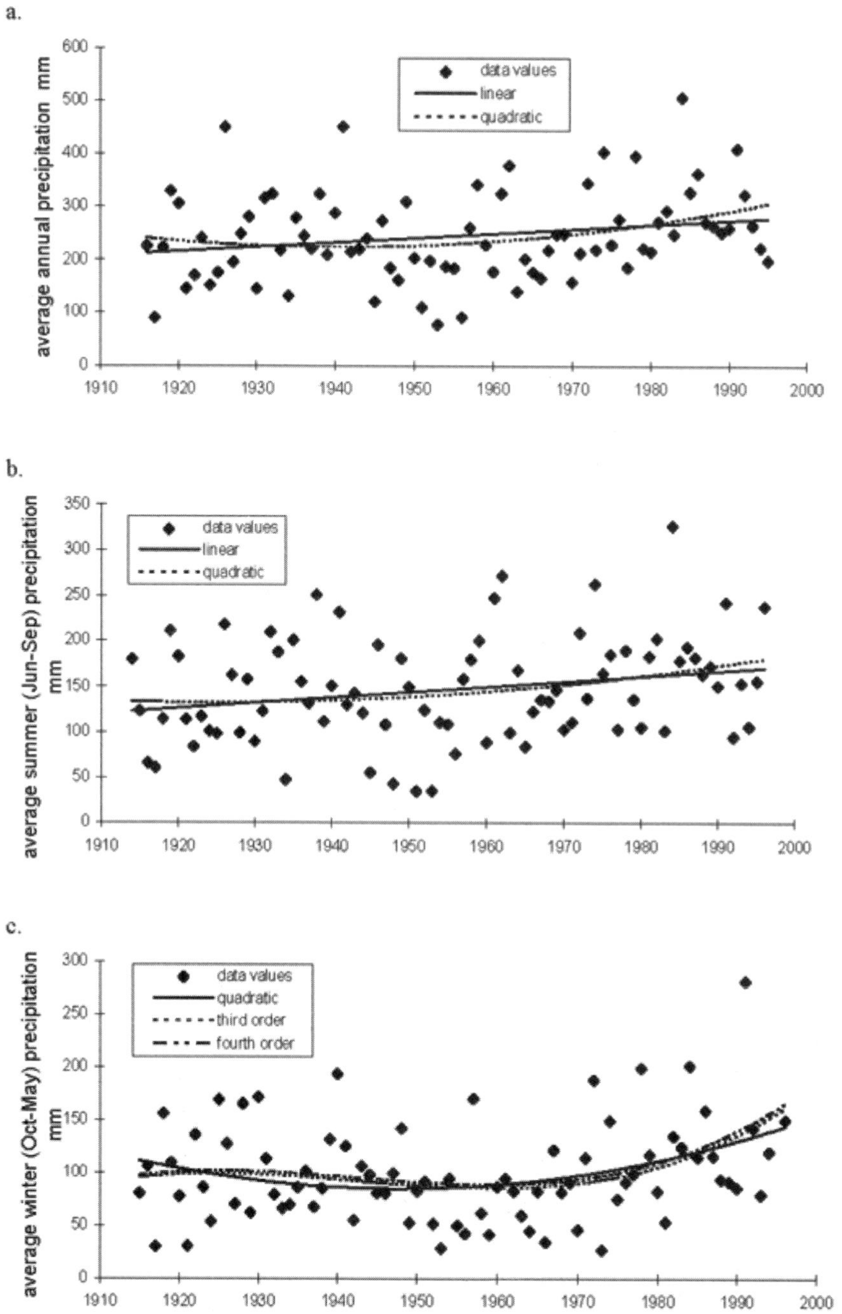

Figure 3-8. Trends in precipitation data from Jornada (headquarters): (a) average annual; (b) average summer (June–September); and (c) average winter (October–May).

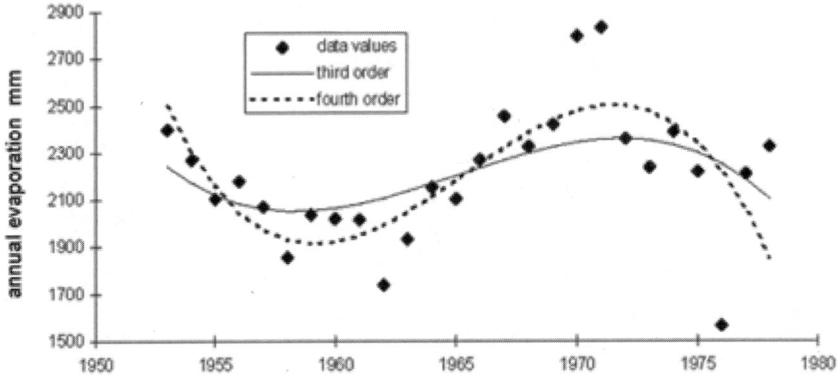

Figure 3-9. Trends in evaporation data from JER headquarters.

5). The linear trend is always positive. Only the areas in the southern, north-central, and eastern portions of the basin show significant linear trends, although there is no clear spatial pattern. Quadratic trends reinforce the pattern from Jornada headquarters, suggesting an increase in the rate of change in recent years. Third-order polynomial trends suggest a different pattern, in that only the Rope Springs gauge follows the same trend as Jornada headquarters, whereas all of the other gauges suggest an oscillation, with rainfall decreasing until the mid-1950s, increasing until the mid-1980s, and decreasing again thereafter. These results suggest that although there are general patterns of long-term variability in the rainfall data, the trends have a complex spatial pattern.

Again, the linear trend was removed from these rainfall sequences to observe whether any shorter term oscillations are present. The results obtained show strong spatial variability in these data. Oscillations on a 3-year cycle are the most common, followed by the 64-year cycle, then the 16, the 6, the 9, and the 32. The 2-, 11-, and 13-year cycles are the least well represented. Only 7 out of the 26 available gauges showed a significant persistence in annual rainfall using a runs test. Although these gauges are located for the main part on the edge of the central part of the basin, again there is no definite spatial pattern. Observation of these series suggests that the main pattern of persistence is the period starting about 1943 or, more usually, 1945 and continuing until 1956 or 1957 when conditions were generally drier than the detrended, long-term median rainfall. This pattern of general aridity is seen in many of the rainfall sequences and is discussed as the drought of the 1950s by Conley et al. (1992). It possibly also includes the drier conditions of the early 1930s. Several of the seven sites also show some persistence of wetter conditions in the late 1930s, early 1940s, late 1960s, and early 1970s. It is likely that these secondary patterns of persistence cause the significance in the runs test in conjunction with the more widespread late 1940s to early 1950s phenomenon. At a seasonal level, only six sites show persistence in values of summer rainfall and four sites in values of winter rainfall. It could thus be concluded that although persistence is important at specific points in space

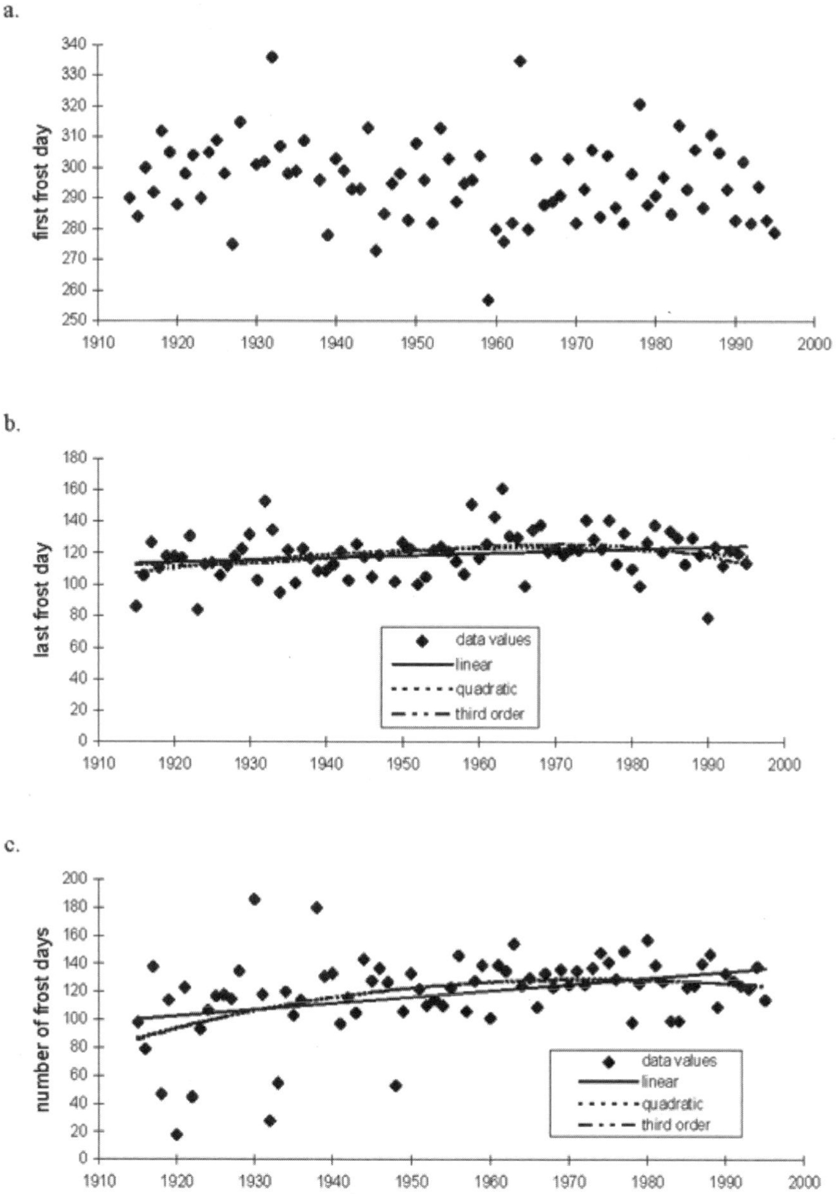

Figure 3-10. Trends in frost-day data from JER headquarters (a) Julian day of first frost; (b) Julian day of last frost; and (c) number of days with frost.

Table 3-5. Long-term (> 40 years) trends in annual precipitation for 28 permanent recording stations in the Jornada Basin.

	Constant	Year	Year^2	Year^3	r^2	p
Antelope	−3928.369	2.113191	0.0682	−0.00827	0.139	0.004
	260031.64	−266.295	48.869		0.172	0.005
	63189540	−96252.8			0.281	0.000
Aristida	−1839.956	1.050213	0.0504	−0.001744	0.063	0.036
	191951.5	−196.566	10.316		0.111	0.020
	13354648	−20330.7			0.128	0.027
Ash Canyon	−3437.851	1.92424	0.0894	−0.007583	0.095	0.018
	342116.01	−349.534	44.823		0.144	0.013
	57993213	−88310.6			0.224	0.003
Brown Tank	−973.2359	0.611896	0.0709	−0.002189	0.024	0.205
	271621.79	−277.417	12.948		0.128	0.011
	16778778	−25530.8			0.159	0.010
Co-op Well	−1761.779	1.005324	0.039	−0.004883	0.050	0.086
	148886.31	−152.221	28.843		0.068	0.134
	37273746	−56793.4			0.133	0.044
Doña Ana	−1723.753	1.000714	0.0379	−0.003451	0.062	0.036
	143941.81	−147.578	20.34		0.092	0.037
	26162753	−39957.2			0.173	0.005
Headquarters	−1398.13	0.839842	0.0248	0.0004916	0.054	0.037
	93343.455	−96.0964	−2.858		0.075	0.049
	−3578673	5539.678			0.078	0.098
Mesquite	−3876.88	2.082954	0.0614	−0.007446	0.160	0.002
	233566.52	−239.424	43.989		0.194	0.002
	56849959	−86618.8			0.307	0.000
Middle Well	−966.313	0.60782	0.0707	−0.00395	0.018	0.268
	271046.68	−276.845	23.309		0.098	0.030
	30052639	−45843.6			0.180	0.004
New Well	−1495.99	0.902292	0.0685	−0.002611	0.040	0.101
	262057.52	−267.899	15.431		0.114	0.018
	19957452	−30397.5			0.149	0.014
Rabbit	−2505.238	1.399172	0.0436	−0.004515	0.101	0.007
	165194.06	−169.61	26.615		0.133	0.008
	34235808	−52285.3			0.241	0.000
Ragged Tank	−777.5242	0.530026	0.0459	−0.001444	0.017	0.266
	175173.13	−179.125	8.5341		0.064	0.091
	11031231	−16806.5			0.081	0.108
Red Lake	−2198.286	1.231195	0.0696	−0.000693	0.096	0.005
	264237.69	−271.096	4.1405		0.224	0.000
	5459924.8	−8237.17			0.229	0.000
Restoration	−3383.717	1.839516	0.0335	−0.007543	0.139	0.003
	126096.21	−129.958	44.501		0.151	0.007
	57350419	−87504.3			0.308	0.000

	Constant	Year	Year^2	Year^3	r^2	p
Road Tank	−1517.61	0.897357	0.0528	−0.005845	0.039	0.097
	201341.32	−206.019	34.441		0.085	0.049
	44273092	−67637.2			0.267	0.000
Rope Springs	−2039.815	1.21825	0.036	0.0007849	0.067	0.023
	135637.64	−139.501	−4.572		0.091	0.028
	−5745201	8877.176			0.095	0.058
Sand Hill	−1889.168	1.082622	0.0485	−0.005321	0.051	0.064
	184666.37	−189.204	31.35		0.084	0.058
	40301082	−61567.9			0.203	0.002
South Well	−2204.621	1.246951	0.0271	−0.00126	0.104	0.004
	101578.42	−104.803	7.4252		0.124	0.007
	9548720.6	−14585.5			0.140	0.010
Stuart Well	−1693.325	0.995332	0.0502	−0.003233	0.051	0.058
	191521.94	−196.084	19.071		0.095	0.033
	24568102	−37493			0.154	0.010
Taylor Well	−2284.302	1.278477	0.0497	−0.007995	0.082	0.026
	189958.25	−194.254	47.214		0.112	0.034
	60978251	−92938.6			0.289	0.000
West Well	−385.0251	0.311672	0.0236	−0.000276	0.007	0.451
	90054.882	−92.128	1.6423		0.025	0.381
	2156024.7	−3259.69			0.026	0.574
Yucca	−977.9326	0.615654	0.057	−0.00285	0.024	0.195
	218039.66	−222.782	16.826		0.094	0.035
	21709147	−33104.9			0.151	0.012
Pasture 15N	−2269.145	1.275022	0.0573	−0.005901	0.068	0.049
	219491.06	−224.272	34.867		0.100	0.056
	45078904	−68669.8			0.178	0.014
Pasture 15S Exclosure	−1478.611	0.867952	0.0462	−0.00586	0.037	0.149
	177008.01	−180.671	34.616		0.061	0.178
	44730372	−68157.3			0.153	0.029
Camp Well	−1760.727	1.019452	0.0289	−0.002415	0.047	0.087
	109832.49	−112.655	14.255		0.058	0.165
	18386505	−28042			0.079	0.178
College Ranch Headquarters	−2621.301	1.457836	0.0624	−0.0042	0.095	0.029
	238243.41	−243.801	24.789		0.143	0.027
	31969217	−48761.2			0.188	0.021
Parker Heights Exclosure	−2245.223	1.2707	−0.008	−0.006079	0.079	0.072
	−34654.55	34.22995	35.843		0.079	0.201
	46136172	−70436.2			0.166	0.072
Selden Well	−1596.798	0.928175	0.0167	−0.002137	0.043	0.099
	62659.448	−64.5265	12.603		0.047	0.229
	16232470	−24774.3			0.066	0.250

and time, it is not a dominant feature of the Jornada rainfall record over the past 80 years.

Effects of the El Niño–Southern Oscillation

A further climatic phenomenon which has been shown to be a significant factor in variability in the region is that of El Niño (see Trenberth 1976; Rasmussen and Wallace 1983; Rasmussen 1985; Nicholls 1988; Diaz and Kiladis 1992 for details on the mechanisms). At the Sevilleta LTER site in New Mexico, Dahm and Moore (1994) found significant differences in winter (October to May) rainfall, with El Niño years having about 1.5 times the medial rainfall and La Niña years having about 0.5 times the medial rainfall. Only La Niña years showed a significant difference at the annual level (0.7 times medial rainfall), although El Niño years had a higher annual rainfall than medial years. There was no significant difference observed in summer (June to September) rainfall in either case.

The patterns of rainfall differences have been assessed for the Jornada using the same definition of El Niño years (1919, 1926, 1940, 1941, 1942, 1952, 1958, 1964, 1966, 1973, 1978, 1983, 1987, 1992, and 1993) and La Niña years (1918, 1939, 1950, 1951, 1956, 1971, 1974, 1976, and 1989) as Dahm and Moore (1994). Of the 47 gauges with sequences sufficiently long to allow analysis, only 5 sites show significantly higher rainfall in El Niño years, and only 8 have significantly lower rainfall in La Niña years, compared to medial years of annual rainfall (table 3-6). In the case of winter (October to May) rainfall, the pattern is much clearer, with 42 sites with significantly higher rainfall in El Niño years and 33 sites with significantly lower rainfall in La Niña years compared to medial years. On the other hand, only seven sites show significantly higher summer rainfall in El Niño years, and three sites show significantly lower rainfall in La Niña years compared to medial years. These results support the conclusion drawn by Dahm and Moore (1994) that El Niño affects the climate in the area by affecting the moisture supply. Thus it is only really significant in the winter months when rainfall is dominantly drawn from the Pacific Ocean. On average, El Niño years have 1.13 times and La Niña years 0.84 times the annual rainfall in medial years. The respective ratios for winter rainfall are 1.59 and 0.63 and for summer rainfall, 1.02 and 0.84. It is possible that the La Niña events in 1950, 1951, and 1956 were at least partly responsible for the period of dry weather in the 1950s, although the earlier onset (as discussed) suggests that it was probably not the principal mechanism. There seems to be a spatial pattern in the extent to which El Niño and La Niña events affect the rainfall within the basin. The effects in El Niño years are particularly pronounced in the higher altitude sites in the east and northeast of the basin, whereas the La Niña years seem to have a stronger effect in the west of the basin, particularly in the summer months. These differences probably also relate to the differences in the moisture supply and the rainfall mechanisms through the year.

More detail on the relationship between El Niño and the rainfall pattern can be obtained by comparing monthly rainfall values with values of the Southern

Table 3-6. Annual rainfall (mm) for 47 long-term precipitation-recording locations in the Jornada Basin analyzed for La Niña years (1918, 1939, 1950, 1951, 1956, 1971, 1974, 1976, and 1989), El Niño years (1919, 1926, 1940, 1941, 1942, 1952, 1958, 1964, 1966, 1973, 1978, 1983, 1984, 1992, and 1993) and medial years (all others between 1915 and 1995)

	Annual Rainfall (January–December)			Winter Rainfall (October–May)			Summer Rainfall (June–September)		
Rain Gauge	El Niño	Medial	La Niña	El Niño	Medial	La Niña	El Niño	Medial	La Niña
Antelope	270.33[a]	224.17[a]	179.46[a]	136.50[d]	86.69[e]	49.65[f]	151.48[g]	136.53[g]	113.11[g]
Aristida	250.74[a]	218.96[a]	176.73[a]	134.87[d]	81.99[e]	55.64[f]	132.61[g]	135.77[g]	108.33[g]
Ash Canyon	388.12[a]	343.42[a]	288.53[a]	206.08[d]	133.24[e]	80.98[f]	199.13[g]	213.71[g]	171.96[g]
Ber	208.56[a]	241.38[a]	228.95[a]	126.94[d]	94.28[d]	66.53[d]	108.31[g]	144.23[g]	134.40[g]
Brown Tank	263.28[a]	222.93[a,b]	189.98[b]	135.59[d]	79.38[e]	58.16[e]	148.18[g]	139.32[g]	116.53[g]
Co-op Well	251.93[a]	208.53[a]	187.95[a]	125.48[d]	76.61[e]	50.40[f]	145.96[g]	129.43[g]	120.14[g]
Doña Ana	291.71[a]	227.01[b]	217.10[a,b]	136.53[d]	85.71[e]	53.43[f]	166.74[g]	142.04[g]	140.39[g]
Exclosure A	238.23[a]	244.89[a]	281.38[a]	145.80[d]	92.38[d]	66.20[e]	123.13[g]	150.92[g]	178.98[g]
Exclosure B	253.39[a]	226.70[a]	258.00[a]	142.88[d]	86.42[e]	58.38[e]	135.34[g]	140.76[g]	162.90[g]
Goat Mountain	334.07[a]	387.21[a]	350.65[a]	237.90[d]	137.71[e]	84.18[e]	198.28[d]	243.69[g]	211.35[g]
Headquarters	290.18[a]	235.90[b]	216.18[a,b]	145.05[d]	93.59[e]	59.88[f]	159.75[g]	144.79[g]	131.37[g]
IBP	235.13[a]	230.68[a]	204.40[a]	162.98[d]	99.42[d,e]	61.63[e]	101.80[g]	131.03[g]	113.08[g]
Mesquite	256.65[a]	214.25[a]	182.83[a]	130.59[d]	83.40[e]	53.73[f]	145.81[g]	128.84[g]	113.31[g]
Middle Well	261.14[a]	223.84[a]	174.39[a]	130.48[d]	84.68[e]	51.46[f]	138.36[g]	137.84[g]	111.25[g]
New Well	316.44[a]	272.35[a,b]	215.63[b]	163.32[d]	99.41[e]	66.65[f]	173.77[g]	172.07[g]	128.70[g]
Northeast Exclosure	235.20[a]	237.52[a]	274.10[a]	140.26[d]	92.04[d,e]	62.53[e]	121.10[g]	144.83[g]	183.15[g]
Parker	263.91[a]	229.93[a]	204.51[a]	140.14[d]	84.02[e]	66.77[e]	151.53[g]	144.10[g]	119.70[g]
Pasture 2	246.97[a]	250.81[a]	218.93[a]	155.13[d]	95.30[d,e]	63.40[e]	114.76[g]	154.14[g]	128.80[g]
Rabbit	273.63[a]	236.65[a]	199.21[a]	145.42[d]	89.88[e]	62.69[f]	147.68[g]	144.98[g]	121.84[g]
Ragged Tank	327.09[a]	247.38[b]	233.64[b]	153.41[d]	96.75[e]	65.59[f]	188.41[g]	151.55[h]	141.33[g,h]
Red Lake	253.40[a]	207.43[a,b]	163.59[b]	132.42[d]	79.56[e]	47.78[f]	135.76[g]	126.46[g]	104.28[g]
Restoration	270.49[a]	226.57[a]	189.56[a]	141.61[d]	83.04[e,f]	60.60[f]	152.46[g]	139.99[g]	115.51[g]
Road Tank	285.87[a]	231.78[a]	228.81[a]	141.47[d]	90.82[e]	58.78[f]	156.28[g]	141.27[g]	148.09[g]
Rope Springs	403.81[a]	338.63[b]	276.46[b]	218.58[d]	128.62[e]	74.56[f]	207.50[g]	211.90[g]	163.16[g]
Sand Hill	275.68[a]	223.41[a]	224.40[a]	147.72[d]	85.93[e]	55.31[f]	153.43[g]	134.80[g]	149.23[g]
South Exclosure	248.00[a]	252.08[a]	278.70[a]	136.43[d]	88.44[e,f]	71.05[f]	136.54[g]	163.02[g]	173.70[g]
South Well	268.87[a]	229.08[a]	224.74[a]	145.56[d]	87.09[e,f]	59.90[f]	146.29[g]	141.45[g]	138.40[g]
St Nicholas	315.07[a]	394.54[b]	382.28[a,b]	203.95[d]	154.80[d,e]	92.85[e]	167.14[g]	243.67[h]	210.13[g,h]
Stuart Well	300.36[a]	249.20[a]	242.38[a]	154.18[d]	95.07[e,f]	66.65[f]	161.15[g]	154.34[g]	149.58[g]
Taylor Well	257.35[a]	227.39[a]	196.96[a]	133.95[d]	81.62[e,f]	59,76[f]	148.03[g]	142.43[g]	119.00[g]
West Well	264.58[a]	222.66[a,b]	172.62[b]	132.58[d]	83.58[e]	53.25[f]	144.00[g]	138.41[g]	106.91[g]
Yucca	272.61[a]	221.61[a]	201.19[a]	136.21[d]	82.79[e]	55.88[f]	149.54[g]	137.60[g]	129.26[g]
Pasture 15N	270.49[a]	241.09[a]	160.26[b]	140.85[d]	90.83[e]	50.00[f]	156.35[g]	144.05[g]	91.56[h]
Pasture Exclosure 15S	260.06[a]	230.98[a]	163.24[b]	139.01[d]	89.38[e]	53.69[f]	146.71[g]	140.08[g]	87.99[h]
Pasture 3N	269.79[a]	241.56[a]	164.16[b]	143.65[d]	91.65[e]	47.63[f]	149.92[g]	147.57[g]	89.77[h]
Pasture 3S	255.82[a]	261.26[a]	215.10[a]	176.15[d]	104.88[d,e]	55.48[e]	117.82[g,h]	151.69[g]	110.40[h]
Pasture 3W-15	228.04[a]	251.74[a]	106.85[a]	149.42[d]	92.68[d]	63.40[d]	128.53[g]	147.85[g]	69.75[g]
Pasture 3W	278.22[a]	270.62[a]	206.68[b]	182.68[d]	114.70[d]	59.73[e]	130.48[g,h]	158.19[g]	104.55[h]
Camp Well	265.82[a]	243.30[a]	180.43[b]	142.82[d]	89.70[e]	47.50[f]	150.59[g,h]	149.60[e]	117.82[g,h]
Creosote Site	254.93[a]	214.98[a]	155.57[a]	120.23[d]	77.88[d]	41.03[e]	153.28[g]	134.19[g]	111.03[g]

(continued)

Table 3-6. *(continued)*

	Annual Rainfall (January–December)			Winter Rainfall (October–May)			Summer Rainfall (June–September)		
Rain Gauge	El Niño	Medial	La Niña	El Niño	Medial	La Niña	El Niño	Medial	La Niña
College Ranch Headquarters	273.68^{a}	235.49^{a}	161.52^{a}	142.93^{d}	84.25^{e}	57.92^{d,e}	168.53^{g}	138.98^{g}	104.32^{g}
Pasture 12 Farm	281.68^{a}	263.80^{a}	145.10^{b}	154.59^{d}	89.20^{e}	64.25^{d,e}	168.61^{g}	160.87^{g,h}	89.35^{h}
Pasture 13W	295.94^{a}	221.54^{a}	133.27^{b}	116.08^{d}	79.43^{d}	41.53^{d}	177.53^{g}	132.21^{g}	90.80^{g}
Pasture 10	302.23^{a}	188.26^{a}	161.45^{a}	129.83^{d}	98.45^{d}	44.35^{d}	195.33^{g}	93.66^{h}	102.35^{g,h}
Parker Heights Exclosure	319.39	247.09	148.10	159.80^{d}	97.98^{e}	61.46^{d,e}	177.33^{g}	142.50^{g}	90.24^{h}
Seldon Well	267.39^{a}	224.04^{a,b}	165.48^{b}	137.42^{d}	90.18^{e}	53.05^{f}	153.93^{g}	134.20^{g}	89.81^{h}
Seldon South	276.56^{a}	236.31^{a}	138.93^{b}	160.64^{d}	96.56^{d,e}	59.10^{e}	153.43^{g}	131.23^{g,h}	84.37^{h}

Values with same letter in each period are not different $p \leq 0.05$

Oscillation Index (SOI). The values of the SOI covering the period of record in the Jornada Basin were obtained from the Climatic Research Unit at the University of East Anglia (available online at fttp://daac.gsfc.nasa.gov) and compared with the monthly rainfall at the Jornada headquarters. Cross-correlation of these two data series from November 1915 to January 1996 shows a weak but significant negative correlation at lags of −2 to 4 months and at 9 months. In other words, the Jornada headquarters' rainfall leads the response of SOI by two months and continues to respond until four months later. Analysis of the other long sequences shows again that the pattern of response is relatively complex. Lags of −3, −2, −1, 3, and 9 months dominate the response within the basin, whereas the direct response at lag 0 is recorded in only a few cases. A number of sites concentrated in the CDRRC area in the west of the basin also show a response between lags of six and nine months. It is possible that the atmospheric fluctuations that tend to precede the El Niño–Southern Oscillation (ENSO) events (Rasmussen 1985; Diaz and Kiladis 1992) also cause the rainfall patterns at Jornada to precede the changes in the SOI. The delayed response at lags of six to nine months is most likely to be related to the change in the source of moisture in the summer months followed by the return to Pacific moisture sources in the next winter rainfall season. Similar correlations between SOI fluctuations and rainfall in southern New Mexico have been demonstrated by a number of authors (Rasmussen 1985; Ropelewski and Halpert 1986, 1987; Andrade and Sellers 1988; Nicholls 1988; Kiladis and Diaz 1989; Molles and Dahm 1990; Redmond and Koch 1991; Diaz and Kiladis 1992; Kahya and Dracup 1993). The impact of major episodic fluctuations in precipitation has been demonstrated by Swetnam and Betancourt (1990, 1992) in terms of its correlation with major episodes of wildfires across large parts of the American Southwest, which has important implications for vegetation patterns.

The other climatic data for Jornada headquarters have also been analyzed to test for significant differences related to El Niño events. The results of this analysis suggest that El Niño events have a small impact on the temperature and evaporation characteristics of the basin. In El Niño years, only February has significantly colder average air temperatures, and January and February have significantly colder maximum air temperatures when compared to medial years. March average temperatures are significantly higher in La Niña years than in medial years, March and June have significantly higher maximum temperatures, and February has significantly lower maximum temperatures. Although Diaz and Kiladis (1992) showed that there was a significant decrease in December–February temperature in parts of New Mexico and Texas relating to ENSO events, the Jornada Basin is on the edge of the zone they delimited as being significant. In contrast, Redmond and Koch (1991) found no significant correlation for October–March temperatures in southern New Mexico. The relationship between SOI and temperature is also very weak in this region according to the analyses of Ropelewski and Halpert (1986); they suggest this is due to the more complex teleconnections associated with the development of the Pacific–North America ridge.

Effects of Other Large-Scale Circulation Patterns

The impact of the North Atlantic Oscillation (NAO) was assessed because of its potential link to fluctuations in summer moisture supplies from the Gulf of Mexico. Long-term trends in the NAO were presented by Trenberth and Hurrell (1995) and also by the University of East Anglia Climate Research Unit and NOAA. These data show no major differences between years with high positive or negative deviations from the NAO and annual precipitation at the Jornada. There are, however, a series of linkages demonstrated by cross-correlation analysis at the monthly level. For the Jornada headquarters' precipitation data from 1914, there are weak but significant negative correlations at a lag of–3 months (i.e., precipitation leading NAO) and at lag 4 months. This would tend to suggest that the effects of the NAO signal occur more in the autumn and spring, rather than in the summer, as originally hypothesized. This may relate more to the generation of anomalous airflow patterns during these periods bringing moisture from the Gulf of Mexico during the winter rainfall season (Trenberth and Hurrell 1995). No significant correlations were found with the temperature or these evaporation data.

The Pacific–North America Index (PNA), defined in terms of pressure anomalies between the Aleutians and the Gulf of Mexico, illustrates the effects of other Pacific circulation patterns. In this case, there are significant positive correlations with the Jornada headquarters' precipitation at lags of −3, −2, and 8 months. The minimum, maximum, and average temperatures have significant positive correlations at lags of −2, −1, and 10 months. The 10-month lag positive correlation is also present in these evaporation data. Redmond and Koch (1991) examined the relationship between October–March PNA and temperature and precipitation at a regional scale. They found an insignificant positive correlation for precipitation and a significant negative correlation ($p < 0.05$) for temperatures. Although the Jornada data do have negative correlations for temperatures at lags 1–7, which would correspond to the signal observed by Redmond and Koch, in no case are they significant.

Implications of Changes within the Instrumental Climate Record

The instrumental record at Jornada shows that there have been significant fluctuations in temperature, precipitation, and by extension, the number of frost days and evaporation. These fluctuations exist on cycles extending from 3 to 64 years. Precipitation fluctuations are reinforced by the occurrence of ENSO events, with significant increases in winter precipitation in El Niño and significant decreases in La Niña years. To a certain extent, these fluctuations are reinforced by teleconnections with the NAO and PNA signals. The coincidence of these larger scale teleconnections with other cycles in the climate may serve to amplify the variations, for example, the repeated La Niña events in the 1950s superimposed on the preexisting drought signal. Reynolds et al. (1999a) show how shrubs on the Jornada can withstand drought by switching growth activity between seasons, which would allow a certain degree of tolerance of such fluctuations. Neilson (1986) suggested that black grama (*Bouteloua eriopoda*) seedling production occurred preferentially when winter drought was followed by high summer rainfalls. In that the summer rainfall has no significant relationship with the large-scale circulation patterns analyzed, such conditions may develop when La Niña events occur in association with otherwise wetter than average cycles. It is possible, then, that longer term fluctuations in vegetation may be triggered by combinations of the various climatic conditions, controlled in the winter season by the ENSO, PNA, and NAO signals and in the summer by the movements of the Intertropical Convergence Zone and the development of monsoonal conditions and their interplay with auto variations in the local climate.

Conclusions

Although there are some elements of persistence in the Jornada climate at the daily scale, the general climate in the basin is characterized by variability at all spatial and temporal scales. There is a clear seasonality, relating to the development of the summer monsoonal system, that controls the amount of rainfall received. The general fluctuations within the years and between years can be ascribed to the highly variable nature of the convective rainfall generated by such systems. As well as such apparently random variability, there are a number of fluctuations that occur on a cyclic basis, with measured cycles of up to 64 years in length present in the instrumental records. Although the mechanisms behind most of these remain unexplained, there is evidence for important quasi-periodic fluctuations relating to variability in global circulation patterns, particularly those represented by the Southern Oscillation and to a lesser extent the NAO and the PNA signal. The Southern Oscillation is particularly important in controlling the amount of winter precipitation, with significantly wetter conditions occurring in El Niño years and significantly drier conditions occurring in La Niña years. According to Neilson (1986), the El Niño years should thus favor shrub vegetation and the La Niña years grass vegetation. The more frequent occurrence of El Niño events at the end of the last century and in the first half of the present century (Anderson et al. 1992) may thus have provided favorable conditions for the increase of shrubland in the light of other land use and environmental changes.

Wainwright (2005) reviews evidence that suggests that this climatic regime seems to have been in place for around the last 4,500 years, albeit with a number of smaller scale oscillations. The longer term variability related to the Southern Oscillation phenomena was well established over this time period, suggesting that both the general trend and shorter timescale variability has been in operation over at least this time scale. There also appear to have been a number of episodes of shorter time scale variability over the last millennium, which are well documented by tree ring analysis. The climate seems to have oscillated between extremes of wet and dry conditions over this time period, although some of the records suggest that the present century has seen the largest magnitude extremes. Before this time period, there is a phase where conditions were warmer than present, again lasting for up to 4,000 or 5,000 years. This broadly corresponds to the early Holocene peak of insolation. Vegetation evidence as well as global change model experiments suggest that this period saw enhanced monsoonal activity, which seems to have led to generally wetter conditions than present. There is evidence for oscillations of even wetter conditions superimposed on this increase both in the Jornada Basin and elsewhere in the Southwest, although this seems to have been more pronounced in the Sonoran Desert than the Chihuahuan Desert. The mechanisms of these changes are unknown. It has been suggested that the Southern Oscillation phenomena were not operative during these periods of warmer conditions, but it is not known to what extent other large-scale phenomena could have controlled such variability. The vegetational evidence suggests continual changes and migrations of the plant species that now characterize these areas throughout this period in response to such changes.

Given the control of the monsoonal circulation patterns on the climate, it is probable that the modern patterns have been in place for the major part of the Pleistocene, although soil carbonate and geomorphic data suggest that there was a gradual transition until around 700ka B.P., when these cycles became fully established. Glacial periods were probably characterized by generally pluvial conditions, although this term should also be understood with respect to available moisture under cooler than present conditions. Estimates of temperature change vary from about 2°C to 6°C cooler than present at the late glacial maximum. It is likely that the patterns of seasonality were highly different from those at present, due to the absence of the summer monsoonal patterns and the Equatorward displacement of the polar jet stream. Summer rainfall was probably slightly less than at present, and winter rainfall was probably substantially higher. Expanded analyses beyond the scope of this chapter (Wainwright 2005) can be viewed at www.ambiotek.com/advances/advemma/indivs/wainwright.pdf.

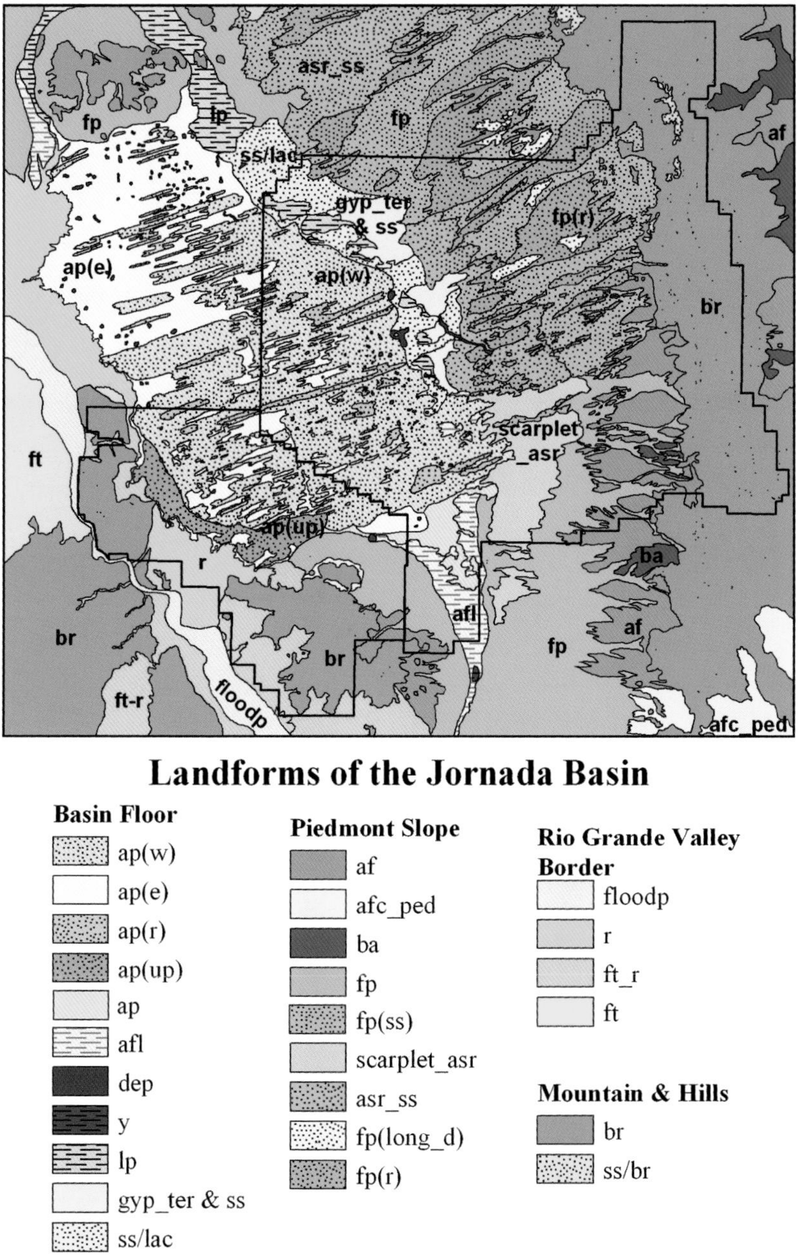

Figure 2.4 Landform map of the Jornada Basin. Map unit definitions are given in table 2-2.

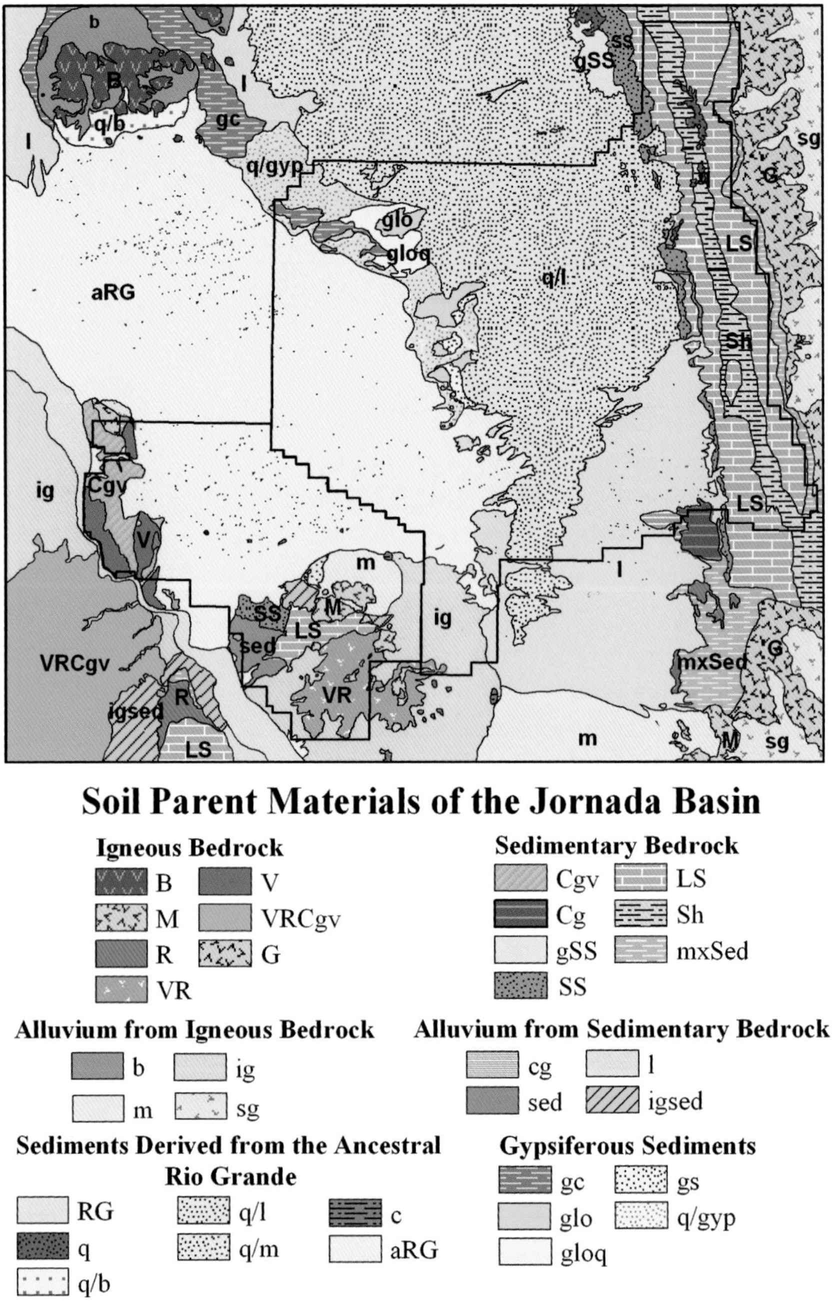

Figure 2.8 Soil parent material map of the Jornada Basin LTER site. Map unit definitions are given in table 2-3.

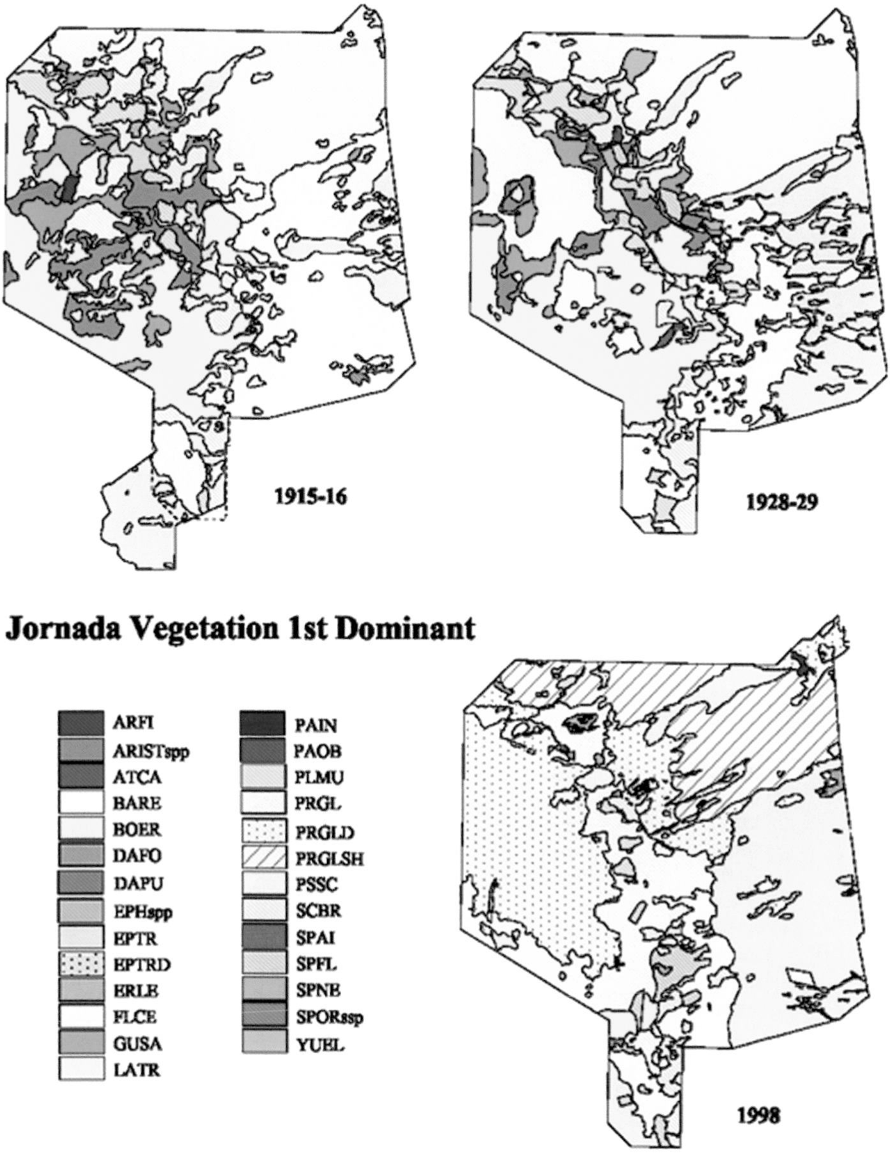

Figure 10.1 Vegetation maps of the first dominant species derived from reconnaissance surveys of the plains portion of the Jornada Experiment Range in 1915–16, 1928–29, and 1998. Species codes are defined in table 10-1. More area was fenced in 1915–16 than at present, but only the area shown by dotted lines and conforming to present boundaries was used in computing areas. The areas dominated by mesquite in 1998 were categorized as mesquite (soil accumulations at base of plants 20 cm or less), mesquite dunes (20 cm to about 3 m in height), or mesquite sandhills (greater than 3 m in height, sometimes 6 m or more).

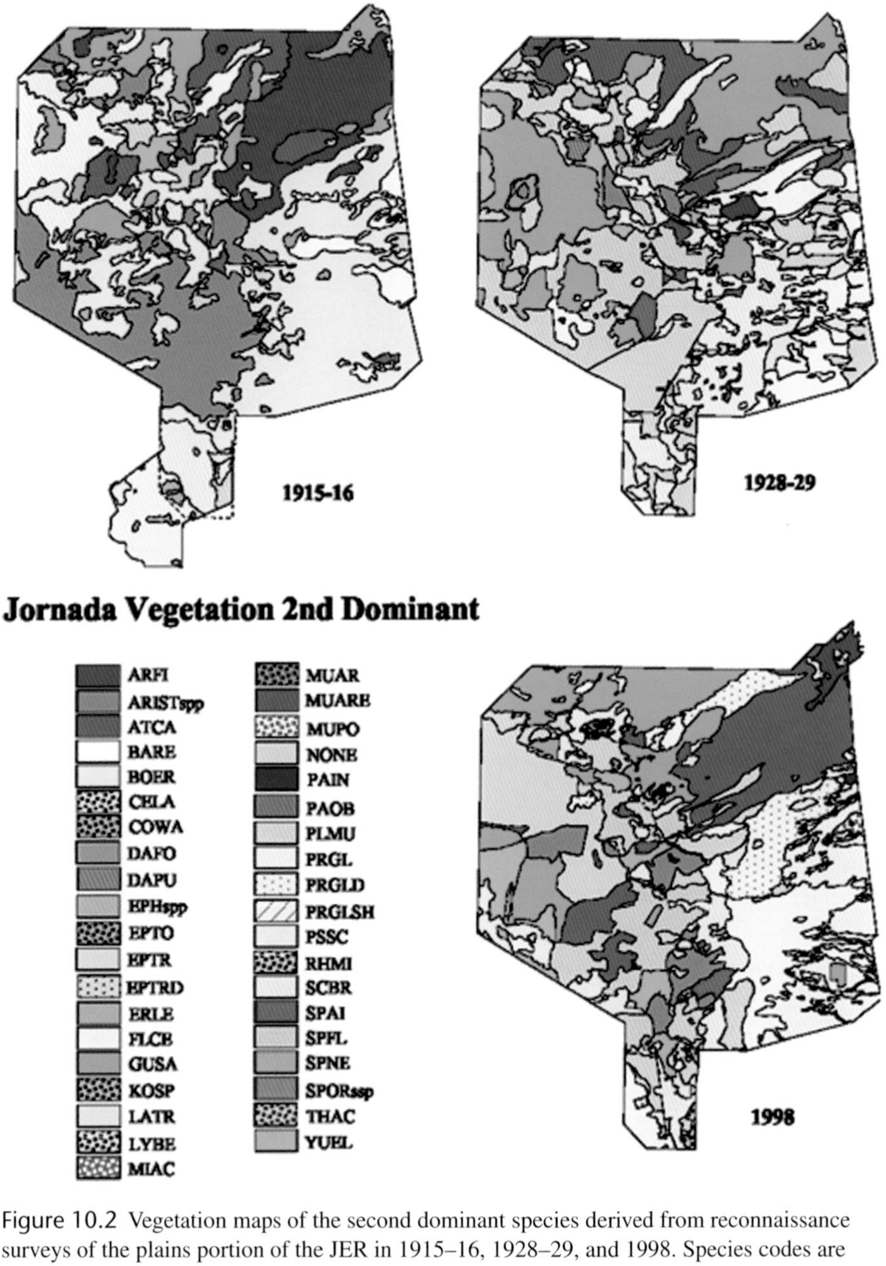

Figure 10.2 Vegetation maps of the second dominant species derived from reconnaissance surveys of the plains portion of the JER in 1915–16, 1928–29, and 1998. Species codes are defined in table 10-1. More area was fenced in 1915–16 than at present, but only the area shown by dotted lines and conforming to present boundaries was used in computing areas. The areas dominated by mesquite in 1998 were categorized as mesquite (soil accumulations at base of plants 20 cm or less), mesquite dunes (20 cm to about 3 m in height), or mesquite sandhills (greater than 3 m in height, sometimes 6 m or more).

4

Soil Development in the Jornada Basin

H. Curtis Monger

Soils of the Jornada Basin are the substrate on which Jornada ecosystems reside and interact. Understanding soils and plant–soil feedback processes have been integral to understanding vegetation change and desertification (Buffington and Herbel 1965; Schlesinger et al. 1990). Formal studies of Jornada soils extend back to 1918 (figure 4). The most detailed study of Jornada soils is the USDA-SCS Desert Soil-Geomorphology Project (Gile et al. 1981), a 400-mi^2 study area that includes the southernmost areas of the Jornada Experimental Range (JER) and Chihuahuan Desert Rangeland Research Center (CDRRC) (refer to figure 2–1). This chapter highlights findings of soil and geomorphology studies, discusses factors and processes of soil development, and lists several ways soils of the Jornada Basin carry a memory of past climates.

Types of Soils

In addition to the Veatch (1918) study and the Desert Soil-Geomorphology Project, other investigations of soil types in the Jornada Basin include three soil surveys by the Soil Conservation Service: the first was of Jornada Experimental Range (SCS 1963), the second was of the White Sands Missile Range that includes the eastern Jornada Basin and San Andres Mountains (Neher and Bailey 1976), and the third was of Doña Ana County (Bulloch and Neher 1980).

The 1918 Map

The 1918 investigation by J.O. Veatch of soils of the Jornada Basin was a reconnaissance study of the Jornada physical landscape. The purpose of the investigation was to make observations on the relation between soils and native vegetation and of the effect of overgrazing on different soil types. Veatch divided the study area into the higher mountain slopes, the foothills, and the Jornada Plain (as he described it, the plain included the currently recognized basin floor and piedmont slope). He recognized that the Jornada Plain was of Pleistocene age and contained extinct lakes with gypsum precipitated from desiccating water. He wrote that little change existed between the soil and subsoil, that "in reality a description of 'soils' here is but little more than a description of the various lithologic phases, appearing at the surface of a recent geologic formation." Still, he made the important observation that development of caliche depends on the age of the soil, greatest in old soils.

The 1918 soil map contains 13 units (figure 4-1). Some of the main characteristics of these units are paraphrased next, beginning with the western units of the Jornada Plain and progressing eastward to the San Andres Mountains. The Jornada Plain west of the gypsiferous playas had three units. The first, West Well Gravelly Sand (WGS), was an area where "whitish limestone-like caliche" of at least 6 feet in thickness occurred at depths of 2–6 feet beneath sandy and gravelly surface soils. In 1918, the WGS unit was predominately covered with a threeawn (*Aristida)* grama (*Bouteloua*) association with no brush except for a few mesquite (*Prosopis glandulosa*) in wind-eroded areas. To its east, the Jornada Red Loamy Sand (RLS) also contained caliche substratum, but this caliche was not exposed (except where excavated by digging animals, such as black-tailed prairie dogs (*Cynrmys ludovicianus*) because the sandy soils had yet to be eroded. The abundant black grama (*Bouteloua eriopoda*) and absence of brush made RLS the most valuable forage land, as well as the most vulnerable to degradation. The third unit, Jornada Sand (RLS[W]), was described as the eroded phase of RLS in which sand of the originally smooth grama surface had been shifted by wind into low mesquite hummocks and ridges. In RLS(W), sand buried caliche to depths of 6–10 feet, whereas in other places caliche was exposed due to wind scouring.

Veatch mapped four units in the gypsiferous playa area. The Lake Bed Clay (LC) unit was described as highly calcareous, chocolate red or grayish clay with scattered gypsum crystals overlying pure gypsum beds of lacustrine origin at depths up to 10 feet. The vegetation in LC was burrograss (*Scleropogon brevifolius)* and tobosa grass (*Pleuraphis mutica*) in 1918, with an entire absence of soapweed and shrubs present on neighboring sandy soils. The unit labeled Gypsum Soil (GY.S) described small areas where gypsum was at or very near the surface, covered only by a thin veneer of silt or very fine sand with only scant amounts of dropseed (*Sporobolus*) and Mormon tea (*Ephedra*). The Middle Well Sand (SG) was deep, loose sands that covered substrata of gypsum. This unit was then nearly barren except for some remaining threeawns and black grama. Jornada Gray Sand (GSL) was a second sand unit that occupied the bottoms and slopes of lake basins and, like SG, overlaid gypsum. This unit differed in color from

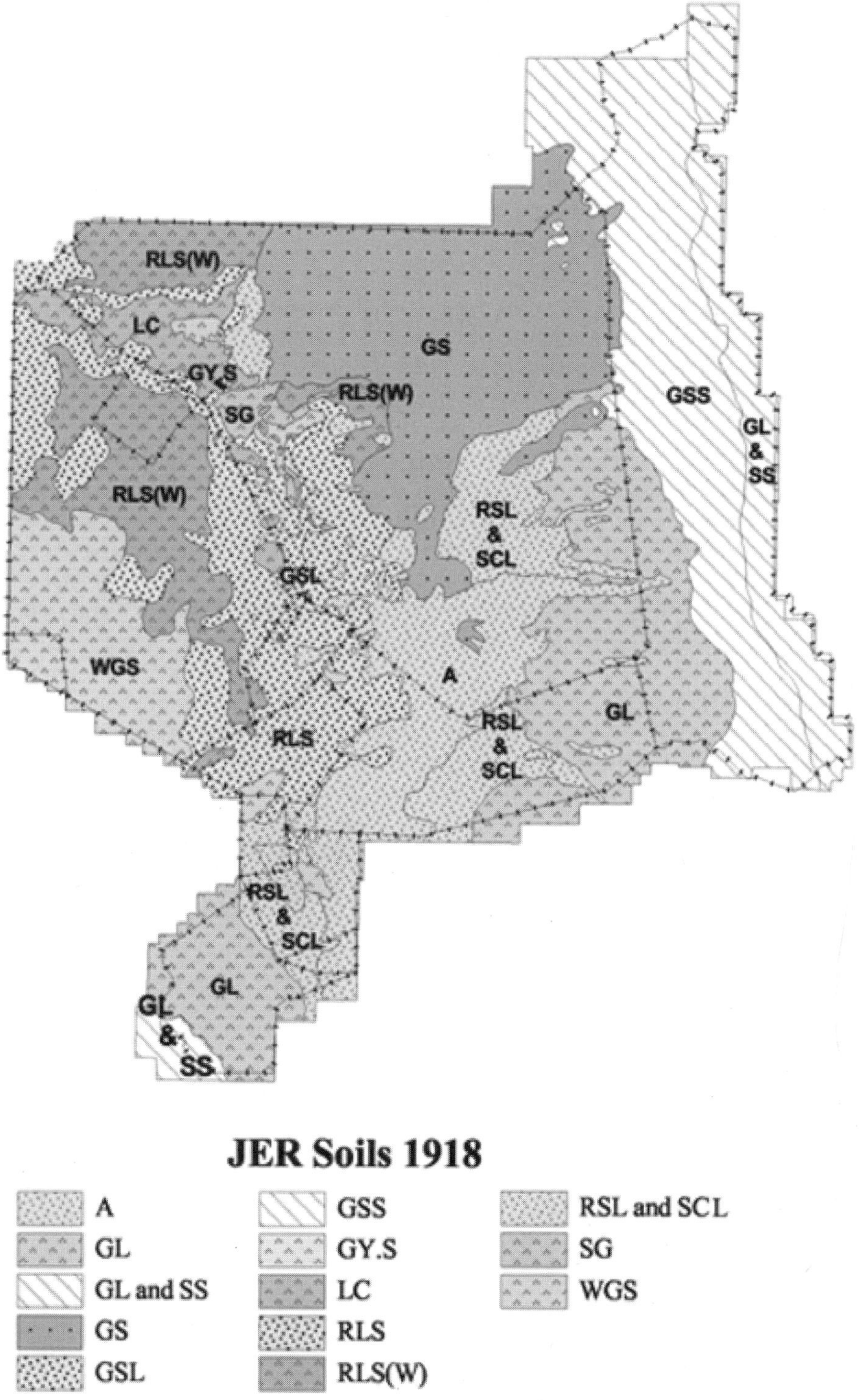

Figure 4-1. The 1918 soil map of JER produced by J. O. Veatch under a cooperative agreement between the U.S. Bureau of Soils and the U.S. Forest Service (the agency that operated JER at that time). The 1918 soil survey contained 13 map units. Abbreviated descriptions of the map units are given in table 4-1.

other sands, which were prevailingly red. It also differed because its sand was derived from lakes rather than the Jornada Plain. The vegetation had a mixed character, transitional from burrograss and tobosa grass of the lake bottoms to considerable black grama, threeawn, and soapweed (*Yucca elata*) on the slopes.

The piedmont slopes contained four units—the clay or adobe soil (A) that extended across the current boundary of the piedmont slope and basin floor. These clay flat soils were described as having some "lime cementation." They ascended toward the foothills by broad, flat steps and scarplets on which numerous patches and low-winding ridges of loose, reddish, wind-laid sand resided. The vegetation was predominantly burrograss, tobosa grass, and tarbush (*Flourensia cernua*) on the flats with soaptree yucca, honey mesquite, Mormon tea, and creosotebush (*Larrea tridentata*) on the ridges. The Goldenburg Sands (GS) were wind-laid deposits overlying various alluvial strata that because of their less-developed caliche were recognized to be younger than similar duned soils of the Jornada Plain whose sands are essentially the same lithology. Vegetation of the GS unit was dominated by mesquite, with common amounts of sand sage (*Artemisia*) and saltbush (*Atriplex*), and only sparse growth of dropseed and threeawn grass. The Jornada Clay Loam (RSL and SCL) was used to map various intermediate piedmont slope positions. The unit had a range of textures, degrees of lime carbonate, and vegetation, which consisted of brush (tarbush, creosotebush, and mesquite) and grasses (burro, tobosa, grama, and threeawn). Higher on the piedmont slopes, the alluvial fan and fan-piedmont landforms were mapped as the Middle Tank Gravelly Soils (GL). These soils contained coarse rock detritus that in places were cemented into conglomerates that restricted root penetration. Vegetation was described as a creosotebush-black brush (assumed to mean tarbush) association of low density, which attributed to the low moisture content of the soil.

Bedrock hills and mountains were mapped using two units. Foothills (GSS) were described as consisting of thin, silty, residual soils developed from underlying Paleozoic sedimentary rock. In places, sand of the Jornada Plain has been blown up on the hills and given rise to mesquite and soaptree yucca vegetation, in sharp contrast to the creosotebush and grass vegetation of the residual bedrock soils. Mountain Slope Soils (GL and SS) consisted of very thin, stony, yet dark soils that bear close relation to the underlying bedrock. From a distance, the mountain slopes were described as appearing nearly barren; however, mixed grasses, mountain mahogany (*Cercocarpus*) , and other shrubs occurred on the slopes while piñon *(Pinyon),* juniper *(Juniperus),* and oak *(Quercus)* were in the more favorable situations.

The 1963 Map

The second soil map of the Jornada Basin (see figure 4-2) was completed by the SCS but was not published except in a slightly modified form in Buffington and Herbel (1965). The 1963 map made several advances. It expanded the 13 units of the 1918 map to 20 units. These units provide added detail to gypsiferous units, subdivisions of basin floor soils, and slopes, such as the delineation of the

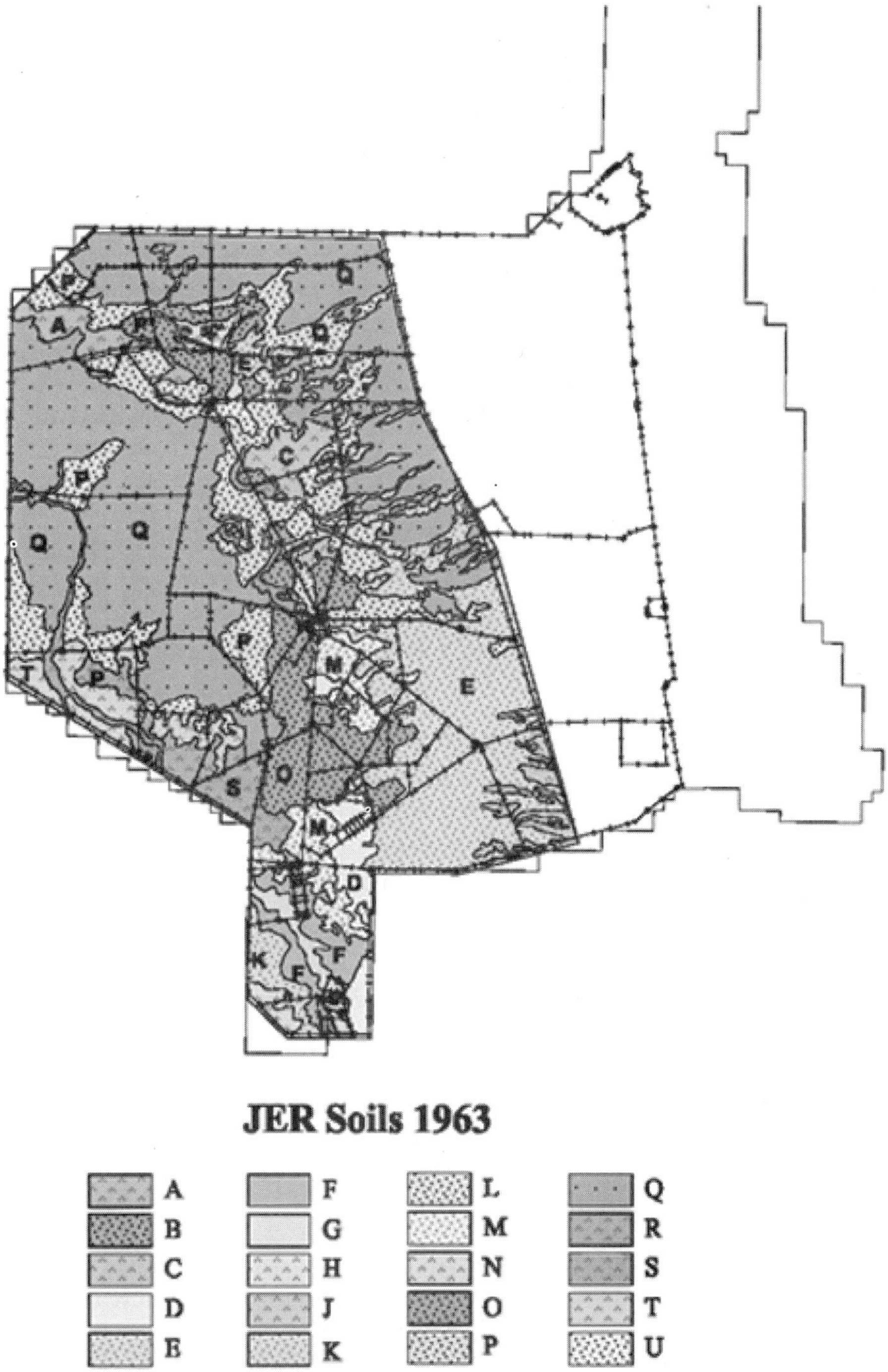

Figure 4-2. Soil map of JER produced by the Soil Conservation Service (1963 unpublished). This soil survey was used, in a slightly modified form, to examine soil–vegetation relationships (Buffington and Herbel 1965). Abbreviated definitions of this map and their correlation to the 1918 map are given in table 4-1.

fault scarp near the western JER boundary. One significant observation is represented by the western boundary of the O unit. This line delineates the boundary between indurated caliche (stage IV and V petrocalcic horizons) to the west and nonindurated caliche (stage II and III calcic horizons) to the east. Recently, this soil boundary was described by Gile (1999), who attributed it to an early middle Pleistocene lake, Lake Jornada (Gile 2002).

The detail and accuracy of the 1963 map make it suitable for many landscape-scale ecological studies today (e.g., Buffington and Herbel 1965). However, it has limitations because it predated soil taxonomy (Soil Survey Staff 1975) and lacks quantitative soil characterization data. The map also carried the Q and P units onto the piedmont slope area of Veatch's Goldenburg Sands. Although these units mark the presence of coppice dunes Q and hummocky soils P on the piedmont slope, they do not convey the general absence of petrocalcic horizons. Nevertheless, the quality of the line work of this map make it the best soils map currently available for the Jornada Basin north of the Desert Project. The unit descriptions of the 1963 map and their correlation to the 1918 map are given in Table 4-1.

The Doña Ana County and White Sands Missile Range Soil Surveys

In 1980, the Doña Ana County Soil Survey (Bulloch and Neher 1980) was published (figure 4-3). It provides a general description of Jornada soils and made the advancement of classifying soils according to the soil taxonomy system (Soil Survey Staff 1975, 1999). This survey also covers both the JER and CDRRC. However, because it was designed to be a general survey, it has limitations for scientific studies requiring detailed soils information.

In 1976, the White Sands Soil Survey was published by the SCS at a scale of 1:100,000 (Neher and Bailey 1976). This survey covers the eastern part of the JER in the area that is jointly administered by White Sands Missile Range and the USDA. This survey also classifies soils using the soil taxonomy system. Unlike most SCS soil surveys, however, this one is unique because the maps do not have an aerial photographic base (because of missile range security).

The Desert Soil-Geomorphology Project

The Desert Soil-Geomorphology Project provides uniquely detailed and quantitative soil information. In August 1957, R. V. Ruhe and L. H. Gile moved to Las Cruces to begin the Desert Soil-Geomorphology Project, which was one of a few regional soil-geomorphic projects established by SCS in the 1950s. At that time, little was known about soils in arid and semiarid regions of the American Southwest away from alluvial valley floors. This study was undertaken to learn more about the morphology, classification, and genesis of desert soils and their relation to late-Cenozoic landscape evolution, as well as to establish principles that could

Table 4-1. Correlation of 1918 and 1963 soil maps of the Jornada Basin area and abbreviated descriptions of 1963 mapping units

1918 Soil Map Units	1963 Soil Map Unit Descriptions (Abbreviated from SCS 1963)
Basin Floor *(Alluvial Plain)*[a]	
West Gravelly Sand (WGS)	(T) Tencee (Simona-Palma complex),[b] gravelly loamy fine sand. Shallow to very shallow loamy sand underlain by caliche at 10 to 20 in. Occurs on bench-like surface higher than land to east. Similar to S except shallower, less B-horizon development, and more surface litter of caliche and blowouts. 0–3% slope.
	(S) Cacique loamy fine sand. Weakly developed textural B [argillic] horizons overlying semi-indurated to indurated caliche at 11 to 48 inches. Surface littered with caliche fragments. Rodent action has mixed caliche. More sloping than T. 1–3% slope.
Jornada Sand (RLS(W))	(Q) Continental loamy fine sand, severely eroded, coppice dunes. Severely altered by erosion. Sand hummocks 3 to 6 ft high, 12 to 20 ft long with interdune blowouts. Semi-indurated to indurated caliche at 3.5 to 5 ft. 0–3% slope.
	(P) Continental loamy fine sand, hummocky. Semi-indurated to indurated caliche at 30 to 60 in. Less eroded than Q, more eroded than O. On piedmont slope, weakly calcareous upper horizons and indurated caliche is absent. 0–2% slope.
Jornada Red Loamy Sand (RLS)	(O) Continental loamy fine sand. Deep, reddish brown sandy soils with moderately well developed textural B [argillic] horizons. Caliche is soft to only semi-indurated; in places it may be indurated. Less wind erosion than P and Q. Different from M by being less calcareous, lighter textured, deeper to caliche, and higher on landscape. 1–3% slope.
	(M) Turney sandy loam. Minor blowouts and sand accumulation of 6 to 12 inches. Weak textured B [argillic] horizon of sandy clay loam. Strong Cca [calcic] horizon that may be weakly indurated. Correlates to mapping unit 56 & 57 of Desert Project. 0–2% slope.
	(R) Palma loamy fine sand, sloping. Sand accumulations on sloping ridges and escarpments. Usually have moderate to strong Cca horizons between 30 & 60 inches. Gypsum at about 36 inches in areas of ancient lakes. 3–6% slope.
Basin Floor *(Alluvial Flat)*	
Clay and Adobe Soil (A)	(G) McNeal (Continental) silt loam, bottom (bolson) phase. Lowest position on landscape. Well developed textural B [argillic] horizon and Cca [calcic] horizon. Subject to runoff water. Correlates with mapping unit 55 of Desert Project. Slope 0–1%.
	(F) McNeal (Continental) loam. Loam to silt loam surface textures. Well-developed textural B [argillic] horizons and strong Cca [calcic] horizons. Soils developed in alluvial parent material of mostly mixed igneous rock. Paleosols are common. Correlates with mapping unit 16V of Desert Project. 0–2% slopes.
	(D) Harvey (Hoban) silt loam. Soils formed in non-gravelly limestone alluvium. Soil do not have textural B [argillic] horizons, but do have moderate Cca [calcic] horizons. Differ from G & F by not having textural B [argillic] horizons. Also, slightly higher landscape position than F. Correlates with mapping unit 51 of Desert Project. 0–1% slopes.

(continued)

Table 4-1. (*continued*)

1918 Soil Map Units	1963 Soil Map Unit Descriptions (Abbreviated from SCS 1963)
Lake Bed Clay (LC)	(A) Skassi (Verhalen) clay. Reddish brown calcareous clay to silty clay loam developed from lake sediments containing some soluble salts and well-rounded, gravel-size gypsum crystals. Underlain by massive gypsum. 0–1% slope.
	(B) Rustler (Russler) loam. Strongly calcareous loam to silty clay loam in older lake basins. Subsoils resemble lacustrine sediments containing gypsum. Associated with shallow gypsum soils on higher positions and very fine textured gypsiferous soils on lower landscapes. 0–1% slopes.
Gypsum Soil (GY.S)	(H) Eroded gypsum land (Cottonwood-Gypsum Outcrop complex). Strongly calcareous, medium-textured soils overlying gypsum beds at 1 to 12 inches. Highly eroded. 0–3% slope.
Middle Well Sand (SG)	(U) Cottonwood fine sand, overblown. Shallow sandy soils over gypsum at 4 to 24 inches. Lacustrine material overblown by aeolian sands with slight depressions & rolling ridges within an old lake basin. 0–3% slope.
Jornada Gray Sand (GSL)	(L) Gomez sandy loam. Weakly developed, strongly calcareous soil with 24 to 40 inch depth to gypsum beds. Soil developed from strongly calcareous lacustrine material deposited by strong winds from adjacent playas or shorelines of ancient lakes. 0–7% slope.
	(C) Cottonwood soils (Cottonwood-Reeves-Hoban complex). Complex of calcareous-gypsiferous soils on a shelf midway between sandy soils higher on landscape and finer soils lower on landscape. Gypsum ranges from soft powder to coarse crystals. Textures range from clay loam to sand. 0–3% slope.
Piedmont Slope *(of San Andres Mountains)*	
Goldenburg Sand (GS)	(Q) Continental loamy fine sand, severely eroded, coppice dunes. (Q also used for RLS(W)). Severely altered by erosion. Sand hummocks 3 to 6 ft high, 12 to 20 ft long with interdune blowouts. Semi-indurated to indurated caliche at 3.5 to 5 ft. 0–3% slope.
	(P) Continental loamy fine sand, hummocky. (P also used for RLS(W)). Semi-indurated to indurated caliche at 30 to 60 in. Less eroded than Q, more eroded than O. On piedmont slope, weakly calcareous upper horizons and indurated caliche is absent. 0–2% slope.
Clay and Adobe Soil (A)	(E) Target-Isacks soils (Dona Ana Complex). Weakly developed, yet strongly calcareous soils on alluvial fans of limestone and calcareous sandstone that extend from the bottom of the Jornada basin eastward to the foothills of the San Andres Mountains. Erosion escarpments range from 6 inches to 7 feet high with narrow sandy ridges parallel to scarps. Surface soils range from silt loam to loamy sand. 0–3% slope.
Jornada Clay Loam (RSL & SCL)	(E) Target-Isaacks soils (Dona Ana Complex). (E also used for Clay and Adobe Soil (A)). Weakly developed, yet strongly calcareous soils on alluvial fans of limestone and calcareous sandstone that extend from the bottom of the Jornada basin eastward to the foothills of the San Andres mountains. Erosion escarpments range from 6 inches to 7 feet high with narrow sandy ridges parallel to scarps. Surface soils range from silt loam to loamy sand. 0–3% slope.

1918 Soil Map Units	1963 Soil Map Unit Descriptions (Abbreviated from SCS 1963)
Middle Tank Gravelly Soil (GL)	(J) Solidad (Cave) gravelly sandy loam. Shallow, violently calcareous soils with weak Cca horizon underlain by weakly cemented limestone cobbles or indurated caliche. Soils developed in alluvial fans consisting mostly of limestone with some calcareous sandstone on long ridges that slope to west. Water erosion is shown by gullies to indurated caliche (calcrete). Surface covered with up to 35% gravels. 2–5% slope.
Piedmont Slope *(of Doña Ana Mountains)*	
Jornada Clay Loam (RSL & SCL)	(K) Daggett gravelly sandy loam. Gravelly, weakly developed regosols. Parent material contains caliche mixed with monzonite and rhyolite gravel. Soils have very weak Cca horizons overlying paleosols with textured B [argillic] and Cca [calcic] horizons. Moderate gully and sheet erosion. Surface covered with gravel more concentrated than in profile. Correlates with mapping unit 13V of Desert Project. 1–3% slopes.
	(F) McNeal (Continental) loam. (F also used for A). Loam to silt loam surface textures. Well-developed textural B [argillic] horizons and strong Cca [calcic] horizons. Lower lying soils developed in alluvial parent material of mostly mixed igneous rock. Paleosols are common. Correlates with mapping unit 16V of Desert Project. 0–2% slopes.
Middle Tank Gravelly Soil (GL)	(N) Solidad (Cavot) very gravelly sandy loam. Soil 10 to 32 inches over either semi-indurated or indurated caliche. Soil paved with gravel and fine cobbles. Soil developed in outwash alluvial deposits of mixed igneous rocks. Profiles free of toxic salts and alkali. Active sheet and gully erosion. 2–5% slope.
Bedrock Hills and Mountains (San Andres Mountains)	
Foothills (GSS and SS)	(V)[c] Rock outcrops. Much of the surface is bare rock. In some areas, sand has accumulated quite high on the slopes and covers or partially covers the rock.

[a] Soils are grouped into landform divisions of figure 2-2.

[b]Names in parentheses are soil names published in Buffington and Herbel (1965). Soil names preceding the names in parentheses are as they appear in unpublished report (SCS 1963 unpublished).

[c] Rock outcrops unit was not recognized in SCS (1963 unpublished) but does appear in Buffington and Herbel (1965).

be used to improve the quality of soil survey in similar settings (Ruhe 1967; Hawley 1975b). The study area of the Desert Soil-Geomorphology Project (informally called the Desert Project) originally encompassed the JER, but it was later reduced to its 400-mi^2 area to focus more effort on a smaller area (Gile personal communication).

The Desert Project made several contributions to the understanding of soil-geomorphic relationships in arid and semiarid climates. In addition to the work on the local Cenozoic stratigraphy (Ruhe 1962; Hawley 1975a), hydrogeology (Hawley and Lozinsky 1992), geomorphic surfaces (Ruhe 1964; Gile et al. 1981), and paleoclimate (Gile 1975c; Hawley et al. 1976), general principles were established that dealt with (1) pedogenic-illuvial origin of carbonate ($CaCO_3$) ho-

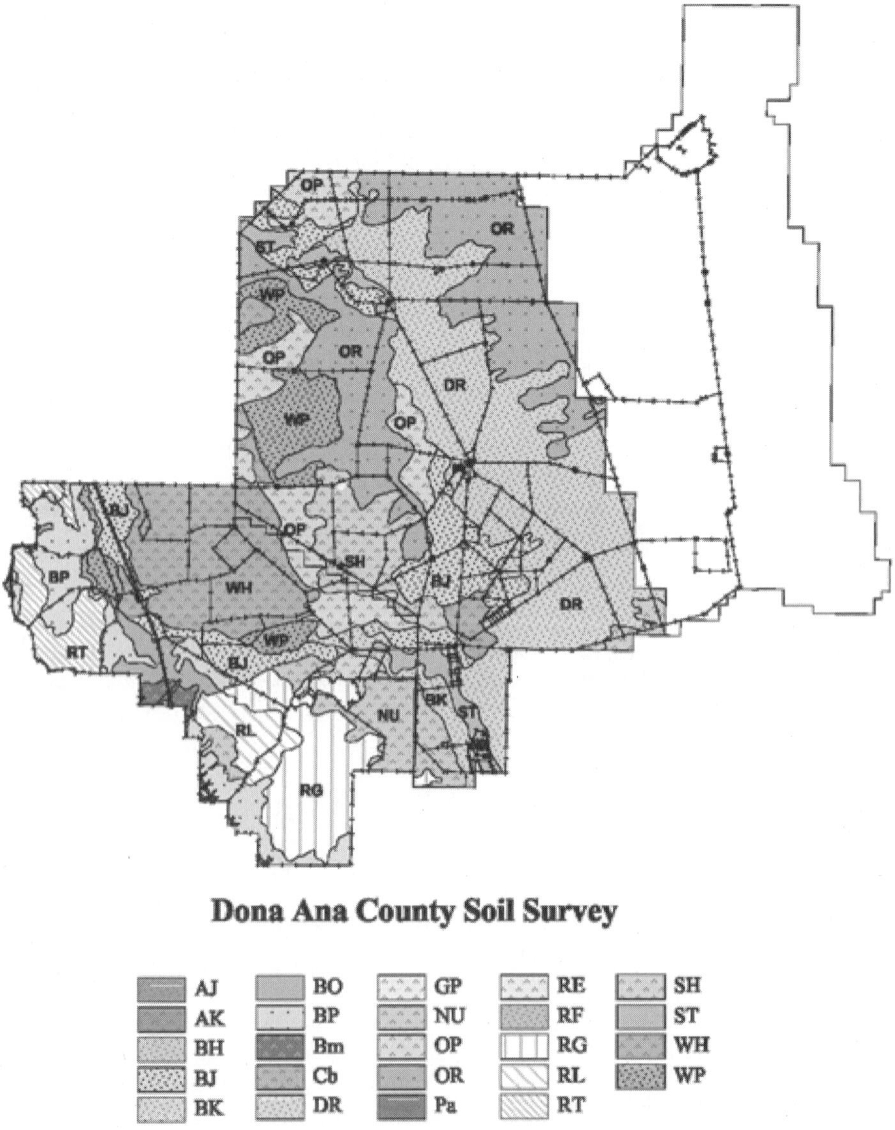

Figure 4-3. The Doña Ana Soil Survey (Bulloch and Neher 1980). This survey provides a general survey of soils at both JER and the CDRRC. Soil series and classification based on the soil taxonomy system are given in table 4-2.

Table 4-2. Classification of soil series mapped for Jornada Experimental Range and Chihuahuan Desert Rangeland Research Center by the 1980 Doña Ana Soil Survey (figure 4-3). The 1975 classification system (Soil Survey Staff 1975) was used in the Doña Ana Soil Survey. Subgroups of Aridisols have since been revised (Soil Survey Staff 1999). Particle-size classes, mineralogy classes, and temperature regimes are the same for both the 1975 and 1999 classification systems.

Series[a]	Family Classification (1975 System)	Revised Subgroup Classification (1999)
Agua	C-l/s or s-k[b] mixed (calcareous), thermic Typic Torrifluvents	no revision
Arizo	Sandy-skeletal, mixed, thermic Typic Torriorthents	no revision
Belen	cl/l, montmorillonitic (calcareous), thermic Vertic Torrifluvents	no revision
Berino	Fine-loamy, mixed, thermic Typic Haplargids	Typic Calciargids
Bluepoint	Mixed, thermic Typic Torripsamments	no revision
Bucklebar	Fine-loamy, mixed, thermic Typic Haplargids	no revision
Caliza	Sandy-skeletal, mixed, thermic Typic Calciorthids	Typic Haplocalcids
Canutio	l-sk, mixed (calcareous), thermic Typic Torriorthents	no revision
Dona Ana	Fine-loamy, mixed thermic Typic Haplargids	Typic Calciargids
Harrisburg	Coarse-loamy, mixed, thermic Typic Paleorthids	Typic Petrocalcids
Lozier	Loamy-skeletal, carbonatic, thermic Lithic Calciorthids	Lithic Haplocalcids
Nickel	Loamy-skeletal, mixed, thermic Typic Calciorthids	Typic Haplocalcids
Onite	Coarse-loamy, mixed, thermic Typic Haplargids	Typic Calciargids
Pajarito	Coarse-loamy, mixed, thermic Typic Camborthids	Typic Haplocambids
Pintura	Mixed, thermic, Typic Torripsamments	no revision
Reagan	Fine-silty, mixed, thermic Ustollic Calciorthids	Ustic Haplocalcids
Simona	Loamy, mixed, thermic, shallow Typic Paleorthids	Typic Petrocalcids
Stellar	Fine, mixed, thermic Ustollic Haplargids	Ustic Calciargids
Upton	Loamy, carbonatic, thermic, shallow Typic Paleorthids	Calcic Petrocalcids
Wink	Coarse-loamy, mixed, thermic Typic Calciorthids	Petronodic Haplocalcids
Yturbide	Mixed, thermic Typic Torripsamments	no revision

[a] Symbols and Mapping-Unit Names for the Dona Ana Soil Survey are AJ: Agua Variant soils, moderately wet; AK: Agua Variant and Belen Variant soils; BH: Belen Variant soils; BJ: Berino-Bucklebar association; BK: Berino-Dona Ana association; BP: Bluepoint-Caliza-Yturbide complex; BO: Bluepoint loamy-sand, 1–5% slopes; Bm: Bluepoint loamy-sand, 1–5% slopes; Cb: Canutio-Arizo gravelly sandy loam; DR: Dona Ana-Reagan association; NU: Nickel-Upton association; OP: Onite-Pajarito association; OR: Onite-Pintura association; Pa: Pajarito fine sandy loam; RE: riverwash; RF: riverwash-Arizo complex; RG: Rock outcrop-Argid complex; RL: Rock outcrop-Lozier complex; RT: Rock outcrop-Torriorthents associations; SH: Simona-Harrisburg association; ST: Stellar association; WH: Wink-Harrisburg association; WP: Wink-Pintura complex.

[b] Abbreviations of particle size classes; c = coarse, f = fine, l = loamy, s = sandy, sk = skeletal.

rizons, (2) atmospheric additions, (3) soil chronology, (4) silicate clay accumulation, and (5) causal factors for soil boundaries.

Pedogenic-Illuvial Origin of Carbonate Horizons

Several hypotheses for the origin of calcium carbonate in soils had been proposed by the late 1950s when the Desert Project began, including lacustrine, fluvial, ascending groundwater, and pedologic origins (e.g., Bretz and Horberg 1949; Brown 1956; see Gile and Grossman 1979 for further discussion). However, except for small outcrops of groundwater carbonate, the following field evidence argued for a pedogenic-illuvial origin of carbonate in Desert Project soils: (1) carbonate horizons are parallel to the land surface; (2) carbonate horizons have upper boundaries within several inches to about 2 feet of the soil surface; (3) carbonate horizons have distinctive morphologies that show lateral continuity and differ markedly from morphologies of overlying and underlying horizons; (4) carbonate horizons occur between horizons containing little or no carbonate; (5) carbonate horizons occur across sediments of various compositions and textures; and (6) carbonate horizons form in a developmental sequence related to time (Gile et al. 1965). This evidence has been useful for understanding the origin of carbonate in other desert sites, such as the proposed Yucca Mountain nuclear waste repository in Nevada (e.g., Kerr 1992; Hill et al. 1995; Monger and Adams 1996).

Atmospheric Additions

Although there were multiple lines of evidence that suggested a pedogenic-illuvial origin for most carbonate horizons in the Desert Project, the source of Ca was an enigma. First, soils with igneous parent materials, especially rhyolite, were only slightly weathered, and second, these sediments contained only small amounts of Ca (Gile et al. 1966; Ruhe 1967). Therefore, an analysis of dust ensued. Dust traps were set out across the Desert Project. Ten years of dust measurements revealed that calcareous dust fell ubiquitously on the landscape, ranging from 0.2 $g/m^2/yr$ in a grassy basin floor area to 1.1 $g/m^2/yr$ in a sandy bare-ground area (Gile and Grossman 1979). Meanwhile, rain as a source of Ca became recognized as a more important source of Ca than dust. Chemical analysis of rain by Junge and Werby (1958) and Lodge et al. (1968) revealed that Ca from rainwater could produce an estimated 1.5 g $CaCO_3/m^2/yr$, assuming 200 mm annual rainfall. Therefore, if ample bicarbonate is generated by roots and microbes, carbonate resulting from Ca in rain could be roughly two to three times greater than carbonate resulting from calcareous dustfall (Gile et al. 1981).

Soil Chronology

Measurements of soil age, which ranges from Pliocene to Holocene in the Jornada region, were made using several techniques. For soils of historical age, land survey notes were used. These notes made it possible to determine which coppice dunes (Gile 1966a) and arroyo sediments were deposited between 1858 and 1922 (Gile and Hawley 1968). For soils of prehistorical age, radiocarbon dates of buried charcoal, which ranged from less than 200 to 9,360 yr B.P., provide the most authoritative dates (Hawley and Kottlowski 1969; Gile et al. 1981). Radiocarbon dates of $CaCO_3$ carbon were also used to help determine soil ages, but calcium carbonate is subject to inputs of modern carbon and is of little value in distinguishing between late Pleistocene and older soil horizons (Gile et al. 1981). For soils of middle Pleistocene age and greater, the Lava Creek B (0.61 to 0.67 Ma) and Bishop (0.76 Ma) volcanic ashes have been important for bracketing soil ages (Hawley et al. 1976; Izett et al. 1992; Sarna-Wojcicki and Pringle 1992). Paleontological remains provided some of the earliest clues to the ages of soils. For example, horses (*Equus*), short-jawed mastodons (*Mammut*), mammoths (*Mammuthus*), and stegomastodons (*Morrillia*) provided evidence that basin-floor soils formed in Camp Rice fluvial sediments were at least Kansan in age (Ruhe 1962; Hawley et al. 1969). Subsequent work refined the age of this faunal assemblage to be early Pleistocene (Tedford 1981).

Tracing soil horizons and geomorphic surfaces laterally provided some of the most conclusive evidence about relative ages of soils. For example, a soil can be determined to be younger than a neighboring soil if its geomorphic surface buries, cuts, or is inset against the neighboring surface. Another indicator of relative soil age is the degree of profile development, such as solum thickness (i.e., depth to the bottom of B horizons) and the expression of carbonate horizons (Gile 1970).

Silicate Clay Accumulation

Like the accumulation of carbonate and formation of calcic and petrocalcic horizons, the accumulation of silicate clay and formation of argillic horizons is an important diagnostic criterion for soil classification. For a horizon to qualify as argillic, clay must be illuvial and have more clay than the overlying eluvial horizon (Soil Survey Staff 1999). Clay accumulation in arid soils, however, was formerly thought to be due to in-place weathering rather than illuviation (Nikiforoff 1937; Brown and Drosdoff 1940). Yet clay in soils of the Desert Project showed coated particles in the B horizon and evidence of having an illuvial origin and meeting the criteria of an argillic horizon (Gile and Grossman 1968).

Other studies focused on the affect of parent material on argillic horizons and on factors that obliterate argillic horizons. Parent material was found to have an important control on clay illuviation because argillic horizons typically do not exist in parent material that contains abundant limestone rocks (Gile et al. 1981). The explanation apparently lies in the flocculating effect that carbonate has on clay movement. Factors that obliterate argillic horizons include (1) landscape dissection and erosional truncation of argillic horizons, (2) engulfment of argillic horizons by pedogenic carbonate, and (3) faunal mixing in which tunnels and mounds made by kangaroo rats *(Dipodomys)*, badgers *(Taxidea)*, and termites (primarily *Gnathamitermes*) destroy the fabric of argillic horizons (Gile 1975a).

Characterization Data of Jornada Basin Soils

Sites where soil characterization has been conducted are shown in figure 4-4. The sites marked Desert Project Pedons contain physical and chemical laboratory analyses performed by the SCS (now the Natural Resources Conservations Service, NRCS). These include analysis of particle size, organic carbon, nitrogen, carbonate, extractable iron, bulk density, shrink-swell potential, water-holding properties, extractable ions, clay mineralogy, pH, and electrical conductivity, among others (Gile and Grossman 1979).

Other sites with NRCS laboratory data are (1) the soil water sites, (2) the NRCS 1999 sites, and (3) the NRCS 2002 sites. Data from the NRCS 1999 and 2002 sites are on the NRCS Web site (http://soils.usda.gov). Data from the soil water sites are in Herbel et al. (1994). Other soil data were gathered at the root excavation sites (Gibbens and Lenz 2001) and carbon isotope sites (Connin et al. 1997a, b).

Processes of Soil Development in the Jornada Basin

When Veatch made the 1918 soil map of the Jornada Basin, Curtis Marbut (1863–1935) was chief of the U.S. National Soil Survey. During that time, Marbut read a German translation of K. D. Glinkas's monograph on the nature of soil science and pedology in Russia (Gardner 1957). This introduced Marbut to V. V. Dokuchaiev (1846–1903) and the Russian concepts of soil development. Of primary importance was the concept that soils are the evolutionary product of five factors—topography, climate, parent material, biota, and time. Since that time, the understanding of soil development in any region has been enhanced by knowledge of the five factors of soil formation (Jenny 1941).

The five soil-forming factors establish a context and provide the external forces for soil development in the Jornada Basin. Within that context, several physical, chemical, and biological processes are active. Soil-forming processes (as contrasted with factors) were grouped into four categories by Simonson (1959): additions, transfers, transformations, and removals (figure 4-5). The cumulative result of these pedogenic processes is the transformation of bedrock or sedimentary deposits into soil with pedogenic horizons. Pedogenic horizons (e.g., A, E, B horizons) are produced by the four soil-forming processes. The C horizon, in contrast, is typically a sedimentary layer little affected by pedogenesis (that is, only slightly modified by the four soil-forming processes). All the processes illustrated in figure 4-5 occur in soils of the Jornada Basin, but they occur at different rates. A few of the more notable results of these processes are described (organic matter accumulation, pedogenic carbonate accumulation, desert pavement and the formation of eluvial horizons, and clay accumulation).

Organic Matter Accumulation

Organic matter plays an important role in nutrient cycling and aggregate stability (chapter 6). Desert soils in the Jornada Basin have low concentrations of soil

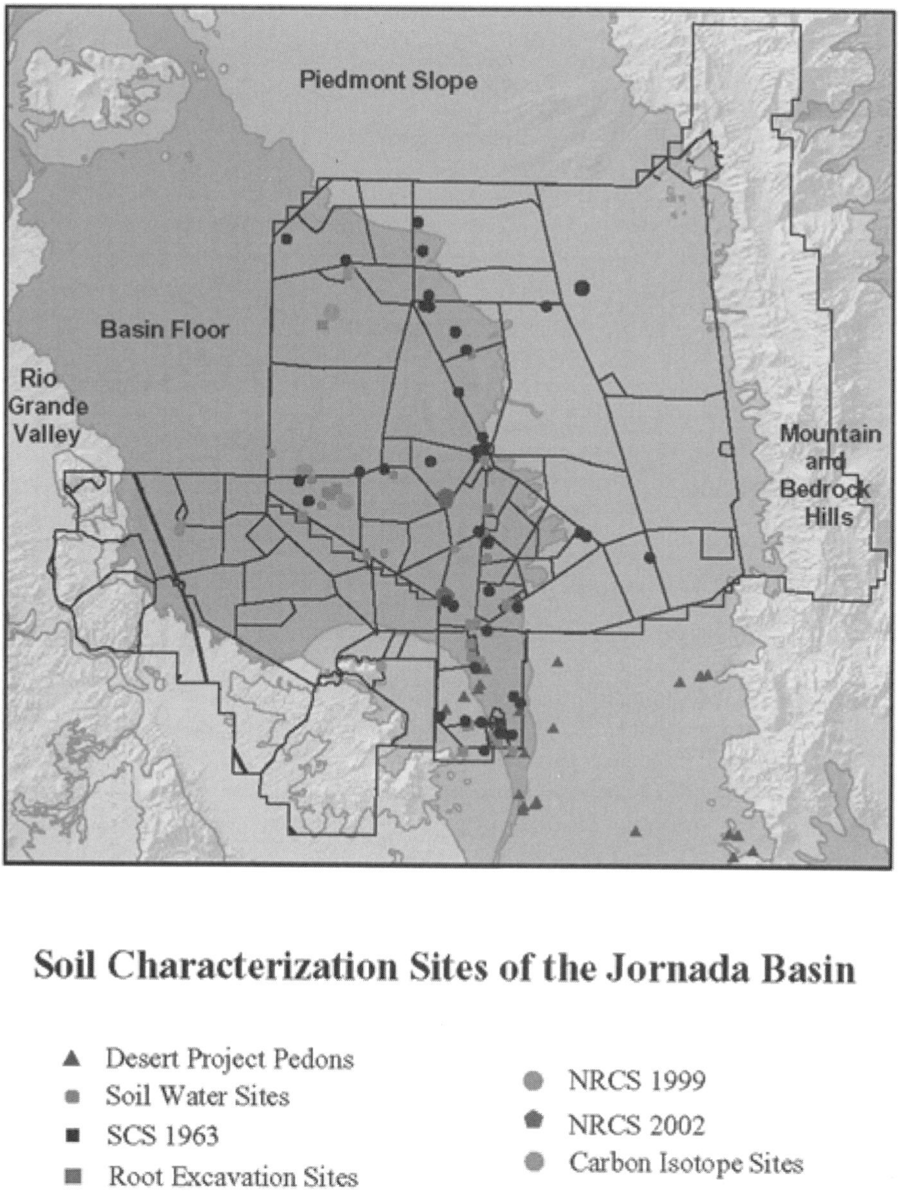

Figure 4-4. Map of physiographic units and locations of sites where soil characterization data have been generated. Publications and Web sites where data are located are in Gile and Grossman (1979), Herbel et al. (1994), NRCS Web site (http://soils.usda.gov), Gibbens and Lenz (2001), and Connin et al. (1997a,1997b).

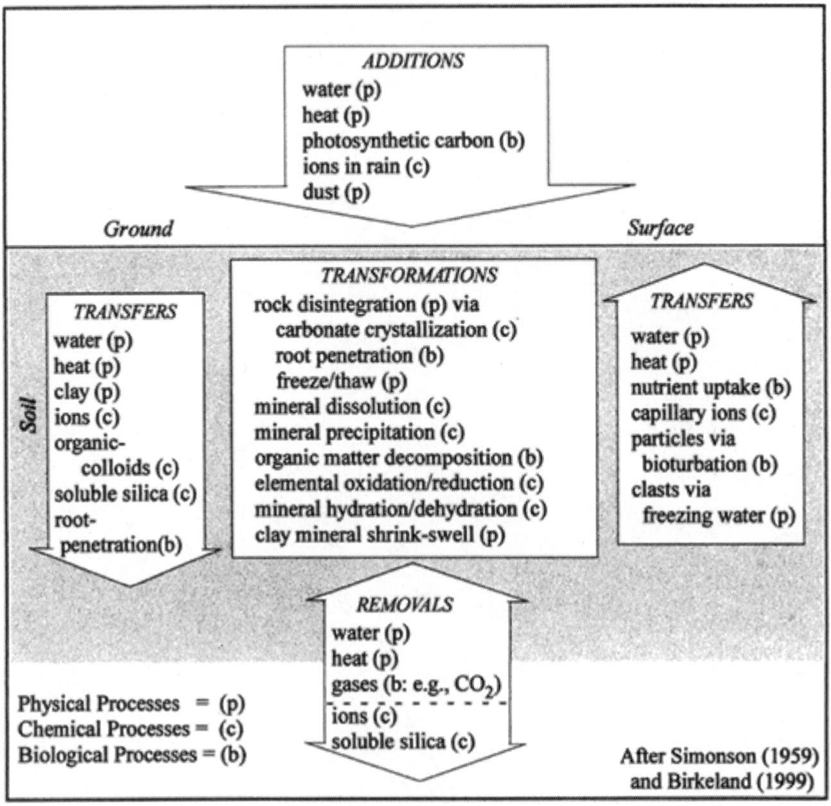

Figure 4-5. Physical, chemical, and biological processes that operate interactively to produce soil horizons. These processes are grouped into four categories: additions, transfers, transformations, and removals (Simonson 1959). All of these processes are active in Jornada soils, though some, like freeze-thaw, are less active than processes such as heat transfer or organic matter decomposition.

organic matter. Even in A horizons most concentrations are only 0.1–0.3% organic carbon. On the high end, soils above elevations of about 1,524 m (5,000 ft) can have up to 1.3% organic carbon in A horizons (Gile et al. 1981). These soils have enough organic matter to classify as Mollisols, rather than Aridisols or Entisols. Downslope, these soils change from Mollisols to Aridisols. Where soils form in alluvium derived from bedrock with granitic texture, such as the sorts around Summerford Mountain, A horizons retain a dark color (giving a false impression of high organic content) as a consequence of thin organic coatings on sand and fine gravel (Encina-Rojas 1995).

Other soils containing relatively high levels of organic matter are located in topographically low areas that receive water from hill slope runoff. In these soils, organic carbon can be as high as 2.21% (Herbel et al. 1994). This is the result of greater plant densities supported by increased water supply and an increased clay content that curtails organic matter decomposition (Deng and Dixon 2002). In addition, soils developed in limestone alluvium have a tendency to contain higher amounts of organic matter than neighboring soils formed in igneous alluvium (Grossman et al. 1995).

On the low end, some sandy soils barren of vegetation in the Jornada Basin contain as little as 0.1% organic carbon. In areas where black grama grass has converted to mesquite coppice dunes over the past 150 years, losses of soil organic matter have occurred, although carbon storage of the ecosystem as a whole has changed little (Schlesinger and Pilmanis 1998). In bare interdune areas, the loss of organic matter is the result of not only decreased organic inputs but also the loss of organic matter by wind and water erosion (Schlesinger et al. 1990).

Pedogenic Carbonate Accumulation

Unlike organic carbon, which reaches an equilibrium concentration within an ecosystem within decades to centuries, pedogenic carbonate accumulates over millennia to progressively higher concentrations. Consequently, carbonate is one of the prominent (if not *the* prominent) pedogenic feature in the Jornada Basin. For example, soils of the southeastern region of the JER, which are of early Pleistocene age (Mack et al. 1996), have pedogenic carbonate amounts that reach 223 kg C/m^2, a value that equals some of the highest concentrations of organic carbon in peat-bog Histosols (NRCS Web site 2005). As in the Mojave Desert (Schlesinger 1985), roots (Gallegos 1999) and microorganisms (Monger et al. 1991a) play a role in precipitating pedogenic carbonate, which in the Jornada Basin has been studied from both soil-geomorphic and ecological aspects.

Soil-Geomorphic Aspects of Pedogenic Carbonate

Several names have been used to describe soil carbonate in arid and semiarid soils: caliche, calcrete, croute calcaire, tosca, caprock, crust, calcic horizons, and petrocalcic horizons (Gile 1961; Goudie 1973; Dregne 1976; Soil Survey Staff 1999). Forms of carbonate accumulation have been described as filamentary, concretionary, cylindroidal, nodular, plugged horizons, and laminar horizons (Gile 1961; Gile et al. 1966). These carbonate accumulations range from nonindurated, which slake when placed in water, to very strongly indurated, which do not slake in water and cannot be scored with a knife. On the low end, soil horizons with carbonate filaments can contain as little as 1% $CaCO_3$ and, depending on texture, have a bulk density of about 1.68 g/cm^3 and an infiltration rate of 12.4 cm/h (Gile 1961). On the high end, laminar horizons contain as much as 93% $CaCO_3$, have a bulk density of 2.22 g/cm^3, and have an infiltration rate of 0.1 cm/h.

Carbonate is an important indicator of soil age because progressively older geomorphic surfaces contain progressively greater amounts of carbonate. Over time, four diagnostic morphogenetic stages of carbonate are formed (Gile et al. 1966). For example, soils formed in sediment deposited by the Rio Grande progress from having no display of carbonate to having stage I filaments, then stage

II nodules, a stage III plugged horizon, and eventually a stage IV laminar horizon atop the plugged horizon. Researchers applying this classification to desert soils elsewhere have added two more stages of carbonate accumulation: stage V and stage VI (Machette 1985). Stage V contains diagnostically thick laminae (> 1 cm) and pisolites; stage VI contains multiple generations of laminae, breccia, and pisolites. All six stages are found in the Jornada region of south-central New Mexico (Gile et al. 1996). The time required to reach a certain morphogenic stage depends on the texture of the soil. Gravelly soils pass through the stages more quickly than fine-textured soils because gravelly soils have lower surface area and less poor space.

Until about 1965, horizons with carbonate accumulation were designated as C horizons with a "ca" suffix or "cam" suffix if indurated, following the procedure of the *Soil Survey Manual* (Soil Survey Staff 1951). Subsequent work in the Desert Project showed that carbonate accumulation was a pedogenic process. Therefore, horizons with significant amounts of carbonate should be designated B horizons. The prominence of carbonate horizons in arid and semiarid regions led Gile et al. (1965) to propose the K horizon as a master horizon (i.e., horizons such as the O, E, A, B, C, and R horizons). The K horizon is based on the presence of K-fabric, which is defined as "fine-grained authigenic carbonate that coats or engulfs skeletal pebbles, sand, and silt grains as an essentially continuous medium" (Gile et al. 1965). The K horizon, by definition, contains 90% or more K-fabric. Although not formally accepted by the NRCS, except that the suffix "k" has replaced the "ca" suffix, the K horizon has been used extensively in quaternary geology studies of the Southwestern United States (e.g., Machette 1985; Birkeland 1999).

Ecological Aspects of Pedogenic Carbonate

Mineralogically, pedogenic carbonate in the Jornada Basin is calcium carbonate in the form of calcite, regardless of whether it formed in igneous or limestone parent material (Kraimer 2003). Calcium carbonate is known to have important chemical influences on plant growth by its control on pH, phosphorous, and micronutrient availability (chapter 6). For rangeland ecosystems, carbonate accumulation also has an important physical influence on plant growth by its influence on water storage via its affect on soil texture. Carbonate crystals are generally in the size range of coarse clay to fine silt, approximately 1–10 microns (Monger et al. 1991b), which increases surface area and microporosity of the horizon impregnated with carbonate. For example, soil-water release curves for carbonate nodules revealed that the nodules store about twice the water as adjacent soil not impregnated with carbonate (Monger 1990). Hennessy et al. (1983b) measured water properties of caliche in the Jornada Basin and found that caliche absorbed appreciable quantities of water and retained it for extended periods, which may have contributed to certain areas of black grama grass underlain by caliche surviving the 1950s drought (Herbel et al. 1972). Moreover, creosotebush on sites underlain by caliche showed less water stress when rainfall was low than sites without caliche (Cunningham and Burk 1973).

Desert Pavement and the Formation of Eluvial Horizons

Gravelly soils of Pleistocene age in the Jornada Basin commonly develop desert pavements. Depending on local circumstances, these gravelly pavements can form as the result of erosion removing smaller particles (Cooke et al. 1993), eolian materials accumulating beneath and lifting gravels (McFadden et al. 1987), or a combination of both processes. Many desert pavements in the Jornada Basin region, especially those of stable geomorphic surfaces, are blackened and reddened by desert varnish on coarse fragments other than limestone (Gile and Grossman 1979).

Thin vesicular A horizons, Av (Birkeland 1999), characterized by having spherical voids disseminated through the horizon, occur just beneath many desert pavements. These apparently form by the entrapment of air that is displaced upward by infiltrating water (Springer 1958). The rate at which vesicular A horizons form can be quite rapid based on the observation that well-developed vesicular horizon formed in tire tracks known to be less than two years old (Gile and Grossman 1979).

The vesicular A horizon is one form of an eluvial zone (i.e., a soil zone from which clay and chemical compounds have been removed by percolating water). Eluvial horizons are designated by the letter E (formerly A2) when eluviation dominates other properties of the A or B horizons (Soil Survey Staff 1999). E horizons are not widespread across the Jornada Basin but do occur where leaching and eluviation have taken place and the horizon has not been truncated by erosion (e.g., Gile et al. 1981), such as the stellar series on the lower piedmont slope of the Doña Ana Mountains.

Clay Accumulation and Mineralogy

The amount of clay in soil affects vegetation by its control on water storage, air and water permeability, nutrient storage, shrink-swell properties, soil hardness, and erodibility. As clay content increases, water storage, nutrient storage, shrink-swell, and hardness generally increase, whereas erodibility and air permeability generally decrease. In the Jornada Basin, soils with the highest clay content occur in playas. The Red Lake Playa site, for example, contains soils that have up to 69% clay (largely smectite). Its water content is as high as 32% at 1/3 bar and 20% at 15 bar tension, and it has a cation exchange capacity of up to 30 cmol/kg (table 4-2). For comparison, a coppice dune along the Camino Real has as little as 5% clay, a water content of 5% at 1/3 bar and 3% at 15 bar tension, and a cation exchange capacity of 5.8 cmol/kg.

Mineralogically, the clay fraction (< 2 microns) in soils of the Jornada Basin contains smectite, illite (mica), kaolinite, vermiculite, mixed-layer clays, palygorskite, sepiolite, quartz, feldspars, and calcite (Vanden Heuvel 1966; Gile and Grossman 1979; Monger and Lynn 1996). These occur in various proportions depending on parent material and age. For example, palygorskite and sepiolite only occur in petrocalcic horizons (Monger and Daugherty 1991a), in contrast to smectite, illite, and kaolinite, which are nearly ubiquitous.

Chemical weathering of monzonite alluvium of the Jornada Basin area has produced kaolinite (Monger and Lynn 1996). Another example of chemical weathering can be found in soils of the basin floors that formed in ancestral Rio Grande alluvium. The sand mineralogy in these soils show the highest degree of chemical weathering in the A and B horizons and lowest degree of chemical weathering in C horizons (e.g., Monger 1990). In petrocalcic horizons, pressure solution resulting from carbonate precipitation is an additional process that hastens the dissolution of sand grains and promotes the neoformation of palygorskite (Monger and Daugherty 1991b).

Processes that Obliterate Soil Development

Alluvial Activity

Alluvial activity in the Jornada Basin consists of two processes: erosion and sedimentation. Water erosion is prominent on the piedmont slopes (chapter 7) but is also important in interdune areas on the basin floor where short-distance (on the order of meters to tens of meters) erosion and sedimentation occur. Erosion obliterates soil development by truncating pedogenic horizons, as exemplified along the Rio Grande, where lowering base levels have initiated landscape dissection (Gile et al. 1969). Figure 4-6 illustrates how progressive erosion and truncation removes pedogenic soil horizons, which, in turn, changes the soil classification from Aridisols having argillic and calcic horizons (Calciargids) to Entisols having no subsurface diagnostic horizons (e.g., Torriorthents).

On piedmont slopes, erosion upslope supplies sediment downslope that progressively fills broad swales and interfan valleys until these landforms are no longer the topographic lows. Subsequently, the locus for further deposition is shifted laterally across the piedmont slope into new topographic lows (Peterson 1981). As a result, alluvial fans and coalescent fan piedmonts (bajadas) are built. In general, erosion in the Jornada Basin produces sediments that are not removed by water from the basin. The exception is the Rio Grande Valley, where the drainage system is integrated with the river.

Small lakes were sites of additional alluvial (lacustrine) deposition in the Jornada Basin. Evidence that these deposits are lacustrine in origin includes the terraces, thick clay deposits, and large gypsum crystals (selenite) found there. Lakes of the Jornada Basin have probably been dry since the end of early Holocene (8,000 years ago) when similar lakes in the northern Chihuahuan Desert became extinct (Hawley 1993).

Eolian Activity

Like alluvial activity, eolian activity consists of two categories: erosion and sedimentation. However, unlike alluvial activity in the Jornada Basin, wind erosion can completely remove particles from the basin ecosystem (chapter 9). During

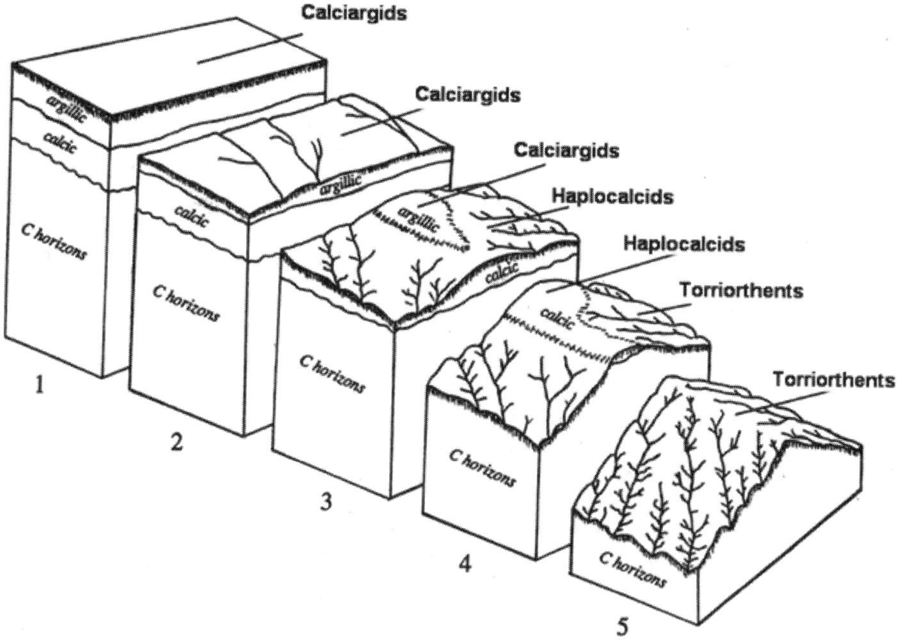

Figure 4-6. Illustration of the effects of erosional dissection on argillic and calcic horizons. Block diagram 1 shows a stable, nondissected landscape with an argillic and calcic horizon overlying C horizons, which classifies as Calciargids at the Great Group level (Soil Survey Staff 1999). With progressive dissection and truncation of these horizons, soils become Torriorthents, as illustrated in block diagram 5 (modified from Gile et al. 1981).

high winds, much of the sand fraction saltates and creeps laterally. If mesquite shrubs are present, a portion of this material is trapped and coppice dunes form (Gile 1966b). Huge amounts of sand have been carried across the basin floor in an east-northeasterly direction by the prevailing wind and blown up the piedmont slope of the San Andres Mountains, over the bedrock outcrops, and into the intermountain valleys. The timing of this process, based on the degree of argillic and calcic horizon development in many of the intermountain sand deposits, predates the arrival of cattle in the late 1800s. However, the precise timing of the eolian deposition and the magnitude of present-day eolian deposition in relation to natural cycles of prehistoric deposition remain to be determined.

Erosion by wind (deflation) is a critically important process in the Jornada Basin. As with landscape dissection, deflation truncates soil horizons and changes soil classification. Large, linear deflational streaks are oriented to the east-northeast in the western portion of the Jornada Basin. In these areas, the soil profile is truncated and petrocalcic horizons are very near the surface (Gile 1999). Some of these features, called streets (Okin and Gillette 2001), occupy areas between coppice dunes, which in some cases are also linear and oriented in the prevailing wind direction.

Faunal Activity

Fauna, such as small mammals, reptiles, and insects, play a major role in obliterating soil horizons by their burrowing activity (chapter 12). For example, faunal burrowing obliterated argillic horizons in soils of the southern Jornada Basin (Gile 1975b). Kangaroo rats can penetrate into carbonate horizons and bring carbonate fragments to the surface (Anderson and Kay 1999), where they are apparent in aerial photographs. In some cases, termites are agents of carbonate movement. In the Jornada Basin, it is common to find termite sheaths that are calcareous in contrast to the noncalcareous soil surface on which they reside. Thin sections of such galleries and of the termites themselves revealed an arrangement of carbonate crystals that led to the hypothesis that termites biomineralize carbonate for use as cement for gallery construction (Monger and Gallegos 2000). However, in other areas of the Jornada Basin, especially where the underlying soils contain argillic horizons, the sheaths are cemented with silicate clays. This evidence, combined with isotopic, mineralogic, and electron microscopy, now indicates that termites transport carbonate from subsoil rather than biomineralizing it for gallery construction (Liu 2002).

Links to Past Climates

Regions that have undergone bioclimate change, such as the Jornada Basin, often contain soil profiles with relict mineralogical and morphological properties generated during climates of the past. These properties can provide evidence about past climates and vegetation, provided other soil-forming factors can be held relatively constant. Although properties like clay mineralogy and elemental composition have been useful for deriving paleoclimatic information (e.g., Birkeland 1999), pedogenic carbonate is particularly valuable because it contains evidence about past climates in at least three forms: the presence of carbonate, carbonate depth, and carbon and oxygen isotopes in carbonate. In addition to carbonates, landscape stability and soil formation provide clues about past climates.

Presence of Carbonate

Presence of pedogenic carbonate is a common feature of arid and semiarid soils. Although pedogenic carbonate can exist in humid climates under certain circumstances of high water tables and calcareous parent materials (Sobecki and Wilding 1983), an analysis of 1,168 soil profiles with carbonate showed that 95% of them exist in climates where the annual precipitation is less than 760 mm (Royer 1999). A typical value of 500 mm (20 inches) has been used as the general boundary between soils with carbonate and soils without carbonate (Birkeland 1999). Therefore, inferences can be made about past climates based on the premise that soils formed in humid climates do not have carbonate. For example, the sequence of paleosols (i.e., multiple buried soil profiles) shown in figure 4-7a–c provide evidence that the climate has not been humid enough to remove carbonate from

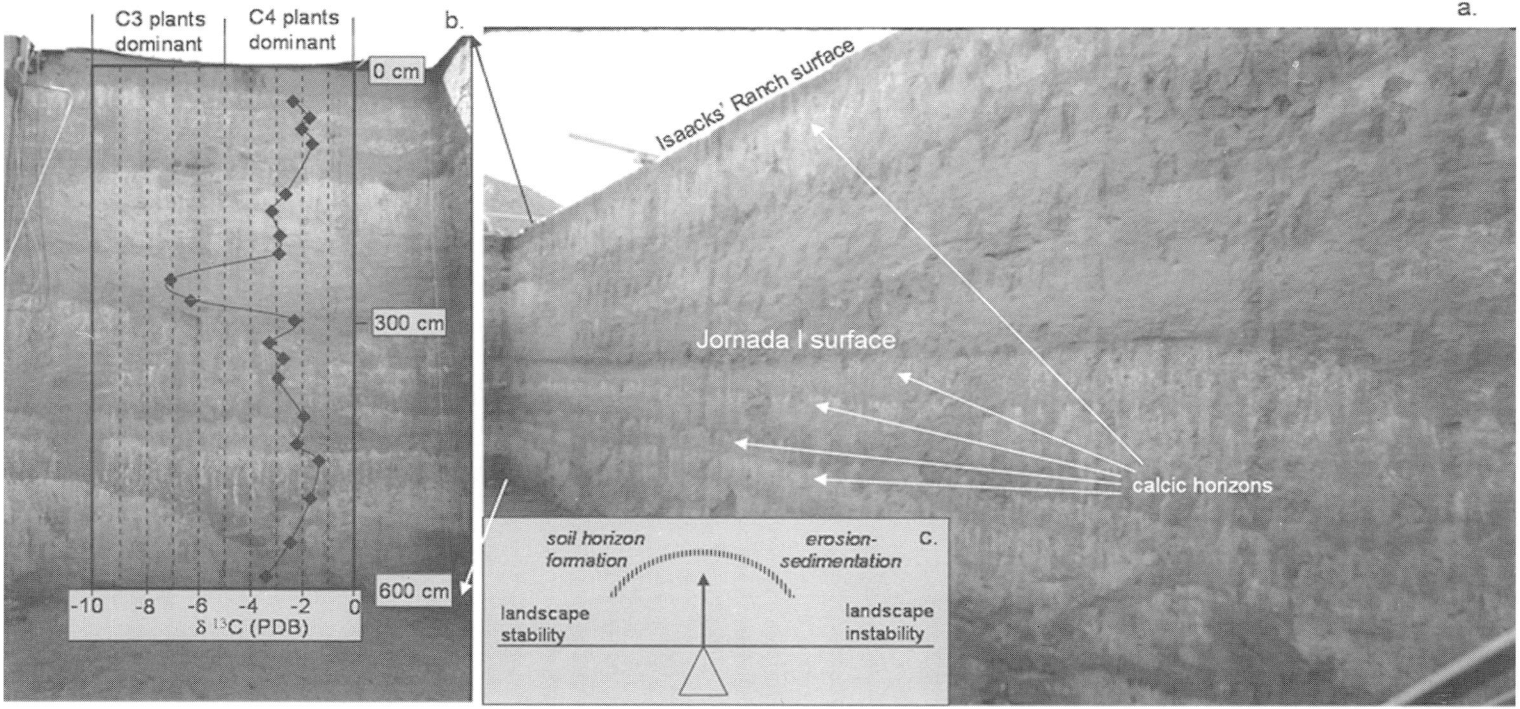

Figure 4-7. (a) North face of a 6-m-deep trench on the piedmont slope northeast of Summerford Mountain along the LTER I transect (Wierenga et al. 1991). The stratigraphic record extends from the current land surface and the Isaacks' Ranch geomorphic surface to below the Jornada I surface, which has an estimated age as old as 500,000 years (see table 2-1). White arrows locate some of the calcic horizons, which are overlain by red argillic horizons unless truncated. (b) West face of the trench showing depths and a graph of the $\delta^{13}C$ values of carbonate. Values of $\delta^{13}C$ greater than −5% are thought to represent vegetation dominated by C_4 plants (Cerling 1984). (c) Illustration of a landscape stability model in which increased landscape stability gives rise to increased soil horizon formation, in contrast to progressive instability, which is characterized by progressive erosion and sedimentation.

the modern soil or any of the buried soils when they were at the land surface. In other words, the climate in the Jornada Basin has probably been arid, semiarid, or at most subhumid for at least the past 500,000 years, which is the estimated age of the stratigraphic section based on correlation to the Jornada I surface at other sites in the region that are dated by K/Ar and paleomagnetic methods (Hawley 1975a; Seager et al. 1984; Gile 1990, 1999; Mack et al. 1993).

Carbonate Depth

Within the confines of dry climates, it has been observed that the depth of pedogenic carbonate is proportional to annual rainfall, that is, greater precipitation corresponds to greater depths to the top of the carbonate horizons (Jenny and Leonard 1934; Arkely 1963; Gile 1975c). However, erosion, runoff, and run-in can confound this general relationship, and it has recently been shown that a statistically significant correlation does not exist between carbonate depth and rainfall, especially for shallow carbonate in arid and semiarid climates (Royer 1999). Nevertheless, within a small region and especially within a single soil profile, carbonate depth is likely to be a function of long-term depth of soil wetting. For example, the vertical, karst-like pipes that cross-cut petrocalcic horizon in middle Pleistocene soils of the Jornada Basin are probably the result of deep water penetration during wetter climates (pluvials) in the Pleistocene (Gile et al. 1981). Similarly, horizons of carbonate filaments that occupy shallow depths in the same middle Pleistocene soils provide evidence of an upward shift in the depth of wetting during subsequent drier climates (Gile et al. 1981; Monger 2003).

In addition to field observations of carbonate depths, models have been useful for understanding the relationship of carbonate depth and climate. McFadden and Tinsley (1985) developed a model that defined soil as a vertical sequence of compartments, each with a specified texture, bulk density, water-holding capacity, mineralogy, pCO_2 content, ionic strength, and temperature. Similarly, the model by Marion et al. (1985) and a later model by Marion and Schlesinger (1994) used stochastic precipitation, evapotranspiration, chemical thermodynamics, soil parameterization, and soil water movement to simulate carbonate and gypsum deposition. For soils in the Jornada Basin, the depth of carbonate was modeled as being about 15 cm deeper during wetter Pleistocene climates than during current arid conditions (Marion et al. 1985).

Carbon and Oxygen Isotopes in Carbonate

Carbon and oxygen isotopes in pedogenic carbonate contain climatic information by recording the relative abundance of C_3 to C_4 plants and the isotopic signatures of meteoritic water (Cerling 1984). In the Jornada Basin and Desert Project, early isotope studies were conducted by Gardner (1984), who showed that soils in different landscape positions contained different isotopic signatures and that repeated dissolution and reprecipitation tended to homogenize isotopic signatures in older soils. Later studies found isotopic shifts across stratigraphic boundaries of middle Holocene age in both alluvial (Cole and Monger 1994; Monger et al. 1998) and eolian paleosols (Buck and Monger 1999). In both cases, shifts in $^{13}C/^{12}C$ ratios indicated a change from C_4 (grass) to C_3 (shrub), which slightly predated sedimentation, and an upward shift in the depth of carbonate deposition. Though relative amounts vary, carbon isotopes suggest that C_4 plants have been dominant during the late quaternary in the Jornada Basin (figure 4-7b). Much younger isotopic shifts in carbonate and organic matter have also recorded the replacement of grass by mesquite within historical sediment (Connin et al. 1997a, b).

Landscape Stability

Landscape stability (i.e., periods when land surfaces are not being rapidly dissected by erosion or rapidly buried by sedimentation) is prerequisite for pedogenic carbonate to develop to a substantial degree. For example, stage I carbonate filaments can develop within 100 years, whereas stage II nodules take about 8,000 years to form (Gile et al. 1981). Stage III and IV petrocalcic horizons might take 75,000 years to form depending on texture. Carbonate horizons are therefore important indicators of sustained soil formation on relatively stable surfaces. Argillic horizons (which typically form above carbonate horizons) also provide important clues that soil formation has proceeded for lengthy time periods. For example, argillic horizons are common in soils of the Isaack's Ranch geomorphic surface just south of the JER, which are about 8,000–15,000 years old (Gile et al. 1981), but are less common in younger soils.

The presence of carbonate and argillic horizons in stacked sequences of paleosols has been interpreted as evidence for alternating periods of landscape stability and instability in the Jornada Basin and across the Desert Project area (Gile and Hawley 1966). For example, paleosols with calcic horizons shown in figure 4-7a could be interpreted using the stability-instability model to be a record of intervals of landscape instability (when erosion upslope deposited sediments downslope) which alternated with periods of landscape stability (when sedimentation ceased and calcic and argillic horizons formed).

However, there are caveats that need to be applied to this model. First, instead of landscapes oscillating between the two stability-instability end members (as illustrated in figure 4-7c), the pointer may spend a lot of time at various places between the extremes. That is, there is probably much time during which erosion rates and sedimentation rates occur slowly enough to allow some carbonate and clay accumulation. A second complication involves the shifting of a main depositional channel, which would have caused a cessation of sediment deposition. Therefore, the hiatus would not be the result of a change from landscape instability to stability but rather to moving loci of deposition.

Conclusions

The 1918 soil map and report of the JER (Veatch 1918) captures the essence of the physical landscape of the Jornada Basin and provides a valuable account of the native vegetation, soil-vegetation associations, and the effects of overgrazing in 1918. The second soil map (SCS 1963) provides accurate delineations but lacks quantitative soil classification because it predated soil taxonomy by about a decade. The most detailed soil map of the Jornada Basin was produced by the Desert Soil-Geomorphology Project (Gile et al. 1981), which covers a 400-mi^2 area around Las Cruces, including the southern portions of the JER and the CDRRC. This map and associated investigations provide large databases of soil properties, detailed geomorphic mapping, and thorough studies of Cenozoic geology, the origin of soil carbonate, the importance of atmospheric additions of $CaCO_3$ dust and Ca^{2+}, soil chronology, the accumulation of silicate clay by illuviation, and factors that cause soil boundaries.

Notable consequences of soil development in the Jornada Basin include the accumulation of organic matter, pedogenic carbonate, and silicate clay. Organic carbon, which has important influences on nutrient cycling and aggregate stability, ranges from 0.1% weight in sand dunes to 2.2% weight in soils in depressions that receive run-in water. Pedogenic carbonate, which has influences on water storage and is an important indicator of landscape evolution, ranges from less than 1% weight in soils of historical age to 90% weight in soils of Pliocene age (Gile et al. 1981, 1996). Silicate clay, which affects both water and nutrient storage as well as permeability and erodibility, ranges from less than 5% weight in coppice dunes to as much as 69% in playas.

Erosion by water and wind are the main processes that truncate soil horizons. Sediments from water erosion are moved laterally downslope but remain within the Jornada Basin. Sediments from wind erosion, though also moved laterally across the landscape, can be removed entirely from the Jornada Basin, especially particles of very fine sand and smaller. Erosion and sedimentation working in concert with Cenozoic tectonic extension has produced the modern terrain of the Jornada Basin.

Jornada soils of Pleistocene age have existed through multiple cycles of bioclimatic change. The presence of carbonate in buried paleosols indicates that the Jornada Basin has had a climate drier than 760 mm of annual precipitation for at least the past 500,000 years. However, dissolution pipes through carbonate horizons, combined with regional paleolake, paleoecology, and isotope studies, indicate that there have been multiple swings between relatively wet and dry climates in the Jornada Basin.

5

Patterns and Controls of Soil Water in the Jornada Basin

Keirith A. Snyder
Katherine A. Mitchell
Jeffrey E. Herrick

This chapter focuses on controls and patterns of soil moisture in the Jornada Basin. First we describe general properties that commonly contribute to soil water heterogeneity; second, we offer a brief overview of soil water research in the Jornada Basin; and last, we describe specific patterns of soil water content and availability observed in the Jornada Basin. Our goal is to describe general patterns of soil water that are likely to occur across the Chihuahuan Desert region.

In arid and semiarid regions, water is typically thought to be the most limiting resource to biological activity (Noy-Meir 1973), though colimitation by water and nitrogen may be a more general rule (Hooper and Johnson 1999; see also chapter 6). The availability of water affects plant productivity, microbial activity, activity of biological soil crusts, nutrient cycling, and organic matter decomposition. It also directly and indirectly affects soil erosion, chemical weathering, and carbonate formation. There are several hypotheses addressing how water availability affects plant productivity in desert environments. Beatley (1974) proposed that various functional types (e.g., shrub, perennial grass, annual forb) have different seasonal rainfall thresholds to trigger phenological responses. The annual productivity of functional types is therefore determined by the timing and amount of rainfall. Westoby (Noy-Meir 1973) proposed the pulse-reserve paradigm to explain population dynamics of desert plants. In this view, a rain event triggers a pulse of production. Some of that production is used to generate new tissue, but part of the production is diverted into reserves. The amount of reserves in part determines the next production pulse, as well as the minimum size of the next trigger event. Rainfall is highly variable both spatially and temporally in arid regions; therefore, understanding patterns of rainfall and interactions between rainfall patterns, soil characteristics, temperature, and topography are critical to predicting ecosystem responses.

The relationship between average annual precipitation and plant productivity across arid regions has substantial predictive ability (Le Houérou 1984). However, for a given site, the relationship between annual precipitation and yearly plant productivity has limited explanatory power (Lauenroth and Sala 1992). At local scales when time is substituted for space, precipitation appears to be a poor measure of water availability and productivity because of the complex effects of differences in rainfall frequency, timing and magnitude, landscape position, soil texture, soil structure, macropores, microrelief, and feedbacks between the vegetation and hydrologic processes such as stem flow, infiltration, percolation, and runoff. Soil moisture is a more direct indicator of available water for biological activity, but accurate data are rarely available at relevant spatial scales due to measurement and scaling limitations (Williams and Bonnell 1988). Recent studies on soil water availability and plant water use have emphasized that the relative availability of different sources of water (i.e., shallow soil water, deep soil water, groundwater) may play an important role in structuring communities and seasonal productivity (Ehleringer and Dawson 1992; Snyder and Williams 2000; Schwinning and Ehleringer 2001). There is also increased recognition that frequency and magnitude of dry periods (interpulse periods) relative to the frequency and magnitude of rainfall pulses affects plant and ecosystem responses in arid systems (Huxman et al. 2004; Loik et al. 2004; Reynolds et al. 2004). Soil is the regulator and interface between plants and precipitation. Therefore, understanding patterns of soil moisture is critical in establishing predictive models of ecosystem function.

The Southwestern desert regions of North America are characterized by a bimodal pattern of rainfall where precipitation is received in both the winter months and during the summer growing season. The ratio of summer to winter rainfall varies across the desert regions. Precipitation in these arid and semiarid regions is highly unpredictable, especially during the summer growing season (chapter 3). Consequently, precipitation and resultant pulses of soil moisture are very heterogeneously distributed in time and space.

As detailed in chapter 10, the Jornada Basin is an area where shrubs have invaded and dramatically changed the landscape and ecosystem processes of areas formerly dominated by grasslands. Walter (1979) proposed the two-layer model to explain the stable coexistence of woody plants and grasses in water-limited environments. According to this model, deep, rooted woody plants access water from deep in the soil profile, whereas shallow-rooted grasses rely on water in shallow soil layers. Reynolds et al. (1999b) found that co-occurring shrub species may partition water temporally with different phenological strategies. These theories highlight the importance of understanding patterns of soil water availability over multiple years, soil types, and landscape positions to explain resultant vegetation patterns.

Causes of Soil Water Heterogeneity

Climate

In the Chihuahuan Desert, region rainfall is bimodal with slightly more than half of the annual rainfall occurring in the three months from July through September and the majority of the rest falling during winter months (especially January and February). Winter precipitation is generally characterized by slow-moving frontal systems. These systems generally produce long-duration and low-intensity rainfall (chapter 3). Winter rainfall often percolates to greater depths because lower rainfall intensities allow more of it to enter the soil and because lower evapotranspirational demands increase the probability that upper soil layers will already be near field capacity when precipitation events begin. Summer rainfall is characterized by localized convective storms, which are generally of short duration and high intensity. These rainfall events generally do not percolate to deep soil layers because of high evaporation and transpiration rates and an increased likelihood of surface runoff. This pattern can sometimes be reversed in the lowest landscape positions, which benefit from run-in during more intense summer storms, generating deep water percolation during the summer months. The seasonal patterns of rainfall create differences in the vertical distribution of soil water, and the stochastic nature of rainfall, particularly during summer months, contributes to the spatial and temporal heterogeneity of soil moisture.

Landscape Position and Soil Properties

Average or seasonal precipitation is a poor predictor of soil moisture in part because the amount of water available for infiltration varies widely as a function of landscape position. Rain gauge records show that rainfall increases with elevation and that there is also a high level of variability at the same elevation in the Jornada Basin (chapter 3). The volume of water available to infiltrate at any particular point on the landscape is further modified by its relative position: lower landscape units generally have a higher probability of receiving run-in, though landscape units at the top of alluvial fans surrounding the basin can benefit from mountain runoff (chapter 7). Runoff is generally higher from steeper slopes.

Landscape position is also a good predictor of soil texture, which affects infiltration capacity, water holding capacity, and bare-soil evaporation rates. In general, infiltration capacity is highest on the sandy basin soils (refer to tables 4-1 and 4-2 figure 4-3, chapter 4), intermediate on the loamy alluvial fans and fan piedmont areas, and lowest on the fine-textured soils of the alluvial flats and lake plains. The low infiltration capacity of the fine-textured soils of the ephemeral playa lakes is balanced by the lack of runoff. Deepwater infiltration into some of these low-lying areas is facilitated by deep cracks that form in soils with 2:1 expanding clays when the soil is dry.

Deep water infiltration also varies as a function of soil profile characteristics. Whereas clay-rich soil horizons are relatively rare on the Jornada, widespread calcic horizons can have a significant effect on deep percolation of water due to their generally low hydraulic conductivity.

Volumetric water content is the proportion of the soil volume that is occupied by water. The maximum volumetric soil water content after gravity drainage is referred to as field capacity. However, water content does not reflect the actual water available to plants because available water is a function of soil water potential. Soil water potential is the summation of soil matric potential, gravitational potential, and osmotic potential, and measures how tightly water is held in the soil (Kramer and Boyer 1995). Soil matric potential is a function of soil texture and structure and reflects how tightly water is bound to adjacent soil particles due to capillary and surface binding properties. Gravitational potential is the force of gravity operating on soil water, whereas osmotic potential is a function of the chemical composition of water and increases with increasing solute concentrations. In nonsaline soils, plant available water is largely determined by soil matric potential and unsaturated soil hydraulic conductivity. Plant available water also varies with plant species characteristics such as root morphology, root length density, and microbial associations.

Plant available water-holding capacity refers to the percentage of the soil volume at soil water potentials less than field capacity that is extractable by plants. Plant available water-holding capacity is generally highest in intermediate-textured soils (loamy) and lowest in sandy soils. It is reduced by the presence of rocks but may actually be increased by the presence of calcium nodules in sandy soils (Hennessy et al. 1983a).

Evaporation rates are affected by landscape position primarily through aspect differences, with higher rates occurring on south-facing slopes. Aspect effects are readily apparent in areas north of the Jornada Basin. Evaporation is generally slower through soils with a coarse-textured surface horizon and those in which the unsaturated hydraulic conductivity is lower for the top few centimeters of soil than for the layers below. This phenomenon may be quite significant in the Jornada Basin, where deposition of eolian sand on top of relatively fine-textured basin soils has created extensive areas of these types of surfaces (tables 4-1 and 4-2, and figure 4-3).

The relationship between soil texture and soil water properties varies with soil structure. Soil structure is formed through modification of the inorganic soil by both abiotic and biotic processes. In arid and semiarid ecosystems, soil surface structure is generated by litter decomposition under plants, microbiotic crusts in plant interspaces, and macroinvertebrates in both plant and interspace microsites (Herrick and Wander 1998). Repeated cycles of wetting and drying help form aggregates, and freeze-thaw cycles can also be important. Macroinvertebrates, especially ants and termites, are extremely important in the Chihuahuan Desert for increasing infiltration through the formation of macropores (Elkins et al. 1986; Herrick 1999). The relationship between microbiotic crusts and soil water is complex and poorly understood (Warren 2001). Development of soil structure below the soil surface is similar to more humid environments, where root decomposition and associated soil biotic activity dominate; however, ants and termites replace earthworms as the dominant macropore-forming organisms in these environments (Herrick 1999). Soil structure-forming processes are self-reinforcing as improved soil structure facilitates greater water infiltration and retention, leading to higher litter production in subsequent years and, therefore, greater substrate availability for soil biological activity. Due to the importance of these plant litter inputs, soil structure tends to be extremely patchy at nearly every spatial scale in the Chihuahuan Desert (Herrick and Whitford 1999).

Vegetation

Vegetation has a number of effects on soil moisture heterogeneity, in addition to modifying soil structure. Increased plant basal cover can increase water percolation depth by slowing runoff, and canopy cover reduces raindrop erosivity and therefore limits soil surface degradation (chapter 7). Shading by plant canopies and increased litter cover below plants reduces soil surface temperatures and resultant evaporation, which in turn can increase soil water under plant canopies relative to interspace areas. There also tends to be a greater occurrence of macropores under plant canopies that can increase infiltration and soil water. However, interception of rainfall by plant canopies and extraction of soil water by plant roots can reduce soil water beneath plant canopies (Breshears et al. 1997, 1998). Shrub canopies and stems tend to channel water to the root crown, where depending on litter cover, soil characteristics, and slope, it will either follow root channels or be an early initiator of surface runoff (Martinez-Mesa and Whitford 1996; see also chapter 7).

Hydraulic redistribution of soil water by plant roots is another important control that plants exert over soil water. The classic example is hydraulic lift, where deeply rooted plants redistribute water from wet, deep soil layers to drier, shallow soil layers (Richards and Caldwell 1987; Caldwell et al. 1998). This is a passive response by plants to changes in soil water potential. It characteristically happens at night when stomata are shut and the gradient between wet and dry soil layers becomes the driving gradient for water movement through the soil-plant-atmosphere continuum. However, the process of hydraulic redistribution may also transfer water downward through the soil in response to differences in soil water potential gradients. Specifically, it has been found that after rainfall events, roots of some species transfer water from wet shallow soil layers to drier deeper soil layers (Burgess et al. 1998, 2000; Schulze 1998; Smith et al. 1999; Ryel et al. 2002). Hydrogen and oxygen stable isotope ratios of plant xylem sap and environmental water samples have been used to determine where in the soil profile plants access water, and studies have shown that not all sources of water are used equally across functional types (Sala et al. 1989; Ehleringer and Dawson 1992; Donovan and Ehleringer 1994; Weltzin and McPherson 1997; Gebauer and Ehleringer 2000; Snyder and Williams 2000). Therefore, shifts in community composition may contribute to variation in soil moisture. These complex factors interact to accentuate the heterogeneity of soil water.

Major Historical and Ongoing Efforts to Measure Soil Water

Beginning in 1957 and continuing to 1976, gypsum blocks were installed to measure soil matric potential (or soil water availability) at 16 sites on the Jornada Experimental Range (JER) (Herbel and Gile 1973; Herbel and Gibbens 1985, 1987, 1989) at depths of 10, 25, 41, 61, 91, and 122 cm (see figure 5-1 for site locations of all studies). The deepest block was placed at the top of any petrocalcic horizon occurring at a depth of less than 122 cm. An additional set of blocks was installed inside 3-m diameter cylinders at three sites, and two more sets were installed at a fourth site for a total of 21 moisture profiles. Data were collected one to three times during rainy periods when soil matric potential exceeded −1.5 MPa and once a month during drier periods. A threshold value of −1.5 MPa was used because this was historically classified as the permanent wilting point for many herbaceous species (Kramer and Boyer 1995). These data were used to calculate the average number of days over the 19-year period when soil matric potential exceeded −1.5 MPa. Because the data are not continuous, a high degree of interpolation between sampling dates was required. Soil matric potentials below −1.5 MPa were not measured. However, many desert soils have soil matric potentials below −1.5, and most desert species are capable of extracting water far below this theoretical wilting point. This study was the first effort at the JER to quantify spatial and temporal variability in soil moisture, and it does allow for coarse comparisons among sites.

The second major effort to measure soil water was part of early Jornada Basin LTER studies. Soil moisture and rainfall were measured every 30 m along a 2.7-km transect in the Jornada Basin (figure 5-1). The transect extended northeast from the base of Summerford Mountain downslope into an ephemeral playa lake. Soil water content was measured every 2 weeks with a neutron probe at depths of 30, 60, 90, 110, and 130 cm from July 1982 to February 1987 and monthly thereafter (Nash et al. 1991, 1992).

The third comprehensive and ongoing effort to measure soil water was established as part of the expanded Jornada LTER program (figure 5-1). Since July 1989, soil water content measurements have been made once a month at 10 depths (where possible) at each of 10 access tubes at each of the 15 LTER NPP (net primary production; chapter 11) sites using a neutron probe (Campbell Model 503DR Hydroprobe). Measurements are made at 30, 60, 90, 120, 150, 180, 210, 240, 270, and 300 cm or to the greatest depth possible to install the access tubing before hitting an impenetrable petrocalcic horizon. The majority of LTER soil water data were collected using the same neutron probe throughout the multiyear period. Another probe was used when the primary probe was being repaired. Both probes were calibrated separately at the same location. A single calibration curve for each probe was used to convert neutron probe counts to volumetric moisture content (cm^3 water/cm^3 soil) for all LTER transect points and LTER NPP locations and depths (Nash et al. 1992). The use of a single calibration curve is appropriate for within-soil comparisons and for comparing temporal patterns of soil moisture variability. Future analysis will incorporate soil-specific calibrations that take into account texture and coarse fragment content. Additional analyses

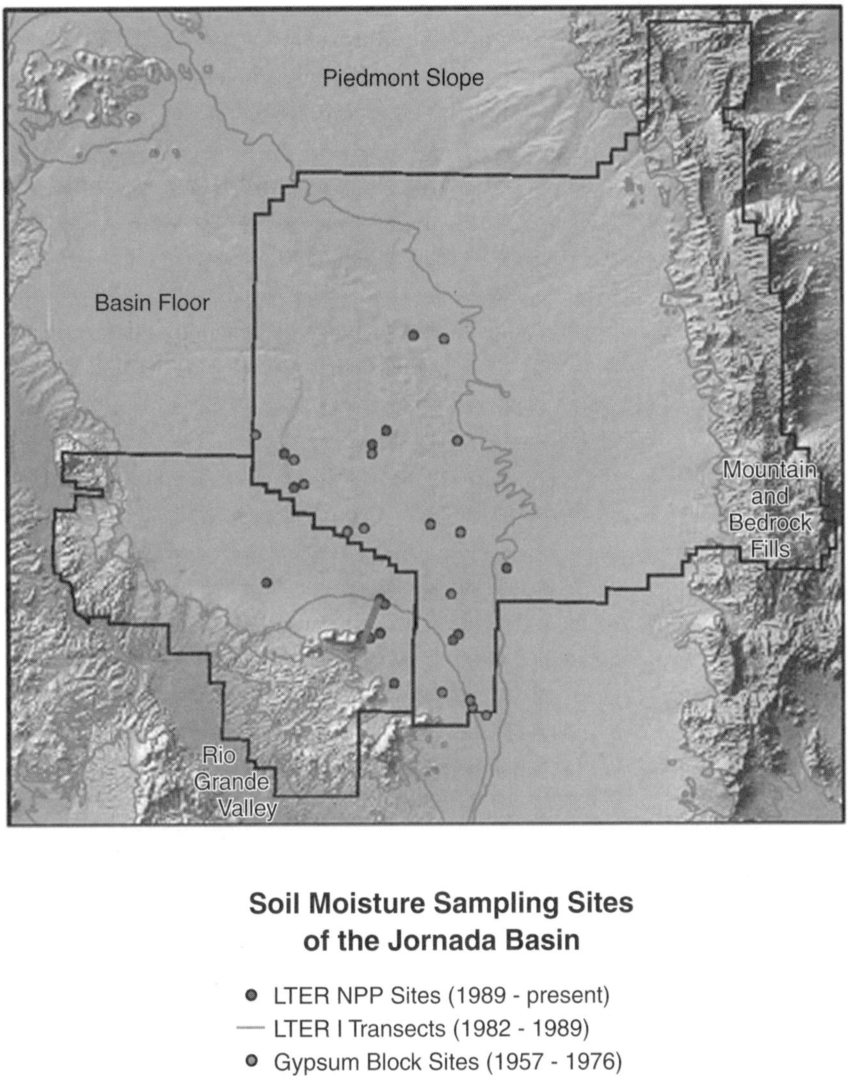

Figure 5-1. Location map of the LTER II net primary production (NPP) sites, LTER I transect, and gypsum block soil moisture study sites.

will also take into account soil- and soil horizon–specific differences in the relationship between volumetric soil moisture content and soil matric potential, which ultimately determines how much of the water is available to plants. Preliminary analyses of the LTER transect data were reported in Nash et al. (1991, 1992).

Patterns of Soil Moisture on the Jornada

The LTER NPP soil water data set was selected for preliminary analysis in this chapter because it represents a much broader range of variability than the LTER transect data. The LTER NPP site selection was based on a replicated sampling design with three sites representing each of the five major plant communities in the Jornada Basin (see table 5-1 for abbreviations and site descriptions). Within each of the 15 sites, data from all 10 tubes were averaged to produce monthly soil water content at each sampled depth for all the years 1990–2001 (Snyder and Mitchell unpublished data). In many cases, neutron probe tubes did not extend to the full depth because a petrocalcic horizon was encountered; therefore, all depths are not represented equally across sites and communities (table 5-2). Soil textures

Table 5-1. The five LTER net primary production (NPP) vegetation communities, the three individual sites within each community, and the soil mapping unit name and texture class for each individual site. Soil information from the Doña Ana County Soil Survey (Bulloch and Neher 1980)

Community/Site	Soil Name	Soil Texture	
Creosotebush			
Caliche (CALI)	Nickel-Upton association	Gravelly sandy loam	(gsl)
Gravel (GRAV)	Nickel-Upton association	Gravelly sandy loam	(gsl)
Sand (SAND)	Nickel-Upton association	Gravelly sandy loam	(gsl)
Grasslands			
Basin (BASN)	Onite-Pajarito association	Loamy sand	(ls)
IBP (IBPE)	Berino-Doña Ana association	Fine sandy loam	(fsl)
Summerford (SUMM)	Onite-Pajarito association	Loamy sand	(ls)
Mesquite			
North (NORT)	Onite-Pintura complex	Loamy fine sand	(lfs)
Rabbit (RABB)	Onite-Pintura complex	Loamy fine sand	(lfs)
West Well (WELL)	Onite-Pajarito association	Loamy sand	(ls)
Playas			
College (COLL)	Stellar association	Clay loam	(cl)
Small (SMAL)	Wink-Harrisburg association	Fine sandy loam	(fsl)
Tobosa (TOBO)	Doña Ana-Reagan association	Sandy clay loam	(scl)
Tarbush			
East (EAST)	Doña Ana-Reagan association	Sandy clay loam	(scl)
Taylor Well (TAYL)	Doña Ana-Reagan association	Sandy clay loam	(scl)
West (WEST)	Stellar association	Clay loam	(cl)

Table 5-2. Number of neutron probe measurements at each depth for each of the 15 Jornada Basin LTER NPP sampling sites (see table 5-1)

Depth (cm)	Creosote			Grassland			Mesquite			Playa			Tarbush		
	CALI	GRAV	SAND	BASN	IBPE	SUMM	NORT	RABB	WELL	COLL	SMAL	TOBO	EAST	TAYL	WEST
30	10	10	10	10	10	10	10	10	10	10	10	10	10	10	10
60	10	10	10	10	10	10	10	10	10	10	10	10	10	10	10
90	10	7	10	10	10	10	10	10	10	10	10	10	10	10	9
120	10	4	10	10	10	10	10	10	8	9	10	10	10	9	7
150	8	1	9	10	6	10	10	10	6	8	10	10	10	7	5
180	5	0	8	10	4	10	10	9	5	8	10	10	10	6	4
210	4	0	7	10	3	10	10	8	4	8	10	10	8	5	2
240	2	0	6	10	2	10	9	7	2	7	10	10	5	5	2
270	2	0	5	10	2	10	9	6	2	4	8	9	1	5	1
300	1	0	3	7	0	8	7	1	0	4	8	5	0	4	0

at the NPP sites range from clay loams to loamy sands (table 5-1). The effect of this wide variability in soil texture on plant available water limits the among-site comparisons that can be made using the data presented here.

General Patterns of Soil Water Content

To analyze general variations in soil moisture between different community types and to understand the amount of variation within a community, noncalibrated volumetric water content (hereafter referred to as water content) was averaged by depth across all sampling dates for each site, and then all sites within a community (n = 3) were averaged to calculate mean soil water content by depth (figure 5-2).

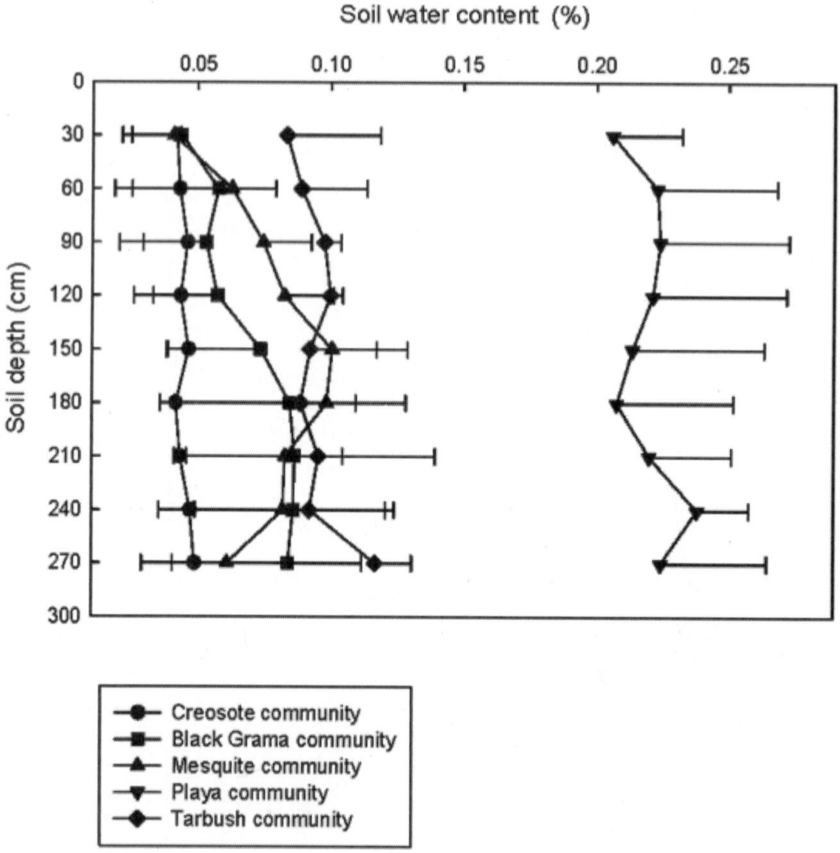

Figure 5-2. Mean soil volumetric water content (cm³ H_2O/cm³ soil, expressed by a %) by depth for each of the five vegetation communities. Data pooled across the three sites within each community and across the entire time period from 1990 to 2001. Error bars are standard deviation from the mean and represent intracommunity variability (n = 3, the amount of variation among the three sites within each community type).

Therefore, the calculated standard deviations indicate the amount of variability between the different sites within a given community. Mean soil water content was highest in the playa community (approximately 23%), intermediate in tarbush (*Flourensia cernua*) (10%), slightly less in mesquite (*Prosopis glandulosa*) and grassland communities (6%), and lowest for creosote community (4%) (figure 5-2). The lower water content in the creosote community is likely due to a combination of higher runoff rates associated with higher slope and a relatively high proportion of the soil volume being occupied by coarse fragments (rocks) at two creosote sites (C-CALI and C-GRAV, acronyms follow table 5-1). The IBP and BASN black grama (*Bouteloua eriopoda*) sites may have lower runoff rates due to better soil structure associated with more continuous plant cover (Neave and Abrahams 2002; Schlesinger et al. 2000), generally lower slope, and coarser-textured soils. Because water available to plants varies with soil texture and there is limited information on texture at these sites, it is not possible to compare plant available water across sites based on water content alone without further calibration of data. In addition to texture, rock content also affects the relationship between soil water content measured with the neutron probe and plant available water. With high rock content, the water that is present is concentrated in a smaller volume.

Shallow versus Deep Soil Water

Neutron probe and gypsum block data were used to compare shallow and deep soil water. Two additional soil water content data sets were generated by averaging neutron probe measurements made at 30 and 60 cm soil depth (hereafter referred to as shallow soil water) and by averaging measurements from 90 to 300 cm soil depth (hereafter referred to as deep soil water). To examine the temporal patterns in water content, time-series plots of shallow and deep soil water for each community are shown for each monthly sampling date for the period 1990–2001 (figure 5-3).

Shallow Soil Water

Average water content in the shallow soil depth for each community type was extremely variable in response to differences in monthly rainfall (figure 5-3). The shallow depths exhibited variable soil water content in response to monthly rainfall. This variability is probably influenced by a combination of soil texture, slope, plant cover, and other factors affecting infiltration, runoff, and run-in. Finer-textured soils at playa sites were more responsive to rainfall events than sandy loams and loamy sands at the black grama and creosote sites, but extremely sandy soils of mesquite were the least variable likely due to generally rapid infiltration in these coarse soils.

The standard deviation from the 12-year average mean water content for each community (n = 141 months, 3 missing observations) includes both within- and among-year temporal variability data from (figure 5-3). Water content of shallow soils was highly variable in playa soils (SD = 0.034). The playa sites are located on soils that are quite high in clay, and low rates of infiltration are expected (table

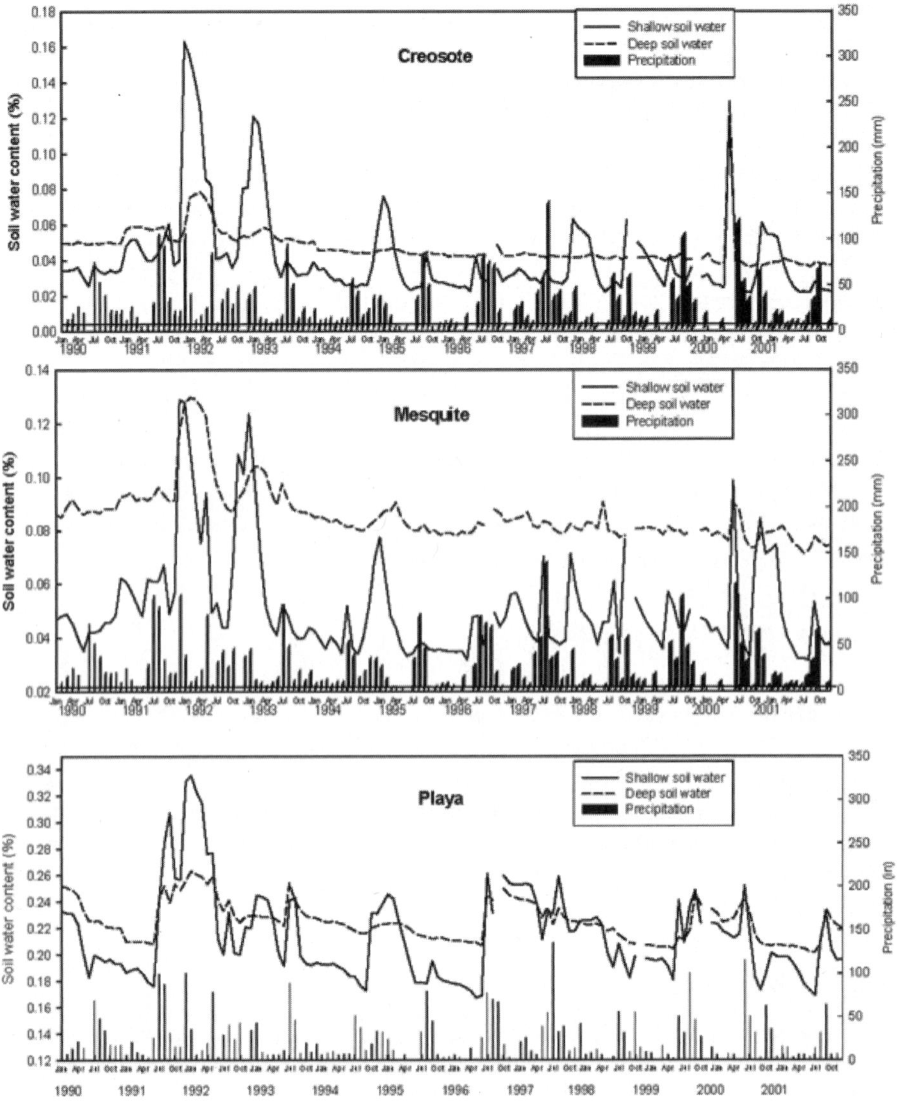

Figure 5-3. Long-term temporal patterns in shallow versus deep soil volumetric water content for each of the five vegetation communities. Soil water at 30 cm and 60 cm depth averaged for "shallow" water content; soil water from 90 cm to 300 cm depth averaged for "deep" water content. Vertical bars along *x*-axis show mean monthly precipitation. (Note: *y*-axis varies by community).

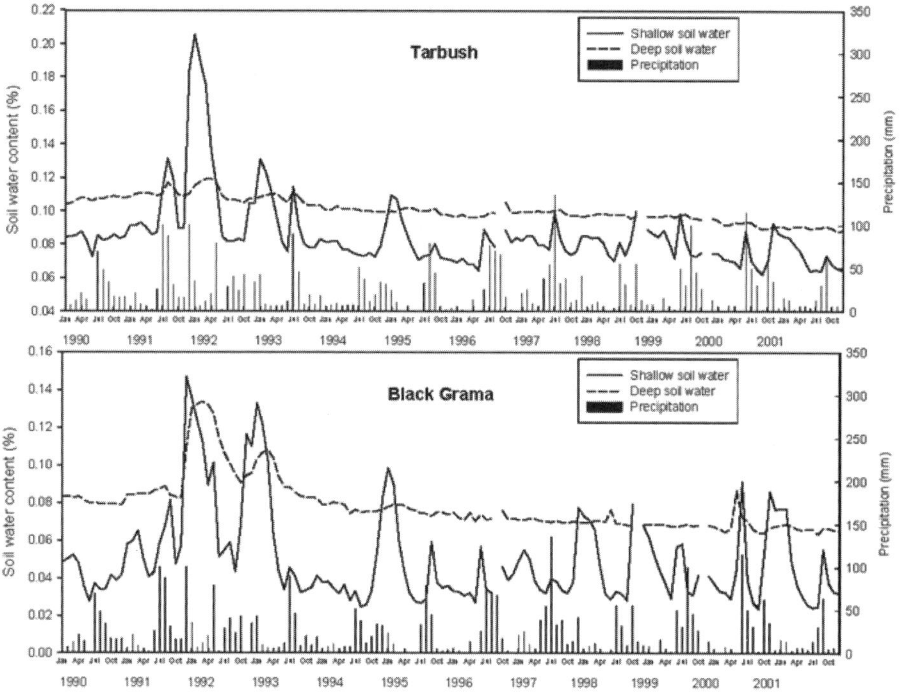

5-1). However, these sites are located in topographical depressions where precipitation and infiltration are enhanced by run-in processes that generally occur after larger rainfall events, creating high temporal variability in soil water content (Herbel and Gile 1973). Creosote (SD = 0.026) and black grama (SD = 0.026) communities showed similar variation in water content, while tarbush had slightly less variation (SD = 0.022). Tarbush sites are located in sandy clay loams and clay loam soils. These sites have a fairly high clay fraction that reduces infiltration, but are located on slopes of less than 1%. Intense rains generally produce significant runoff, whereas most of the rainfall from less intense events infiltrates into the soil. Shallow soil water was least temporally variable in the mesquite community (SD = 0.020). Mesquite sites are located on loamy fine sands and loamy sands that appear to have rapid infiltration and percolation, low water-holding capacity, and therefore low variability in soil water contents at monthly time scales.

Gypsum block data also illustrate that soil matric potential in shallow soil layers (30 cm and 60 cm) varied in response to rainy periods, and illustrate the importance of soil texture and landscape position (Herbel and Gibbens 1987, 1989). Soils of the narrow, level basin floor and an adjacent fan piedmont exhibited consistently different patterns of soil water availability than soils of the broad basin floor. All the coarse-textured soils (loamy sands and sandy loams) in the broad basin had a sandy surface layer that appears to have increased infiltration. The majority of these sites had a higher probability of soil matric potential less negative than –1.5 MPa in comparison to the finer-textured clay-loam soils of the narrow valley that had a lower probability of soil matric potential less negative than –1.5 MPa. For a clay-loam soil dominated by tobosa (*Pleuraphis mutica*), run-in processes doubled the probability that soil moisture at 25 cm was above –1.5 MPa, in comparison to plots where run-in had been experimentally excluded by 3-m diameter cylinders inserted to a depth of 15 cm. These data further support the hypothesis that the lower variability in the NPP tarbush data are due to infiltration limitations associated with relatively fine soil texture (table 5-1) or degraded soil structure (Herbel and Gibbens 1989).

Deep Soil Water

NPP soil water data, which were averaged for all depths greater than and including 90 cm, were very stable and fairly invariant through time at the community level. Average soil water contents of deep soil were generally greater than shallow soil water contents (figure 5-3). Inspection of soil water by individual depths within the mesquite and black grama communities (figure 5-4) confirms that soil water content increases and is less variable with depth. This probably reflects actual differences in plant water availability, at least at the mesquite sites and the BASN and SUMM black grama sites, because texture is relatively consistent throughout the profile for the dominant soils at these sites and because there are relatively few coarse fragments in these soils (although the presence of calcium carbonate nodules at greater depths may increase water-holding capacity at greater depths in some sites; Hennessey et al. 1983b). In general, water content variability at 90 cm decreases substantially in comparison with water content variability at 30 and 60 cm (figure 5-4). Less variability in soil water at and below 90 cm indicates that rainfall rarely infiltrates to these depths, although the aggregated monthly time scale of these measurements affects the ability to detect changes in soil water content that result from individual rain events. Playa, creosote, and tarbush communities had fairly consistent water content throughout the soil profile, but variability again decreased with depth. The amount of soil water did not consistently decrease with depth in any community type. The neutron probe data show consistent soil water at depths of greater than 1–2 m in this arid system, where groundwater is approximately 100 m below the soil surface.

Recharge of Soil Water

The presence of water at depth leads to questions about how deep soil water is recharged in these systems. Research at the nearby Walnut Gulch Experimental Watershed in Arizona found that transmission losses in ephemeral channels may be up to 80% (Renard 1970) and are the primary contributor to groundwater recharge. Less is known about the recharge of unsaturated vadose zone water for the Jornada Basin (see chapter 7). Transmission losses from smaller rills that flow fairly frequently are likely to contribute to vadose zone recharge (chapter 7), as big channel flows are rare in the Jornada Basin. Time-series techniques (cross-correlation and autocorrelation) were used to determine the relationship between

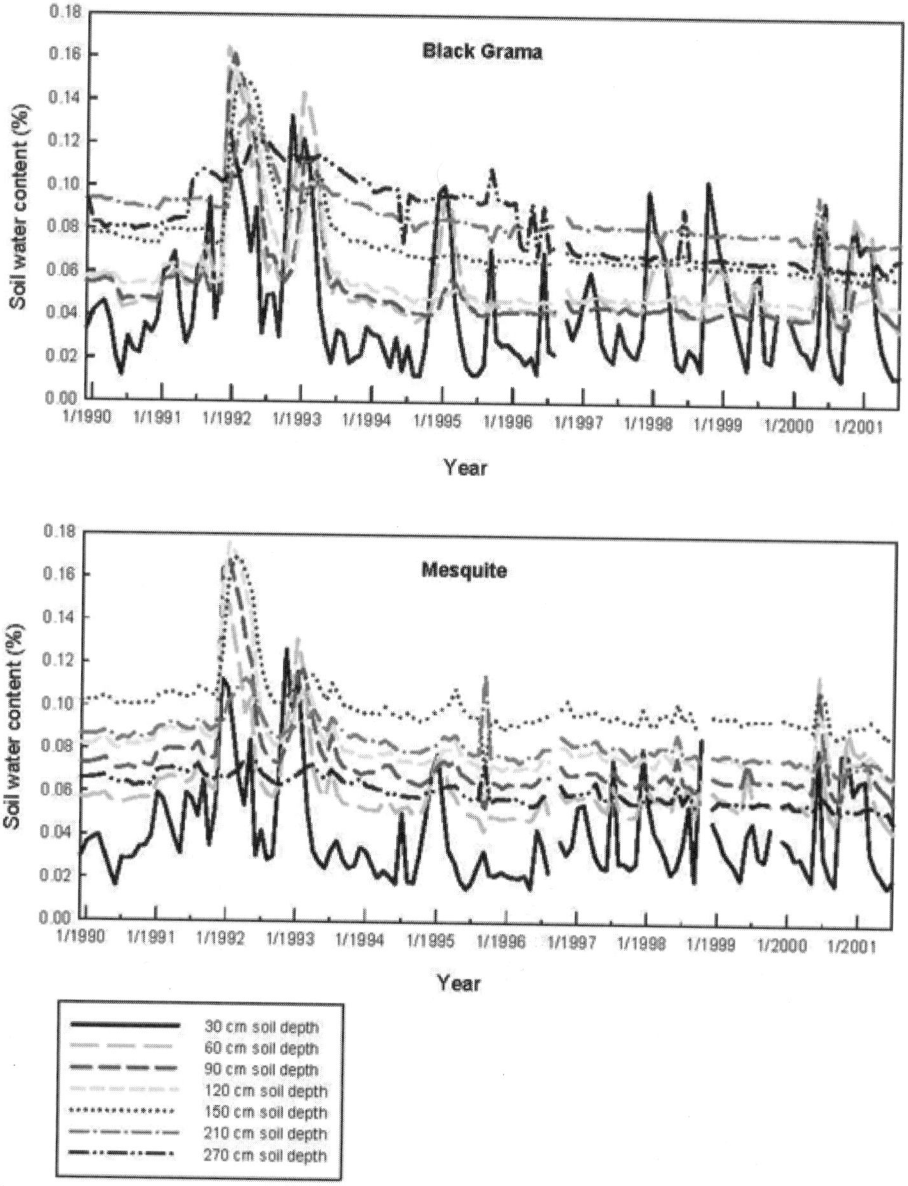

Figure 5-4. Seasonal and interannual patterns of soil water content by soil depth for the period 1990 to 2001 in the (top) black grama grassland community and (bottom) mesquite community (the three individual sites for each community were averaged monthly).

soil water content down to 130 cm and rainfall for the LTER transect (Nash et al. 1991). Rainfall samples were collected at one-week intervals, and soil water content was measured at two-week intervals for the period (July 1982 to March 1986). The soils along the 2.7-km transect were highly variable and so were divided into six relatively homogenous segments based on soil texture, morphology, vegetation, and soil water content (table 5-3) (Nash et al. 1991). Rainfall was slightly higher in the upper piedmont segment, and soil moisture increased moving downslope to the playa. Rainfall events preceded peaks in soil moisture by 4 weeks at 30 cm depth and 11 weeks at 130 cm depth. Cross-correlations of rainfall and soil water content were used to determine the lag time for soil water content to respond to rainfall for the entire sampling period. The response time to rainfall was short (i.e., less than 2 weeks) at 30 and 60 cm depths for all segments of the transect, but water content was slower to respond at deeper depths and lagged behind rainfall as much as 10 weeks at 130 cm. Furthermore, in upper and lower piedmont and playa soils rainfall reached a depth of 110 cm in two weeks. Upper and lower piedmont soils had the lowest clay fraction, whereas playa soils had an extremely high clay content. Therefore, changes in water content could not be predicted based on soil textural differences alone. Vertic clays in the soil at the College playa may facilitate deep percolation of water at the beginning of precipitation events through the formation of large, deep, vertical cracks (Nash 1985). The movement of water to 130 cm depth in nonplaya soils with an intermediate amount of clay and an increase in clay content with depth took four to six weeks. Time to reach maximum soil water content was substantially longer. Playa soil reached maximum infiltration at all depths in 8 weeks, but all other sites took between 14 and 24 weeks to reach maximum water content.

At a finer scale, variability in the structure and depth of petrocalcic horizons also affects deep soil water recharge. These horizons limit deep infiltration due to extremely low permeability caused by infilling of pores by calcium carbonate. Depth to the top of these horizons can vary by as much as 100% across distances of less than 10 m (chapters 2 and 4). Deep infiltration can occur through relatively carbonate-free "pipes," which form where the petrocalcic horizon is penetrated by animal burrows, strong roots, or anything that allows water to preferentially move through the horizon. Increased water flow dissolves more carbonate, further increasing permeability.

Another potential mechanism of deep water recharge is the poorly understood process of internal release of water from indurated caliche. Calcium carbonate ($CaCO_3$) is common in soils of the Jornada (chapters 2 and 4). Hennessy et al. (1983b) found in a laboratory study that calcium carbonate nodules derived from Jornada soils have a saturated water content of 27% by volume. Water loss from saturated nodules to the atmosphere was slow, but a majority of water was eventually released. It remains to be comprehensively tested in field settings, but buried calcium carbonate may release water to dry soils or provide a source of water that is accessible to plants. Herbel and Gibbens (1989) found some evidence for upward movement of water inside steel cylinders designed to exclude run-in on clay loam soils of the Stellar and Reagan series (table 4-2 and figure 4-3).

Another tentative (though untested) hypothesis is that some portion of unsat-

Table 5-3. Correlation of soil type, vegetation type, and soil water content along the LTER transect (taken from Wierenga et al. 1987). Zones of similar soil water content were determined using split-moving window analysis (Hotelling Lawley trace *F*-values for dissimilarity measure). Illustration below the table compares soil zonation, vegetation zonation, and the water content during 1982.

Soil Water Zone Breaks	Soil Series Zonation		Vegetation Type Zonation		
Station	Stations	Soil Mapping Unit	Stations	Vegetation Zones	Vegetation
6	1–6	Dalby (f)	1–7	*Playa*, grassland	*Panicum obtusum* (Paob)
	7	Headquarter variant (fl)			
11	8–10	Headquarter (fl)	8–10	*Playa fringe*, shrub-land	*Prosopis glandulosa* (Prgl)
19	11–25	Bucklebar (fl)	11–57	*Lower basin slope*, grassland	*Aristida longiseta* (Arlo)
42	26–45	Berino (fl)			
	46–55	Onite (cl)			
56	56–70	Doña Ana (cl)	58–72	*Upper basin slope*, shrubland	*Larrea tridentata* (Latr)
73	71–89	Aladdin (cl)	73–81	*Lower piedmont slope*, grassland	*Erioneuron pulchellum* (Erpu)
81			82–89	*Upper piedmont slope*, grassland	*Bouteloua eriopoda* (Boer)
	90–91	Rockland (r)	90–91	*Rocky slope*, shrub-land	*Ericamera Laricifiolia* (Erla)

Notes: f = fine, fl = fine loamy, cl = coarse loamy.

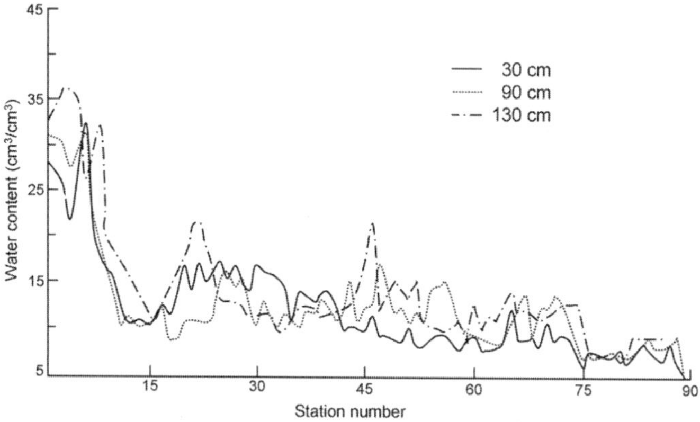

Variation in water content along the transect at depths of 30, 90 and 130 cm during the week of 30 April-6 May 1982.

urated soil moisture in some landscape units is a result of recharge processes that occurred on millennial timescales. The increased shrub component, especially the rapid encroachment of mesquite, is a relatively new phenomenon within the past 150 years, though increased shrub dominance has been found at other times in the geologic record in response to more arid climates (Monger 2003). Thus some of the water used by shrubs at deeper depths may be "old" water. The issue of deep diffuse recharge (i.e., in interdrainage areas) has not been resolved (Phillips 1994; Scanlon et al. 2005; Small 2005). An eight-year lysimeter studies by Scanlon et al. (2005) in the Mojave Desert found that recharge below the plant-rooting zone did not occur in the vegetated lysimeter due to vegetation feedbacks (increased biomass in response to more rainfall), even in El Niño (wet winters) years. Soil water storage was greater and at deeper depth in the nonvegetated lysimeter. Deep soil cores were used to measure soil water potential and chloride concentrations. Chloride bulges, which indicate a region of low soil moisture flux, were right above 5 m. Soil water potentials increased with depth and were remarkably consistent below 5 m. Scanlon et al. (2005) interpret these data to indicate that moisture below 5 m was derived from the Pleistocene, and currently there is gradient for water to move upward from deep soil layers. Mesquite frequently have roots at depths up to 5 m (Gibbens and Lenz 2001) and exceeding 5 m (Phillips 1963) and consequently may be using "old" water. However, focused recharge (e.g., beneath ephemeral streams, in playas and burrow pits) may be more important in some landscape units.

The Role of Seasonal Precipitation

Winter Soil Moisture

Soil water data from NPP sites indicate that recharge of deep soil moisture takes place in above-average precipitation years, especially if this precipitation falls during winter months. This is partly due to the frontal nature of winter precipitation, which is characterized by storms of low intensity and longer duration that enhance infiltration. In addition, evaporative losses are less due to lower temperatures, and transpiration losses are reduced due to the majority of plants being dormant in these cooler months. Both 1991 and 1992 were above-average precipitation years, and an example from the black grama community illustrates that extremely high rainfall during the early growing season of June and July, when convective storms dominate, had little effect on soil moisture below 60 cm (figure 5-5). However, rainfall events in December 1992 appeared to recharge deep water down to 240 cm. The playa community had different patterns with both summer rainfall and winter rainfall recharging soil water at depth (figure 5-5). This illustrates the importance of topographic position. Playa sites are in topographic depressions that have substantial run-in in both seasons. Substantial run-in has been observed to cause prolonged flooding at College playa and the Small playa; flooding has not been observed at the Tobosa playa.

Reynolds et al. (1999b) used rainout shelters to determine the effect of summer

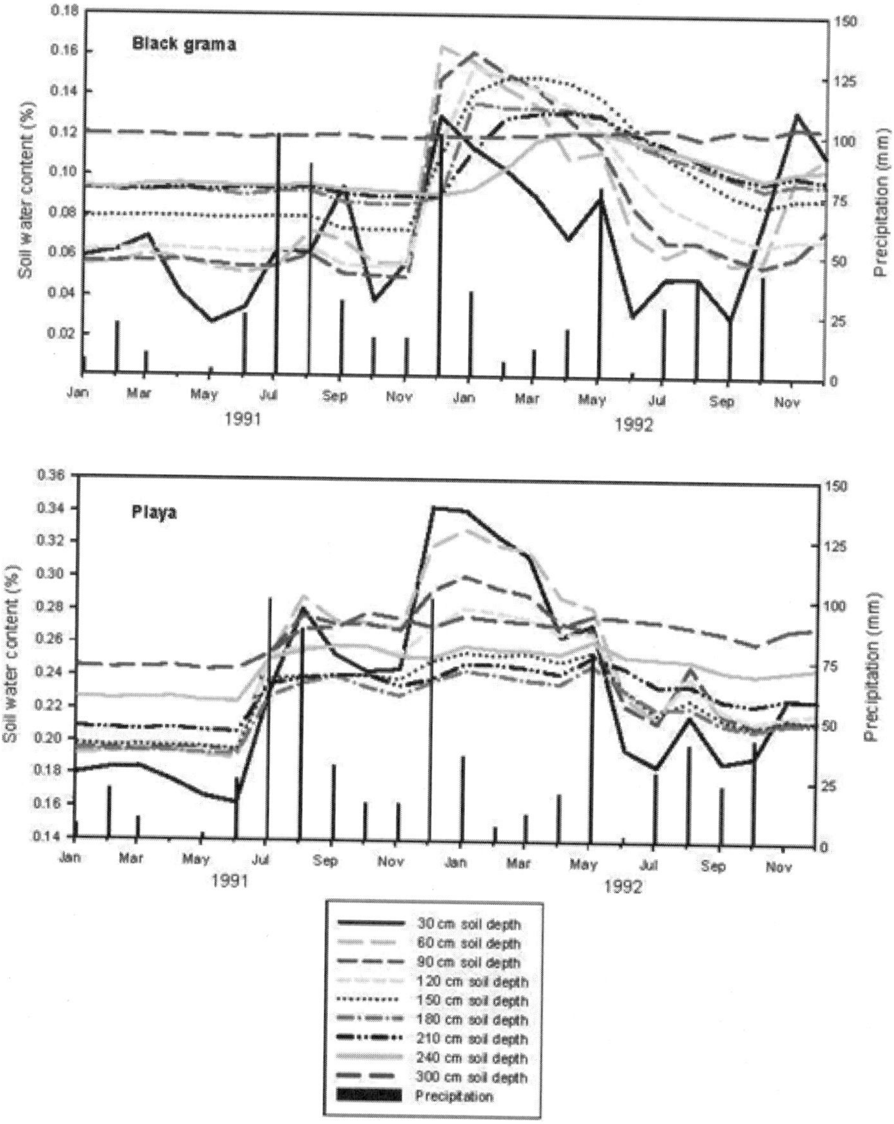

Figure 5-5. Differences in soil moisture response to winter versus summer precipitation. Soil water by depth for (top) black grama the grassland community and (bottom) the playa community for the period January 1991 through December 1992. Vertical bars along *x*-axis show mean monthly precipitation. Winter 1991 was an unusually wet year; figure illustrates recharge at depth with winter rain.

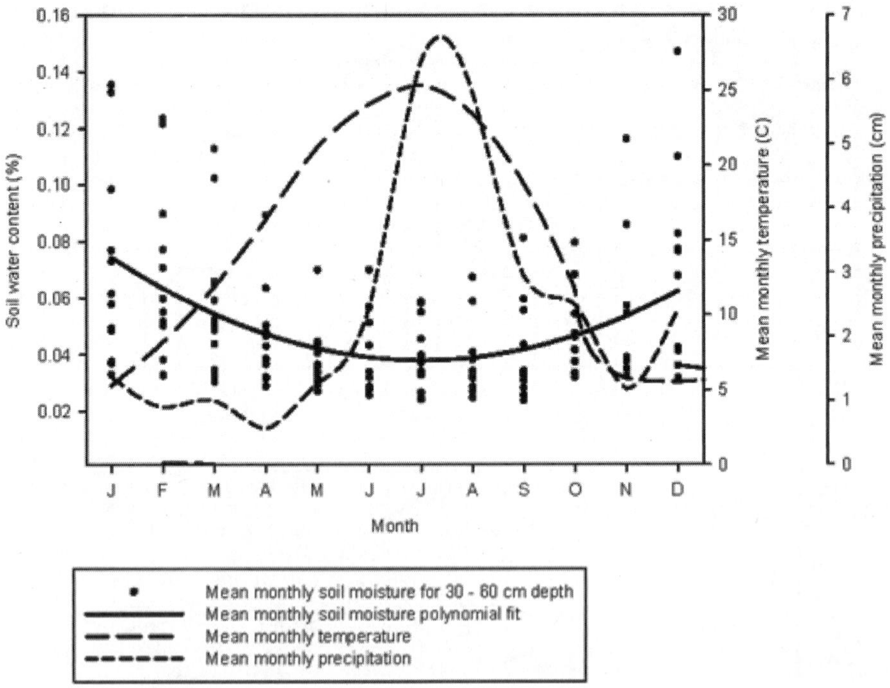

Figure 5-6. The relationship between mean monthly soil moisture, temperature, and precipitation averaged over the 1990–2001 period. Data are from the Jornada LTER meterological station. Also shown are means monthly shallow soil water (30–60 cm depth) data for the black grama grassland community for the same period.

and winter drought on creosote and mesquite. After three years of summer drought, there was little difference in soil moisture in comparison to control plots for either species (one site with a significantly larger rainout shelter and differing soils and vegetation did not follow this pattern). Conversely, imposed winter drought resulted in substantially lower soil water contents at depths down to 90 cm relative to controls plots; however, extrapolation is limited by the fact that winter drought was only imposed at one site. Rainout shelter studies in these types of landscape are further limited by the difficulty of excluding blowing rain during intense summer storms and the effects of shading on reducing summer evapotranspiration.

Although rainfall is greatest during the summer monsoon months, especially July and August, shallow soil depths are generally wetter in winter months than in summer (figure 5-6). Gypsum block data illustrated that the probability of wet soil was as great or greater during winter months than during the summer growing season on both loamy sands of the basin floor and clay loams of the fan piedmont (Herbel and Gibbens 1987, 1989). Data from NPP sites show a consistent trend of available shallow soil water right after winter that then declines throughout the typically dry spring periods as air temperatures increase (figure 5-6). It appears that there is shallow water in late winter and early spring that is potentially not fully used by plants. This may represent an unused resource space that could be at risk for invasion by winter annuals, such as red brome (*Bromus rubens*) that has invaded much of the Mojave, Great Basin, and Sonoran Deserts (Hunter 1991) or early growing, warm-season perennials, such as Lehmann lovegrass (*Eragrostis lehmaniana*).

Summer Precipitation

During the summer growing season, soil matric potentials of sandy loams at 10 cm were not greater than −1.5 MPa if daily precipitation was less than 13 mm (Herbel and Gibbens 1987). Likewise, it took greater than 20 mm of daily precipitation to raise the water potential of clay loams at 10 cm depth to greater than −1.5 MPa (Herbel and Gibbens 1989). Although these data indicate that fairly large rain events are needed during the summer months to change soil moisture at depths greater than 10 cm, the importance of smaller rainfall events on near-surface soil layers, sandy soils (Reynolds et al. 1999b), and vegetation is potentially not trivial. Again, only soil matric potentials above −1.5 MPa were measured, so changes in soil matric potential at more negative water potentials were undetected. Small storm events account for a large proportion of precipitation events in semiarid regions (Sala et al. 1992; Hochstrasser et al. 2002). Blue grama *(Bouteloua gracilis)* was found to have a significant increase in leaf water potential and conductance in response to simulated 5 mm rainfall events on the shortgrass steppe in Colorado (Sala and Lauenroth 1982). The importance of moisture in near-surface soil layers is likely to be extremely important to shallow-rooted species.

Vegetation and Soil Moisture

Consistent patterns of soil chemistry and vegetation type have been found to exist for four vegetation groups on the Jornada Basin (Stein and Ludwig 1979). However, the relationship between soil water content, soil texture, and vegetation type is less clear.

Transect data from LTER were used to determine the correlations between soil moisture, soil texture, and type of vegetation (Wierenga et al. 1987). A multivariate, moving, split-window technique was used to delineate different zones along the transect. On the basis of soil texture, nine distinct zones were found that correlated well with the soil survey. Seven distinct vegetation types were identified that were generally correlated with a soil-mapping unit (i.e., soil series plus surface textural characteristics) (table 5-3). Seven distinct zones of soil water were found that were in general well correlated with vegetation and soil texture. The exception to the general pattern of distinct zones was in the vegetation zone, which spanned a majority of sampling stations (stations 15–60) and was historically dominated by black grama, but at the time of the study, black grama cover was greatly reduced. The area was dominated by threeawn (*Aristida longiseta*), a perennial grass, soaptree yucca (*Yucca elata*), and globe mallow (*Sphaeralcea subhastata*). This same vegetation zone encompassed three soil-mapping units (two fine loams and one coarse loam) and three distinct changes in soil moisture that corresponded roughly to changes in soil series. These data indicate that soil texture and soil moisture are related at a landscape scale to vegetation patterns; however, within vegetation types there may still be substantial difference in soil moisture and changes in soil-mapping units. Analysis of plant cover and soil water from LTER using two geostatistical analysis methods reached the conclusion that volumetric soil water content could not fully explain vegetation patterns along the transect (Nash et al. 1992).

Gypsum block data showed substantial changes in soil moisture under different vegetation on the same soil type. For example, on a Stellar soil series comparison of soil water under tobosa grass showed much higher soil matric potential than the same soil under burrograss. This pattern was attributed to difference in near-surface soil structure; a silty platy structure may have decreased infiltration of soil water under burrograss (Herbel and Gibbens 1989). Runoff from burrograss plots was three times higher than tobosa grass under simulated rainfall events (Devine et al. 1998). It is also plausible that feedbacks between vegetation type and soil type have positively reinforced this decline in soil matric potential. For example, feedbacks between carbon exudates of roots and soil aggregate formation affect both infiltration capacity and the water-holding capacity of soils. However, on sandy loams and loamy sands, Herbel and Gibbens (1987) came to the conclusion that the type of vegetation, shrub or grass, did not appear to influence soil matric potentials.

Soil water content averaged across all months (n = 141 months) at up to 10 measured depths (see figure 5-2) was used to obtain the total water content of the soil profile (figure 5-7). Measured content at each available depth was assumed to represent water content ± 15 cm around the measured point (see table 5-2), and values were summed to obtain content of the profile. There is a general relationship between total profile soil water content and soil surface texture; finer surface soil textures were associated with higher total water content. This is expected based on their higher water content at soil water tensions that may be inaccessible to plants. However, when the playa sites are excluded, the variation in cm of total water showed as much variation within a community as across communities.

There is also tremendous variation in soil water content by depth between sites within a given community type and with depth (figure 5-8). Variability in water content is highest near the surface and decreases with depth. If the playa sites are excluded, there appears to be nearly as much variation with depth and within a community as across all communities (figures 5-2 and 5-8). Even though the three sites within a community type were selected to include minimum and maximum biomass of each community type (see Huenneke et al. 2002; chapter 11), we still expected soil water contents to be more similar within a community type than across community types. Similarly, extreme variability in soil texture and depth were found over a small spatial scale dominated by a single vegetation type, mesquite (Gile et al. 1997). Our analyses strongly suggest that community com-

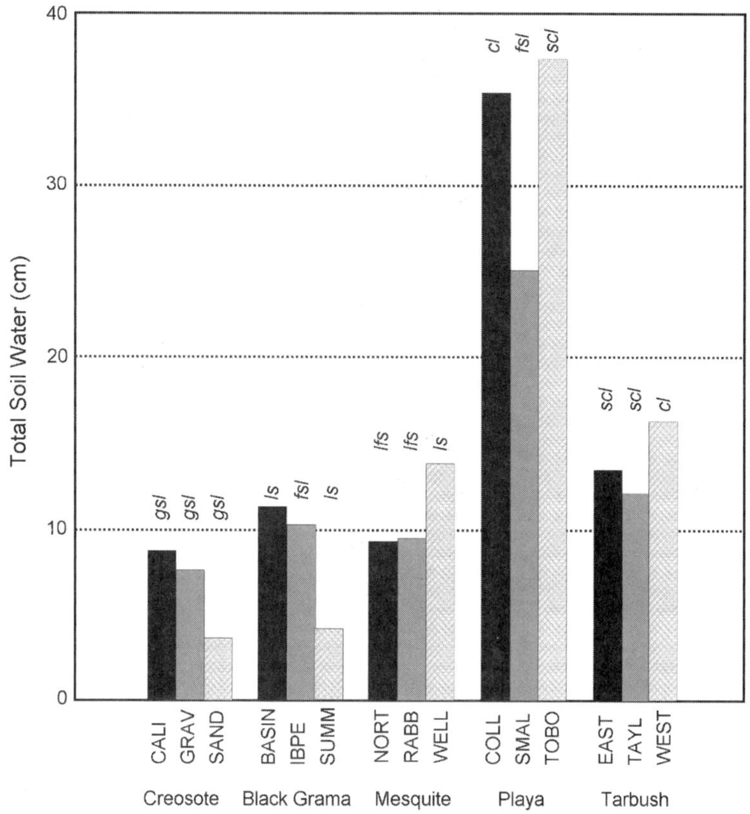

Figure 5-7. Total soil water in the soil profile for each of the 15 individual NPP sites. (Soil water content at each depth of 10s multiplied by 30 cm, summed, up to and including soil water at 270 cm.) Four-letter notation on the x-axis is for each of the three sites within each of the five vegetation types (creosote, black grama, mesquite, playa, and tarbush) follows chapter 11, figure 11-1. Creosote at the gravel site only has data to 150 cm; therefore soil water was estimated for depths from 180 cm to 270 cm. Average of all sampling dates from July 1989–March 2001 (n = 141). Above each site there is notation of the soil texture class from table 5-1.

position and surface soil texture are not the only controlling factors for soil water and that landscape linkages must also be considered.

Interspaces versus Beneath-Plant Canopies

There may also be variation in soil water within a site due to feedbacks between hydrology and vegetation. Much research in the Jornada Basin has contributed significantly to our understanding of the processes behind "resource island formation," whereby spatially homogenous grasslands are converted to patchy shrub-

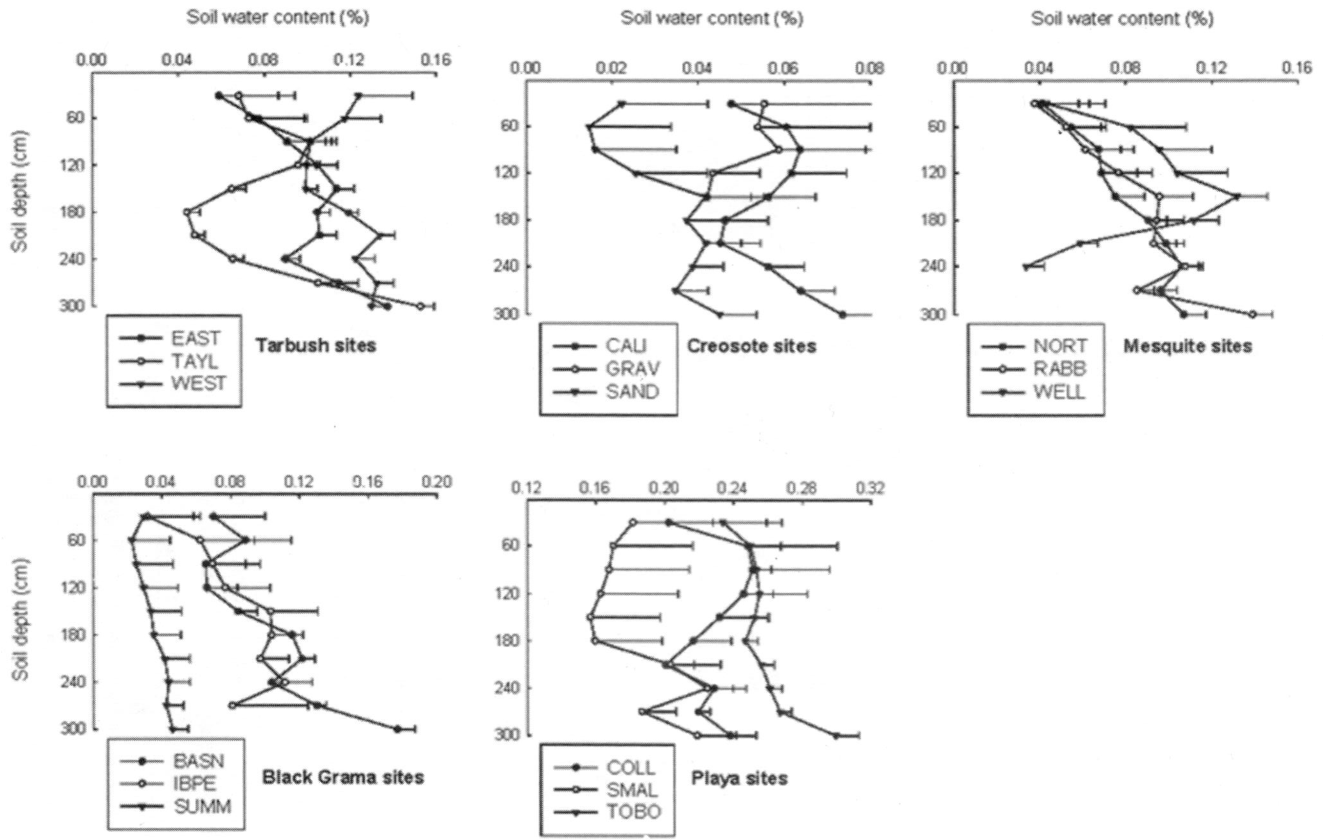

Figure 5-8. Mean soil water content (± 1 SD, n = 141) by depth for each site within each of the five vegetation communities for the period 1990–2001. Individual site names correspond to table 5-1.

land communities (Schlesinger et al. 1990). Shrub communities are characterized by a heterogeneous spatial distribution of resources which become increasingly concentrated below and around shrubs while the bare interspaces between shrubs have reduced amounts of litter, nutrients, and inputs from animal activity (Schlesinger et al. 1990; Cross and Schlesinger 1999). It has been proposed that increased interception of precipitation by shrub canopies leads to increased stem flow and, subsequently, increased depth of infiltration below shrubs (Martinez-Meza and Whitford 1996). In keeping with this view, Reynolds et al. (1999b) hypothesized that large shrubs should have greater soil moisture relative to small shrubs. Contrary to their hypothesis, small mesquite shrub islands were found to have greater soil moisture in summer (measured biweekly at depths > 30 cm) than large shrub islands. Furthermore, there was no indication that large shrub islands of either creosote or mesquite had greater soil moisture storage than young islands. Hennessy et al. (1985) compared soil properties under mesquite dunes with bare interdunal spaces. Dune soils with established mesquite had greater infiltration, greater hydraulic conductivity, and less evaporation than interdunal soils. However, gravimetric water content at 15 cm did not differ between dunes and interdunes and was closely related to rainfall. At 30 cm in depth water content, measured with a neutron probe every two weeks, was significantly less in vegetated dunes than in interdune areas. Similarly, preliminary analysis of NPP soil water data indicates that water content measured in neutron probe tubes in interspaces of the mesquite sites is consistently greater than water contents measured in neutron probe tubes under mesquite canopies (Snyder and Mitchell unpublished data). Because neutron probes measure water stored in nearby plant roots, as well as nearby soil, actual soil water content may potentially be even lower under shrubs (Hennessy et al. 1985). Although water may infiltrate faster and deeper below shrubs, limited evidence suggests there is not more water stored in shrub islands. This is likely due in part to plants having greater root volume in canopy areas and, consequently, using more water from the canopy area. Shrub islands may be areas of greater infiltration and percolation, but plants may quickly take up this water; to detect these differences, continuous measurements of soil water may be necessary.

Conclusions

Patterns of soil water appear to vary greatly within a community type as well as across the landscape, and the relationship to community type and production appears to be complex. However, some patterns did emerge from previous Jornada research and our current data analyses presented herein. Shallow soil water is highly variable and cannot be completely explained by temporal differences in precipitation and spatial differences in soil texture. There is water at depth in these systems, and this water is less temporally variable. Recharge of unsaturated soil moisture at depth is a poorly understood process but appears to occur mostly in wetter-than-average winter months. There is little evidence to support predictions that shrub islands have more stored soil water than barren interspaces, but these islands do appear to be areas of greater infiltration.

More detailed calibration of existing soil water content data and developing relationships between soil water content and soil water potential would allow for better comparisons across different soil types and would resolve whether water stored at depth is available to plants. More detailed measurements at shallower depths (< 30 cm) are needed to determine the importance of small, frequent rainfall events on plant productivity and potential competitive interactions between grasses and shrubs. Because of the need for detailed, continuous measurements on soil water availability and plant composition and production, it is impractical to attempt this for every soil series in the Jornada Basin. Instead, care should be taken to select soil series that appear to be most susceptible to vegetation change and erosion. Sandy loams appear to be a particularly susceptible soil type—where temporal interactions are greater, vegetation change is dynamic and enhanced information on the feedbacks between soil water and vegetation is particularly warranted.

6

Nutrient Cycling within an Arid Ecosystem

William H. Schlesinger
Sandy L. Tartowski
Sebastian M. Schmidt

Low quantities of soil nitrogen limit plant growth in the Chihuahuan Desert (Ettershank et al. 1978; Fisher et al. 1988; Lajtha and Whitford 1989; Mun and Whitford 1989) and in other deserts of the world (Wallace et al. 1980; Breman and de Wit 1983; Sharifi et al. 1988; Link et al. 1995). Indeed, although deserts are often regarded as water-limited systems, colimitation by water and N may be the more general rule (Hooper and Johnson 1999; Austin and Sala 2002). In a broad survey of desert ecosystems, Hooper and Johnson (1999) found evidence for colimitation by water and N even at the lowest levels of rainfall. In arid ecosystems, water is delivered in discrete events separated by drier periods, which restrict biological activity and uncouple plant uptake of nutrients from decomposition. Local variations in net primary production in arid and semiarid ecosystems are largely determined by processes that control the redistribution of water and soil nutrients across the landscape (Noy-Meir 1985; Schlesinger and Jones 1984; Wainwright et al. 2002; see also chapter 11). In this chapter we focus on the N cycle in different plant communities of the Jornada Basin with the recognition that after water, N is the most likely resource to determine the plant productivity of this ecosystem.

Where arid environments are dominated by shrubby vegetation, the distribution of soil properties is markedly patchy with strong accumulations of plant nutrients under shrubs and relatively infertile soils in the intershrub spaces (Noy-Meir 1985). These islands of fertility are particularly well described in the Chihuahuan Desert and other areas of the American Southwest. Local accumulations of nutrients under vegetation are also documented for desert habitats on other continents, including Europe (Gallardo et al. 2000), Africa (Gerakis and Tsangarakis 1970; Belsky et al. 1989; Wezel et al. 2000), Australia (Tongway and Ludwig

1994; Facelli and Brock 2000), and South America (Rostagno et al. 1991; Mazzarino et al. 1991, 1998; Gutierrez et al. 1993). In the Jornada Basin, Schlesinger et al. (1996) used geostatistics to compare the scale of soil heterogeneity in arid habitats dominated by shrubs and in adjacent areas of arid grassland. A nearly random distribution of extractable N was found in grassland soils, but in areas dominated by creosotebush (*Larrea tridentata*) the distribution of soil N was patchy at a scale close to the average size of shrubs (figure 6-1).

The patchy habitat created by shrubs also determines the biodiversity of animals at higher trophic levels, including lizards and birds (Pianka 1967; Naranjo and Raitt 1993). Patchy distributions of soil microbial biomass (Mazzarino et al. 1991; Gallardo and Schlesinger 1992; Kieft 1994; Smith et al. 1994; Herman et al. 1995), nematodes (Freckman and Mankau 1986), and microarthropods (Santos et al. 1978) reflect the heterogeneous distribution of soil nutrients in desert shrublands. Indeed, most ecosystem function in shrub deserts is localized under vegetation, whereas the adjacent shrub interspaces are comparatively devoid of biotic activity (see chapter 12).

Greater microbial activity under shrubs is manifest in high rates of N mineralization and nitrification (Charley and West 1977; Mazzarino et al. 1991; Smith et al. 1994). These microbial processes have the potential to produce gaseous byproducts—NH_3, NO, N_2O and N_2—that are lost to the atmosphere. In some deserts, the emission of these gases is an important part of the biogeochemical cycle (West and Skujms 1977; Westerman and Tucker 1979; Virginia et al. 1982), but in most cases, the shrubs act to conserve N by its immobilization in the litter and microbial biomass of soil mounds (Peterjohn and Schlesinger 1991; Schlesinger

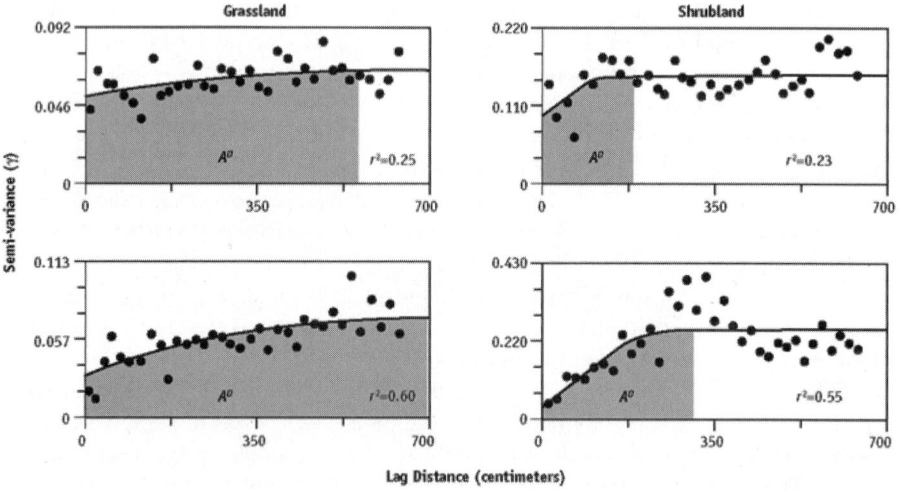

Figure 6-1. Spherical model semivariograms for the distribution of available N in soils of grassland and shrublands at the JER (n = 2 each) in the Chihuahuan Desert of New Mexico. The range of spatial dependence or autocorrelation is designated as Ao in each panel (modified from Schlesinger et al. 1996).

and Peterjohn 1991; Gallardo and Schlesinger 1992; Zaady et al. 1996). Indeed, the local nutrient accumulation under shrubs may exceed the original nutrient concentrations in grassland soils that the shrubs have invaded (Kieft et al. 1998; Cross and Schlesinger 1999).

The spatial heterogeneity of biotic activity in shrublands controls movements of water and soil materials in desert ecosystems. Total ground cover is the most important variable influencing runoff and sediment production on desert rangelands in southern New Mexico (Wood et al. 1987) and other arid and semiarid regions (Zobisch 1993). In the Jornada Basin, vegetation aerial cover can be up to 50–60% in grasslands versus 30% in creosotebush and mesquite (*Prosopis glandulosa*) shrublands (Schmidt unpublished data). When shrubs replace grasslands, the rate of erosion increases and the surface soil materials are progressively lost from the barren shrub interspaces, especially for sand-textured soils (Bull 1979; Abrahams et al. 1994, 1995; Gutierrez and Hernandez 1996; see also chapter 9). When shrubs are widely spaced, the barren intershrub soils are also subject to wind erosion that redistributes soil materials across the landscape (Snow and McClelland 1990; Stockton and Gillette 1990; Okin and Gillette 2001). Losses of soils and soil nutrients are closely tied to the degradation and desertification of desert grasslands in southern New Mexico and other arid and semiarid regions of the world.

Nutrient Cycling in the Jornada Basin

Simultaneous limitation of plant growth by water and N (response to water added alone and to N added alone) is common in the Chihuahuan Desert (Ettershank et al. 1978; Gutierrez and Whitford 1987a). For example, Fisher et al. (1988) found that the growth of creosotebush nearly doubled with experimental additions of 100 kg N/ha, with the greatest plant growth seen when N and water were added together (figure 6-2). Depending on the site and circumstances, pulses of primary production related to variable rainfall may shift between water and nutrient limitation of plant growth (Mun and Whitford 1989). During years of high winter rainfall in the Jornada Basin, the decomposition of an abundant growth of spring annuals can immobilize soil N, leading to deficiencies that subsequently limit the growth of creosotebush during the summer (Parker et al. 1984a). These interactions between water and N may contribute to the poor correlation between annual net primary production and recent precipitation in semiarid habitats worldwide (Le Houerou et al. 1988; see also chapter 11).

Intersystem Nutrient Flux

N inputs from the atmosphere average around 2.5 kg/ha/yr in the Jornada Basin (Schlesinger et al. 2000) with about half as NH^{+}_{4} and half as NO^{-}_{3}. Deposition of N in rainfall dominates over the deposition of N in dust and particles. The N in atmospheric deposition is supplemented by rather meager inputs, mostly < 1

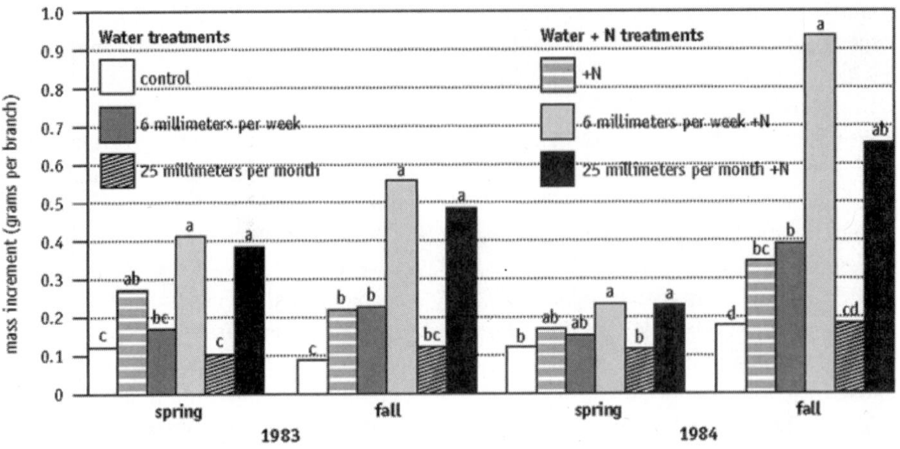

Figure 6-2. Effects of nitrogen fertilization and two patterns of irrigation on seasonal mass increments in vegetation in creosotebush shrubs in the Jornada Basin. Within each year and season, significant differences between treatments are indicated by different letters (modified from Fisher et al. 1988).

kg N/ha/yr, from asymbiotic N-fixing bacteria in soil crusts (Loftis and Kurtz 1980; Hartley and Schlesinger 2002). In the Jornada Basin, the only appreciable rates of soil fixation, ranging up to 10 kg N/ha/yr, are found in tarbush (*Flourensia cernua*) communities, especially in soils with a low N/P ratio (Hartley and Schlesinger 2002). Herman et al. (1993) also report N-fixing bacteria in the rhizosphere of black grama (*Bouteloua eriopoda*), but their contribution to the N economy of the grassland habitats remains unknown. Field experiments show that most asymbiotic N fixation appears to be due to heterotrophic soil bacteria, and the rate of fixation is stimulated by experimental additions of carbon and water (Hartley and Schlesinger 2002). The low rate of N fixation in Jornada desert soils contrasts sharply with high inputs by this process in the cold desert steppe of the Great Basin (West and Skujms 1977; Belnap 2002) but compares favorably to the results of Rundel and Gibson (1996), who report similar low rates of asymbiotic fixation in creosotebush habitats in the Mojave Desert of southern Nevada.

Symbiotic N fixation is largely confined to habitats dominated by mesquite. At the Jornada, Jenkins et al. (1988) report nodules on mesquite roots from 13 m depth, complicating attempts to estimate overall N inputs to this ecosystem. Based on studies in other ecosystems, we might expect N fixation in mesquite habitats to range from 40 (Rundel et al. 1982) to 150 kg N/ha/yr (Johnson and Mayeux 1990), depending on plant cover. From measurements of $\delta^{15}N$ in its foliage, Lajtha and Schlesinger (1986) estimated that mesquite in the Jornada Basin obtains 48% of its N from symbiotic fixation, accounting for about 20 kg N/ha/yr.

In the face of these inputs of N, the Chihuahuan Desert persists in an N-deficient state, owing to soil erosion by wind and water and to the microbial production and loss of N-containing gases to the atmosphere. Wind erosion may remove up to 14 kg N/ha/yr from mesquite habitats, assuming a soil N concentration of 0.1% (Gallardo and Schlesinger 1992) and a net soil loss of 1,400 g/m^2/yr (chapter 9). Additional losses of 1–5 kg N/ha/yr are associated with the suspended and bedload sediments carried in runoff waters. Only a small fraction of the runoff loss occurs in forms that are available to plants. For instance, Schlesinger et al. (2000) report runoff losses of dissolved N totaling 0.15 kg N/ha/yr in black grama grasslands and 0.33 kg N/ha/yr in creosotebush shrublands. A surprising fraction of the dissolved N loss, nearly 70% on bare soils, is carried in dissolved organic forms (DON). Relatively little N is lost to groundwater in the Chihuahuan Desert, in contrast to large apparent losses of NO^{-}_{3} to groundwater in other North American deserts (Peterjohn and Schlesinger 1991; Jackson et al. 2004; Seyfried et al. 2005).

Volatilization of NH_3 to the atmosphere is rather small (0.03–0.35 kg N/ha/yr) (Schlesinger and Peterjohn 1991), and losses of nitric oxide (NO) associated with nitrification (figure 6-3) range from 0.15 to 0.38 kg N/ha/yr (Hartley and Schlesinger 2000). Peterjohn and Schlesinger (1991) report bursts of denitrification after wetting events, potentially leading to losses of 7.2 kg N/ha/yr from these ecosystems. A significant fraction of this loss may occur as N_2O, which Hartley (1997) estimated to return up to 3.9 and 6.2 kg N/ha/yr to the atmosphere from grasslands and shrublands, respectively. By comparison, Guilbault and Matthias (1998) report the loss of 0.4 kg N/ha/yr as N_2O from soils of the Sonoran Desert.

In summary, new inputs of N are limited and a variety of processes lead to the loss of N from desert soils and to a deficiency of N for plant growth, especially in wet years in the Jornada Basin. Nutrient losses from shrubland habitats exceed

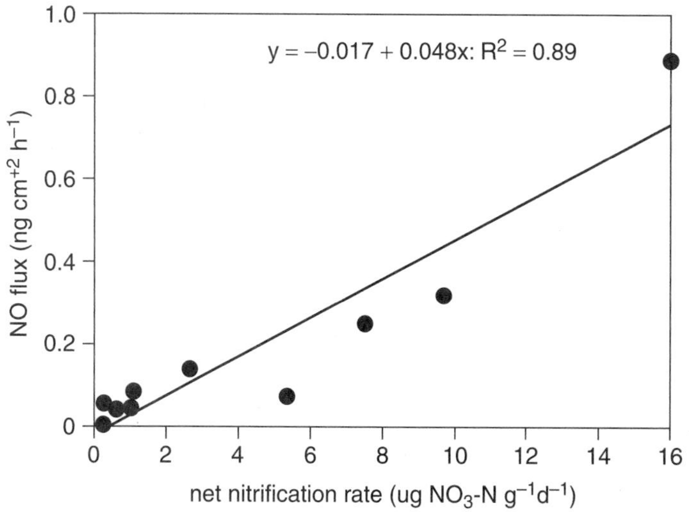

Figure 6-3. Relationship between soil NO emission and potential net nitrification rates, measured in the laboratory, in a variety of soils from the Jornada Basin (from Hartley and Schlesinger 2000).

those from grasslands, but much of the apparent loss may simply result in a local redistribution of soil nutrients on the landscape.

Intrasystem Nutrient Cycling

Plant Nutrition

Plant nutrient uptake by Chihuahuan Desert shrubs is closely tied to the availability of soil water. In watering experiments, creosotebush showed a rapid uptake of soil N, whereas mesquite showed little response (BassiriRad et al. 1999). This difference may result from the relatively shallow rooting system in creosotebush and a greater reliance by mesquite on N-fixation at depth (Lajtha and Schlesinger 1986; Ho et al. 1996). In fourwing saltbush (*Atriplex canescens*) Sisson and Throneberry (1986) found the highest levels of leaf nitrate reductase during the seasonal periods of greatest leaf water potential, consistent with patterns of soil N availability, which show a dominance of NO_3 over NH_4 during the summer wet season. Creosotebush shows a distinct preference for NO_3 as a N source, consistent with its growth during the summer (BassiriRad personal communication). That plant uptake affects soil N dynamics is clearly evident in the greater accumulations of inorganic N that are found in the soils of rainfall-exclusion experiments (Reynolds et al. 1999a).

Although some desert plants have been reported to have especially high leaf N contents (El-Ghonemy et al. 1978), the N content of plants in the Jornada Basin is not significantly different from that of many forest species (Killingbeck and Whitford 1996). As for most plants, photosynthesis in creosotebush is positively correlated to leaf N content (figure 6-4), but N-use efficiency declines significantly with increasing leaf N contents (figure 6-5; Lajtha and Klein 1988). N-use efficiency is inversely correlated to water-use efficiency, so both C acquisition and water conservation are greater in desert plants with high leaf N contents (Lajtha and Whitford 1989). However, plants with high leaf N contents are also more attractive to insect and mammalian herbivores, who enhance the return of N to the soil through their feeding (Lightfoot and Whitford 1989, 1990; Day and Detling 1990; Frank and Evans 1997).

Mycorrhizae may contribute to the nutrient-uptake capacity of many species in the Jornada Basin (Herman 2000). Abundant fungal endophytes with mycorrhizal traits have been found on fourwing saltbush (Barrow et al. 1997) and black grama (Barrow 2003) in the Jornada Basin, and creosotebush is reported to harbor mycorrhizae in the deserts of southern California (Bethlenfalvay et al. 1984). The mycorrhizae on fourwing saltbush are dark septate fungi with the ability to solubilize rock phosphate (Barrow and Osuna 2002). Mycorrhizal production of oxalic acid may be particularly important to phosphorus mobilization in high carbonate soils (Jurinak et al. 1986).

The widespread deposition of pedogenic calcite in arid soils affects the availability of phosphorus, and much of the total P pool in the soil is bound to calcium minerals (Marion and Babcock 1977; Lajtha and Bloomer 1988). Lajtha and

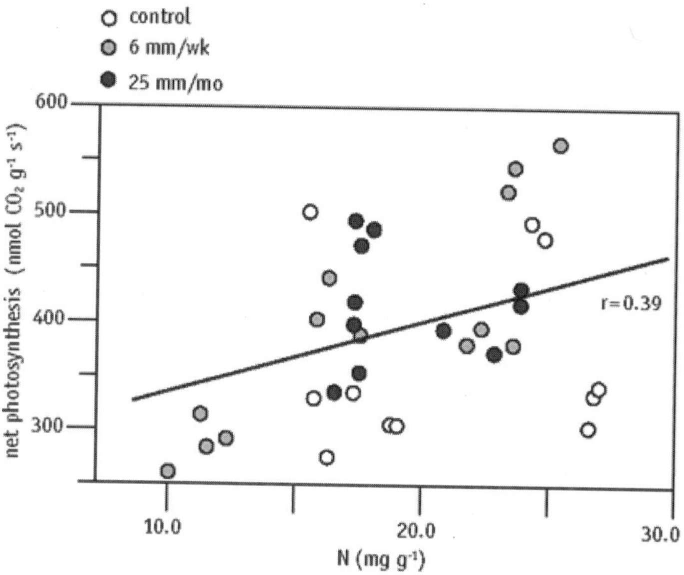

Figure 6-4. Relationship between net photosynthesis and leaf N in creosotebush in the Jornada Basin, among shrubs receiving different irrigation treatments (from Lajtha and Whitford 1989).

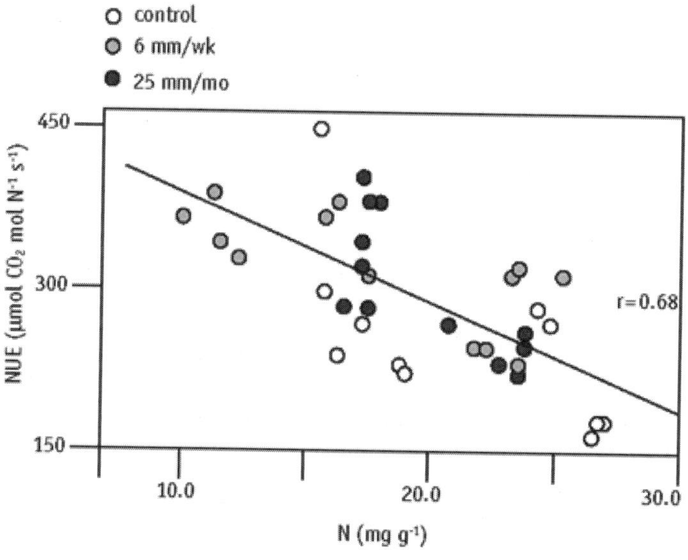

Figure 6-5. Relationship between leaf nitrogen-use efficiency (NUE) and leaf N concentration in creosotebush (from Lajtha and Klein 1988).

Schlesinger (1988) report decreased uptake of inorganic P in creosotebush in response to experimental additions of $CaCO_3$ to soils. In the Chihuahuan Desert only a small amount of phosphorus is found in organic forms, which may be of special importance to plant phosphorus nutrition (Cross and Schlesinger 2001). Often, organic and bicarbonate-extractable P (forms easily available for plant uptake) are concentrated beneath the canopy of shrubs, whereas Ca-bound P is greatest in the shrub interspace (Charley and West 1975; Cross and Schlesinger 2001).

In creosotebush, resorption of N and P before leaf abscission is positively related to the nutrient status of foliage (Lajtha 1987; Lajtha and Whitford 1989). For creosotebush, especially high resorption efficiencies for phosphorus (72–86%) may be a response to limited P availability in the high carbonate soils of the Jornada Basin (Lajtha 1987). For comparison, N-resorption efficiency in creosotebush is about 50–60% (Lajtha 1987; Lajtha and Whitford 1989), not unlike the global mean for terrestrial plants (Aerts 1996). The N-resorption efficiency of mesquite is exceptionally low (36%), perhaps reflecting its reliance on N fixation as a source of N (Killingbeck and Whitford 2001).

Soil Processes

Whitford et al. (1981b) noted that the rate of decomposition of plant litter in the Chihuahuan Desert was greater than predicted by the simple correlation with actual evapotranspiration, as promulgated by Meentemeyer (1978) in his seminal study comparing decomposition in various terrestrial biomes. Later, Schaefer et al. (1985) found that lignin content, C/N ratio, and the lignin/N ratio in plant litter also failed to predict decomposition in the Jornada Basin, despite the success of these variables as predictors of decomposition in a variety of other ecosystems (Melillo et al. 1982). Unexpected high decomposition rates of surface litter were thought to be related to its photo-oxidation by ultraviolet light and to the abundant activity of microfauna and termites (*Gnathamitermes*) in desert soils. Despite several attempts to demonstrate high rates of photo-oxidation of surface litter materials, the importance of this process remains equivocal (Moorhead and Reynolds 1989a; MacKay et al. 1994). Meanwhile, the importance of soil microfauna to litter decomposition and soil N content has been clearly demonstrated by using insecticides to inhibit their activity in litterbags (Santos and Whitford 1981; Santos et al. 1981; Brown and Whitford 2003). Johnson and Whitford (1975) found that termites consumed about 50% of the surface litter in creosotebush and mesquite communities (see also Whitford et al. 1982; Silva et al. 1985; Whitford 1991; see also chapter 12). Photo-oxidation of litter appears more important in the Patagonian steppe of Argentina where termites are less abundant (Amy Austin, personal communication).

Because much of the plant litter in the Jornada Basin is processed by the activities of termites, rainfall timing and amount is not as strong a predictor of decomposition in the Chihuahuan Desert (Santos et al. 1984; Whitford et al. 1986; Kemp et al. 2003) as it is in other desert ecosystems (Strojan et al. 1987). Because surface litter is subject to rapid drying after rainfall, the rate of decomposition of litter in local patches where it accumulates, such as under shrubs, is greater than in adjacent areas of lesser accumulation where it dries quickly (Whitford et al. 1980b, 1982; Parker et al. 1984a). For most species the decomposition of buried litter and roots, largely mediated by microarthropods (Santos et al. 1984), is faster than the rate of disappearance of surface litter, which shows greater fluctuations in moisture content (Schaefer et al. 1985).

Litter quality, especially N content, appears to play only a limited role in determining rates of decomposition in desert habitats of the Jornada Basin (Schaefer et al. 1985) and elsewhere in the American Southwest (Murphy et al. 1998). Experimental additions of N had little effect on the decomposition of black grama or creosotebush litter (MacKay et al. 1987a; see also Mun and Whitford 1998).

Among habitats of the Jornada Basin, soil microbial biomass is related to the content of soil organic carbon and extractable N (NH_4 + NO_3) (Gallardo and Schlesinger 1992). Fertilization with N increases microbial biomass in grassland soils, whereas additions of C have little effect. In shrublands, fertilization by C increases microbial biomass and decreases extractable N and P, which are immobilized during microbial growth (Gallardo and Schlesinger 1995). In many areas of the Jornada Basin, where shrubs have invaded upland grasslands, the proportional net decrease of soil organic C exceeds that for soil N, so that soil C/N ratios decrease and C becomes limiting for microbial biomass as desertification proceeds (Gallardo and Schlesinger 1992, 1995; Kieft 1994).

Completing the nutrient cycle, the release, or "mineralization," of N from soil organic materials is closely tied to fluctuations in soil moisture. Fisher et al. (1987) found that small, frequent experimental applications of simulated precipitation (6 mm/week) caused greater rates of N mineralization than a larger, infrequent event (25 mm/month), followed by periods of drought. The mineralization of NH_4^+ proceeds even at relatively low soil water potentials, presumably due to fungi and bacteria whose activity may extend to soil water potentials of −6 to −8 MPa (Whitford 1989). Nitrifying bacteria are particularly sensitive to drought (Wetselaar 1968), so the concentration of soil NH_4^+ builds up during the dry season with rapid conversion to NO^-_3 when the summer rains begin (Fisher and Whitford 1995; Mazzarino et al. 1998; Reynolds et al. 1999a; Hartley and Schlesinger 2000). In wet soils, nitrifying bacteria can deplete the pool of NH^+_4 more rapidly than it is restored by mineralization, so the nitrification rate, as measured by NO production, declines with repeated watering (Hartley and Schlesinger 2000).

N mineralization, approximated from estimates of plant uptake, ranges from 28 to 64 kg N/ha/yr in creosotebush habitats (Whitford and Parker 1989) where net primary production (NPP) averages about 139 g/m^2/yr (chapter 11). This estimate of N mineralization is considerably higher than 6.5 kg N/ha/yr reported for a sparsely vegetated creosotebush desert in Nevada, where NPP ranges from 20 to 60 g/m^2/yr (Rundel and Gibson 1996). Our estimate of N mineralization is much lower than 149 kg N/ha/yr reported for habitats dominated by creosotebush in Argentina (Mazzarino et al. 1991).

Nutrient Budgets for the Jornada Basin

Comparative N budgets for representative plant communities in the Jornada Basin are shown in figure 6-6. Preparation of these budgets is complicated by the different scales of spatial heterogeneity in biomass and soil characteristics in each community and the need to extrapolate to landscape-scale values using assumptions about the relative cover of vegetation and bare ground. The data in these diagrams are compiled from many sources and constrained to yield a mass balance for the annual intrasystem cycle of the plant community. These should be regarded as rough estimates of nutrient cycling in the Jornada Basin and an impetus for future work.

In all habitats, the pool of N in vegetation is dwarfed by that held in the upper meter of the soil profile with percentages in vegetation ranging from 0.8% in grassland to 2.3% in mesquite shrublands. Despite inputs from N fixation, the soil N pool under mesquite (3,760 kg N/ha) is less than that in grassland (4,260 kg N/ha), whereas that in creosotebush is slightly higher (5,000 kg N/ha). In grassland and creosotebush habitats, the intrasystem cycling of nutrients dominates over new inputs of nutrients from outside the system. For instance, the internal cycle in grasslands provides about 50 kg N/ha/yr for plant uptake versus new inputs of < 3 kg N/ha/yr. In contrast, in mesquite the input of N from N fixation provides 48% of the N uptake (Lajtha and Schlesinger 1986). If estimates of N uptake by plants are used as a proxy for mineralization rates, grasslands have a mineralization rate that is roughly double that of the shrubland habitats that have replaced them over much of the Jornada Basin. Mesquite, through its reliance on N fixation, is less dependent on N derived from soil microbial mineralization and shows the lowest rate of N uptake from the soil.

The rapid rate of N mineralization in grasslands is associated with a higher rate of biomass turnover in grasslands than in the shrubland habitats. Overall, the biomass turnover is quite rapid in Chihuahuan Desert ecosystems, with mesquite having the longest mean residence time for N in biomass—about two years. In all habitats, turnover of N in roots is greater than or equal to that of aboveground components, and the percentage turnover of N in roots is greatest in grasslands, where the mean residence time for N in roots is < 1 year. In each habitat, only a small fraction of the total annual demand for N is satisfied by internal plant cycling (i.e., nutrient resorption) which increases nutrient-use efficiency.

The budgets presented in figure 6-6 reflect the local geomorphology that typically characterizes each community. For example, creosotebush tends to dominate coarse soils on the bajada slopes of the Jornada Basin, so estimates of the runoff losses of N are derived from field studies in plots located on those slopes (Schlesinger et al. 2000). Runoff losses are also included for adjacent plots in grassland, but many of the grasslands in the Jornada Basin are found on relatively flat ground with little long-distance overland flow. The budget for mesquite does not include a runoff component because mesquite dunelands normally show only a local redistribution of materials between shrub mounds and the local shrub interspace (Parsons et al. 2003). Eolian losses of N are shown only for mesquite, which is

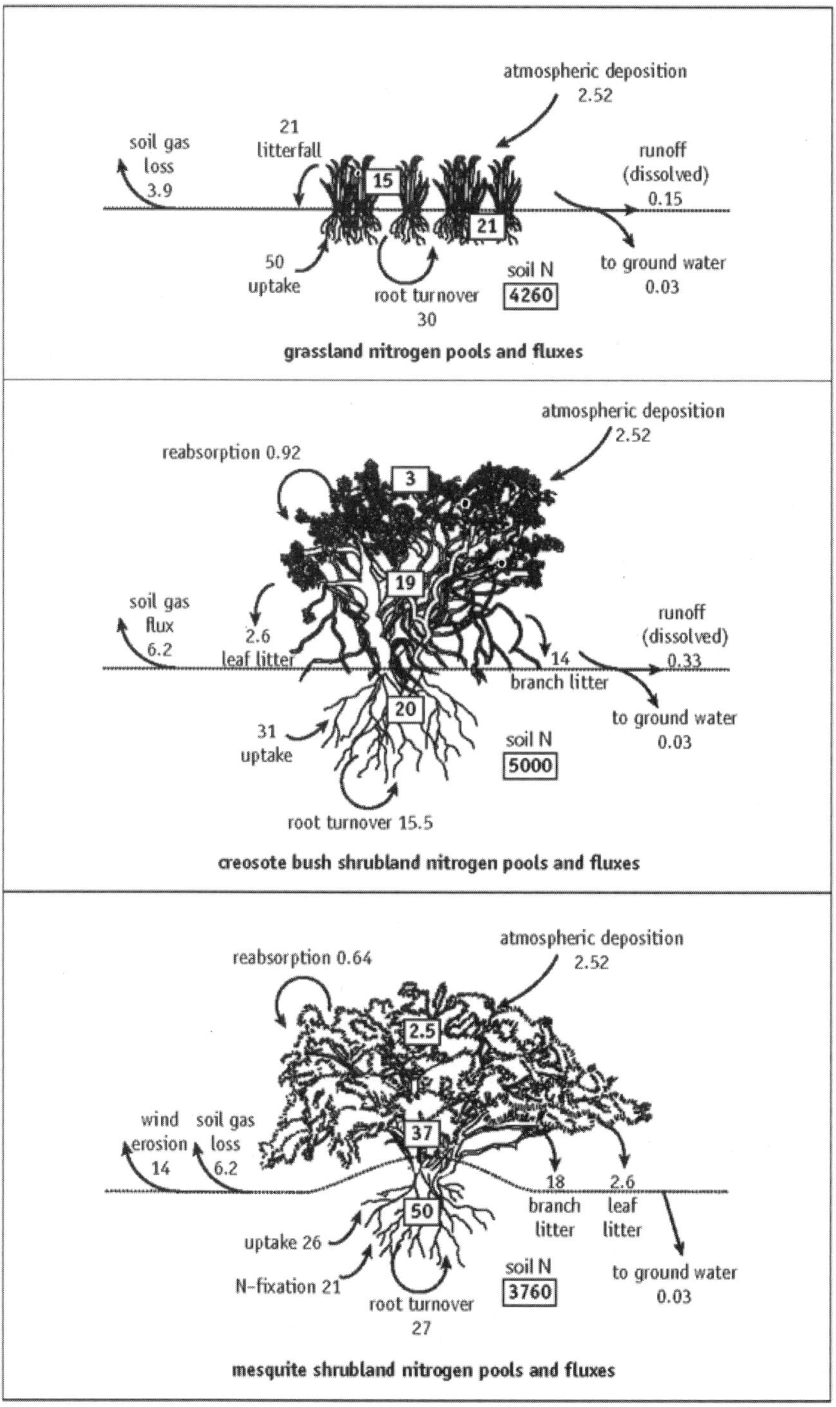

Figure 6-6. Nutrient budgets (pools within boxes in kg/ha; fluxes associated with arrows in kg/ha/yr) for grassland, creosotebush shrubland, and mesquite shrubland, as synthesized from the data in this chapter and other sources.

the only major habitat showing net removal of materials by wind in the Jornada Basin (chapter 9).

Given similar topography, there is greater N loss through runoff in shrub-invaded systems (Schlesinger et al. 2000). These systems are more susceptible to runoff because there is less total plant cover to intercept rainfall and to prevent erosion and scouring by wind and water. With larger plants and gaps between plants, flow paths are connected over longer distances (Howes and Abrahams 2003). There are also greater gaseous losses of N in shrublands through higher denitrification rates (Peterjohn and Schlesinger 1991; Hartley 1997). Thus our comparison of the N cycle in shrubland and grassland communities shows that areas invaded by shrubs have greater N losses even though shrubland landscapes appear to sequester equivalent or greater amounts of soil N (Jackson et al. 2002). Desertified areas may be more sensitive to further disturbance and nutrient loss because the ability of shrubs to conserve nutrients is reduced compared to that of the uninvaded grassland.

Desertification of Grasslands and Remediation of Shrublands

Redistribution of Soil Nutrients with Desertification

In southern New Mexico, desertification is associated with the loss of grassland, dominated by black grama, and the invasion of desert shrubs, primarily mesquite and creosotebush (Buffington and Herbel 1965; see also chapter 10). Huenneke et al. (2002) report similar levels of NPP in grassland and shrubland communities but much higher spatial variation in the distribution of NPP and biomass in shrublands (chapter 11). On a landscape-scale, the mass of soil nutrients in shrublands is similar to or higher than that in grasslands (Kieft et al. 1998; Cross and Schlesinger 1999; Hibbard et al. 2001; Jackson et al. 2002), but the spatial distribution of the soil nutrients contrasts strongly between these communities. Table 6-1 shows only a minor (4%) change in the estimated N pool of the Jornada Basin ecosystem between 1858 and 1963. Thus, desertification is not so much associated with a loss of biotic productivity as with the redistribution of soil resources on the landscape (Schlesinger et al. 1990, 1996) that increases the scale of patchiness (Hook et al. 1991; Tongway and Ludwig 1994). Lower NPP is the expected and traditional outcome of arid land degradation, but changes in the spatial distribution of soil resources may be a more effective index of desertification.

Resource islands develop as a function of shrub age (Facelli and Brock 2000; Shachak and Lovett 1998). Large shrub mounds are partly erosional and partly depositional features (Abrahams and Parsons 1991; Abrahams et al. 1995). In the Jornada Basin, ^{137}Cs profiles reveal that soil has accumulated under mesquite shrubs and been removed from interdune areas. In grasslands, the ^{137}Cs data show much less soil redistribution (Ritchie et al. 2003). In an area recently invaded by shrubs in southeastern Arizona, the soil mound contained buried remnants of the surface (A) horizon of the former grassland soil and between the shrubs, the A horizon had been eroded (Parsons et al. 1992). Rainsplash, mediated by the dis-

Table 6-1. Net nitrogen balance at the JER by vegetation type, 1858–1963

Vegetation Type	Biomass Nitrogen (kg N/ha)	Soil Nitrogen (kg N/ha)[a]	Total Ecosystem Nitrogen (kg N/ha)	Areal Extent (ha)[b]		Regional N Pool (10^8 kg)	
				1858	1963	1858	1963
Grassland	36	4260	4300	33,800	0	1.45	0
Mesquite shrubland	90	3760	3850	15,500	37,800	0.6	1.45
Creosotebush shrubland	42	5000	5040	400	7500	0.02	0.38
Tarbush shrubland	82	7870	7950	8700	13,100	0.69	1.04
Total				58,400	58,400	2.76	2.87

[a] To 1 m depth.
[b] from Buffington and Herbel (1965).

sipation of raindrop energy in the shrub canopy, results in a net transport of soil fines from interspaces to shrub mounds (Parsons et al. 1992). Wind erosion also redistributes soil materials across the landscape where they are caught by shrub canopies and accumulate in the soil mounds beneath shrubs (Coppinger et al. 1991). Digging by rodents, especially kangaroo rats (*Dipodomys*), redistributes soil materials in desert landscapes, leading to patches of fertility that may become preferred sites for the establishment of annual plants and shrub seedlings (Moorhead et al. 1988; Mun and Whitford 1990; Chew and Whitford 1992; Whitford 1993; Ayarbe and Kieft 2000). The capture of soil particles adds C and N to the soil mounds that develop beneath desert shrubs (Shachak and Lovett 1998).

Once established, shrubs further enrich the nutrient content of soils beneath their canopy through autogenic, biological processes that help ensure the persistence and regeneration of the shrub ecosystem (Schlesinger et al. 1990). Biotic processes leading to the development of islands of fertility include plant uptake of essential nutrients from the soils of the interspace followed by the deposition of litter in the localized areas beneath shrubs. Thus shrubs appear to "mine" nutrients from the soils of the interspace (Garner and Steinberger 1989), which may support a cryptobiotic crust of algae, fungi, and soil bacteria that fix N (West 1990; Evans and Johansen 1999). Shrubs, such as acacia (*Acacia* ssp.) and mesquite, which maintain symbiotic, N-fixing bacteria in their rooting system, directly contribute to the accumulation of N beneath their canopy (Garcia-Moya and McKell 1970; Gerakis and Tsangarakis 1970; Tiedemann and Klemmedson 1973; Virginia and Jarrell 1983; Lajtha and Schlesinger 1986; Wright and Honea 1986). Despite occasional observations of hydrophobic layers in the soils under shrubs, including creosotebush (Adams et al. 1970), infiltration rates are typically higher under desert shrubs as a result of better soil crumb structure and a lower impact energy of raindrops (Lyford and Qashu 1969; Bach et al. 1986; Rostagno 1989;

Shachak and Lovett 1998; Schlesinger et al. 1999; Wainwright et al. 1999a, 2000; see also chapter 5). A number of desert shrubs also funnel nutrient-rich stem flow waters to the soil beneath their canopy (Navar and Bryan 1990; Mauchamp and Janeau 1993; Martinez-Meza and Whitford 1996; Whitford et al. 1997; see also chapter 8).

In the Jornada Basin, extractable N, P, and K are strongly concentrated under creosotebush shrubs, whereas Na, Li, and Sr are more concentrated in the soils between shrubs, suggesting that physical processes, such as runoff, lead to localized accumulations of some nonessential elements in intershrub spaces and biological processes are more important in concentrating biologically essential elements under shrubs (Schlesinger et al. 1996). In similar Chihuahuan Desert habitats at the Sevilleta National Wildlife Refuge, 300 km north of the Jornada Basin, Cross and Schlesinger (1999) found no spatial variations in the concentration of various soil nutrients in grassland habitats but strong accumulations of N, P, and K under creosote bushes (figure 6-7). Frequency distributions for the occurrence of soil nutrients at the Sevilleta show that N and P are concentrated in shrub islands at levels above those found in adjacent grasslands so that the islands of fertility are not simply a remnant left after erosion (see Kieft et al. 1998).

Concentrations of nutrients in the islands of fertility are greatest at the soil surface and attenuate with depth (Nishita and Haug 1973; Charley and West 1975; West and Klemmedson 1978; Rostagno et al. 1991). High surface concentrations of K have been attributed to the deposition of illite in eolian materials (Singer 1989), but surface accumulations of K are undoubtedly enhanced by nutrient cycling and mineral weathering under shrubs (Rostagno et al. 1991; Kelly et al. 1998). In contrast, the concentrations of nonessential ions (e.g., Na and Cl) or nonlimiting elements (e.g., Ca and ${SO^{2-}}_4$) tend to increase with depth due to periodic leaching (Yaalon 1965; Schlesinger et al. 1989). The depth to the peak concentration of various solutes in desert soils follows the global pattern reported by Jobbágy and Jackson (2001) from shallowest to deepest in the order $P < K < Ca < Mg < SO_4 = Na = Cl$. This vertical pattern mirrors the horizontal pattern extending from shrubs to the shrub interspaces (Schlesinger et al. 1996).

When shrubs are removed by cutting, herbicides, or fire, N, P, and other soil nutrients are lost from former islands of fertility. Elimination of the local biogeochemical cycle associated with shrubs allows physical processes to disperse soil nutrients across the landscape. The redistribution of N seems to be more rapid than that of P, likely due to more rapid gaseous and soluble losses of N and retention of P by adsorption to soil minerals. Thirteen years after the removal of mesquite there was a significant loss of soil N from former shrub islands, but there were no significant changes in P or S over the same period (Tiedemann and Klemmedson 1986). Similarly, in Australia, Facelli and Brock (2000) found that P-rich spots persisted for 50 years after the death of Western myall (*Acacia papyrocarpa*), but N was lost rapidly from former shrub islands. At the JER, Virginia (unpublished data) observed a degradation of the N pool in shrub islands within 15 years after spraying mesquite with herbicides.

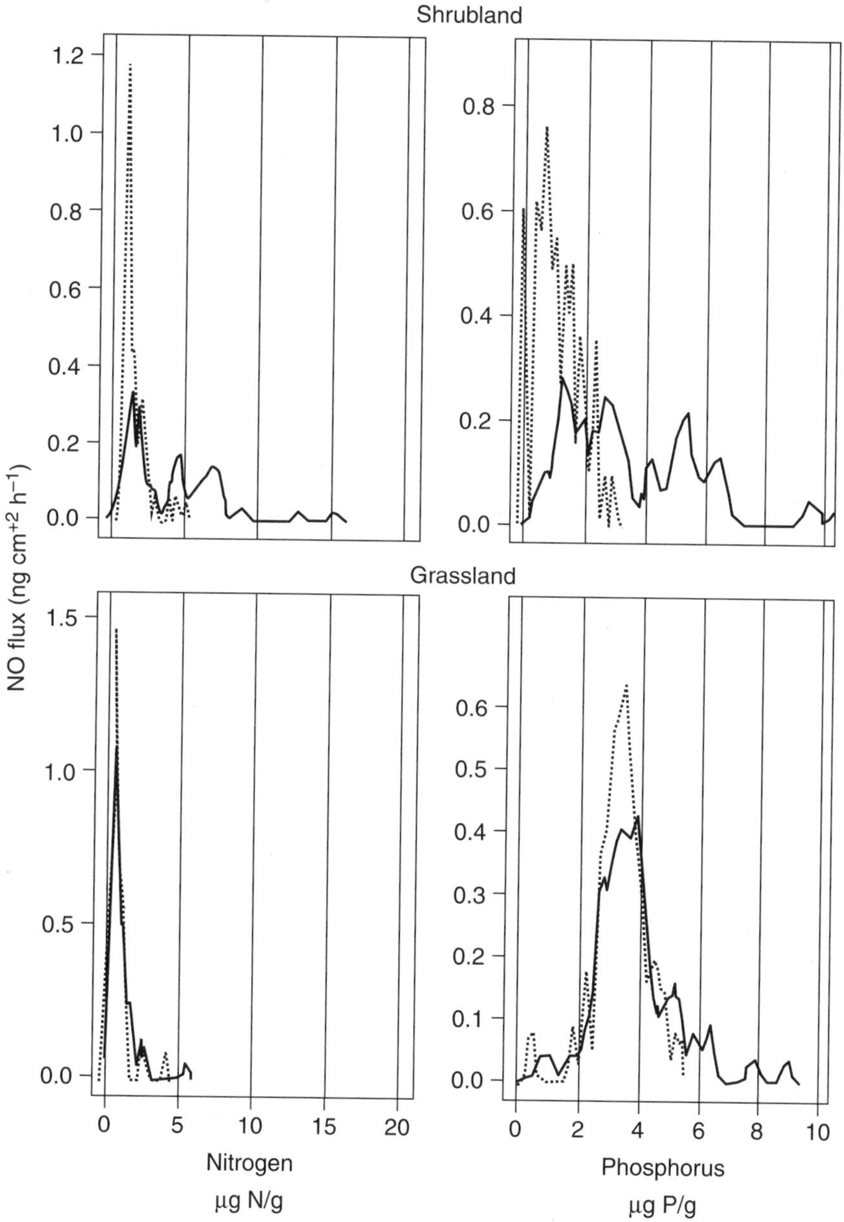

Figure 6-7. Frequency histogram for the concentration of available N and water-soluble P in soils from adjacent grassland and shrubland sites at the Sevilleta National Wildlife Refuge, New Mexico. In each graph, the solid line is for samples taken under vegetation and the dashed line is for samples taken between vegetation (from Cross and Schlesinger 1999).

Recovery of desert shrub vegetation on cleared areas is most rapid when the original soil conditions, such as the islands of fertility, remain intact. Wallace et al. (1980) found more than twice as much shrub biomass regenerated on bare, undisturbed desert soils compared to plowed, disked, or scraped soils after 20 years of plant succession in the Mojave Desert. When land managers wish to reestablish shrub-dominated vegetation on soils that have been homogenized by human activities, such as cultivation or construction, they must consider creating heterogeneity in soils by artificial means (Boeken and Shachak 1994; Shachak and Lovett 1998). Similarly, conversion of shrubland to grassland involves consideration of redistributing resources into finer scale patchiness. Otherwise, transitions of shrub-dominated states on these lands can be extremely slow (e.g., Carpenter et al. 1986; McAuliffe 1988; see also chapter 14).

Conclusions

In the Chihuahuan Desert, water and N limit net primary production. Water is delivered in pulses, interspersed with dry periods, which uncouple soil nutrient cycling processes from plant uptake and allow more opportunity for redistribution or export of nutrients. Water, wind, and animals transport nutrients, creating strong spatial patterns of resources, influenced by the interactions of climate, topography, soils, and vegetation. In desert shrublands, redistribution of resources creates distinct islands of fertility around shrubs, separated by depleted interspaces.

N inputs in the JER are relatively small compared to those in other deserts. Biological soil crusts are poorly developed, but symbiotic N fixation is significant in mesquite shrublands. Outputs of N are relatively large, especially via wind and water erosion from shrublands. Leaching below the root zone is infrequent and much less important than gaseous losses. Plant uptake of nutrients is enhanced by mycorrhizae and fungal endophytes and controlled by temperature and water availability. Decomposition is rapid, but not strongly related to water availability or C/N ratio of the substrate. Termites are especially important in speeding the decomposition of surface litter and some photo-oxidation also occurs. Decomposition of subsurface litter is mainly by fungi and bacteria, mediated by microathropods, and are more strongly influenced by soil moisture.

In grasslands, N cycling is relatively faster with lower outputs and proportionally greater internal fluxes between plants and soil. In contrast, shrublands have greater aboveground biomass, deeper roots, smaller internal fluxes, and greater outputs. Wind is more important in exporting N from sandy mesquite shrubland, whereas water is more important in exporting N from upslope creosotebush shrubland. N fixation by mesquite reduces dependence on soil N and allows increased N concentration in biomass, providing more opportunity for N accumulation under the shrub and more N export in litter and soil.

Conversion of grassland to shrubland in the northern Chihuahuan Desert is accompanied by redistribution of nutrients at the scale of the plant/interspace, which often promotes increased output of nutrients at the patch scale and sometimes, longer distance transport of nutrients at the landscape scale (chapters 7 and 9). Restoration or remediation of landscapes in the Chihuahuan Desert (Chapter 14) could be improved by understanding, accommodating, and possibly managing the redistribution of soil resources.

7

Biogeochemical Fluxes across Piedmont Slopes of the Jornada Basin

Athol D. Abrahams
Melissa Neave
William H. Schlesinger
John Wainwright
David A. Howes
Anthony J. Parsons

This chapter is an overview of recent studies of the movement of water, sediment, and nutrients across a principle piedmont slope, or bajada, of the Jornada Basin. Bajadas are extensive, gently sloping surfaces formed by the coalescence of alluvial fans and are a major landscape component of the basin and range province (see chapter 2, figures 2-3 and 2-6). Over the past four decades a considerable body of research has elucidated the form and function of alluvial fans (Bull 1977; Blair and McPherson 1994; Harvey 1997), but less attention has been paid to bajadas. In particular, the bajadas most neglected are those where channels converge and diverge at irregular intervals downslope. This type of bajada is found at the base of Summerford Mountain, the northernmost peak of the Doña Ana Mountains on the western edge of the Jornada Basin (see chapter 2, figure 2-5b). For convenience, this bajada is hereafter referred to as the Summerford bajada. The research has involved rainfall simulation experiments on small plots, monitoring of two small watersheds on this bajada, and computer modeling of the processes operating in these watersheds and over the bajada as a whole. A detailed understanding of the hydrology and hydraulics of overland flow on this bajada requires a numerical model of the rainfall-runoff process. The objective of this chapter is to detail the model and draw conclusions from model simulations about hydrologic transports of sediment and nutrients across this bajada. Because these piedmonts are important surfaces in this desert (chapter 2) an understanding of their hydrologic and biogeochemical dynamics is crucial to understanding landscape dynamics in the basin and throughout arid regions.

The Summerford Bajada

Geology and Soils

Summerford Mountain is a steep-sided, rocky inselberg (i.e., isolated mountain) that rises 380 m above the surrounding bajada to an elevation of 1,780 m. The mountain is composed of monzonite porphyry of Oligocene age (Seager et al. 1976) and has a fringing bajada on its northern and eastern sides. This study focuses on the bajada to the east, which extends 2.5 km to the basin floor (figure 7-1) at an average gradient of 4%. The soils on the upper bajada at the base of Summerford Mountain are Mollisols or, more specifically, Torriorthentic Haplustolls (chapter 4). The surface horizon is a sandy loam (79% sand, 13% silt, 8% clay) with an abundance (> 30%) of fine to medium quartz pebbles weathered from the monzonite. Downslope a calcite-rich paleosol is exposed over a limited area. This soil is classified as a Typic Haplocalcid. The caliche horizon is at or just below the surface, so infiltration rates are low and shrubs are small. Over the remainder of the bajada the soils are Typic Haplargids. The surface horizon has a sandy loam or loamy sand texture (70–85% sand, 10–20% silt, 5–10% clay) and contains variable amounts of gravel, which become concentrated on the surface as an erosional lag. The gravel cover increases from 5% to > 80% from northwest to southeast across the bajada (figure 7-1). The Typic Haplargids become finer down the bajada with the surface horizon having a loamy texture (64% sand, 24% silt, and 11% clay) adjacent to the basin floor. Scattered across the

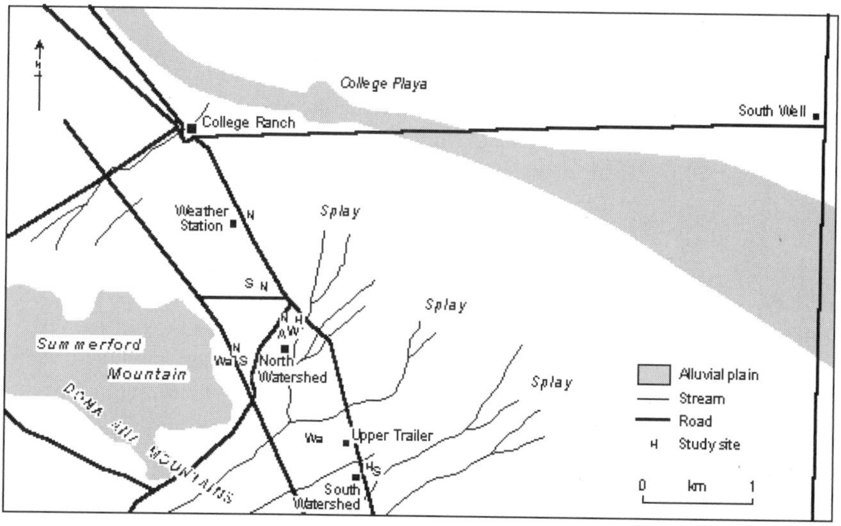

Figure 7-1. Map of the study area showing the locations of the field experiments reported by Abrahams et al. (2003) (A), Howes and Abrahams (2003) (H), Neave and Abrahams (2001, 2002) (N), Schlesinger et al. (1999, 2000) (S), Wainwright et al. (1999a) (W), and T. J. Ward (Wa).

lower bajada are large poorly defined areas of undifferentiated sandy sediments or Entisols that have been deposited by the sandy washes splaying out over the surface.

Vegetation

Vegetation surveys (chapter 10, figures 10-1 and 10-2) indicate that prior to 1915 the Summerford bajada was covered with black grama (*Bouteloua eriopoda*) grasses (Buffington and Herbel 1965). Today, a grassland community is found only on the Mollisols at the base of Summerford Mountain. The community is dominated by bunch grasses, notably black grama, threeawn (*Aristida*) species, and Lehmann lovegrass (*Eragrostis lehmanniana*), an exotic introduced to control erosion along the power line that crosses the bajada. In the middle of the bajada a shrubland community dominated by creosotebush (*Larrea tridentata*) is well established (Stein and Ludwig 1979). Creosotebush extends to the bottom of the bajada, where tarbush (*Flourensia cernua*) replaces it as the dominant shrub on the finer-textured soils. North and east of Summerford Mountain, degraded grassland covers the lower part of the bajada. In the grasslands about 40% of the ground surface is bare, but this percentage is closer to 70% in the degraded grassland, which is believed to represent the transition from grassland to shrubland. Both the creosotebush shrubland and the degraded grassland are underlain by soils with a subsurface argillic horizon. The presence of all three vegetation communities in close proximity to one another was a major reason for selecting this particular area for study.

Surface Crusts and Animal Disturbance

The Summerford bajada is characterized by the widespread development of surface crusts. These crusts consist of a thin surface layer 1–3 mm thick that is both stronger and less permeable than the underlying soil. The crusts may be either physical or biological in character. Most biological crusts in the study area are due to the presence of filamentous cyanobacteria. Crusts of this type cover much of the ground surface in the shrubland and degraded grassland but are patchier in the grassland. They are best developed where the soil surface is relatively moist, such as under shrubs and grasses, though their distribution is quite uneven. They are least developed where erosion rates are relatively high, such as in rills.

In contrast to biological crusts, physical crusts form when raindrop impact blocks soil pores by compressing the soil surface and/or by entraining clays that then get carried into the pores by infiltrating water (McIntyre 1958; Moore 1981; Kidron et al. 1999). Physical crusts often develop where biological crusts have been disrupted. Cyanobacteria may recolonize a disturbed area within two years and reach their predisturbance biomass within five years (Kidron personal communication). In the meantime, a physical crust may form during a single storm where the soil is suitable. Poesen (1992) showed that the optimal soil textures for the development of physical crusts are sandy loams and loamy sands. These textures are found in the surface soils over most of the bajada. As a result, physical crusts develop readily in these soils and have a major impact on surface runoff.

Another distinctive feature of the Summerford bajada is the degree to which the ground is disturbed by faunal activity. Small mammals are largely responsible for this disturbance. These animals dig and scratch the soil in search of food (plant stems, roots, seeds, and insects). The resulting holes are typically 20–30 mm deep but may have depths of 200 mm. In addition, there are burrows of indeterminate depth and extent. Following a storm, the animals dig up the surface until the next storm obliterates their diggings. Inasmuch as digging disrupts the surface crust (both physical and biological) and scatters loose sediment over the ground surface, it might be expected to have a profound effect on the movement of water and materials across the bajada, particularly in the degraded grassland and shrubland (Neave and Abrahams 2001).

Drainage

Runoff on the bajada surface arises both from infiltration-excess rainfall on the bajada itself and from sand-bedded streams issuing from Summerford Mountain that debouch onto the head of the bajada. All sand-bedded streams eventually splay out over the bajada surface, and their flow becomes dispersed. The smaller of these streams terminate within the grassland community at the base of Summerford Mountain, whereas the larger ones survive to the lower bajada. The latter streams have alternating single-channel and multichannel reaches that are equivalent to Bull's (1997) "discontinuous ephemeral streams." Along the channel margins are frequent mesquite (*Prosopis glandulosa*) and white thorn (*Acacia constricta*) shrubs. The splays of these channels are typically grassy and elevated a meter or so above the bajada surface.

Infiltration-excess runoff is highest in the shrubland and concentrates downslope to form an extensive network of rills. These rills generally head within 5 m of the local divide, and like the larger sandy washes, they have alternating single- and multichannel reaches. Elsewhere, the multichannel reaches are replaced by reaches with no clearly defined channelized flow that are herein called beads because they are reminiscent of beads spaced out along a necklace. At their upstream end they are similar to the splays of the larger sand-bed streams. However, they differ at their downstream end, where their flow coalesces into a single channel. Although the sequences of beads and single-channel reaches tempts one to think of them as coherent hydrological units, the functional relationship between a bead and the downstream channel is unclear. An alternative view is that a bead is equivalent to a sand splay on one of the larger sand-bed streams and that the single-channel reach at the bead's downstream end is a hydrologically independent unit that forms as a result of the concentration of infiltration-excess runoff downslope of the bead. The single-channel reaches are typically less than 2 m wide and incised into the bajada surface by as much as 0.8 m. Apart from their uppermost reaches, these reaches are characterized by sandy beds.

In the grassland, the rills are smaller (typically 0.3 m wide) and more widely spaced (about 20 m) than in the shrubland, and they do not form an integrated network. Instead, individual rills extend downslope for 15–20 m before they dissipate, choked by grass and deposited sediment. In the degraded grassland, rills are rare and virtually all drainage is inter-rill.

Ground Surface

In inter-rill areas, the nature of the ground surface is strongly influenced by the vegetation. Thus in the grassland the ground surface is characterized by large clumps of grass, which give rise to a pronounced microtopography with an amplitude of about 0.2 m and a wavelength of about 0.5 m. The degraded grassland, which is found on the gentler slopes of the lower bajada, has smaller and more widely spaced clumps of grass and an almost planar ground surface between the clumps. As a result, the microtopography has amplitude of no more than 0.1 m and a wavelength of 0.25–0.5 m. Surface runoff in the grassland and degraded grassland is clearly dispersed by the grass. Although there are obvious threads of flow and features that resemble small rills (termed "prerills" by Roels 1984), they rarely persist over distances greater than 5 m. In the degraded grassland, the microtopography is subdued and threads of flow are difficult to follow.

In contrast, in the shrubland the inter-rill areas are characterized by broad, shallow swales stretching between mounds topped by shrubs. The microtopography has amplitude of about 0.10 m and a wavelength ranging from 1.5 to 3 m. The mounds grow, at least initially, as fine sediments are deposited beneath shrub canopies by eolian and/or rainsplash processes (Carson and Kirkby 1972; Parsons et al. 1992). Once developed, the mounds divert surface runoff into intershrub areas where it may erode the surface sufficiently to form a rill. Where there is a significant proportion of gravel in the surface soil, a gravel lag accumulates in intershrub areas and impedes erosion in general and rill formation in particular.

Modeling Runoff Events

Howes and Abrahams (2003) developed an event-based, two-dimensional runoff model to simulate overland flow within two small shrubland watersheds—north watershed (figure 7-2a) and south watershed (figure 7-2b)—and made detailed field surveys of these watersheds using a total station. The principal difference between the watersheds is that south watershed is incised about 1 m below the general surface of the bajada and has a surface gravel cover that averages about 30%. North watershed is not incised and is largely free of surface gravel. Both watersheds have gradients of about 4%. In the model, grids of 1 m $\times$ 1 m cells, in which each cell is classified as either intershrub or shrub, represent the watersheds. The rills are treated as intershrub surfaces. Flow is routed from cell to cell by numerically solving the two-dimensional kinematic wave equation using a predictor-corrector, finite-difference scheme (Davis 1988). The flow velocity is computed using the Darcy-Weisbach flow equation, and infiltration is computed using the Smith-Parlange equation (Smith and Parlange 1978). Parameter values

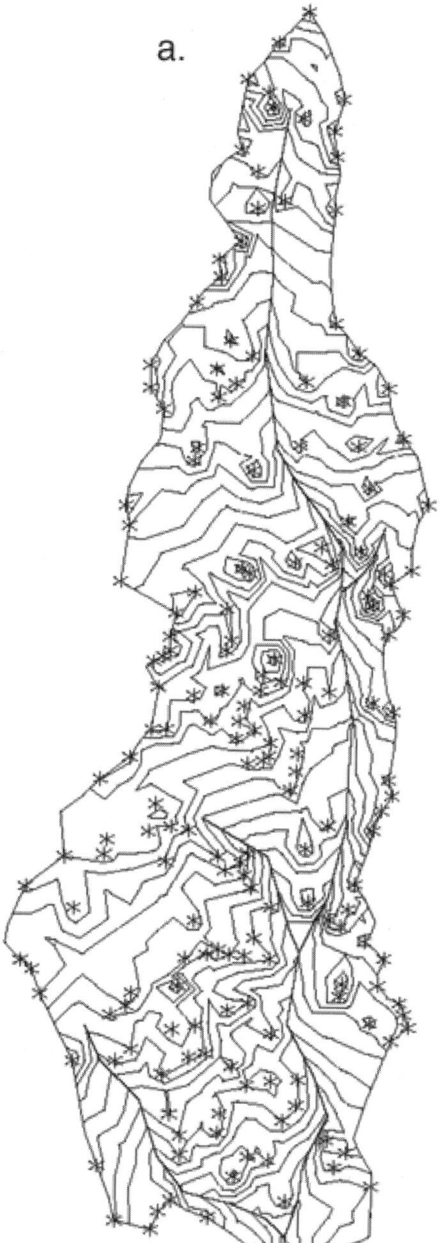

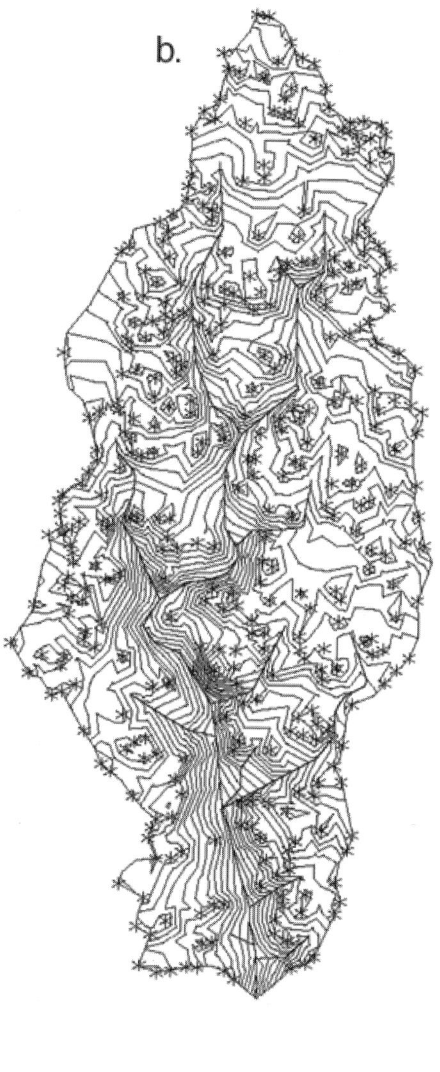

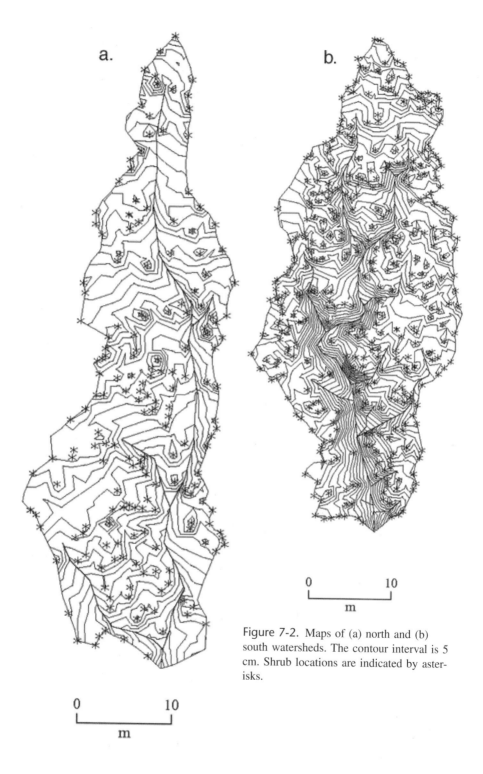

Figure 7-2. Maps of (a) north and (b) south watersheds. The contour interval is 5 cm. Shrub locations are indicated by asterisks.

were obtained for the shrub and intershrub surfaces in each watershed from rainfall simulation experiments and a field survey. Rainfall data were provided by tipping bucket rain gauges located adjacent to the outlets of the watersheds where flow discharge was recorded at calibrated supercritical flumes. Such flumes are designed to accelerate the flow as it passes through, thereby ensuring that there is no deposition on the floor of the flume that might corrupt the stage-discharge rating curve.

Figure 7-3 shows simulated and observed hydrographs for the two watersheds for a storm on July 30, 1997. In the initial simulations, the model underpredicted the runoff from both watersheds as a result of overpredicting the saturated hydraulic conductivity K_s. This overprediction is presumed to be due to the surface crust in intershrub areas being better developed at the time of the storm than at the time of the rainfall simulations, that is, late June when the crust was degraded by animal activity. Model performance was improved by reducing the intershrub K_s values to account for the crusting (Moore 1981; Bosch and Onstad 1988; Rawls et al. 1990). The second pair of simulated hydrographs shown in figure 7-3 was generated by reducing the intershrub K_s from 58.10 mm/h to 34.00 mm/h for the north watershed and from 22.63 mm/h to 16.00 mm/h for the south watershed. The relative reduction in K_s is less for south watershed because gravel on intershrub surfaces promotes runoff and at the same time renders the soil less susceptible to crusting than the north watershed intershrub surface that is free of gravel. Given that the formation of a physical crust has the potential to alter the infiltration characteristics of a watershed from storm to storm, more work is required to develop a procedure for estimating the development and destruction of surface crusts and their effect on K_s for individual rainfall events.

The 2D model provides an investigative tool that can be used to study the hydrologic processes operating in shrubland ecosystems. For example, field studies have indicated that lateral movements of water and nutrients are important in desert ecosystem function (Schlesinger and Jones 1984; Noy-Meir 1985), but the lack of a detailed runoff model has meant that a quantitative investigation of lateral movement of water and nutrients at the scale of the individual shrub has not been possible. Howes and Abrahams (2003) demonstrated how their 2D model could be used to model the lateral movement of water at this scale.

Using the 2D model, Howes and Abrahams (2003) conducted a study of the relative importance of run-in infiltration (i.e., infiltration of overland flow in a cell whose infiltration capacity has not been satisfied) and rainfall infiltration (i.e., infiltration of rain falling directly into a cell containing a shrub) in supplying water to shrubs. Wet and dry antecedent soil moisture conditions in each watershed were simulated for a range of rainfall conditions typical of the bajada. Model cells were assigned a value of K_s at random from a log normal distribution to represent spatial variability in infiltration. The mean and coefficient of variation of the distribution were based on the data from the rainfall simulation experiments that were conducted to parameterize the model. The simulation results were expressed in terms of the mean run-in percentage, which is the depth of run-in infiltration expressed as a percentage of the total depth of infiltration in a cell.

It was found that the mean run-in percentage for all shrub cells varied with

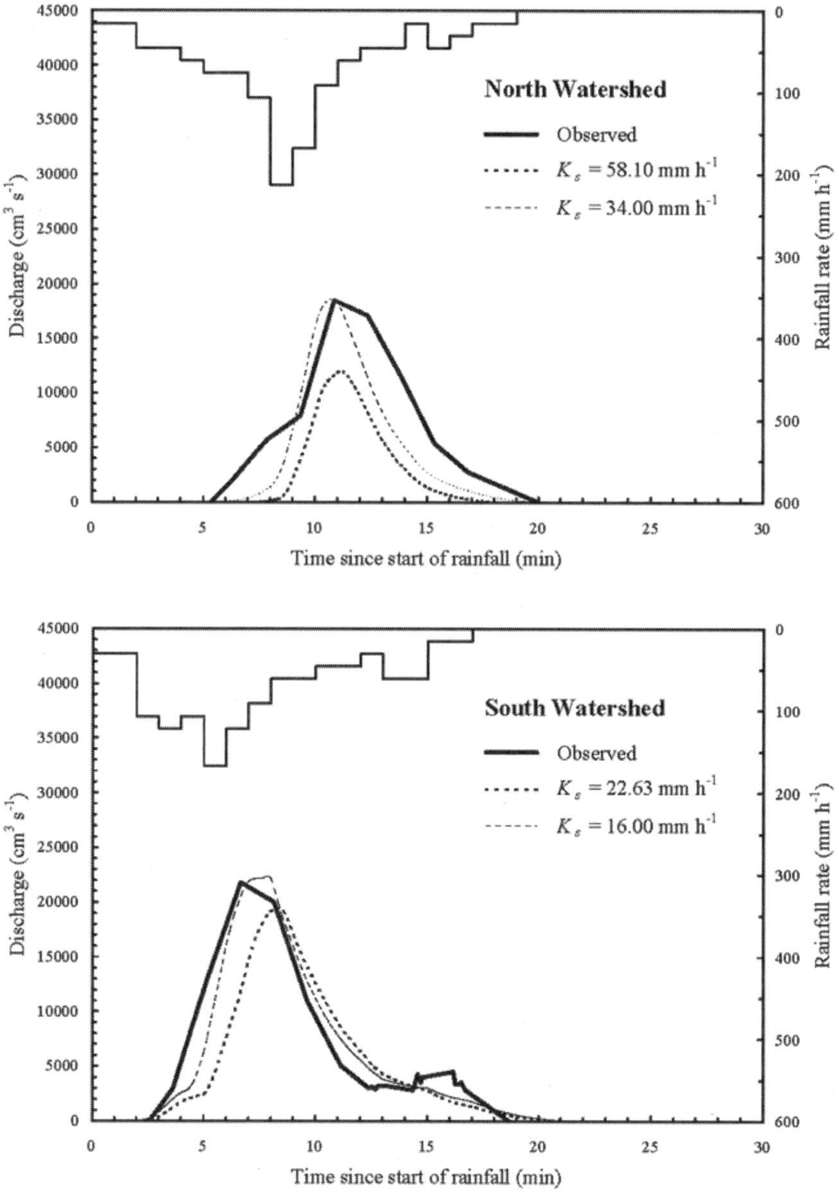

Figure 7-3. Simulated (*K* values) and observed (measured) hydrographs for the July 30, 1997, storm in (top) north watershed and (bottom) south watersheds.

mean rainfall rate in a similar manner in both watersheds (figure 7-4). The highest mean run-in percentages were associated with an initially wet soil, a temporally variable rainfall pattern, and low rainfall rate. Run-in infiltration was slightly more important in north watershed, being equal to 3–12% of the total infiltration, compared to 2–8% for south watershed. This difference was attributed to both the higher values of K_s and the lower shrub mounds in the north watershed (figure 7-2). The computed mean run-in percentages are sufficiently large to suggest (1) that lateral flows must be taken into account in studies of shrubland ecosystems and (2) that a much broader study of the role of run-in infiltration in supplying water to shrubs would be worthwhile.

Transmission Losses

The ability of ephemeral channels to carry water and sediment depends, among other things, on transmission losses through their beds. In our investigation of transmission losses in the discontinuous channels on the bajada surface, we have so far examined only the rills in the shrubland community. This examination has been conducted on both single-channel reaches and on a bead. Sedimentological analysis of the beds of single-channel shrubland rills and sandy washes reveal a

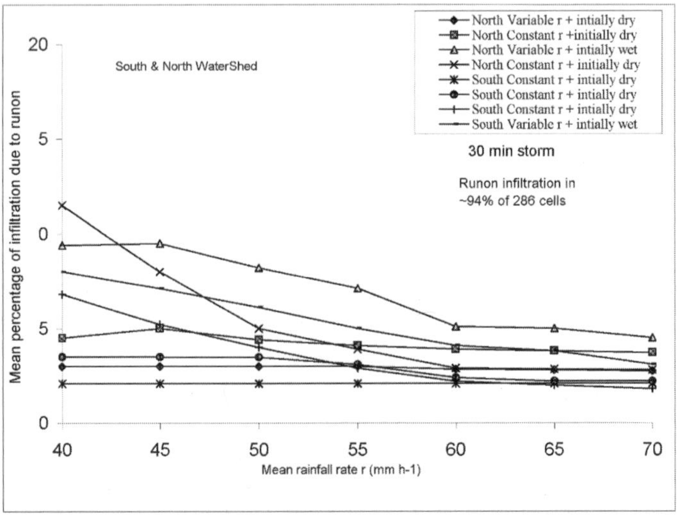

Figure 7-4. Graph of mean percentage of infiltration due to run-on against mean rainfall rate for all shrubs in north and south watersheds predicted by the 2D model. In the north watershed, run-on infiltration occurred in 98% of shrub cells and accounted for 3% to 12% of the total infiltration under shrubs. In the south watershed, run-on infiltration occurred in 94% of shrub cells and accounted for 2–8% of the total infiltration under shrubs.

similar structure. The top 30 cm consists of loose, coarse sand and gravel below that is a finer, more indurated layer. This sedimentological similarity suggests that the transmission losses of the sandy washes may not be dissimilar to those measured in the rills.

Rills

Experiments were conducted on 10 rill reaches (Parsons et al. 1999). Water simulating rill flow was introduced at the upper end of each reach, and the outflow from the reach was measured by taking timed volumetric samples. Transmission losses at equilibrium averaged 9.24 mm/min, which is about an order of magnitude greater than final infiltration rates measured in bare inter-rill areas between shrubs. This difference may be due to several factors, the most likely being that surface sealing occurs in inter-rill areas but not in the rills and hydraulic heads are greater in rill flow than in inter-rill flow.

Support for the notion that hydraulic head accounts for greater infiltration loss in the rills is obtained from estimates of downstream changes in discharge through the rill reaches. These estimates were made by measuring depth and width of flow at four cross-sections within the rill reaches and assuming uniform velocity throughout the reach. Figure 7-5 shows the pattern of discharges obtained from these measurements. The pattern is clearly nonlinear and shows that transmission loss decreases downhill as the hydraulic head decreases. This pattern implies that where rill flow extends beyond the spatial limits of the generating storm, most of the flow will be lost through the bed close to the edge of the storm.

Beads

The significance of beads for runoff from the bajada has been investigated using supercritical runoff flumes to monitor discharges in natural events. Three flumes were installed in north watershed and three in south watershed (figure 7-1). In south watershed, flumes were set up on two tributary rills, and a third flume was set up just downstream of their confluence. All rills consist of a single channel. The contributing areas for these flumes are 889.7, 431.3, and 1,833.9 m^2. In contrast, north watershed contains two tributary channels that feed into a bead. Flumes were installed on these rills, just upstream of the bead. A third flume was installed just downstream of the point where a clearly defined channel emerges from the bead. The contributing areas for these flumes are 775.2, 3,379.3, and 5,936.1 m^2. Serendipitously, the ratio of the sum of contributing areas of the two upstream flumes to the downstream flume is 1.4 in both cases. The flumes are equipped with stilling wells that record maximum depth of flow in a storm event. Using the ARS calibration for the flume design (Smith et al. 1981), these depths were converted to peak discharges.

Between January 26, 1995, and July 28, 1997, 49 rainfall events were recorded in north watershed and 45 in south watershed. Of the rainfall events in north watershed, runoff was recorded on 18 occasions in one or both of the upstream flumes and on 12 occasions in the downstream flume. For south watershed, the

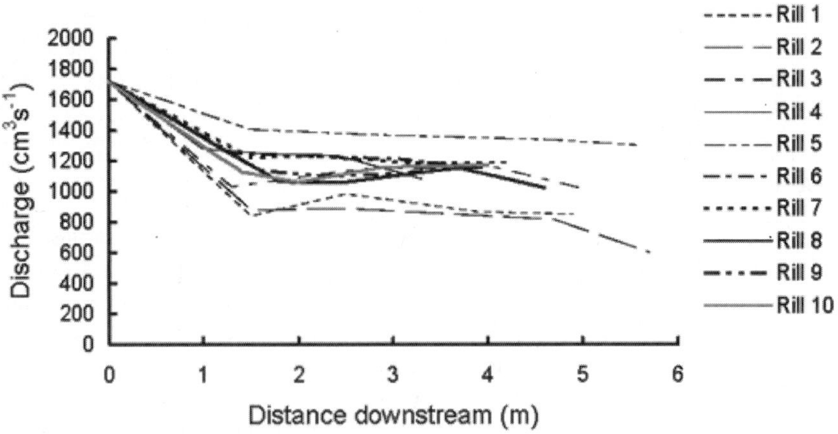

Figure 7-5. Rill discharge as a function of flow distance for 10 rills in the shrubland community of the Summerford bajada study area.

respective figures are 17 and 11 occasions. Whereas those events that were recorded upstream but not downstream in the south watershed are typically of very small magnitude (< 850 cm^3/s); in north watershed, the threshold is much greater ($> 26{,}200$ cm^3/s). This difference reflects the high infiltration capacity of the bead in north watershed.

Figures 7-6a and 7-6b show the relationships for south watershed and north watershed between peak discharge at the watershed outlet and the sum of the peak discharges from the two tributaries. Assuming that the peak discharge is a reasonable surrogate for total discharge, in both north and south watersheds the ratio of the output to the sum of the inputs might normally be expected to be approximately 1.4 (the ratio of the sum of the tributary areas to the entire watershed area). However, given the values for transmission losses just reported, a more reasonable ratio might lie between 1.1 and 0.7. For the single channels of south watershed, most ratios are within or above this range, and only at very small discharges is this not the case (figure 7-6a). Discounting these very small discharges, there is no clear relationship between the ratio of output to input and the input discharge. In contrast, in north watershed two-thirds of the observed ratios fall below 0.7, and there is a clear trend for the ratio to increase with discharge (figure 7-6b). This result suggests that the bead is absorbing water well in excess of the transmission losses in single channels. Thus beads appear to be important sinks for water on the bajada and, consequently, also for sediments and nutrients. The ecological importance of these sinks needs further investigation.

Synthesis: Modeling Bajada Hydrology

To extrapolate results from plot experiments to the overall bajada scale, a series of numerical simulations has been carried out. The model used is based on that

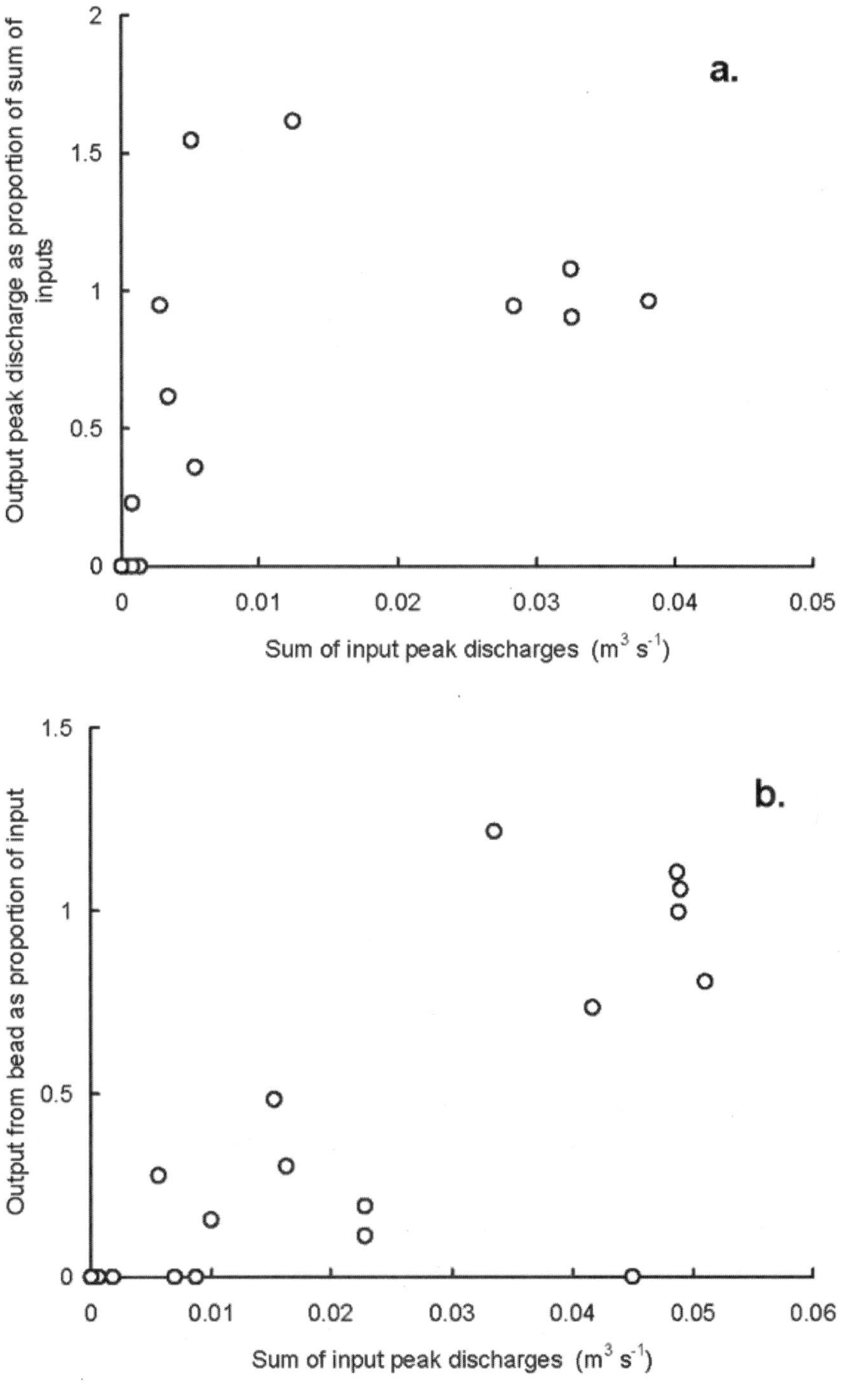

Figure 7-6. Comparison of peak inflow and output discharges for (a) south watershed and (b) north watershed for runoff events between January 16, 1995 and July 28, 1997.

developed by Scoging (1992) and Parsons et al. (1997). Overland flow is generated using a Hortonian infiltration-excess method, based on the Smith and Parlange (1978) model, to allow for moisture storage. This technique allows for run-in infiltration that affects the amount of moisture available to plants in the bajada soils. Flow routing uses the steepest descent method based on the topography, and flow rating is calculated using the Darcy-Weisbach friction factor.

Input data for the model are derived from a number of sources. The topography is derived from the 30-m digital elevation model (DEM) produced from the 1:24,000 maps from the U.S. Geological Survey (USGS). The model was run on the extract of the USGS 30-m DEM that covers the bajada area to the north and east of Summerford Mountain. Infiltration and friction factor parameters for the model were distributed according to the spatial distribution of the vegetation communities present on the bajada, using measurements derived from the field experiments already described. The presence of rills is accounted for by weighting the infiltration rates and friction factors according to the mean rill area in the cells belonging to different vegetation types. The infiltration-weighting factor is by the rate of transmission loss and the friction factor weighting according to measured values from rills at Walnut Gulch (Abrahams et al. 1996). The storm simulated occurred between 15:59 and 16:15 on July 30, 1997. The total rainfall at the north watershed was 19.6 mm, with 1-minute intensities ranging from 15 to 167.4 mm/h.

The results of the model simulation were different patterns in resultant soil moisture according to the different habitats. Summerford Mountain itself had the lowest soil moisture values because of rapid runoff. On the degraded grassland, the predicted soil moisture clearly followed the gradient of the storm cell, suggesting little connectivity down the bajada. There was a distinctive break between the degraded grassland and the shrubland area. The shrubland had more diffuse down-bajada changes in soil moisture and lower soil moisture values for equivalent rainfall. Both of these were the result of the higher runoff rates on the shrubland and suggested that the shrubland is spatially more interconnected. It is difficult to reach firm conclusions about the grassland because it occupies such a small zone at the foot of Summerford Mountain, although it seems to behave in the same way as the degraded grassland. Overall, the results suggest that both the spatial pattern of the rainfall cell and the location of the different habitats control the spatial redistribution of soil water on the bajada.

Sediment

The initial focus of this project was on inter-rill areas because the movement of water, sediment, and nutrients begins in these areas. Although our studies are beginning to shift to the scale of small watersheds, rills, and sandy washes (Parsons et al. 1999; Howes and Abrahams 2003), the study of sediment and nutrient transport continues at the inter-rill scale (Schlesinger et al. 1999, 2000; Neave and Abrahams 2001, 2002). Consequently, the remainder of this chapter examines the processes and controls of sediment and nutrients at the intershrub scale.

Detachment and Transport

Soil erosion involves both the detachment and removal of soil particles. In interrill areas, detachment is accomplished predominantly by raindrops. Where overland flow depth is less than about three raindrop diameters (Kinnell 1991), raindrops falling at terminal velocity are capable of splashing particles with diameters up to 12 mm in all directions (Kotarba 1980). The trajectories of such particles are longer in the downslope direction, so there is a net downslope flux with the rate proportional to the sine of the slope. The rate of splash transport decreases as overland flow becomes deeper, and it decreases as vegetation, litter, and gravel covers increase. Owing to the scarcity of vegetation, splash transport is generally more important in desert landscapes than in humid ones. Nevertheless, it remains a relatively minor transport process, accounting for only 5–25% of the sediment transported by overland flow (Abrahams et al. 1994).

The main role of rainfall is to detach or loosen soil particles and lift them into the flow, which then transports them downslope. Raindrops not only loosen soil particles but also compact the soil surface and produce a crust (Moore and Singer 1990). The former process is more common early in a rainfall event, whereas the latter process dominates later in the event. Shallow overland flow generally does not exert sufficient shear stress on a soil surface to detach soil particles. However, where the soil is highly erodible and/or overland flow becomes concentrated along particular flow paths, the depth and velocity of flow may increase to the point where flow detachment dominates raindrop detachment, leading to the formation of rills and gullies. This is specifically the situation in the shrubland where shrub mounds divert overland flow and concentrate it in intershrub areas.

Sedigraphs of Simulated Runoff Events

Graphs of sediment concentration against time since the start of rain (hereafter called sedigraphs) for the rainfall simulation experiments suggest that sediment transport by overland flow on the Summerford bajada is generally detachment-limited. Where animal digging is limited, sedigraphs are usually monotonically decreasing (figure 7-7). This form is attributed to a reduction in the availability of detached sediment as the flow event proceeds. At the start of runoff, sediment concentration is high because the discharge is low and the flow is transporting loose sediment that has accumulated on the ground surface since the last event or that has been detached by rainfall prior to runoff. As discharge increases, the availability of detached sediment decreases, and the sedigraph declines monotonically (Neave and Abrahams 2001).

In contrast, where animal digging has disturbed a significant proportion of the plot surface, the sedigraph may be almost any shape depending on the severity and distribution of the disturbances over the plot surface. Monotonically increasing, convex-upward, and oscillating sedigraphs are all common (figure 7-8). These shapes reflect the time it takes the flow to transport loose sediment from the

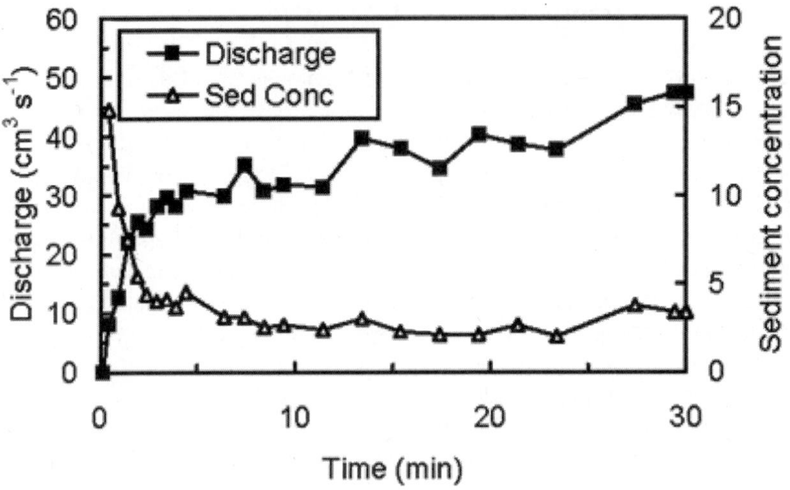

Figure 7-7. Hydrograph (discharge) and sedigraph (sed. conc.) for a 30-min simulated rainstorm at target intensity of 144 mm/h^{-1} on a runoff plot with little animal disturbances of soil surface.

various disturbance sites within the plot to the plot outlet. Thus both the shape of the sedigraph and the rate of sediment transport appear to be determined by the availability of detached sediment rather than the transport capacity of the flow, and the availability of such sediment are strongly influenced by animal disturbance of the soil surface (Neave and Abrahams 2001).

Controls

There is an extensive body of literature (reviewed by Weltz et al. 1998) on the controls of soil erosion and sediment yield on rangeland hill slopes. The principal controls of sediment yield are generally the same as those of runoff (already discussed): namely, plant and surface cover variables (standing biomass, life form class, canopy cover, and ground cover) and soil properties (bulk density, soil texture, soil organic carbon, and aggregate stability). Although the general character of these controls is similar from place to place, there are often important differences in detail. Rainfall simulation provides a powerful method not only for identifying these controls but also for investigating the pathways whereby they affect sediment yield.

Sediment yield, S_Y (kg/m^2), from a 30-min simulated rainfall experiment is equal to the product of water yield, W_Y (m), and sediment concentration, S_C (kg/m^3). Thus, the factors controlling S_Y may be thought of as acting through W_Y and/or S_C. The controls of W_Y were investigated earlier, and the findings are incorporated into figure 7-9. This figure depicts a causal model based on regression analyses of data obtained from the rainfall simulation experiments. The purpose of the model is to identify the surface properties controlling S_Y. The properties

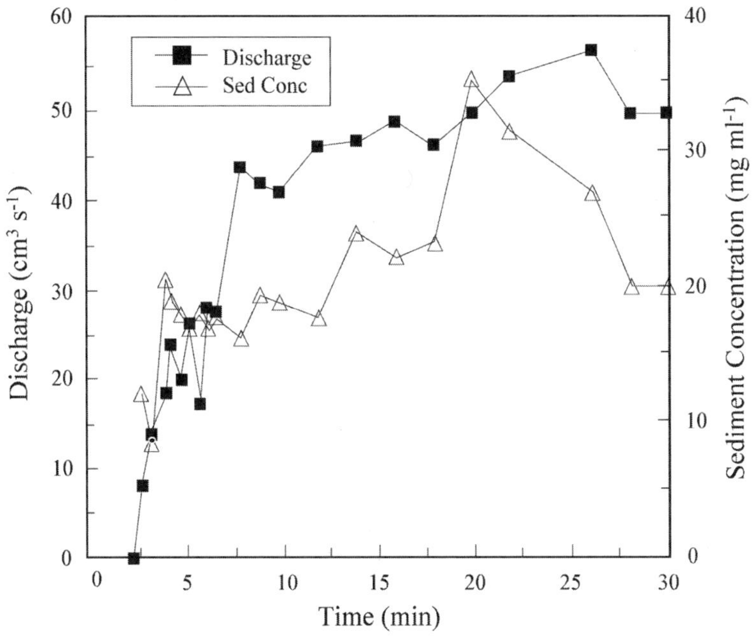

Figure 7-8. Hydrograph (discharge) and sedigraph (sed. conc.) for a 30-min simulated rainstorm at target intensity of 144 mm/h on a runoff plot with extensive animal disturbance of soil surface over its upper zone.

considered in the analysis are gravel cover, *percentG* (particle diameter > 2 mm); fines cover, *percentF*; ground vegetation cover, *percentV*; litter cover, *percentL*; the percentage of the surface disturbed by animals, P_A; and the mean diameter of the disturbances, D_A (mm). In figure 7-9 (1) only correlations that are significant at the 0.10 level are represented by arrows, (2) the arrowhead signals the direction of causality, (3) the strength of the causal relation is indicated by the standardized regression (beta) coefficient beside each arrow, (4) because $S_Y = W_Y S_C$, a regression of S_Y against W_Y and S_C always gives $r^2 = 1$, and (5) the r^2 values for the relations between either W_Y or S_C and the surface properties are presented on the right-hand side.

Figure 7-9 shows that surface properties affect S_Y via their effect on W_Y in all cover types. In the grassland and shrub cover types this is the only way surface properties influence S_Y, because S_Y is independent of S_C. In contrast, in the intershrub and degraded grassland cover types, S_Y is related to S_C, which in turn is controlled by D_A. The positive correlations between D_A and S_C and between S_C and S_Y reflect the strong influence of animal digging on the movement of sediment in these two cover types (Neave and Abrahams 2001).

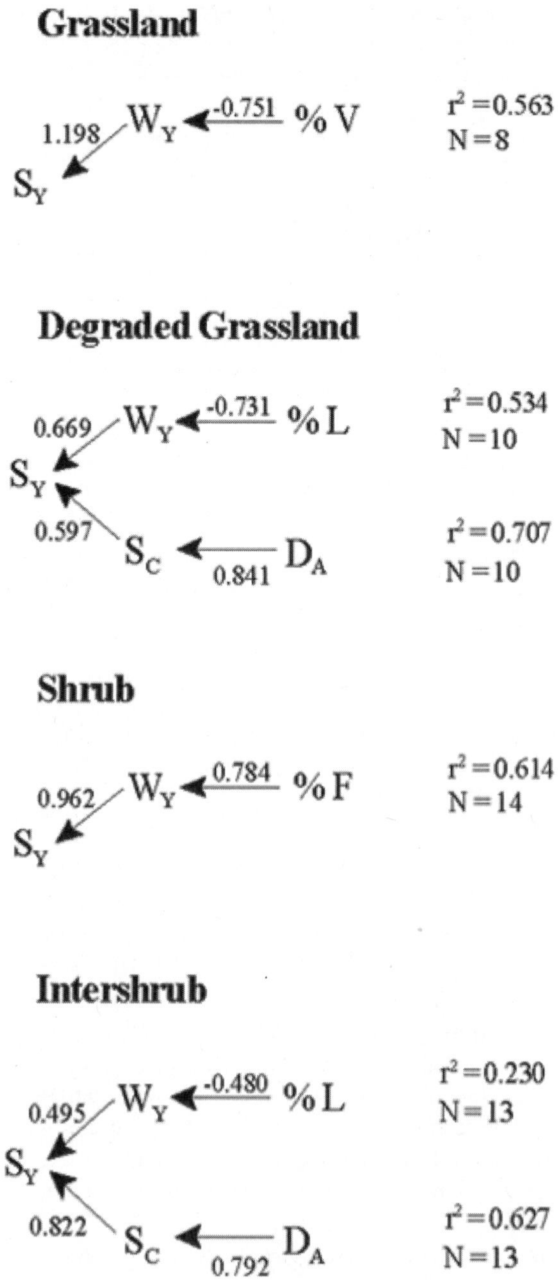

Figure 7-9. Causal diagrams showing how surface properties control sediment yield (S_Y) through water yield (W_Y) and sediment concentration (S_C). The surface properties are vegetation cover ($\%V$), litter cover ($\%L$), fines cover ($\%F$), and mean diameter of the animal disturbances (D_A). The numbers beside the arrows are standardized partial regression (beta) coefficients. All coefficients are significant at the 0.10 level. N is the sample size, and r^2 is the coefficient of determination.

Rainfall Duration

The rainfall simulation experiments on which figure 7-9 is based were run at a target intensity of 144 mm/h for 30 min. Such a storm has a recurrence interval in excess of 100 years. An intensity of 144 mm/h is not unusual, but it is unusual for this intensity to persist for 30 min. The data were therefore examined to ascertain whether the relations displayed in figure 7-9 hold for shorter durations. The analyses revealed that all the significant relations in figure 7-9 are established after 15 min of rainfall and that some relations were established even earlier. The relations were weaker for the 5- and 10-min durations because runoff begins at different times on different plots, and the time to runoff has a major effect on the water and sediment yields during the first few minutes of a storm (Neave and Abrahams 2001).

Natural Runoff Events

Sediment yields obtained from the rainfall simulation experiments cannot be used to estimate annual sediment yields because the simulations were performed at a single intensity and duration, whereas natural rainfalls vary in intensity and duration. T. J. Ward (unpublished data) measured sediment loss from inter-rill areas during natural rainfall events. Because his data represent the rainfall regime, they may be used to calculate mean annual sediment yields for the grassland and shrubland. For this calculation, it was necessary to combine the data from the high-cover and low-cover plots in each community, and this in turn required that each plot type be assigned a weight. Based on the assumption that half the grassland has high cover and half low cover, each plot type was assigned a weight of 0.50. In the shrubland, the high-cover and low-cover plots correspond to shrub and intershrub areas, respectively. Because shrub canopies occupy about 38% of the shrubland surface (Schlesinger et al. 1999, 2000), the high-cover plots were assigned a weight of 0.38, and the low-cover plots had a weight of 0.62.

The mean annual sediment yield and volume-weighted mean sediment concentration are highest for the low-cover shrubland (intershrub areas) and lowest for the high-cover grassland. The differences are large, with the sediment yield for the low-cover shrubland being 7.9 times that for the high-cover shrubland. The sediment yield and the sediment concentration are also higher for the shrubland than for the grassland, with the sediment yield for the former exceeding that for the latter by a factor of 2.7. The mean annual runoff for the shrubland is 3.4 times that for the grassland, and the volume-weighted mean sediment concentration for the shrubland is 1.4 times that for the grassland. Thus the higher sediment yields in the shrubland are due to both higher runoffs and higher sediment concentrations, with the runoff having a larger effect than the sediment concentration.

Nutrients

Our hypothesis for the desertification of piedmont grasslands in the Jornada Basin suggests that as the plant cover in grasslands declines, greater rates of hill slope erosion remove nutrients from exposed soils, thereby reinforcing the development of resource islands under invading shrubs. To test this hypothesis, we measured the nutrient concentrations in runoff during 24 rainfall-simulation experiments on areas of black grama grassland and creosotebush shrubland on the Summerford bajada (Schlesinger et al. 1999). The objective was to compare the runoff of nitrogen (N) and phosphorus (P) from these habitats to assess whether greater losses of soil nutrients are associated with the invasion of grasslands by shrubs.

Volume-weighted mean concentrations of total dissolved N in runoff (total N loss divided by total water loss) were 1.72 mg/L, 1.44 mg/L, and 0.55 mg/L for the grassland, shrub, and intershrub plots, respectively. Weighted by the average cover of shrub (38%) and intershrub (62%) areas on the landscape, the mean nitrogen concentration was 0.77 mg/L in the runoff from shrublands. Thus the concentration of total dissolved N in the runoff from grasslands was 2.23 times greater than that from the shrubland. Volume-weighted mean concentrations of dissolved organic nitrogen (DON) were 1.00 mg/L, 0.87 mg/L, and 0.41 mg/L for the grassland, shrub, and intershrub plots, respectively. Weighted by the average cover of shrub and intershrub areas, the mean concentration of DON in the runoff from shrublands was 0.52 mg/L. Thus, the DON concentration in the runoff of grasslands was 1.9 times greater than in the runoff from shrublands.

For grassland and shrub plots, N yield was always more highly correlated with water yield than with N concentration, whereas the reverse was true for the intershrub plots. The standard deviation of water yield was always greater than the standard deviation of N concentration for grassland and shrub plots, but less than the standard deviations of these variables on intershrub plots. This suggests that intershrub plots were similar in their ability to generate runoff but different in their ability to supply N compared to grassland and shrub cover types. Given that intershrub plots were virtually devoid of ground vegetation and litter, and whereas the grassland and shrub plots had varying covers of plant matter, the greater hydrological uniformity of the intershrub plots is not surprising. It is less obvious, however, why the intershrub plots were more variable than the grassland and shrub plots in their supply of N in general and of organic N in particular. There were no significant correlations between our measures of surface properties in these plots and their yield of runoff or N.

Bolton et al. (1991) suggested that a useful statistic for comparing nutrient losses in arid lands is the volume-weighted mean concentration (e.g., mg N/L) divided by 100, which is equivalent to the loss of N (kg/ha) per ml of runoff. Over a 6-year period, 1989–94, an average of 14.6 mm/yr of runoff (6% of incident precipitation) was measured on eight 2-m by 2-m runoff plots in two grasslands of the Jornada Basin. Using the volume-weighted mean concentration of 1.72 mg N/L for total dissolved N in the runoff from grasslands, nitrogen losses were calculated to be 0.25 kg/ha/yr.

To estimate the annual loss of N from shrublands, the volume-weighted mean concentrations of total dissolved N in runoff (1.44 mg N/L for shrub plots and 0.55 mg N/L for intershrub plots) were weighted by the runoff and proportional land area of shrubs (38%) and intershrub areas (62%) and then multiplied by a runoff estimate of 56.2 mm/yr (18% of incident precipitation) determined on four

2-m by 2-m plots monitored in creosotebush shrubland during the same 6-year period as in the grasslands. The calculated N loss was 0.43 kg/ha/yr. The higher loss of dissolved N from shrublands than from grasslands stems from the 3.8 times greater runoff measured in shrublands during the long-term field studies.

Despite the sparseness of arid land vegetation, DON was greater than 50% of the total dissolved N measured in the runoff from grassland and shrub plots, accounting for nearly 70% of the N lost from the barren soils of the intershrub plots. The total yield of N during each rainfall simulation declined monotonically (figure 7-10), suggesting that the decline in the concentration of N over time was

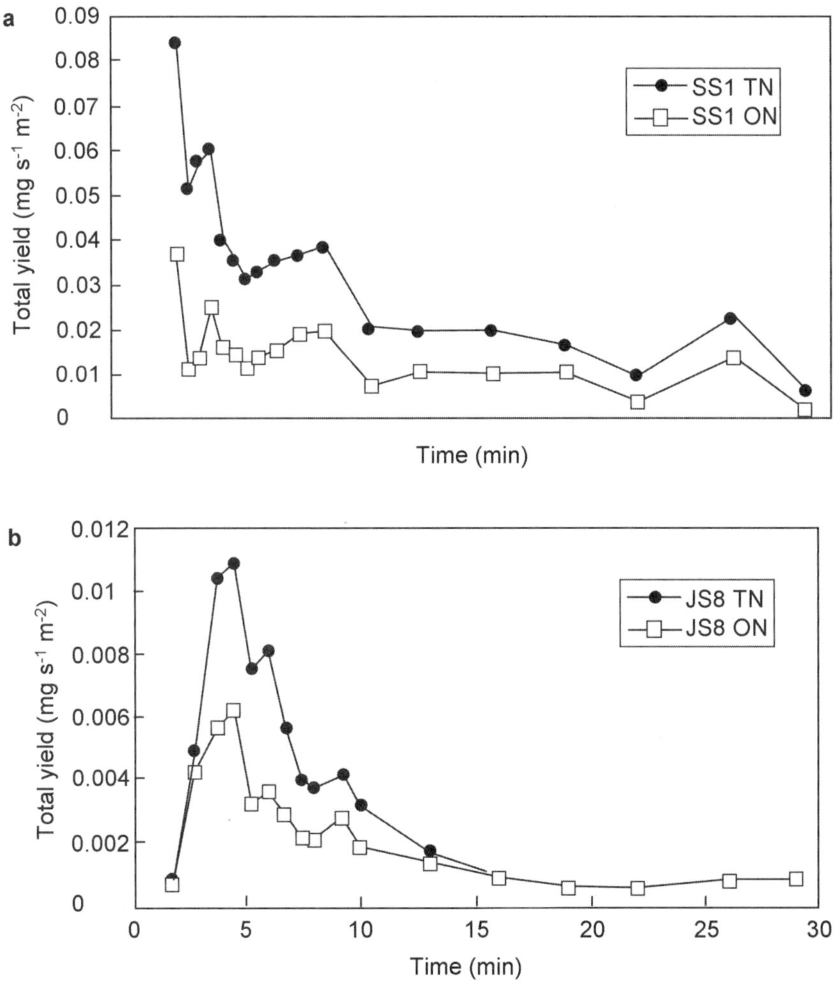

Figure 7-10. Yields of organic nitrogen (ON) and total dissolved nitrogen (TN) in discharge as a function of time since the start of simulated rainfall from (a) grassland and (b) shrubland plots.

not simply due to dilution by the increasing runoff but to depletion of the pool of available N in the soil. The decline was greater for dissolved inorganic forms of nitrogen (NH_4 and NO_3) than for DON, so DON became an increasing fraction of the total N yield as duration of runoff increased. These high concentrations of DON may be seasonal; the rainfall simulation experiments were performed at the end of the dry season, when a large quantity of soluble organic N compounds may have accumulated undecomposed in the soil.

The nutrient losses in runoff during the rainfall simulation experiments were compared to the losses of dissolved nutrients (N, P, K, Ca, Mg, Na, Cl, and SO_4) in runoff measured during natural runoff events of varying intensity and duration on grassland and shrubland plots in the Jornada Basin (Schlesinger et al. 2000). Across all plots at each site, runoff was logarithmically related to precipitation volume, with r^2 ranging from 0.44 to 0.48 (figures 7-11a and 7-11b). An analysis of variance showed that the mean slope of this relationship, calculated from the slope of the regressions for the individual runoff plots in each area ($N = 4$), was not different between grassland and shrubland plots. However, the intercepts of the relationship were lower for both grasslands than for the shrubland, suggesting that runoff commenced at a lower threshold of storm size in the shrubland than in the grasslands. The annual runoff coefficient (total depth of runoff expressed as a percentage of the total depth of precipitation received) averaged 19% in the creosotebush shrubland and 6% in the grasslands.

Within each area, plant cover did not significantly affect the slope of the relationship between runoff and precipitation, although the difference between high- and low-cover plots is nearly significant in the grassland. The volumetric runoff coefficient averaged 8.5% in low-cover plots and 4.1% in high-cover plots during a 5.5-year period of collections in the grasslands. During the 7-year record in the creosotebush shrubland, the annual runoff coefficient was 16.4% in high-cover plots versus 20.8% in low-cover (intershrub) plots. As in the rainfall simulation experiments, the data from the field plots show that a loss of grass cover leads to greater runoff from dryland ecosystems (see Abrahams et al. 1995; Gutierrez and Hernandez 1996; Castillo et al. 1997). Discharge commences earlier in the shrubland than the grassland. Specifically, an average of 2.72 mm of rain is required to initiate runoff in the shrubland, whereas 3.25 mm is required in the grassland. Total discharge during a storm increases with the volume of rainfall received, but the slopes of these relationships differ little between the grassland and shrubland (figure 7-11) or between high- and low-cover plots (figure 7-11b).

The concentrations of dissolved constituents declined with increasing runoff volume in all habitats. The best relationships between concentration and volume were always logarithmic, reflecting a rapid dilution of dissolved constituents with increasing discharge (figures 7-12a–c). The yield of dissolved constituents in each runoff event was calculated by multiplying the concentration (mg/L) by the runoff volume (L). Annual losses of dissolved forms of plant nutrients in runoff were always higher in the creosotebush shrubland than in the grasslands. Losses of total dissolved nitrogen ranged from 0.28 to 0.41 kg/ha/yr in the shrubland (figure 7-12a) versus 0.11–0.21 kg/ha/yr in the grassland plots (figures 7-12b and 7-12c). DON made up 10–30% of the loss of total dissolved N in each habitat. Higher

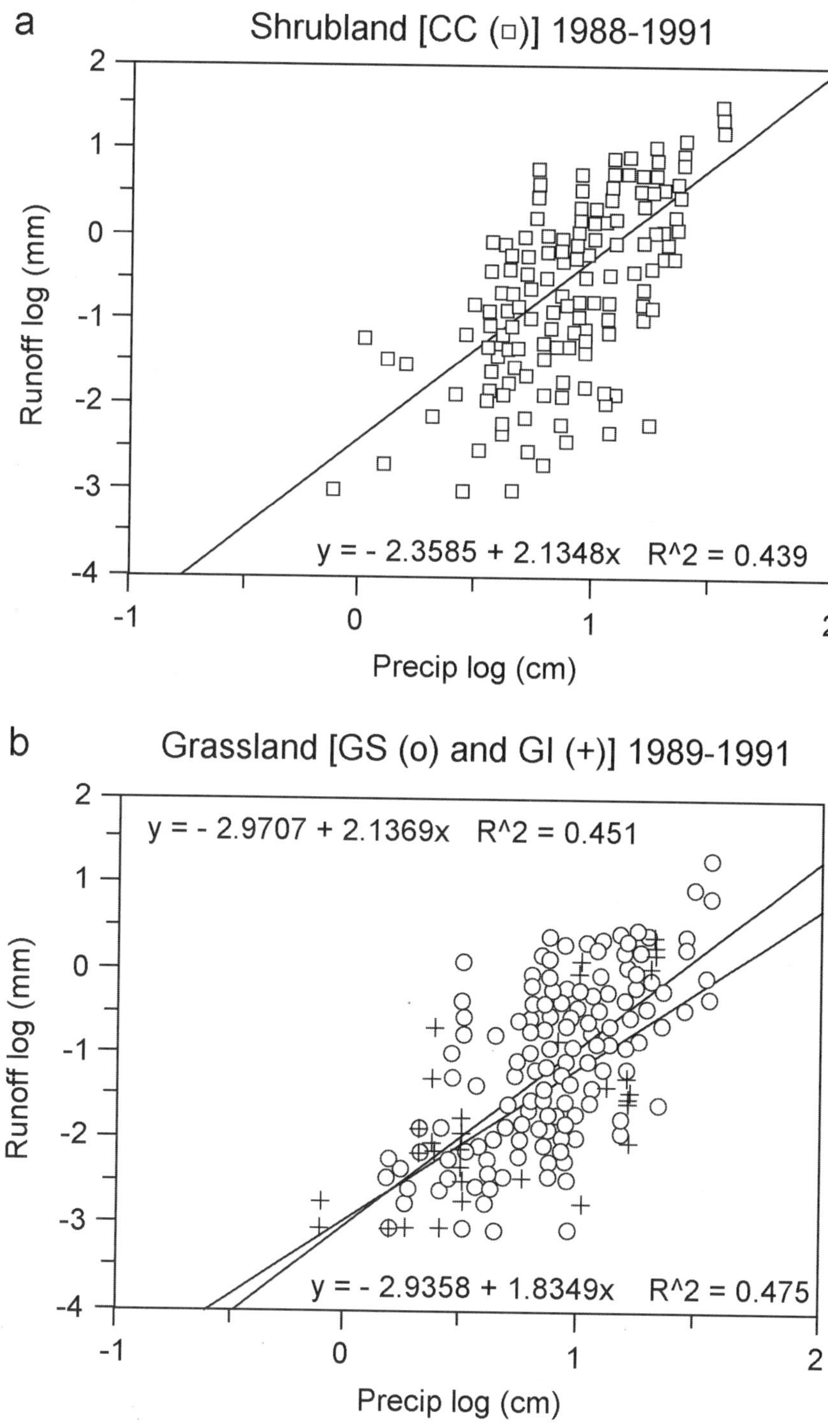

Figure 7-11. Total runoff during natural rainfall events from plots in (a) shrubland and (b) grassland as a function of total rainfall during the event. CC denotes creosotebush shrubland on the Summerford bajada, GS denotes grassland on the Summerford bajada, and GI grassland about 10 km to the north.

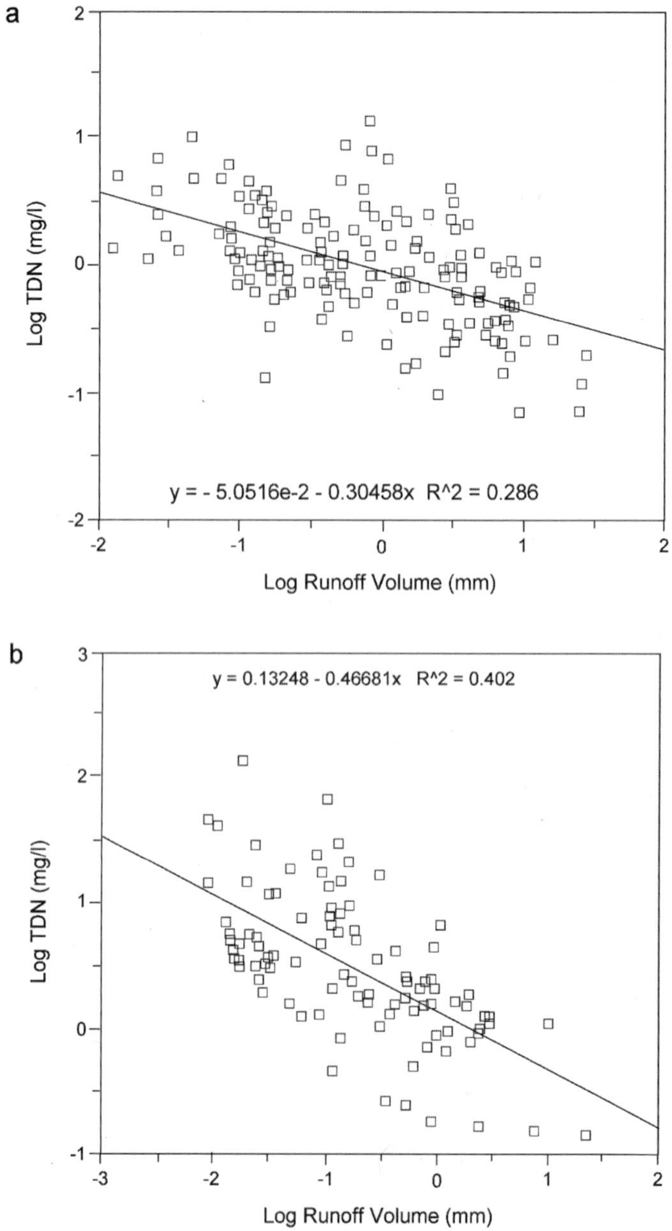

Figure 7-12. Concentration of total dissolved nitrogen (TDN) against runoff during natural rainfall events for (a) creosotebush (CC) and (b) grassland (GS) plots on the Summerford bajada and for (c) grassland plots (GI) 10 km to the north (1988–91).

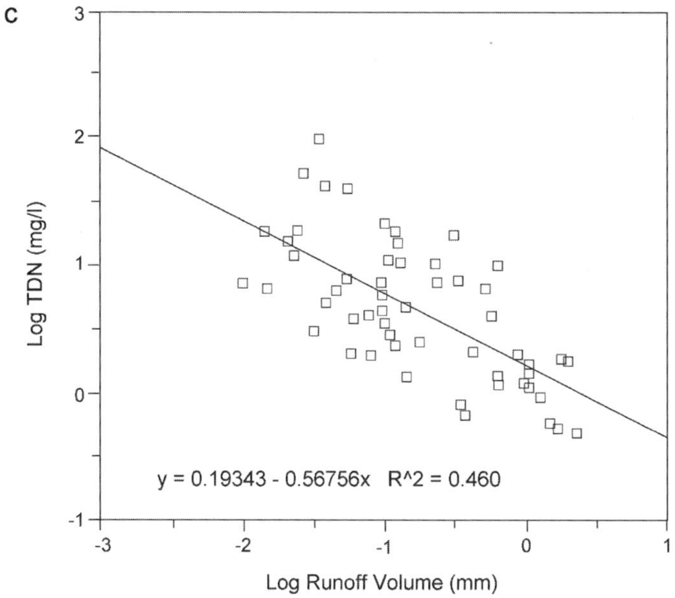

overall discharge, rather than higher nutrient concentrations in runoff waters, accounts for the greater losses of nutrients from shrublands. The weighted mean concentration of total dissolved N was 0.71 mg N/L in the creosotebush shrubland and 1.39–3.24 mg N/L in the grasslands. These concentrations compare favorably to those measured in shrubland and grasslands during the rainfall simulation experiments.

In both the shrubland and grassland plots, the losses of N found were less than those estimated from the rainfall simulation experiments (0.43 and 0.25 kg/ha/yr, respectively; Schlesinger et al. 1999). The mean runoff for 1988–91 was below the long-term average (1989–94) that we used to make a long-term extrapolation of N yield from the rainfall simulation experiments.

Although the mean N losses in runoff are greater in the shrubland (0.33 kg ha/yr) than in the grasslands (0.12 and 0.19 kg/ha/yr), the transport of dissolved N compounds in runoff cannot be wholly responsible for the depletion of soil N that is associated with desertification in this region (Schlesinger et al. 1996; Kieft et al. 1998). The N losses in runoff are lower than the inputs in atmospheric deposition; so all habitats show a net gain of soil N if one considers only atmospheric deposition and runoff.

Conclusions

This chapter summarizes our characterizations of the movement of water, sediment, and nutrients across the Summerford bajada. The lateral transfer of water and materials is accomplished by both channelized flows in rills and washes and unchannelized flows over inter-rill surfaces. The focus of the work has been on inter-rill flows, which are strongly influenced by raindrop impact. Rainfall simulation is a powerful tool for investigating such flows, and we have made extensive use of it in these characterizations. Analyses of rainfall simulations (including unpublished data provided by Ward) and natural rainfall events have led to the following findings.

1. Creosotebush shrubs and the areas (interspaces) between them are hydrologically very different. In particular, under equilibrium conditions interception by the shrub canopy causes the mean kinetic energy of the subcanopy rainfall to decline by 30%, which reduces surface sealing and contributes to the formation of shrub mounds by differential splash. These mounds, which are the topographic representation of a resource island, divert overland flow and concentrate it in intershrub areas, where flow detachment lowers the surface and may ultimately scour a rill. Stem flow is another process contributing to the formation of resource islands. An average of one-sixth of the rainfall intercepted by the canopy is funnelled to the base of the shrub. Direct and released throughfall also contribute water to resource islands, and the disposition of this water (i.e., infiltration and surface runoff) is controlled by the density of subcanopy vegetation. On plots devoid of such vegetation, runoff averages 77% of rainfall intensity. In contrast, where there is dense subcanopy vegetation, this proportion ranges from 0 to 16% as rainfall intensity increases from 0 to 20 cm/h.
2. Whereas runoff from shrub areas is strongly influenced by subcanopy vegetation, runoff from intershrub areas is largely controlled by surface crusting. Biological and physical crusts coexist, and their effects are generally difficult to separate. Physical crusts, however, are more dynamic than biological crusts: They can form during a single storm and be effectively destroyed in a matter of weeks by animal activity. As a result of crusting, runoff from intershrub areas is about three times that from shrub areas, and runoff from degraded grassland is about three times that from grassland.
3. Runoff coefficients for the grassland and shrubland are 6% and 19%, respectively, whereas maximum runoff in the shrubland is 2.2 times that in the grassland. These differences between grassland and shrubland are largely attributed to the fact that almost two-thirds of the shrubland community consists of barren intershrub areas with crusted surfaces and low infiltration rates. Most runoff from the shrubland is generated in these intershrub areas and is largely confined to these areas as it flows downslope.
4. Rainfall simulation experiments demonstrate that animal digging and scratching, which are most pronounced in the interspaces between shrubs and in the degraded grassland, have a major effect on sediment yields. Sediment yields from the shrubland are 2.7 times those from the grassland. Because the mean annual runoff for the shrubland is 3.4 times that of the grassland while the sediment concentration is only 1.4 times that for the grassland, it is evident that the higher sediment yields for the shrubland are

due to both higher sediment concentrations and higher runoffs, with the higher runoffs having the greater effect.

5. The foregoing numbers suggest that as shrubland has replaced grassland on the Summerford bajada, intershrub areas have experienced greater rates of erosion, which have removed nutrients and promoted the development of resource islands. Rainfall simulation experiments indicate that although the mean concentration of total dissolved N is lowest in bare intershrub areas and highest in grassland, total N yield is highest from intershrub areas and lowest from grassland due to the higher runoff from the intershrub areas. Surprisingly, in all three cover types, more than half the N transported in runoff is carried in dissolved organic compounds. Analyses of the data for natural rainfall events indicate that average nutrient losses from the shrubland is 0.33 kg/ha/yr, which is more than twice the value of 0.15 kg/ha/yr obtained for the grassland. Moreover, these data confirm that the greater nutrient losses from the shrubland are due to higher runoff rather than higher nutrient concentrations in runoff. These findings pertain to inter-rill areas, where runoff exceeds run-in and erosion processes dominate.

8

Water and Energy Balances within the Jornada Basin

Vincent P. Gutschick
Keirith A. Snyder

This chapter describes general characteristics and components of the energy and water balances in arid regions, with specific examples from the Jornada Basin. Various research efforts to characterize the energy and water balances and resultant carbon dioxide fluxes in the Jornada Basin are detailed. We provide a brief overview of how plant physiology interacts with energy and water balances in this region, and characterize general abiotic conditions and some physiological traits of plants in this arid region.

Energy Balance Characteristics of the Jornada Basin

The surface of a landscape may be considered as a layer with some amount of vegetation. More general descriptions divide the vegetation, like the soil, into layers, but the concern here is energy balance at the interface with the atmosphere. The net energy balance of the land surface is determined by (1) inputs (radiant energy), (2) outputs (reflection [i.e., albedo], emission of longwave radiation, convective heat transfer to the atmosphere [i.e., sensible heat flux], evapotranspiration of water [i.e., latent heat flux], and conduction of heat into soil), and (3) changes in heat storage. The balance of these terms is adjusted as the surface temperature comes into steady state or nearly so. Increased solar input will drive surface temperatures higher until longwave emission and other losses come into a new balance. The net energy input, as inputs minus outputs, may be stated formally as an energy-balance equation

$$\text{Rate of heat storage} = S = Q^{+}_{SW} + Q^{+}_{TIR} - Q^{-}_{TIR} - Q^{-}_{E} - Q^{-}_{H} - Q^{-}_{S}, \quad (8\text{-}1)$$

where the superscript + indicates an input, and − indicates an output or loss, and all terms are expressed as flux density in units of W/m². Q^{+}_{SW} is the energy added to the surface layer by solar radiation from above. Q^{+}_{TIR} is the thermal infrared radiation emitted by gases in the atmosphere, principally water vapor and CO_2, whereas Q^{-}_{TIR} is the thermal infrared radiation emitted from components of the Earth's surface and lost back to the atmosphere. Q^{-}_{E} is the latent heat flux from the heat of vaporization of water vapors resulting from soil evaporation (E) and plant transpiration, generally measured as the composite evapotranspiration flux (ET). Q^{-}_{H} is the sensible heat transfer from surface to atmosphere by convection. Q^{-}_{S} is conduction of heat into soil. Heat storage (S) in the vegetation is only significant in thick, voluminous vegetation such as closed stands of trees (absent on the Jornada) or at the single-plant level for cactus phyllodes (Nobel 1978). An alternative formulation of the energy budget is

$$\text{Rate of heat storage} = S = R_n - LE - SH - G, \tag{8-2}$$

where the first three terms of equation 8-1 are represented as total net radiation, R_n, and the other terms are analogous to equation 8-1. LE is latent heat flux, SH is the sensible heat flux, and G is soil heat conduction.

The Jornada Basin is characterized by low mean cloud cover, moderately high elevation (about 1,350 m), moderately low aerosol burden, and low latitude (32.5°N), such that solar energy input is large. Details of the climate are described in chapter 3. Surface temperatures are high on summer days, despite rather reflective soils (about 35% of shortwave energy is reflected). Cooling by evapotranspiration is often low, as water is frequently depleted from soil surfaces and from vegetation. Mean annual precipitation is 245 mm with more than 54% of total rainfall falling between July and September. Mean annual temperatures range from 3.8°C in January to 26.0°C in July. The clear skies and low water vapor content, which produce low absorptivity for thermal infrared radiation leaving the surface, allow strong radiative cooling. Thus daily fluctuations in surface and air temperatures are very large, commonly exceeding 20°C even without passage of weather fronts. The large diurnal and seasonal ranges of temperature strongly condition the performance and survival of plants.

On large spatial scales, the Jornada and adjoining arid regions are usually a net source of sensible heat flux to the atmosphere, a weak source of latent heat flux or ET, and a modest sink (above the atmosphere) for net radiation (total radiant energy flux density in minus total radiant energy flux density out; Hartmann 1994). The net heat flux out is the net result of an input of heat from compressive heating of the air subsiding here from the tropics minus the output from hot surface to the cooler air transported over hundreds of km. On the scale of tens of km, the sensible heat flux helps drive convective storm activity in summer. On larger scales, the heat flux contributes to setting regional atmospheric circulations that determine climatic patterns.

Energy leaves this surface in several ways. One is by reflection of solar radiation, commonly accounted in the solar input as net absorption. Vegetation absorbs strongly in the photosynthetically active radiation (PAR) band from 400 to 700 nm wavelength, which reduces the landscape's reflectivity (i.e., albedo) of solar radiation. Changes in the composition of vegetation and the amount of vegetation change the albedo of a landscape. For example, shrub encroachment into former grasslands generally results in higher albedo, because bare soil usually increases, and it is generally lighter in color and thus more reflective (Kurc and Small 2004). However, individual shrub canopies are often darker in color than some grass species or subshrub species depending on the time of year, and this can also alter reflectance (Franklin et al. 1993).

Thermal infrared radiation (TIR or Q_{TIR}) is emitted by the surface in close proportion to the fourth power of the absolute (Kelvin) temperature and emitted by gases in the atmosphere. The TIR gain from the sky and loss from the surface actually make up the major part of the annual energy budget of the surface (about 290 and 390 W/m^2, respectively). TIR fluxes are less variable than other energy exchanges. TIR emission from the surface changes by only 5 W/m^2 per 1°C shift in surface temperature when surface temperatures are near 15°C. In contrast, sensible and latent heat fluxes vary by hundreds of W/m^2 diurnally, weekly, or seasonally. As mentioned, the low water vapor content of the skies leads to low Q_{TIR}^{+} in dry regions, and the effective radiative temperature of the sky is many tens of degrees below air temperature. Commonly on a summer day with an air temperature of 35°C, the sky radiative temperature is below 0°C. This leads to radiative deficits at night that cause "radiation frosts" in winter (Leuning 1988; Jones 1992): Horizontal surfaces (such as leaves) may stabilize at temperatures up to 5°C lower than air temperature.

Sensible heat flux (Q_H or H) constitutes a net loss when surface temperatures exceed air temperature, and the rate is proportional to that temperature difference multiplied by the surface boundary-layer conductance. The conductance rises with wind speed and also with surface roughness, which is in part determined by vegetation structure. During sunlight hours, soils are commonly sources of sensible heat transfer to the air, whereas plant leaves vary. Fast-transpiring leaves are sinks for sensible heat flux, and water-stressed, slow-transpiring leaves are sources of sensible heat flux. In the Jornada Basin because vegetation cover is low, soil heat transfer commonly dominates sensible heat flux.

Another energy loss is conduction of heat into soil (Q_S or G). Hot soil surfaces send heat into soil. There is a characteristic lag in the diurnal cycle of four hours, so that peak heat flux outward occurs about four hours before dawn. Soil heat flux is very modest in thicker vegetation, such as crops, because soil is shielded from much radiant exchange by high canopy cover of vegetation. In deserts, soil heat transfer can be much larger, on the order of 50 W/m^2, and is also quite variable spatially depending on whether soil heat flux is being measured in open interspaces, partial canopy cover, or full canopy cover (Kustas et al. 2000).

With physical models of heat propagation (e.g., Campbell and Norman 1998), it can be predicted how peak soil temperature decreases with depth in the soil. This peak decreases by a factor of $1/e = 0.37$ times the daily amplitude over a characteristic damping depth of $D = \sqrt{(2K/\omega)}$, where K is the thermal diffusivity and w is the angular frequency of the cycle (2π per day or 7.27×10^5/s for the daily cycle). The depth at which peak soil temperature begin to dampen in dry Jornada soils is commonly around 10 cm, when volumetric water content is below

10%, as is typical on the Jornada (see chapter 5). The surface soil temperature in the summer may range between 20°C at night and 62°C in the day. Thus the mean temperature in the top 10 cm is close to 41°C, and below 10 cm in depth soil temperature begins to damp with depth to eventually reach the long-term mean of 15°C. The amplitude is one-half the total swing, or 21°C at the surface, damping with depth z as $\exp(-z/D)$. Roots may tolerate 50°C repeatedly, though shallow-rooted cacti may go higher, to 56°C in one study (Jordan and Nobel 1984). To damp to a peak temperature of only 50°C, soil must be found where the amplitude is only 9°C. This occurs at about 8.5 cm in dry Jornada soils. The decline of temperature with depth from the annual cycle is superimposed on the diurnal cycle. This falloff is about 1°C per 10 cm. It shifts the depth at which 50°C is expected upward by about 1 cm. High soil temperatures result in high evaporation from bare soil, and both increased root respiration costs and some soil temperatures may be lethal to plants. However, water in root tissues, shading by plant canopies and heterogeneity of soil water complicates the soil temperature profile.

Another energy loss is latent heat flux resulting from the heat of vaporization from ET (Q_E or LE). Tall vegetation with access to the groundwater and much exposure to wind may give fluxes up to 500 W/m², even in arid regions in riparian zones (Schaeffer et al. 2000). An average value of $\bar{Q}_E$ can be calculated from the average annual precipitation of 245 mm on the Jornada Basin. Evaporating 245 kg H_2O/m² requires 610 MJ. Spread out over 3.1×10^7 seconds in a year, this is an average flux density of 19 W/m², assuming all evaporation occurred equally throughout the year. The amount of energy leaving though the latent heat flux term is low compared to global standards and in comparison to other components of the energy budget. For example, average incoming Q_{sw} is approximately 209 W/m². Plants affect the partitioning of latent heat flux into its component fluxes of evaporation and transpiration, by their (partial) control of ET via their regulation of both stomatal conductance and leaf area development which determine the transpiration component of ET. If ET is curtailed, due to soil water deficits, while solar inputs are not changed, the other losses must increase, with surface temperatures rising to strike the new balance. Indeed, the temperature difference from surface to air is a useful measure of water stress, as well as of sensible heat flux (in crops, Jackson et al. 1981; over arid lands and other sparse vegetation, Humes et al. 1994; Shuttleworth and Gurney 1990; Kustas et al. 1994).

Water Balance Characteristics of the Jornada Basin

The Jornada Basin is a region with low precipitation and high potential evapotranspiration (PET) due to the characteristics of the energy budget and consequently low hydrologic yield (chapter 3). A simplified version of the water budget for the region is

$$\text{Precipitation} + \text{Run-in} = \text{Evapotranspiration} + \text{Runoff} + \text{Recharge.} \quad (8\text{-}3)$$

The basin, by virtue of its geomorphology, is closed hydrologically (see chapter 2). There is little runoff other than locally, and there are no perennial stream flows. Lateral subsurface flows in aquifers or in the vadose (unsaturated) soil zone are essentially absent on all but small spatial scales, such as in the playas and then only episodically. Soil hydraulic conductivity is generally low given the predominantly low water content, and terrain relief is slight except for some localized hill slopes. Hydrologic recharge of deep soil is minimal in most years (Phillips 1994), but may occur under certain rainfall conditions (Small 2005; see also chapter 5), and there is great spatial variation in recharge rates due to local geomorphology (Scanlon et al. 1999). Playas, fissures, gullies, and burrow pits have recharge rates up to 120 mm/yr (Scanlon and Goldsmith 1997) but cover a very minimal portion of surface area. Thus, essentially all precipitation is lost through soil evaporation (*E*) and plant transpiration. Despite the lack of water inputs, salinity is rare in soils of the Jornada Basin. However, calcium concentrations can exceed limits tolerated by calcifuge vegetation.

Evapotranspiration is approximately 95% of total precipitation inputs. In fact, Dugas et al. (1996) report that ET is 100% ± 12% for the Jornada Basin. However, even though almost all water is lost through ET, whether water is lost through soil evaporation or plant transpiration pathways is important to ecosystem processes because evaporation and transpiration drive different processes. Evaporation (E) largely determines soil processes (soil respiration, soil water recharge), whereas transpiration drives plant productivity (Huxman et al. 2005).

Conversion of grassland to shrubland has resulted in changes in both the horizontal distribution of plant cover and the horizontal and vertical distribution of plant roots. In the horizontal dimension, grass cover is rather uniform, whereas shrub cover is patchy. For example, on the mesquite (*Prosopis glandulosa*) dunes of the Jornada, crown cover (defined as to the edge of the crown, ignoring the partial nature of light interception within the crown boundaries) is 15–30% (Rango et al. 2000). Roots systems of 11 shrub species were mapped in the Jornada Basin (Gibbens and Lenz 2001). Maximum root spread was found above petrocalcic and calcic horizons at depths above 1 m and ranged from 1 to 3 m from the center of the plant for tarbush (*Flourensia cernua*) and creosotebush (*Larrea tridentata*) and was greater than 6 m for some mesquite shrubs. Ten of the 11 shrub species had roots that penetrated the petrocalcic and calcic horizons and grew to depths of 5 m (roots were not mapped below this depth). Roots of 11 grass species were found to extend radially between 0.5 m and 1.4 m in sandy soils but did not penetrate petrocalcic and calcic horizons and did not extend to depths greater than 1.6 m. The conversion from grassland to shrubland may affect the water balance of the Jornada Basin because deep-rooted shrub species may transpire deep water to the atmosphere that was previously below the root zone of grass and forb species. Additionally, many shrub species, particularly the evergreen creosotebush, have longer periods of phenological activity than grasses, which extend the period of time that water is transpired to the atmosphere (Reynolds et al. 1999b). The total amount of water lost will also depend on the rates of transpiration by various plant species. However, greater transpiration by individual shrub species may be offset by evaporative water losses in larger open interspaces. The partitioning of ET depends on rainfall characteristics, plant species composition, plant phenology, and photosynthetic pathway of vegetation cover, soil and air temperatures, and nutrient availability. Empirical estimates on the relative contribution of E and transpiration in semiarid regions have found disparate results with the ratio transpiration/ET ranging from 7–80% (see review in Reynolds et al. 2000; Dugas et al. 1996). Modeled estimates of plant transpiration showed high variability (1–60% of ET) that was largely in response to rainfall patterns (Reynolds et al. 2000).

Additionally, shrubs can redistribute water horizontally through their effects on overland flow (see chapter 7) and can also affect the ET component by altering interception losses and stem flow. Whitford et al. (1997) found that stem flow in creosotebush was 16.8% ± 1.9% of bulk precipitation. Stem flow effectively concentrated rainfall amounts around the bases of shrubs and may allow for deeper infiltration in the plant rooting zone than would be predicted based on soil texture and storm size. However, this effect varies by storm size and events smaller than 6 mm were not found to cause measurable stem flow. Canopy interception is another mechanism whereby plants may interact with the hydrologic cycle. Tromble (1983) found that 20% of artificially applied rainfall was intercepted by creosotebush leaves. When scaled to the average crown cover of native creosotebush stands (approximately 30.5% crown cover) 22% of bulk precipitation could be lost through interception. In terms of small storm sizes, there may be disproportionate effects on the amount of intercepted precipitation. In the Jornada Basin the majority of storms are less than 6 mm. If an average storm size of 3 mm is used, nearly 12% of 3 mm rainfall may be lost through this pathway in creosotebush communities (Tromble 1983).

Changes in the surface energy balance or in vegetation cover and composition may produce interactions that can potentially modify regional or local climates (Anthes 1984; Lyons et al. 1993; Sud et al. 1993; Pielke et al. 1998). For example, clearance of vegetation, all else being equal, increases the albedo in arid zones with light-colored soils, thus decreasing the average sensible heat flux to the atmosphere. Charney (1975), Charney et al. (1976), and Otterman (1974, 1975) proposed that the lower sensible heat flux leads to lower convection and cloud formation in the troposphere, producing a feedback that would stabilize desertification by reducing precipitation. Empirical proof has not been found (Le Houerou 1996), even though general circulation models show albedo effects acting at least on the scale of continents (Garratt 1993). More sophisticated climate modeling with global change models has revealed more complex interactions. For example, in the African Sahel coupling of ocean surface conditions to land surface conditions is critical for understanding precipitation characteristics. (review by Lau 1992; Myneni et al. 1996).

On the spatial scale of the Jornada Basin (2,500 km^2) and with its absence of excess upwind surface water, there may be little effect of vegetation change on precipitation. ET is the ultimate source of virtually all precipitation globally. However, vegetation change on the Jornada may have little effect on regional precipitation because total accumulated ET is generally low and the recycling ratio (fraction of precipitation derived from local evapotranspiration; Eltahir and Bras

1996) is low. The recycling ratio clearly increases with spatial scale (becoming 1.0 over the globe) and is highest for any choice of spatial scale in Amazonia. Although coupled regional atmospheric models may find subtle differences in latent and sensible heat flux from vegetation change (Beltrán-Przekurat et al. 2005), it is unclear if these small changes can produce changes in weather patterns. On larger scales, the feedback of depleted water supply to depleted atmospheric humidity to depleted rainfall is expected (Bravar and Kavvas 1991). Consequently a drought cycle must be broken by a large-scale disturbance in atmospheric circulation.

Measurement of Energy Balance and CO_2 and H_2O fluxes in the Jornada Basin

The interactions of the energy and water balance with the composition of vegetation and soil processes determine the amount of carbon dioxide that is either taken up by vegetation or released by the terrestrial ecosystem to the atmosphere. Consequently some efforts to measure these balances have also included measurements of CO_2 flux. Surface energy balances and CO_2 and H_2O fluxes are commonly measured using one of two micrometeorological methods: the eddy covariance method, a direct measurement of flux (Baldocchi et al. 1996) and the Bowen ratio method, an indirect or calculated flux measurement (Dugas 1993; Dugas et al. 1999). The Bowen ratio technique has commonly been used in studies of the Jornada Basin. This technique measures gradients in both air temperature and air humidity over a vertical span above the canopy. The ratio of sensible heat flux to latent heat flux (H/LE) is in proportion to the ratio of these gradients. If measures of the net radiation flux density, R_n (the sum of the first three terms in equation 8-1), and soil heat flux density are taken, then LE can be computed, for example, as $(R_n - G)/(1 + H/LE)$. The method is commonly estimated as having an error band of 50–100 W/m^2 (Bertela 1989; Dunin et al. 1989). Errors in measuring net radiation are dominant (Kustas et al. 1998), latent heat flux is used to determine ET flux, and if additional CO_2 sensors are used, then CO_2 fluxes can be determined.

The first effort to determine how energy balance components varied over five distinct vegetation types during a two-year period used the Bowen ratio energy balance (BREB) technique (Dugas et al. 1996). Mini-soil lysimeters were employed to partition ET into soil E and plant transpiration. The communities measured included tobosa (*Plueraphis mutica*), black grama (*Bouteloua eriopoda*), tarbush, creosotebush, and mesquite (see chapters 10 and 11 for community descriptions). For all communities there were high percentages of bare soil, and high percentages of solar radiation incident on the soil surfaces resulting in high soil heat flux. Net radiation was similar in all communities with the exception of mesquite where midday R_n was 20% lower likely due to a highly reflective soil surface. Total accumulated ET was $\pm$ 35 mm of accumulated precipitation, indicating that nearly all precipitation was lost as ET. ET rates were comparable among all community types, with the exception of the tarbush community, where it was 50% higher. This was attributed to both increased herbaceous cover in this shrub community and its typical downslope position, which makes it the recipient of runoff water. The proportion of soil E to total ET ranged from 0.3 to 0.6, increasing in creosotebush and mesquite communities probably because of large open interspaces between shrubs. The ratios are biased toward high soil E rates because the mini-lysimeters were measured following precipitation events. In general, differences in energy balance components were minimal, reflecting changes in the size of interspaces and water availability. These minimal differences did not appear to be directly attributable to differences in plant community types.

Results of Dugas et al. (1996) were supported by research at the Sevilleta LTER site in New Mexico, approximately 340 km north of the Jornada Basin (Kurc and Small 2004). The Sevilleta study used BREB methodology to quantify difference in energy balance components over three summer monsoon seasons in a nearly monospecific black grama grassland site and a monospecific creosotebush site. Midday ET rates were similar between these two sites, indicating that shallow soil water is the primary source of water for ET losses and direct evaporation is likely a large component of ET. Midday ET rates were similar because other components of the energy balance varied between the two sites. The similarity in ET rates was attributed to differences in midday available energy. Midday available energy was higher at the grassland than the shrubland by 20% due to differences in net radiation and soil heat flux. Net radiation was greater at the grassland because emitted longwave radiation was decreased in the grassland, due to a 3% lower albedo of the grassland site. Differences in soil heat flux also contributed to greater midday available energy. Soil heat fluxes at the shrubland site at midday were 30% greater than the grassland, probably due to the 30% greater amount of bare ground in the shrub community.

Shorter intensive campaigns in the Jornada Basin have used the more direct method of eddy covariance (Hipps et al. 1999) in which vertical fluxes caused by atmospheric turbulent eddies are directly detected. These studies have been limited to upland grassland and mesquite-dominated communities. Precipitation events were not coincident with the studies, so that only the rapid decline of transpiration (or ET) during the two weeks following rainfall events was characterized. This study also compared BREB methodology with eddy covariance (EC) methodology and found that BREB measurements were in agreement with EC measurements for the grassland site, but that in mesquite coppice dunes BREB measurements were unable to resolve small gradients in LE fluxes during dry conditions.

Concerns over global climate change have lead to regional networks to measure net ecosystem exchange of carbon (NEE). Though much research has been done in forested systems, far less data exists in semiarid and arid rangelands. The next major effort to provide long-term measurement of NEE and ET in the Jornada Basin was part of a USDA ARS network of Bowen ratio towers in 11 rangelands sites. The Bowen ratio tower was installed on the Jornada in 1996. A compilation of the first six years of data (Mielnick et al. 2004) summarized NEE fluxes over this period (figure 8-1).

These data indicate that for the majority of days in a year, the Jornada Basin hovers around zero, or there is a slight source of carbon efflux to the atmosphere.

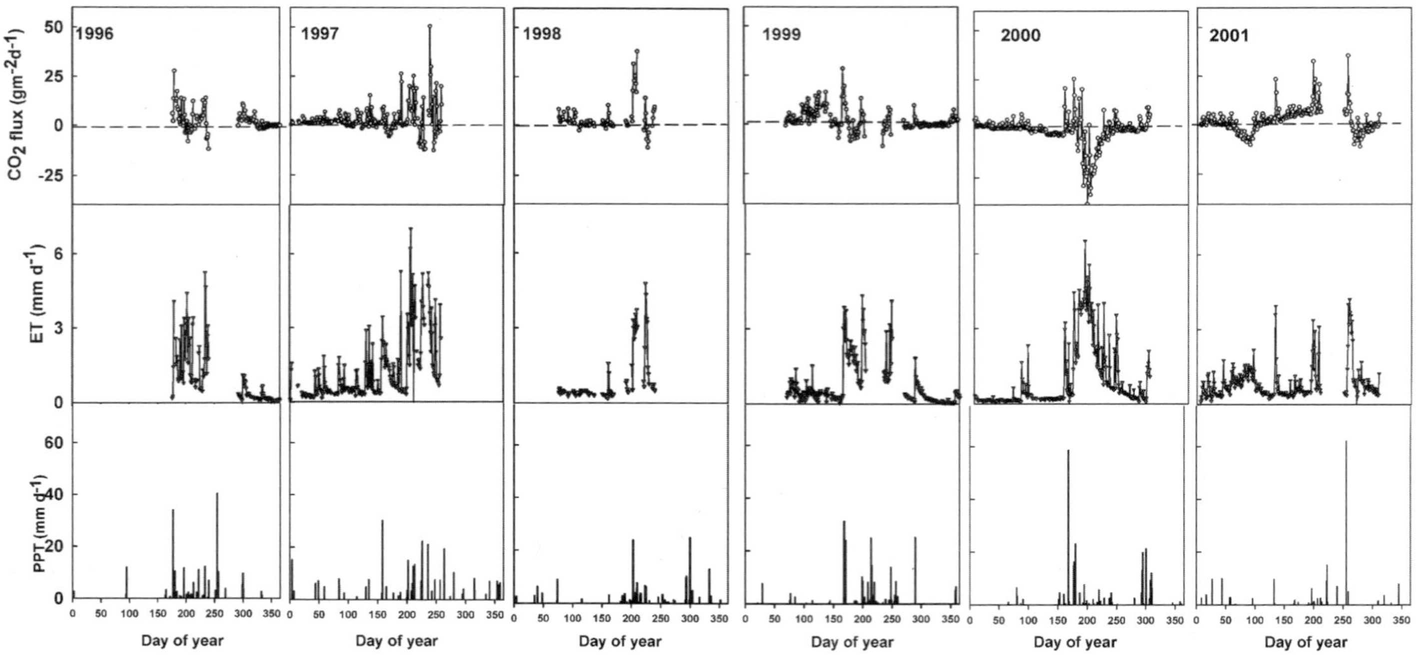

Figure 8-1. Daily total CO_2 flux, evapotranspiration (ET), and precipitation (PPT) for 1996 through 2001. (Negative CO_2 flux indicates uptake). Redrawn from Mielnick et al. 2004.

When rainfall events occur, there is an initial burst of carbon efflux, which is likely due to increased heterotrophic activity of soil microbes. However, if the rainfall is of large magnitude, the system may shift to net carbon uptake due to increased photosynthetic assimilation of plants and perhaps assimilation by biological soil crusts (Cable and Huxman 2004). The analysis by Mielnick et al. (2004) found that when averaged through time the Jornada Basin was a fairly significant source of carbon efflux through this time period (+145.3 g $C/m^2/yr$). Daily ET rates were at a maximum in July through September, and maximum ET rates ranged from 5 mm/day in dry or average precipitation years, but increased to 7 mm/day in the two wet years. Average ET rates ranged from 0.15 mm/day in December to 2.15 mm/day in August. Average yearly ET was 299 mm, indicating that as expected all precipitation was lost to the atmosphere via this pathway.

Although the interpretation by Mielnick et al. (2004) indicates that on average the Jornada Basin was a net carbon source during this time period, there are several potential limitations to applying this estimate to the heterogeneous Jornada Basin as whole. The instrumented site was a grassland site. This type of vegetation occupies only a small percentage (7%) of total vegetation cover in the Jornada Basin (Gibbens et al. 2005; see also chapter 10). Additionally, because of the nature of continuous data, periods of missing data existed. During a 6-year period, 1,243 days of actual data existed of the 2,190 total days possible (57% of the total days in a 6-year period), and the data averaging techniques provide no indication of the daily variance in CO_2 flux. So a small daily positive efflux number was multiplied by a large time period, which produced a large efflux estimate. Estimates of variance may indicate that small positive effluxes are not significantly different from zero. Last, even though this is a long period of record for CO_2 data, in terms of semiarid and arid ecosystem dynamics the period of record may not adequately reflect the rainfall variability. For example, during this period of measurement, if the Jornada Basin had high rainfall distributed throughout the growing season, the source/sink outcome may have been different. The Jornada Basin is characterized by pulses of rainfall and intervening dry periods, consequently data gaps or data-filling techniques based on previous periods of measurement may miss important pulse events that quickly change the CO_2 fluxes of these systems. These flux data support the conclusion that these grasslands are a potential carbon source. However, they also illustrate potential limitations of these methods in a highly pulse-driven environment, such as that of the Jornada Basin.

A simple reanalysis of these data illustrates the effect of variance on these estimates and the effect of eliminating one dry year and adding another wet year. We used daily data from 1996, 1997, 1999, 2000, and 2001 (1998 was excluded because the BREB tower was only working for 70 days). No attempt was made to fill missing data values. The idea was only to estimate the variance associated with daily CO_2 flux when the tower was functioning properly. Mean daily C flux ($\pm$ 1 SD) was 0.36 $\pm$ 1.7 (g $C/m^2/d$) and was significantly different than zero ($p < 0.0001$). On a daily basis this site exhibited high variability and could be either a source or sink, though on average this site was a small source of CO_2 for the period of record. To examine how another wet year would affect these results, 2001 data values from a dry year were eliminated and data from the wet year of 2000 were used twice in the same analysis. This produced a mean daily carbon flux of 0.07 ± 1.8 (g C/m²/d) and this mean value was not significantly different from zero ($p > 0.16$). These simple analyses demonstrate the importance of temporal variability and the need for longer term records of CO_2 flux that encompass the decadal variation in rainfall patterns (Snyder and Tartowski 2005). The difference in variability in CO_2 flux between the Jornada and a site in Niwot Ridge, Colorado, characterized by a coniferous forest with relatively steady-state declines in soil water availability through time, was addressed using composite flux duration curves (see Huxman et al. 2004). Composite flux duration curves were constructed from the probability distribution of CO_2 fluxes for multiple years and are analogous to stream flow duration curves (Potts and Williams 2004). The comparison between the two sites showed that in terms of CO_2 fluxes, the Jornada fluxes were more frequently near zero and were punctuated by infrequent positive and negative values of NEE. Niwot Ridge had a gentler slope, indicative of less variable NEE responses. CO_2 fluxes of the Jornada Basin site are strongly driven by large rainfalls that occurred infrequently during the period of record for the Mielnick et al. (2004) study. Although these large rainfalls are inherently infrequent, the period of CO_2 record may need to be longer to capture desert rangeland dynamics that are largely event driven.

Interactions of Plants with the Energy and Water Balance

The characteristics of the energy and water budgets in the Jornada Basin produce high incoming solar radiation, high variability in surface and soil temperature, and frequently low water availability, which also limits nitrogen availability. Consequently, plants of the Jornada Basin reflect a diverse array of adaptive strategies. Obviously, the major water-related challenge for plants is drought. Drought stress originates from both low soil water content and from high evaporative demand (i.e., atmospheric drought; Schulze 1986). Both types of drought stress are prominent in the Jornada Basin. Drought stress on the Jornada covers multiple time scales, from weeks between shallow wetting rain events in the monsoon to years of reduced precipitation (chapter 3). Specific plant strategies in controlling their physiological behavior, allocation, and phenologies are diverse and discussed elsewhere (see Gutschick 1987). In general, there are four basic strategies: drought escape, drought avoidance, drought tolerance, and drought endurance (Gutschick 1987). Most annual plants on the Jornada complete their life cycles outside the times of drought, therefore achieving drought escape. In drought avoidance, some plants, largely perennials, retain access to water supplies, by being phreatophytic (Neilson 1986) or by accessing deep water, and partially decoupling physiologic function from highly variable summer precipitation. Drought tolerance is the ability to tolerate low tissue water status and maintain continued physiological function. Common drought tolerance mechanisms are osmotic adjustment to maintain cell volume (and turgor in plants) and accumulation of solutes that protect membranes, (Jones 1992 discusses these mechanisms in plants). Other organisms show drought endurance, ceasing function to help maintain viability in drought. Cryptogamic cyanobacteria in soil crusts are an extreme example in that they tolerate drying to perhaps −100 MPa in water potential and temperatures up to 70°C.

Important correlates of summer drought are high temperatures and high radiation loads, which may have independent effects that must be understood to resolve plant tolerance to drought. Several direct interactions with drought are important. At the time of lowest water availability, relief of high leaf temperatures by transpirational cooling is least affordable. Jornada plants reduce the temperature load by having small leaves that have effective leaf–air heat transfer, so that even leaves in full sun with minimal transpiration do not rise far above air temperature. Among perennials, tarbush has the largest leaves with a minor axis of about 2 cm, and these leaves are mostly drought-deciduous and show the high temporal variation in response to rainfall patterns.

Dominant grass and shrub species are characterized by two different photosynthetic pathways—C_4 and C_3, respectively. C_4 plants possess an effective biochemical pump for concentrating CO_2. Consequently they can maintain higher intracellular concentrations of CO_2 (c_i) at lower leaf conductances and smaller stomatal apertures. Reduced stomatal aperture reduces the amount of water lost to atmosphere in exchange for gaining CO_2. Therefore the efficiency of water use, the ratio of carbon gain per unit water loss, is generally higher for C_4 species relative to C_3 species. The temperature optimum and light saturation point is also higher for the C_4 plants than for C_3 plants. These characteristics of C_3 plants may make them more adapted to dry environments, but as CO_2 levels increase, the relative advantage may become less important. Also, conservative use of water may not be the best strategy if competing species extract water at high rates.

Survival of long and severe droughts is only one aspect of drought fitness. Drought recovery is equally important in ensuring that plants respond rapidly enough and fully enough to use evanescent water resources. Assimilation and transpiration in creosotebush continue at low levels in severe drought. Net CO_2 assimilation remains in the range 1–3 μmol/m²/s in creosotebush as late as 4–6 h after sunrise, dropping to near zero later. However, dark respiration is significant, typically 1.5–3 μmol/m²/s. This respiration may be derived from continuous repair of photodamage; chlorophyll fluorescence shows strong down-regulation of photosystem II, but recovery is strong after 2 h in the dark and overnight. It is plausible that protective complexes form, as they appear to do in other species prone to photodamage (Ball et al. 1991; Gilmore and Ball 2000). Such complexes may explain protection in creosotebush, although the elaborate spectroscopic measurements have yet to be done. In any event, recovery of all function on relief of drought is rapid and complete. Photosynthetic capacity, stomatal conductance, leaf water potential, leaf relative water content, and even subjective greenness and leaf area index all recover within 10–14 days of sufficient cumulative rainfall.

Some of the dominant shrubs of the Jornada Basin, particularly creosotebush and mesquite, may be adapted to severe water stress by their hydraulic architecture. In particular, their xylem vessels resist catastrophic cavitation to very low water potentials. We have measured water potentials to our instrumentation limit of −6 MPa in creosotebush during summer months over multiple years. Narrow xylem vessels are correlated with this resistance, particularly for freezing-induced cavitation (Davis et al. 1999); so, too, are narrow interconduit pits in tension-induced cavitation (Pockman et al. 1995). However, these vessels bear the cost of low hydraulic conductance. This feature does not appear to limit ability to grow and use water resources when they are available. Leaf photosynthetic rates reach high levels on an aerial basis (15–20 mmol/m^2/s) after recovery from drought in creosotebush and, readily, 30 μmol/m^2/s in mesquite.

Empirical evidence exists, in a variety of systems, that drought fitness (DF) and water-use efficiency (WUE)[3] are negatively correlated (Thomas 1986; Grieu et al. 1988). A more mechanistic relation may be afforded by the observations for creosotebush. Continued respiration for photodamage control or repair requires continued assimilation and transpiration. This reduces season total WUE by requiring water expenditure at times of inherently low WUE (high vapor pressure deficit). Continuous recovery processes in drought-active plants may enable plants to quickly use evanescent water resources when they become available.

Conclusions

The Jornada Basin, like other arid regions, is a net source of sensible heat flux to the atmosphere, a weak source of latent heat flux, and a modest sink for net radiation. High solar radiation, high air and soil temperature, and low and variable precipitation strongly condition the water relations of plants in the Jornada Basin. Evapotranspiration is appoximately 95% of total precipitation inputs. Trade-offs between drought fitness and water use efficiency may be necessary to ensure rapid use of pulses of rainfall after intervening dry periods. In addition, characteristics of rainfall such as the size and frequency of rainfall events may elicit different responses from plant species that employ different drought strategies and fitness. The C_4 grasses are primarily active during the warmest months and, being shallow-rooted, rely principally on summer rains for their survival. Because summer rainfall is even more variable interannually than is total precipitation, the productivity of C_4 grasses is highly variable. The C_3 shrubs are remarkably resistant to a paucity of summer rainfall. These physiological differences between grasses and shrubs in this basin illustrate the importance of plant community composition on energy and water balances.

Carbon dioxide fluxes have been intermittently measured within grassland communities within the Jornada Basin. These flux data support a conclusion that these arid grasslands are a potential carbon source. However, CO_2 fluxes in the basin are strongly driven by large rainfalls, and a longer record of CO_2 flux in different vegetation communities is required before we understand these carbon dynamics.

9

Eolian Processes on the Jornada Basin

Dale Gillette
H. Curtis Monger

In arid and semiarid lands, soil erosion by wind is an important process that affects both the surface features and the biological potential of the ecosystem. The eolian flux of soil nutrients into or out of an ecosystem results in enrichment or impoverishment of its biological potential. In the Jornada Basin, wind erosion is the only significant mechanism for the net loss of soil materials because fluvial processes do not remove materials from the basin. Vigorous wind erosion leads to topographic changes, altering the growing conditions for plants and animals. Examples of such changes in topography are the formation of sand dunes or the removal of whole soil horizons. Our goal in this chapter is to describe the construction of a mathematical model for wind erosion and dust production for the Jornada Basin. The model attempts to answer the following questions:

1. Which soils are affected by wind erosion?
2. How does wind erosion occur on Jornada soils?
3. Does changing vegetation cover lead to a change in the source/sink relationship?
4. Is the Jornada a source or sink of eolian materials? If it is a source, what materials are lost?
5. How does wind erosion change the soil-forming process?

We will provide provisional answers for the questions and outline work that will more clearly define these answers.

Airborne dust has a significant residence time in the atmosphere and acts to modify the radiative properties of the atmosphere, mainly by back-scattering the incoming solar radiation (Andreae 1996). Changing land uses in arid and semiarid areas (e.g., overgrazing and cultivation) can drastically alter the dust emissions to

Table 9-1. Physical relationships needed for the wind erosion model.

1. Effect on $u*_t$ of aerodynamic roughness height (z_0), vegetation cover, and size distribution of loose soil.
2. $u*_t$ for crusts (biological and rain-physical).
3. Effect of soil moisture on $u*_t$.
4. Aggregation/disaggregation: destruction of soil crust by sandblasting.
5. Particle supply limitation effect.
6. Total particle mass flux (q) as a function of $u*$ and $u*_t$.
7. Owen effect (increase of z_0 with $u*$ above $u*_t$).
8. Ratio of vertical flux of dust to horizontal flux of coarse particles (F_a/q).
9. Effect of vegetation on wind erosion.

the atmosphere (Tegen et al. 1996). The climatic effects of soil-derived dust were investigated in an experiment in central Asia (Golitsyn and Gillette 1993). Using measured size distributions for emitted dust (Sviridenkov et al. 1993) and various real and imaginary indices of refraction (Sokolik et al. 1993), Sokolik and Golitsyn (1993) calculated climatic effects. Atmospheric dust decreased the total radiative balance of the underlying surface and at the same time induced general warming of the underlying surface–atmosphere system due to a decrease in the system albedo over the arid zones. Because of the climatic effects of dust, it is important to understand the mechanisms that determine the flux of dust in arid and semiarid locations. We need to understand whether humans are having an effect on the global flux of dust and consequently on that part of climatic change caused by a change of atmospheric burden of dust.

Assessing dust emissions requires a framework for identifying the importance of various mechanisms. This is a complex task because wind erosion involves nonlinear and threshold processes. Because the interactions are nonlinear and highly interactive, mathematical or physical modeling is desirable to predict the consequences of land use and to reconstruct past conditions affecting wind erosion. Some erosion processes are governed by universal relationships so well established (Greeley and Iversen 1985) that there is no reason to verify them. One such process is erosion resistance parameterized by threshold friction velocity ($u*_t$) for particles on smooth surfaces. Threshold friction velocity is a measure of the minimum wind force needed to sustain wind erosion; it is the minimum friction velocity ($u*$) to sustain wind erosion. Friction velocity is defined in equation 9-1. For other parameters in the wind erosion model (table 9-1), it is highly desirable to verify their application in the Jornada Basin.

Measurements of Threshold Friction Velocity and Small-Scale Aerodynamic Roughness Height, z_0

The soils affected most by wind erosion are those having the lowest threshold friction velocities ($u*_t$). We tested Jornada soils for threshold friction velocity using a portable wind tunnel described by Gillette (1978). The tunnel has an open-floored test section so that a variable-speed turbulent boundary layer can be developed over a flat soil containing small-scale roughness elements, such as pebbles and small aggregates of soil. The wind tunnel has a two-dimensional 5: 1 contraction section with a honeycomb flow straightener and an expanding rectangular diffuser attached to the working section in a configuration similar to that of Wooding et al. (1973). In field studies at the Jornada, the working section was 231 cm^2 in cross-section and 2.4 m in length, and the wind tunnel was placed in areas free of vegetation. Wind data were obtained 20 cm from the end of the working section at the midpoint of the tunnel width and at eight different heights spaced approximately logarithmically apart from 2 mm above the surface to 10 cm. The Pitot tube anemometer was calibrated against the NCAR reference wind tunnel and was corrected for the air density change caused by elevation above sea level. Data for the wind profiles were fitted to the function for aerodynamically rough flow.

$$U = (u*/k)\ln(z/z_0) \qquad (9\text{-}1)$$

where k is von Karman's constant (set to 0.4), U is mean wind speed, z is height above the surface, $u*$ is friction velocity, and z_0 is aerodynamic roughness height.

The threshold wind speed for wind erosion was defined to be that speed at which we observed small but sustained movement of particles across the soil surface. After slowly increasing the wind to the threshold of particle motion, we measured two sets of wind speed profiles. The following threshold profiles were obtained for each site. (1) For crusted soils, we measured the threshold for loose particles on the surface and for the destruction of the crust. For clay-rich soils of playa and other low elevation soils where there were no loose particles on the surface, only the breakup of the soil surface was measured. (2) At sites without crusts, the threshold for loose surface particles was measured. (3) At all sites, soils were disturbed using either livestock hoof impact or one pass of a 3/4-ton truck moving at a speed of about 8 km/h. These disturbances always created loose particles on the surfaces. Measurements were made of the loose-particle threshold immediately following the disturbance. For each site, two replicates of the threshold measurements were obtained. Sites were chosen to be representative of generic soil classifications made for the Jornada Experimental Range (JER): sandy, silty, clayey, and gravelly. Rock (nonerodible) surfaces were not tested.

A plot of all results from wind-tunnel measurements of $u*_t$ versus z_0 (aerodynamic roughness length) is shown in figure 9-1. The plot clearly shows that the lowest values of $u*_t$ are found for disturbed soils and undisturbed, noncrusted sandy soils. The values for $u*_t$ for sandy soils and disturbed soils (i.e., the lowest $u*_t$) versus z_0 values shown in figure 9.1 are shown in figure 9-2. The property that all the soils of figure 9-2 had in common was an abundant supply of loose surface particles of size $80 < d < 120$ μm. This size is mobilized at a minimum threshold friction velocity of about 21 cm/s for smooth surfaces. Consequently, even though the soil size distribution varied from place to place, this variation did not affect the value of $u*_t$ for a smooth surface. For disturbed and sandy soils,

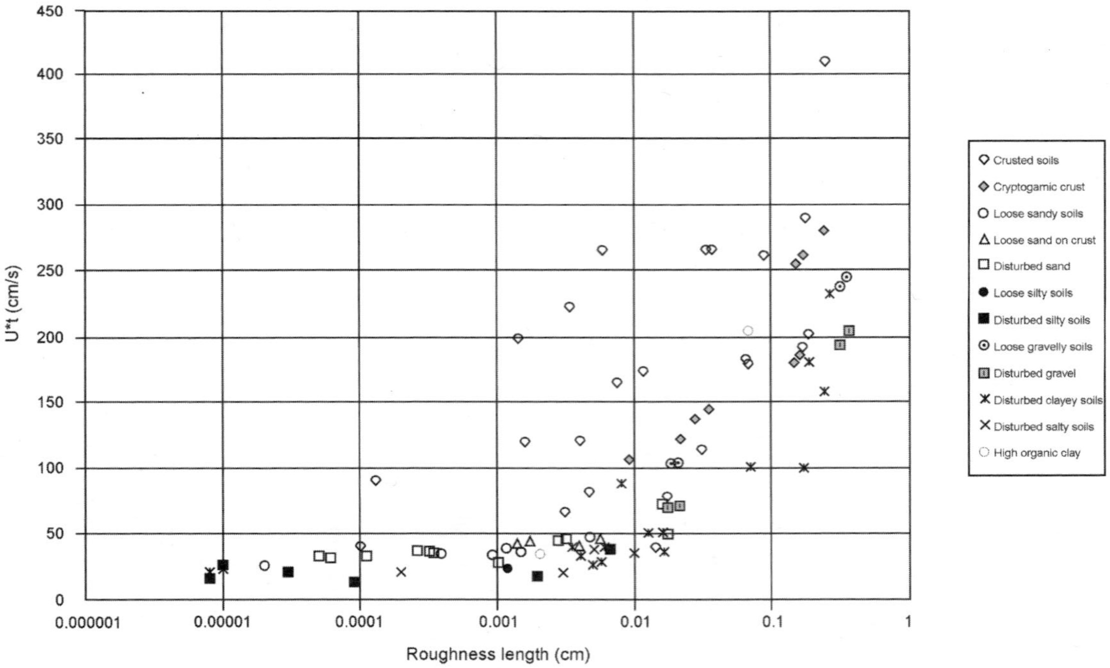

Figure 9-1. Values of threshold friction velocity, $u*_t$, versus aerodynamic roughness length, z_0 (from Marticorena et al. 1997).

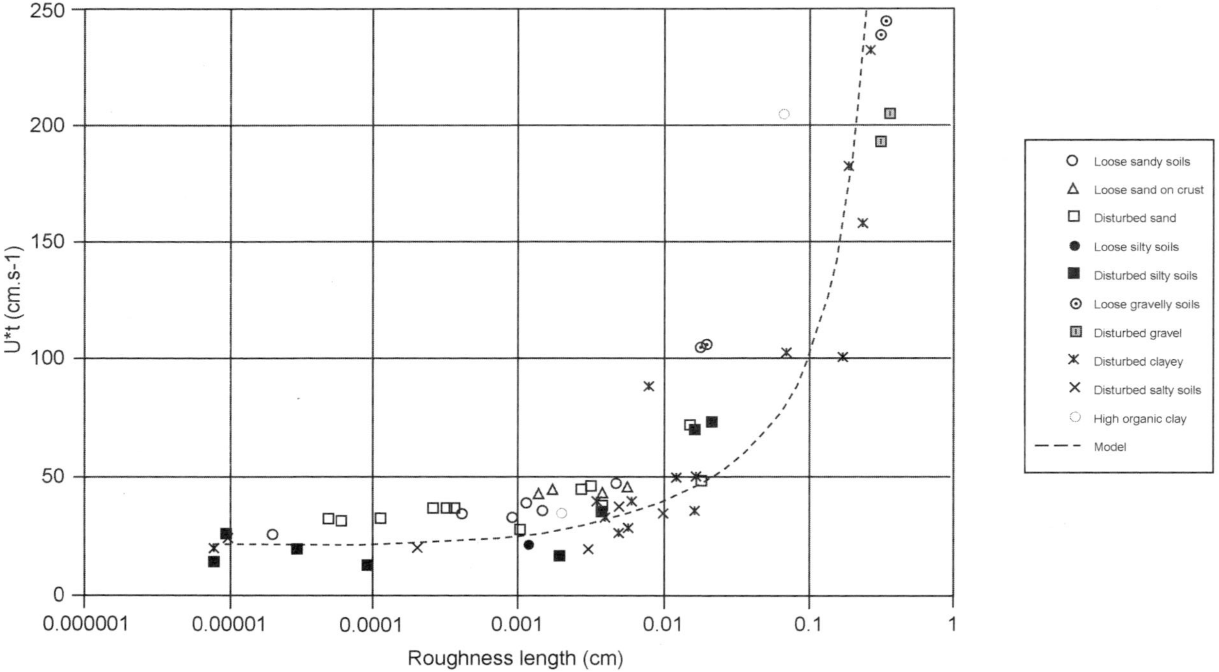

Figure 9-2. Values of threshold friction velocity, $u*_t$ versus aerodynamic toughness length, z_0 for undisturbed sandy soils and disturbed soils, not including gravels and cyanobacteria lichen crusts. The curve is a model of $u*_t$ versus z_0 assuming abundant particles $80 < d < 120$ μm and a momentum partitioning scheme described by Marticorena et al. (1997).

roughness of the surface z_0 controls $u*_t$. This roughness acts to absorb part of the momentum of the wind.

Cyanobacteria lichen crusts (CLC) and physical rain crusts (PRC) contain surface agglomerations of small soil particles bound by physical or biological agents. CLCs occur in open shrub and grass communities in arid and semiarid environments around the world as some combination of nonvascular microphytes (West 1990). Microphytes entangle and hold soil particles together, binding particles to form a crust that should resist wind displacement (Campbell 1979). For areas where vascular plants are sparse, CLCs help stabilize the soil against the wind (Williams et al. 1995). Belnap and Gillette (1997) showed that threshold velocities for sandy desert soils in southeastern Utah are significantly increased by CLCs, which protected soil otherwise at risk. PRCs are not biological in origin but bind particles together with silica, salt, gypsum, and clay, often forming crusts following individual rain events (Williams et al. 1995).

For both PRCs and CLCs, loose particles on the surface are usually much larger than size $80 < d < 120$ μm. Because the smooth threshold friction velocity of pieces of the loose surface crust was quite high compared to the size $80 < d < 120$ mm particles, both CLCs and PRCs had quite high threshold velocities and would erode only in unusually high winds. CLCs increase the threshold in two ways: The crusting roughens the surface, and the biological fibrous growth aggregates soil particles even after the crust is dry, as well as when the biological material is dead. Consequently, the CLCs are effective in the protection of the soil against wind erosion when undisturbed. When disturbed, the CLC loses some but not all of its protective qualities because disturbance smoothes the roughness and breaks the brittle aggregates. Threshold friction velocities for CLCs and PRCs at the Jornada were higher than the typical wind speeds recorded at the Jornada; consequently, areas of undisturbed CLCs and PRCs are areas without wind erosion (Belnap and Gillette 1998).

How Does Wind Erosion Occur on the Jornada Soils?

To estimate wind erosion at the Jornada, various physical relationships need to be developed or verified. At a study site (~ 1 ha) near the center of the JER, vegetation was cleared and a permanent array of instruments installed to measure erosion of a typical Jornada sandy soil subject to disturbance by grazing. The site can be expected to represent the most erodible of the Jornada soils.

The study site is roughly semicircular, having a diameter of > 100 m. Three meteorological towers of 2 m height were located on a line parallel to the dominant direction for wind erosion, southwest. Towers were located 50 m (west), 80 m (middle), and 110 m (east) from the southwestern edge of the clearing. Instrumentation on each of the three towers was as follows: wind speed at 0.2, 0.5, 1.0, and 2.0 m heights; air temperature at 0.2 m and 2.0 m heights; particle collectors at 0.05, 0.1, 0.3, 0.5, 0.6, and 1 m heights; and fast-response particle mass flux sensors at 0.05, 0.1, 0.2, and 0.5 m heights. In addition, markers were set in the soil so that increases or decreases in the height of the crust surface could be measured.

Data on the following relationships were collected:

- Owen effect (increase of z_0 with $u*$ above $u*_t$),
- q (total particle mass flux) as a function of $u*$ and $u*_t$ including particle supply limitation,
- $u*_t$ change by soil moisture,
- aggregation/disaggregation: destruction of soil crust by sandblasting, and
- ratio of vertical flux of dust to horizontal flux of coarse particles (F_a/q).

Owen Effect

The Owen effect is a feedback mechanism by which airborne sediment increases the drag coefficient of the surface. Before sand-sized particles are injected into the air (saltation), the air contains momentum that is transported by turbulent eddy transfer. During saltation, the sand grains interact with the air and transfer a part of the wind momentum to the ground. This occurs because the saltating particles strike the ground in a parabolic trajectory (implicit in the definition of saltation). These particles carry momentum absorbed at heights near the tops of their trajectories; this momentum transport is more efficient than that by air eddy transport. Consequently, momentum from greater heights in the air layer is required to replace the particle-transported momentum. This sequence of momentum transport is very similar to that caused by an increase of aerodynamic roughness length of the surface (z_0). The result is the formation of an internal boundary layer with a higher value of z_0. The increased $u*$ caused by saltating particles grows upward in the wind profile toward the height of the preexisting boundary layer. The Owen theory specifies that an increased z_0 will match the effect on the wind profile of the saltating particles. The effect is most easily detected as an increase of the ratio of friction velocity $u*$ to mean wind U. The Owen effect causes an amplification of sediment movement in a steady wind stream caused by nonhomogenous threshold friction velocity by locally increasing the drag coefficient. For a region having a uniform wind, the Owen effect can cause friction velocities to increase beyond those for areas that are not eroding.

Total Particle Mass Flux as a Function of u* *and* u*$_t$, *Including Particle Supply Limitation*

Although a generalized function for soil particle flux has been given by Iversen and White (1982), it is worthwhile to verify the formula with data from the Jornada. Figure 9-3 shows the response of four Sensit detectors at the middle tower at the Jornada test site on October 22, 1995. The data are categorized as "before 1500" (3 P.M.) and "after 1500." Sensit response is proportional to mass flux and these data were taken at 5-, 10-, 20-, and 50-cm heights. The before 1500 data show that the mass flux increases with the cube of wind speed in approximate agreement with the formula of Iversen and White (1982).

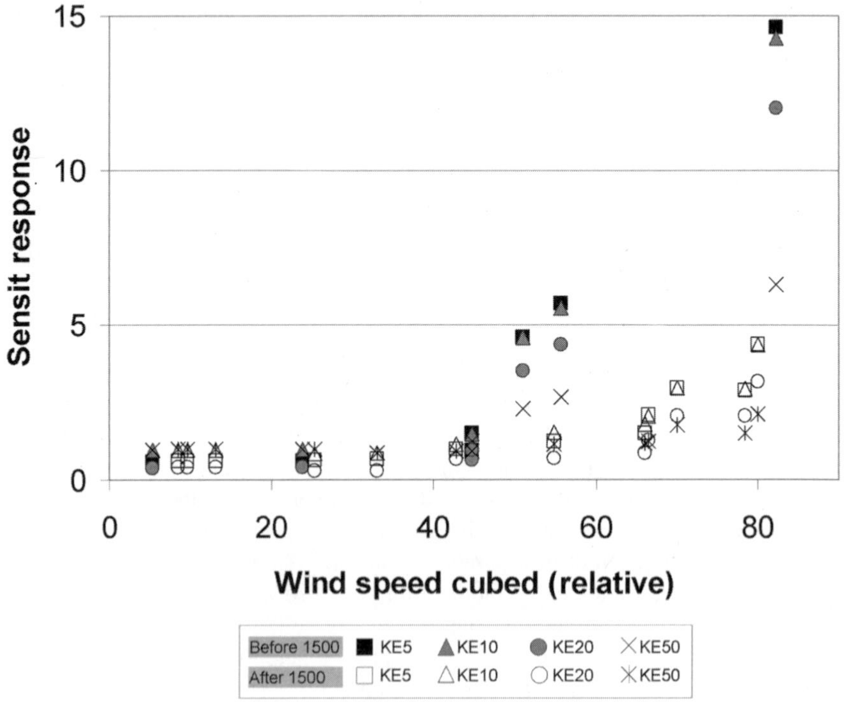

Figure 9-3. Thirty-minute mass flux at a cleared, sandy soil site recorded at the middle tower versus $u*^3$ for October 22, 1995, before and after 1500 h. The mass flux shows a linear relationship with the cube of friction velocity until 1500 h, then decreases.

The mass flux data after 1500 show a decrease of slope. Data obtained before 1500 represent the particle flux from a layer of loose material; those after 1500 represent the new lower flux rate reflecting crust material that has a small mass of loose particles. Consistent with this interpretation were observations that the sample site had a thin, loose cover of sand with several large bare patches of crust. The sharp decrease of the slope of mass flux from one constant value to another lower value probably coincided with the removal of a loose particle layer, leaving behind the supply-limited crust.

Soil Moisture Effect on Threshold Friction Velocity

Soil moisture can increase the threshold friction velocity of a soil (Chepil 1956; Bisal and Hsieh 1966; Saleh and Fryrear 1995). McKenna-Neuman and Nickling (1994) showed that sand grains are held together by the capillary effect of soil moisture. Qualitative observations of the effect of soil moisture on $u*_t$ at the Jornada are shown in figure 9-4 for August 7, 1996. Following a drop in air temperature and shift in wind direction at 1500 h, erosion flux increases just before the arrival of rain. Following the rain, soil moisture at 6 cm increases

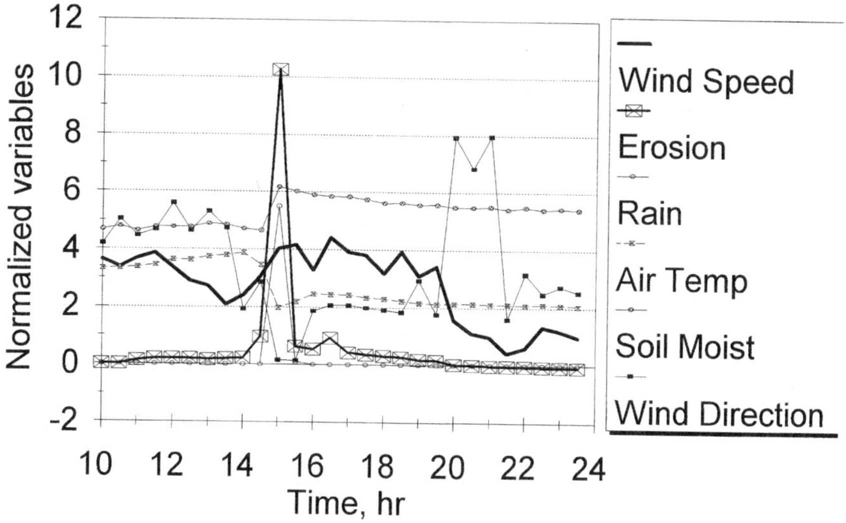

Figure 9-4. Simultaneous half-hour mean measurements at the west vegetation-cleared site on August 7, 1996, of wind speed at 2 m, sand flux at 10 cm height, rainfall, air temperature at 20 cm height, soil moisture at a depth of 6 cm, and wind direction. All measurements are in relative units to show a qualitative picture of wind erosion conditions.

quickly. Although the wind reaches a higher value at 1630 than at 1500, there is greater mass flux at 1500. At 1830 h, the mass flux is zero even though the wind speed is about the same as it was at 1500 (the time of maximum wind erosion). This shows that threshold friction velocity increases with moistening of the soil. However, the figure also shows that the soil dries out quickly, so the effect of soil moisture is short-lived in this desert location.

Crust Formation/Disaggregation

The largest cause of the temporal and spatial variability of threshold friction velocities is the aggregation and disaggregation of soils that can change both $u*_{ts}$ (threshold friction velocity for smooth surfaces) and z_0. Aggregation is a mechanism by which individual particles in the surface sediment become effectively larger particles or crusts. Aggregation or crust formation usually occurs with the drying of a moistened surface. Drying is a complicated process, and the aggregation that occurs is affected by the composition of the soil, evaporation rate, and temperature. Destruction of the crust and surface aggregates depends on sandblasting, freeze-thaw cycles, formation of crystals, and temperature of the sediment. Observations of crust destruction on the Jornada site are shown in figure 9-5. This figure shows the abrasion of the Jornada crust versus total sand passage over the surface. Both the vertical flux of kinetic energy on the crust and the sand passage are proportional to the cube of the wind speed. Because soil aggregates are known to abrade as a func-

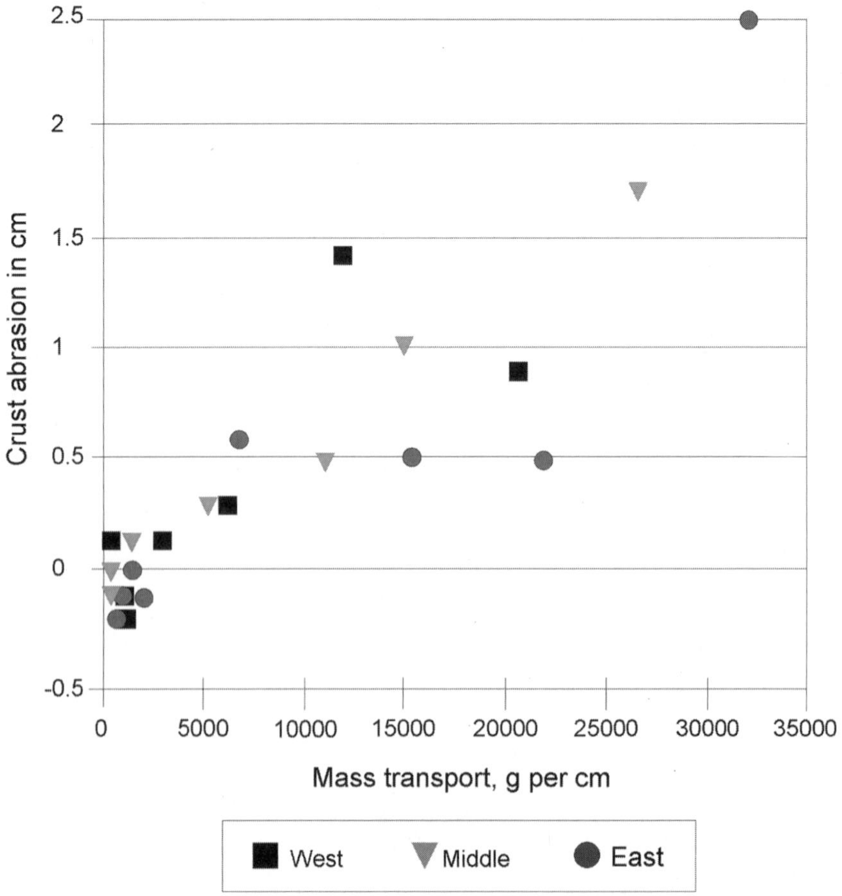

Figure 9-5. Crust abrasion plotted against total sand movement for the three towers (west side, middle of site and east side) at the vegetation-cleared site for August 1995 to March 1996.

tion of the kinetic energy of the sandblasting (Hagen et al. 1992), the linear relationship of the crust abrasion to mass transport q is not surprising.

Relation of Vertical Flux of PM_{10} Dust to Total Particle Mass Flux

Shao et al. (1993) modeled the vertical flux (F_a) of wind-eroded particles smaller than 10 mm (PM_{10}) and related that flux to the saltation-particle mass flux. This model states that F_a is proportional to the vertical flux of kinetic energy of saltating grains. The vertical flux of kinetic energy of saltating grains is proportional to the horizontal flux of the saltating (hopping) grains, as expressed by the Shao et al. (1993) model in equation 9-2:

$$F_a = Km_d(g/\psi)qf(V_H/u*)_2, \qquad (9\text{-}2)$$

where ψ is binding energy, K is a constant, m_d is mass per particle, $f(V_H/u*)$ is a nondimensional function and $q = \int_0^\infty CV_H(z)dz$ where V_H is horizonal speed of the sand grains.

Owen's (1964) theoretical analysis of saltation showed that $V_H/u*$ may be regarded roughly as a constant. The value of g (acceleration of gravity) is also roughly constant for the Earth's surface. The model predicts that F_a/q is a function only of m_d/ψ, the ratio of the mass per saltating particle to the binding energy. Mass per saltating particle reflects the size distribution of the saltating material; coarse sand would have higher mass per particle than fine sand. Binding energy of the PM_{10} particles to larger grains or in aggregates could vary with the following: differences in the texture of the soil or sediment from which they are eroded (i.e., the particle size distribution of the source), chemical composition, clay mineralogy, salt, organic matter content, and a variety of physical properties of the source material, including the (changing) size distribution of soil aggregates as affected by wetting, drying, freezing, thawing, and erosive processes such as sandblasting. Gillette et al. (1997a, b) presented data in figure 9-6 that shows F_a/q_{tot} results for the measurements made in Texas and California.

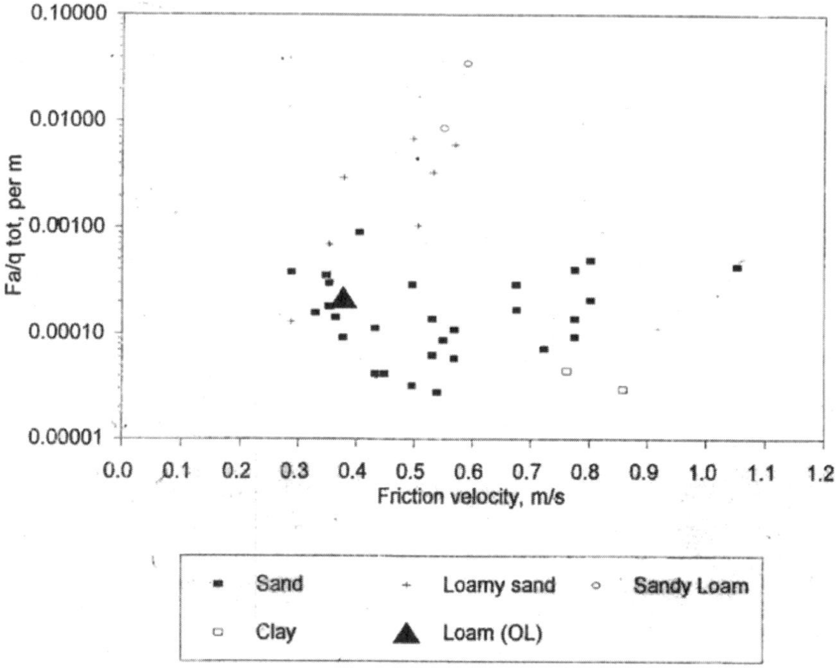

Figure 9-6. Ratio of vertical flux (Fa) of particles smaller than 10 μm (PM10) to horizontal flux of total particle mass (q) versus friction velocity for experiments for which different soil textures were found (from Gillette and Chen 2001).

F_a/q_{tot} for sandy soils does not appear to be a function of friction velocity, but we do not have enough data for other soil textures to evaluate a relationship with $u*$. Clay soils seem to produce less fine particle flux for a given total mass flux, whereas loamy sand soils produce more; the datum from the loam soil is in the middle of the data for sandy soils. The size distributions for q_{tot} values used in figure 9-6 (Gillette and Chen 1999) have a modal value for saltating particles in a range of 120 ± 20 m for all the soils except the clay. The mode of the clay size distribution was roughly three times that for the other soils, which corresponds to a mass about nine times larger. The large modal value for clay corresponds to aggregated particles, not the size of individual clay platelets. Thus m values are highest for clay, intermediate for sand and loam, and lowest for loamy sand.

Effect of Vegetation on Wind Erosion at the Jornada

Airborne particle fluxes for typical vegetated areas of the Jornada LTER were measured at 15 vegetated sites and at 3 vegetation-free sites on sandy soil. The effect of vegetation was demonstrated by calculating the ratio of the fluxes for the vegetated sites to the mean flux for the three vegetation-free sites. Measurements of particle fluxes at the 15 vegetated sites began in January 1998. These are the sites used for measurements of aboveground net primary production (ANPP; chapter 11)—five vegetation categories having three sites each. The vegetation categories are black grama (*Bouteloua eriopoda*) grassland, creosotebush (*Larrea tridentata*), tarbush (*Flourensia cernua*), playa grassland (characterized by *Pleuraphis mutica*), and mesquite (*Prosopis glandulosa*). Results for January 30 through April 30, 1998 (a period of active wind erosion), are summarized in table 9-2. All three of the mesquite sites were strong sources during this period. Of the five vegetation groups at the Jornada, only the mesquite group is a large-scale dust emitter. The mesquite sites contributed dust at a rate roughly a tenth as large as that from a disturbed bare soil site. Although the other four groups

Table 9-2. Ratios of the mean particle movements at the vegetated sites to that for the vegetation-free site for January 30–April 30, 1998. Three different plots for each vegetation type were used.

Vegetation Type	Mean Ratio	SD
Tarbush	0.0006	0.0002
Black grama grassland	0.0007	0.0003
Playa grassland	0.002	0.002
Creosote	0.002	0.001
Mesquite	0.09	0.06

can emit dust during circumstances such as prolonged drought, heavy disturbance, or energetic dust devil activity, they would not normally be significant contributors compared to disturbed bare soil and mesquite sites.

Vegetation acts as a noneroding roughness element that affects the erosion threshold in two ways: (1) directly covering part of the surface and thus protecting it, and (2) absorbing part of the wind momentum that is not then available to initiate particle motion. This momentum partitioning leads to a decrease of the wind shear stress acting on the erodible surface and consequently reduces the erosion efficiency.

Gillette and Stockton (1989) used a scheme developed by Marshall (1971) to calculate the effect of momentum partitioning on threshold friction velocity. Soils having the lowest $u*_t$ were the soils most sensitive to changes of vegetation cover. Ratios of threshold (bare)/threshold (vegetated) versus a measure of the geometry of vegetative cover, L_c, were given by Gillette and Stockton (1989). A formula summarizing these data is

$$u*_t = u*_{t(bare)} \, e^{kLc/2}, \tag{9-3}$$

where k is a constant ($\sim$ 14). The measure of vegetative cover L_c is the ratio of the frontal silhouette area of plants, as seen by the wind, divided by the total land area, including the plants and bare soil. The experimental data represented by equation 9-3 is for solid hemispheres that were randomly placed in the erodible material. Becase the lowest threshold velocity for smooth sandy soils is roughly 20 cm/s and friction velocities larger than 100 cm/s are very rare, a value for L_c of 0.23 should provide good protection for the soil from wind erosion. Assuming a hemispherical model for plant geometry, the fraction of the ground covered by plants (looking down onto the ground) is $2L_c$. Calling this ratio A_c, this model would predict good protection for A_c larger than 0.46. The black grama grassland and playa sites had A_c values of the order 1, so excellent wind erosion protection should be expected. Creosote bush sites do not show significant fluxes in table 9-2 and were visually estimated to have L_c values of about 0.25. Values of this order would be expected to give good protection even on the sandy soils.

All the mesquite sites possessed sandy soils that correspond to vegetation-free threshold friction velocities of 20–30 cm/s. Unlike other vegetation types on sandy soil, however, the mesquite sites have significant erosion fluxes. The mesquite sites had A_c values of 0.44, 0.30, and 0.4. A_c values equal to or larger than 0.4 would normally be associated with moderate or better wind erosion protection. The observation of wind erosion where little would be expected according to equation 9-3 led us to suspect that mesquite is a class of vegetation that is not well described by equation 9-3.

One assumption in equation 9-3 is that individual shrubs are randomly oriented in the field. Visual inspection of mesquite at the Jornada suggests a nonrandom orientation of the mesquite plants superimposed on a random orientation, such that a preferred orientation along the direction of strongest wind erosion (southwest–northeast) could exist. Strong areas of wind erosion along with strong areas of deposition were observed along with the orientation.

Marticorena and Bergametti (1995) advanced a physical formulation of the drag partition between the roughness elements and the erodible surface from an approach developed by Arya (1975). The overall threshold friction velocity $u*_t$ is expressed as the product of the threshold friction velocity for a smooth surface $u*_t$ (the air–particle interaction of a particle of diameter D_p for a smooth surface having aerodynamic roughness length z_{0s}) and an efficiency factor f_{eff} that expresses the fraction of the wind stress that is available to act on the particle for a surface having aerodynamic roughness z_0 that is rougher than the smooth surface z_{0s} (Marticorena et al. 1997). Marticorena and Bergametti (1995) showed that

$$u*_{ts}/u*_t = 1 - (\ln[z_0/z_{0s}]/\ln[0.35(10/Z_{0s})0.8]).^{e} \qquad (9\text{-}4)$$

Thus, a rough surface increases the overall threshold friction velocity above that for a smooth surface that has the same particle size distribution. The relationship of u_t versus z_0 was confirmed by data of Marticorena et al. (1997). For a sandy soil having a smooth threshold friction velocity of about 20 cm/s, roughness lengths larger than about 0.2 cm would result in overall friction velocities of more than 150 cm/s. Consequently, surface roughness leading to roughness lengths larger than 0.2 cm provides good protection against wind erosion.

The tarbush sites have measured z_0 values that were larger than 0.2 cm. In addition, the bare silty soils between plants had threshold velocities larger than 100 cm/s. In some cases, the soils between tarbush plants had CLCs. For the soils with CLCs, threshold friction velocities were largely beyond the power of our equipment. The above-surface conditions combined make tarbush areas at the Jornada unlikely source areas for dust. Equation 9-4 was not applied to the creosotebush, black grama grassland, or playa sites because we had no measurements of z_0 at those sites. Values of z_0 could be roughly estimated for mesquite sites because for areas of abundant roughness, a linear relationship exists between z_0 and height of the roughness elements (bushes). However, application of equation 9-4 to the mesquite data of table 9-2 also leads to poor predictions. Values of z_0 for the mesquite sites are estimated to be in excess of 1 cm. Strong protection predicted by equation 9-4 for such heights was not observed for mesquite.

For application of equation 9-4, it seems that there needs to be a clear distinction of local z_0 and the z_0 appropriate for an ensemble of vegetation and smooth soil. In mesquite areas, there are local areas of several meters wide by tens of meters long that may have very small z_0 values. However, when taken as a whole, the area containing mesquite shrubs may have z_0 well in excess of 1 cm. When wind conditions are right, local areas with small z_0 values are activated in the mesquite dunes even though the area as a whole would be predicted by equation 9-4 to not have wind erosion.

In short, the Jornada possesses a kind of vegetation (mesquite-dominated) that is a relatively poor protector of the soil. Use of existing theory on the protective effect of mesquite does not predict the rather significant amounts of wind erosion observed at the three Jornada mesquite ANPP sites. For the other kinds of vegetation at the Jornada ANPP sites (see chapter 11), however, the theory seems to agree with observations that protection of the soils is adequate.

Is the Jornada a Source or a Sink of Eolian Materials? What Materials Are Lost If It Is a Source of Eolian Materials?

From the above work on vegetation versus wind erosion in each major type of vegetation, it seems that mesquite areas and disturbed areas are significant source areas of dust. Areas having other kinds of vegetation might occasionally be source areas but are more likely to be depositional areas for dust. To assess whether the Jornada is a net source or sink of eolian materials, we need to know the proportion of land covered by mesquite. We must also know the large-scale atmospheric deposition at the Jornada (i.e., how much dust is deposited at the Jornada that comes from large distances). To make this estimate, we need a large-scale emissions/transport/deposition model. Until the model is ready, however, estimates of the vertical flux of eolian material that is carried beyond the borders of the Jornada Basin will be deduced from historical data.

Historical Deposition Rates of Eolian Material

In 1962, seven dust traps consisting of 30- by 30- by 5-cm pans filled flush with 1-cm diameter glass marbles were placed at a height of 90 cm (Gile and Grossman 1979). They were placed in the field each year for 11 years from February through June (the dusty season of the year). The traps were sampling in the dustiest time of the year, and the catch for that interval was considered the total year deposition. Results of deposition averaged for 11 years (1962–72) are shown in table 9-3 along with the size distribution of the particulate material. Using an average bulk density of 1.4 g/cm^3 soil would accumulate at a rate of 2.4 cm per 1,000 years. Thirty-four percent of the deposited material is silt-sized and 24% is clay-sized material.

Table 9-3. Mass deposition in the Jornada Basin, 1962–1972 (after Gile and Grossman 1979).

Trap No.	Deposition (g/m^2/yr)	Particle Size Distribution in % by Mass (mm)				
		2–0.25	0.25–0.1	0.1–0.05	0.05–0.002	<0.002
4	15.7	2	14	14	43	27
5	26.3	2	20	22	34	22
6	58.6	4	27	21	26	22
Average	33.5	2.7	20.3	19	34.3	23.7

Historical Mass Vertical Flux of Eolian Materials

Gibbens et al. (1983) estimated the long-term gross erosion rates and net soil loss rates for three sites at the JER. For one of these sites, Hennessy et al. (1986) determined the size distribution for the soil material that had been lost to the site. This site was labeled the natural revegetation exclosure. Mean rates of net soil loss for 1933–80 were established by measuring the change of level on grid and transect stakes. The mean rate of soil loss per year for the deflated areas at the site was 5,200 $g/m^2/yr$, whereas the mean rate of net soil loss (for the entire area) for the same site was 1,400 $g/m^2/yr$. Hennessy et al. (1986) determined that there was almost no net loss of sand from the site as a whole; silts (84%) and clays (16%) accounted for all the soil lost. The mean rate of deposition for particles smaller than 50 mm (silt and clay) for the three sites cited in table 9-3 is 19.43 $g/m^2/yr$. This rate of deposition is much smaller than the mean source rate at the natural revegetation exclosure site of 1,400 $g/m^2/yr$. Because the area of sandy soil covered by mesquite vegetation is a significant fraction of the JER (more than 10%) and the rate of emission is 73 times the rate of deposition, the Jornada is almost certainly a source area for dust.

For prolific sources of soil dust like the centers of farm fields during dust storms, parts of the Sahara Desert, and parts of Owens (dry) Lake, California, the flux is limited by the momentum from the wind. Gillette and Chen (2000) have found that the supply-limited, vegetation-free source at the Jornada is approximately one-third as emissive as supply unlimited sources. The mesquite-covered, sandy soils of the Jornada would therefore be expected to have emissions rates of the order of 3% (one-third of 9%) of the above prolific dust producers (see table 9-2).

Gillette et al. (1974) found that long-distance transport of emitted soil particles occurs for particles whose sedimentation velocity is less than one-tenth of the vertical root mean square velocity. Practically speaking, this corresponds to particles smaller than about 10 mm. Patterson and Gillette (1977) showed that the typical size distribution of wind erosion particles smaller than 10 mm is roughly log normal with a number mode at 0.48 mm and geometric standard deviation of 2.2. Gillette et al. (1978) showed that aircraft-obtained size distributions of dust in southeastern New Mexico dust were quite similar to what Patterson and Gillette (1977) described as typical size distribution for dust storms. Pinnick et al. (1985) determined size distributions for blowing dust at White Sands Missile Range (less than 100 km from the Jornada). They also concluded that the dust distributions were similar to what Patterson and Gillette (1977) described as bimodal log normal distributions having "about the same mode radii." Pinnick et al. (1993) gave size distributions parameters for the log normal distribution of particles between 0.2 and 30 mm for a dust storm at Orogrande, New Mexico (also less than 100 km east from the Jornada), and those parameters (0.7 μ and 2.2) were very similar to observations by Patterson and Gillette (1977).

Erosion Rates Measured on Vegetation-Free, Sandy Soil Locations

Figure 9-7 shows the abrasion of the crusted surface at the instrumented and cleared site at three measuring locations spaced at 30-m intervals. Figure 9-8 shows the mass fluxes at the same three locations at different points in time. Note in January 1996 an erosion event removed 0.8, 1.7, and 2.5 cm of material from the surface. These large values compare to a mean annual lowering of the soil surface by 1.97, 2.9, and 3.3 cm/yr. Using an average bulk density of 1.6 g/cm^3 measured from seven crust samples obtained at the location, the annual mean masses of surface material lost may be calculated. Multiplying these annual mean mass losses by the ratios of silt and clay to total soil mass for the west, middle, and east sites (0.197, 0.149, and 0.118, respectively) gives the annual mean vertical loss of silt and clay particles. These loss estimates are 6,140, 6,904, and, 6,106 g/m^{2}yr.

These emissions are larger than the annual mean net emission rate of 1,400 g/m^2/yr reported by Gibbens et al. (1983) above; however, the instrumented site is bare, and the sites evaluated by Gibbens and Beck (1988) were protected by mesquite plants. The ratio of the long-term loss of silt and clay reported by Gibbens et al. (1983) for the mesquite site to the 2.3-year average for this bare site is 0.22. This ratio is more than twice the mean ratio of 0.09 reported in table 9-2 (the ratio of erosion at the mesquite sites to erosion at the vegetation-free sites), but we consider this to be fair agreement.

From visual inspection of large areas of sand deposition near the vegetation-

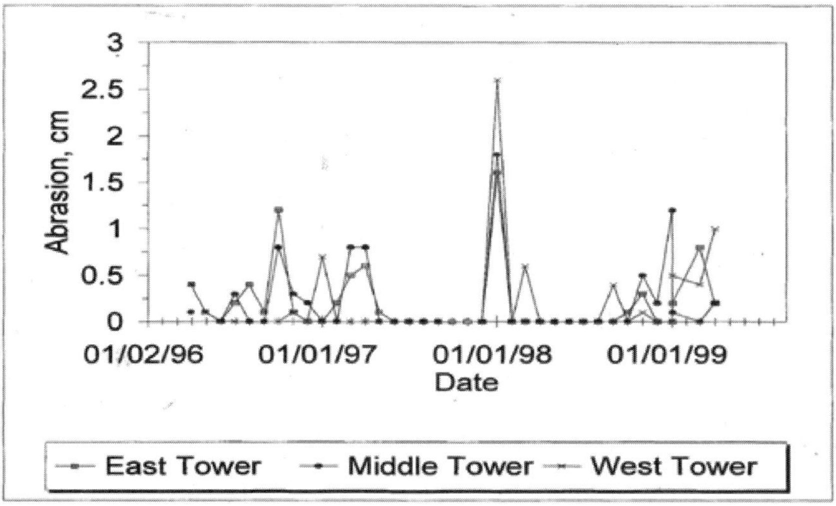

Figure 9-7. Abrasion of the surface crust (progressive distance of the crust surface from an unchanging height above ground) for the three towers at the vegetation-cleared site (from Gillette and Chen 2001) by month from 1996 to 1999.

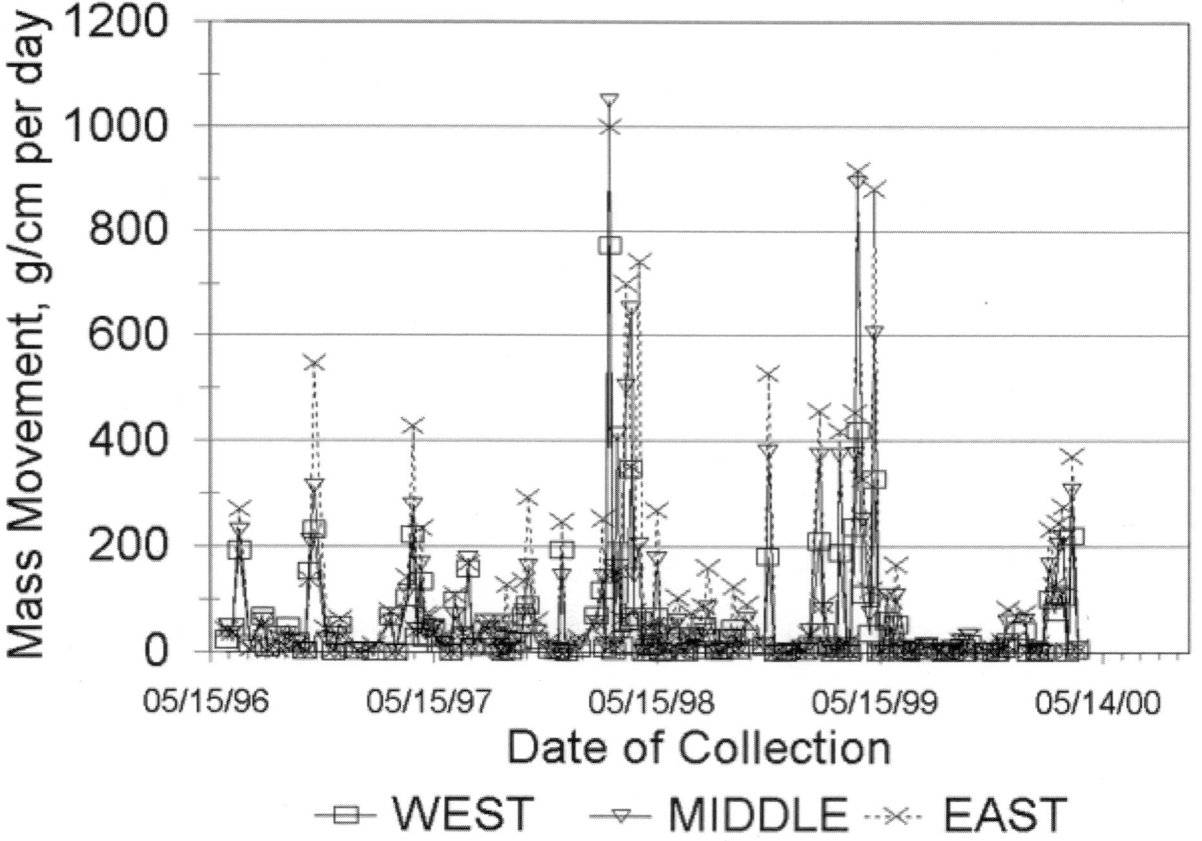

Figure 9-8. Total particle mass movement for the three towers at the vegetation-cleared site (airborne particles moving parallel with the wind and perpendicular to the ground) for a 4-year period in units of g/cm (from Gillette and Chen 2001).

free site, much of the eroded sand was deposited downwind within about 300 m of the point of emission. The loss of silt and clay from the vegetation-free site is about 5–10 times higher than the loss of silt and clay from the mesquite areas. However, the area of disturbed land similar to the vegetation-free site (for example roads, cattle containment areas, and other disturbed areas) is small, so that the important source of windborne dust is the sandy soil covered by mesquite.

How Does Wind Erosion Change the Soil-Forming Process?

Wind is a major geomorphic force in the Chihuahuan Desert and is responsible for many of the soil patterns in sandy areas (Gile 1966a). The Jornada region contains soils that when traced laterally can be seen in their eroded state, their partially eroded state, and their uneroded state (figure 9-9a). Continued tracing of these soils into areas where eolian sediments have accumulated, such as coppice dunes, shows the effects of progressive burial (figure 9-9b).

As described by Simonson (1959), there are four soil-forming processes that transform material at the Earth's surface. These are losses, additions, translocations, and transformations. In erosional settings (figure 9-9a) losses are dominant and responsible for the truncation of many soil profiles in the Chihuahuan Desert (Monger 1995). Because the A horizon (i.e., topsoil) of any soil contains much of the humified organic matter, N, available P, microbial population, and the seed bank, the erosion of this layer results in a loss and redistribution of these biotic components (e.g., Schlesinger et al. 1990). Furthermore, because water-holding capacity is largely controlled by organic matter and silt content (Herbel et al. 1994), removal of these constituents will decrease a soil's ability to retain water. Organic matter is also important for aggregate formation (Brady and Weil 1996). Its decline therefore leads to a decline in infiltration, which in turn leads to greater runoff (Bull 1991), causing hill slope erosion in addition to wind erosion and the perpetuation of bare ground. This bare ground is not only susceptible to further erosion but also a thermally harsh environment in which seedlings do not easily become established (Davenport et al. 1998).

For soils with calcic and petrocalcic horizons, truncation of the profile leaves those horizons near or at the surface. When this happens, the horizons degrade physically as the result of root growth and burrowing animals (Gile 1975b) and degrade chemically by dissolution by percolating waters. Theoretically, this situation would cause the calcic or petrocalcic horizons to change from a reservoir of atmospheric CO_2 to a source of atmospheric CO_2 unless the dissolved products make it to the groundwater or the ocean. Another important aspect of eroded calcic and petrocalcic horizons is their role in water storage. Microporosity causes water to be held more tenaciously than water is held in soil material with larger pores (Hennessy et al. 1983b). Consequently, calcic and petrocalcic horizons are important for preserving sources of water below the layers during droughts for both grasses (Herbel et al. 1972) and shrubs (Cunningham and Burk 1973).

Also in the erosional settings illustrated in figure 9-9a, the other three soil-

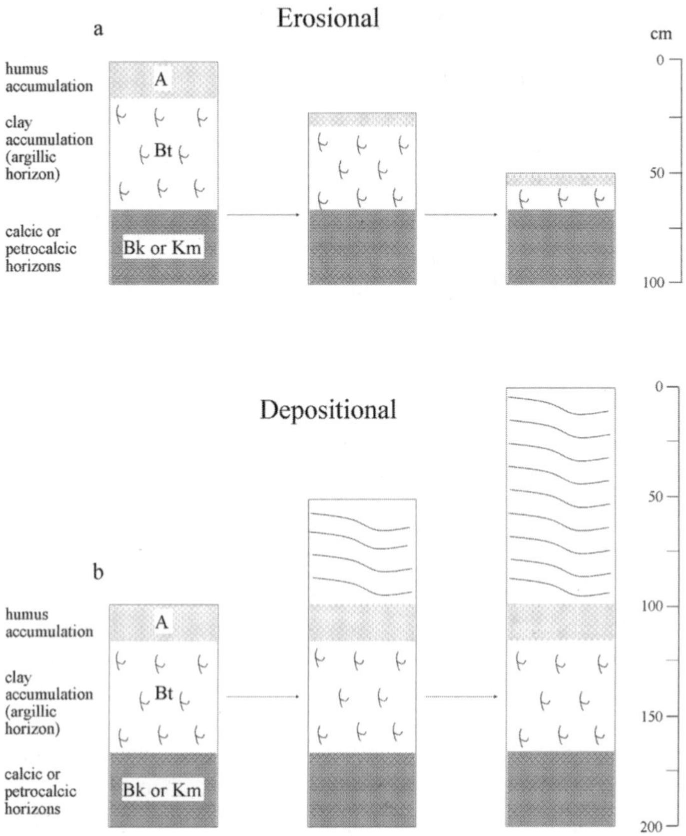

Figure 9-9. Illustration of the effects wind erosion has on sandy soils of the Jornada region. (a) Erosional setting illustrating the truncation of the soil profile and exhumation of the calcic or petrocalcic horizons. (b) Depositional setting illustrating the progressive burial of an existing soil profile by sandy sediments, such as in the formation of mesquite coppice dunes.

forming processes are very much modified by the losses process. Because of increased bare ground, additions to the soil profile, in the form of dry dust, wet dust, ions in rain, and organic matter, would not be retained as readily as on noneroded soils covered with vegetation. Because of greater runoff, translocations of particles and ions into and through the soils are curtailed in bare soils (Gile et al. 1969). For the same reason, transformations, such as chemical weathering and carbonate formation, would diminish in comparison with vegetated noneroded soils.

In the depositional settings (figure 9-9b), the four soil-forming process are dominated by the additions process. In this case, aggradation of the land surface lifts the depth of wetting progressively higher in the soil profile. Consequently, the zone of carbonate formation, the zone of clay accumulation, and the zone of organic matter accumulation rise with additions of accreting sediments. Based on the coarse texture of coppice dunes (Gile et al. 1981), most of the additions are in the form of saltating sand particles. However, suspended particles, organic matter, and associated nutrients also accumulate, as well as ions in rain funneled into the soil along plant stems (Whitford et al. 1997). Cumulatively, these additions produce one form of the islands of fertility described by Schlesinger et al. (1990; 1996).

Losses in the depositional setting are mainly derived from silt winnowed from sand (Hennessy et al. 1986). Little evidence of translocations has been found in coppice dunes (Gile 1966b; Gile and Grossman 1979). The absence of translocated material, however, reflects the young age of the deposits (Buffington and Herbel 1965). In some cases, 86.9 cm of sandy sediments have accumulated in 45 years (Gibbens et al. 1983). Similarly, transformations are minor in coppice dunes because of their young age. However, some biomineralization of calcite on mesquite roots and associated fungal hyphae has occurred, as well as decomposition of plant litter.

Conclusions

Based on the physical principles that the wind erosion model incorporates, we can give provisional answers to the questions asked in the first section.

1. From our measurements at the JER, undisturbed sandy soils populated by mesquite and disturbed soils of all types would be expected to be erodible. Other soils are erodible, but only at very high winds that would be experienced only rarely. The most vulnerable soils are the sandy soils that make up about half of all the Jornada Basin soils. The silt soils, playas, and gravelly soils are less vulnerable to wind erosion, and clay soils are less vulnerable than sand soils but more vulnerable than silt/playa/gravel.
2. Wind erosion at the Jornada is governed by several physical mechanisms. Beyond a threshold wind speed, mass flux increases at about the cube of the wind speed. Other important variables affecting wind erosion are length of bare soil between vegetation, soil crusting (biological and physical) and soil moisture, which has a small but measurable effect in reducing dust flux.
3. Change of vegetation cover leads to a change of the source/sink relationship. Grass is one of the most effective protectors of the soil with respect to wind erosion. When grass is replaced by mesquite plants surrounded by large areas of bare soil, wind erosion increases dramatically. Wind erosion also increases dramatically when grass cover is reduced by drought or disturbances. Vegetation patterns are very important determinants of wind erosion on all soil types.
4. It is highly likely that the Jornada Basin is a net source of eolian materials. Historical studies show that soil loss rates for mesquite areas having sandy soils (1,400 $g/m^2/yr$) are much greater than the average dust deposition of

19.4 g/m^2/yr. Historical soil losses are roughly consistent with current soil loss rates.

5. Wind erosion changes the soil-forming process in both erosional and depositional settings. In erosional settings, soil profiles are truncated, organic matter and nutrients are removed and redistributed, infiltration decreases, water erosion increases, and bare ground is perpetuated by physically harsh conditions in intershrub spaces. In depositional settings, sandy sediments from which silts have been winnowed accumulate and progressively bury existing soil profiles. Consequently, the zones of carbonate accumulation, clay accumulation, and organic matter accumulation are lifted with the accreting sediments. Soil resources are more readily trapped and retained in depositional settings.

10

Plant Communities in the Jornada Basin: The Dynamic Landscape

Debra P. C. Peters
Robert P. Gibbens

Plant communities of the Jornada Basin are characteristic of the northern Chihuahuan Desert both in structure and dynamics. Although a number of plant communities can be differentiated, five major vegetation types are often distinguished that differ in plant species cover and composition, as well as other factors, such as animal populations, soil properties, and elevation. These five types are black grama (*Bouteloua eriopoda*) grasslands, playa grasslands, tarbush (*Flourensia cernua*) shrublands, creosotebush (*Larrea tridentata*) shrublands, and mesquite (*Prosopis grandulosa*) shrublands. Similar to many other parts of the Chihuahuan Desert, these plant communities have experienced major shifts in vegetation composition over the past 50–150 years (York and Dick-Peddie 1969). The most dramatic changes in vegetation and associated ecosystem processes have occurred as a result of a shift in life form due to woody plant encroachment into perennial grasslands (Grover and Musick 1990; Bahre and Shelton 1993).

This encroachment of shrubs has occurred in many arid and semiarid regions of the world, including the Western United States, northern Mexico, southern Africa, South America, New Zealand, and Australia (McPherson 1997; Scholes and Archer 1997). A number of drivers have been implicated in these grass–shrub dynamics, including various combinations of livestock grazing, small animal activity, drought, changes in fire regime, and changes in climate (Humphrey 1958; Archer 1989; Allred 1996; Reynolds et al. 1997; Van Auken 2000). The causes of shrub invasion are quite variable and often poorly understood, although the consequences consistently lead to the process of desertification (Schlesinger et al. 1990). This chapter describes the characteristics of each vegetation type and the documented changes in each type at the Jornada Basin. We then discuss the key drivers influencing these dynamics.

Vegetation Characteristics

Vegetation in the Chihuahuan Desert region has been classified as desert-grassland transition (Shreve 1917), desert savanna (Shantz and Zon 1924), desert plains grasslands (Clements 1920), desert shrub grassland (Darrow 1944), and shrub-steppe (Kuchler 1964). Desert grassland is often used as a general descriptive name for the area (McClaran 1995), although landscapes at the Jornada and throughout the northern Chihuahuan Desert often consist of a mosaic of desert grasslands, Chihuahuan Desert shrublands, and plains-mesa sand scrub (Dick-Peddie 1993). Because of this combination of vegetation types, the flora is diverse (Brown 1994). A total of 77 families, 285 genera, and 490 species have been identified from the region (Allred 2003). Predominant families include the Asteraceae (59 genera, 93 species), Poaceae (35 genera, 78 species), and Fabaceae (18 genera, 36 species). A complete species list for the Jornada Basin can be found online at http://jornada-www.nmsu.edu.

Five major vegetation types can be distinguished at the Jornada based on the dominant species and the associated species composition. Each vegetation type is often associated with a particular soil type and geomorphic position; ecotones between vegetation types in time and space can also be found (Wondzell et al. 1996). A variety of winter and summer annuals are seasonally abundant in all communities (nomenclature follows Allred 2003).

Black Grama Grasslands

The dominant grass species on sandy or gravelly upland sites is black grama. Black grama dominance is highest on deep, loamy soils (Paulsen and Ares 1962). A weak calcium carbonate layer is often present below the surface. On sandier soils, dropseeds (*Sporobolus* spp.) and threeawns (*Aristida* spp.) as well as forbs are more abundant. In wet years, black grama cover can reach up to 75% of total vegetative cover, whereas in dry years, black grama can average 44% of the total (Paulsen and Ares 1962). Black grama is an important forage species year round (Wright and Streetment 1958; Paulsen and Ares 1962). This long-lived (35–40 years) C_4 grass grows in open stands of individual ramets or sets (Canfield 1939; Wright and Van Dyne 1976). Recruitment by seed occurs infrequently as a result of low and infrequent production of viable seeds, few seeds stored in the soil, and a restrictive set of microenvironmental constraints on seedling establishment (Minnick and Coffin 1999; Peters 2000, 2002b). Vegetative spread occurs through the production of stolons (Nelson 1934).

Associated species in black grama grasslands include mesa dropseed (*Sporobolus flexuosus*), purple threeawn (*Aristida purpurea* Nutt.), and spike dropseed (*S. contractus*). These species often dominate where black grama has been eliminated. A number of perennial and annual grasses and forbs, as well as perennial shrubs, occur in these communities (Schmutz et al. 1992; Dick-Peddie 1993). Associated shrubs include soaptree yucca (*Yucca elata*), broom snakeweed (*Gutierrezia sarothrae*), ephedra (*Ephedra trifurca*), and scattered mesquite.

Playa Grasslands

Communities found in low-lying areas with heavy, clayey soils and run-in water are typically dominated by tobosa (*Pleuraphis mutica*), side oats grama (*Bouteloua curtipendula*), and alkali sacaton (*Sporobolus airoides*). Soils are often impervious to deep water infiltration with slow infiltration rates and show little change in texture or structure with depth. A cemented calcium carbonate layer is rarely present. Up to 80% of total cover in wet years and 62% in dry years can be attributed to tobosa on these sites (Paulsen and Ares 1962). Tobosa is a highly productive, drought-tolerant C_4 grass that is palatable only during the growing season (Paulsen and Ares 1962). This short-lived species (7 years) maintains its dominance through time by expanding vegetatively through the production of rhizomes (Neuenschwander et al. 1975; Wright and Van Dyne 1976; Gibbens and Beck 1987).

Burrograss (*Schleropogon brevifolius*) and ear muhly (*Muhlenbergia arenacea*) are also found on heavy soils that may or may not receive run-in water. These species can compose up to 60% of total cover on these sites. Success of burrograss is related to its ability to begin growth under cool temperatures and to spread rapidly by stolons and seeds. Production of many seeds with high viability followed by dispersal by wind, water, and animals, as well as minimal constraints on seedling establishment, lead to the dominance by this species on heavy soils (Paulsen and Ares 1962).

Tarbush Shrublands

Communities dominated by tarbush are often found on clay loam soils with some gravel near the surface (Paulsen and Ares 1962). These sites may receive some run-in water. Tarbush is a deciduous C_3 perennial shrub with a tar-like odor as a result of secondary compounds in its leaves (Estell et al. 1998). Tarbush has an extensive root system and produces seeds that are wind or water dispersed (Mauchamp et al. 1993; Gibbens and Lenz 2001). Herbaceous species in these communities are similar to the playa grasslands, where tobosa and burrograss are the common grasses.

Creosotebush Shrublands

Communities dominated by creosotebush typically occur on bajada slopes and alluvial fans. Soils vary from well-drained sands to shallow stony soils underlain by cemented calcium carbonate at shallow or deep depths (Gardner 1951). Creosote makes up 28–45% of total cover in these communities (Paulsen and Ares 1962). Creosote is an evergreen, drought-resistant C_3 perennial shrub that can live up to 400 years in the Chihuahuan Desert (Miller and Huenneke 2000). Creosotebush produces many secondary compounds that may influence the growth and survival of other species (Knipe and Herbel 1966). Associated species in these communities include bush muhly (*Muhlenbergia porteri*), fluff grass (*Dasyochloa pulchella*), and black grama, as well as a variety of forbs.

Mesquite Shrublands

Communities dominated by honey mesquite have been referred to as a "moving dune complex" (Campbell 1929). Mesquite is a deciduous, thorny, long-lived (200 years) C_3 shrub. This species is a facultative phreatophyte with very deep and laterally extensive root systems (Gibbens and Lenz 2001). Mesquite occurs on most soil types but is particularly prevalent on sandy soils. Most mesquite-dominated soils are deep sands with a calcium carbonate layer at variable depths. Short, multistemmed mesquite plants accumulate blowing sand until a mound forms around each plant. Interspaces between plants are scoured of loose soil resulting in sparse herbaceous cover. In some sites, the calcium carbonate layer is exposed at the surface, whereas coppice dunes, or nabkhas, can be > 3 m high (Langford 2000). Mesquite typically makes up 30–55% of total cover on these sites (Paulsen and Ares 1962). Associated plants in dunes include saltbush (*Atriplex canescens*), broom snakeweed, dropseeds, threeawns, and a variety of forbs.

Vegetation Dynamics

In the Jornada Basin, major changes in plant communities have been documented using a series of vegetation maps and land surveys. The first study to analyze change in vegetation was the classic paper by Buffington and Herbel (1965). General Land Office Survey notes made in 1858 were used to delineate areas covered by shrubs. Collectively, mesquite, tarbush, and creosotebush occurred on 42% of the Jornada Experimental Range (JER) in 1858, and no area was free of shrubs by 1963. Early land survey notes also showed that the increase of shrubs was widespread in southern New Mexico (York and Dick-Peddie 1969).

More detailed vegetation maps have been prepared for the JER for three time periods following the early land surveys: reconnaissance surveys by the U.S. Forest Service in 1915–16 and 1928–29 and by local Jornada researchers in 1998. The goal of the early field surveys was to determine the number of forage acres using five steps: (1) The Jornada was surveyed by section, and vegetation types were outlined on a map; (2) total plant foliar cover was estimated to the nearest 10% within each vegetation type; (3) percentage composition of species making up each of three broad categories of "weeds," "grasses and grass-like plants," and "shrubs" was estimated relative to the total for that area (100%); (4) the palatability rating of each species was multiplied by its percentage cover and summed for all species to obtain a forage-acre factor; and (5) cover was multiplied by the forage-acre factor to result in the number of forage acres in that area.

In 1999–2001, a hand-colored, original 1915–16 vegetation map at a scale of 1 inch = 1 mile and a 1928–29 vegetation map at a scale of 2 inches 1 mile were digitized, and field data were entered into a database. Dominant species for each vegetation type were obtained from the field forms or determined from the percentage composition. Because the early surveys usually did not distinguish between species of threeawns or species of dropseeds, comparisons of communities over time are limited for these species.

In 1998, a vegetation map of the Jornada was again made using a field survey. Transparent overlays on large-scale (4.5 inches = 1 mile) color infrared prints from aerial photographs taken in 1996 were used as a mapping base for an intensive ground reconnaissance; vegetation was mapped to ~ 8 ha minimum area or grain. Vegetation types were defined in terms of dominant species with usually no more than four species used to characterize any given type. Digitization of polygons delineated on the overlays permitted construction of vegetation maps based on primary and secondary dominant species (figures 10-1, 10-2), and percentage area calculations for the first four dominance classes through time (figures 10-3, 10-4). Although changes in rank may not reflect all changes in cover, the main results are not affected.

Perennial Grasses

One of the striking features of vegetation change from 1915–16 to 1998 is the large reduction in black grama–dominated grasslands (figures 10-1, 10-2). Although black grama was the first or second dominant on 28% of the Jornada in 1915–16, this species dominated only 4% of the area in 1998 for a net loss of 24% (table 10-1). Black grama occurred on 48% of the area as either a first, second, third, or fourth dominant in 1915–16 (figure 10-3a). By 1928–29, most decreases in area of black grama occurred in the second through fourth dominance classes. In 1998, very little area (< 0.06%) contained black grama as a third or fourth dominant. Many sites where black grama is still a major component of the community were treated for shrub control prior to 1990. Records from 1×1 m^2 quadrats sampled from 1915 until the 1970s indicate that much of the decrease in black grama cover occurred after the severe drought of the 1950s (Gibbens and Beck 1988; Yao et al. 2002a).

Threeawns were much more prominent in 1915–16 than in 1998 (figures 10-1, 10-2), mainly on areas now dominated by mesquite. Those species were a first or second dominant on 23% of the area in 1915–16 and only 4% in 1998 for a net loss of 19% (table 10-1). In 1915–16, these species were mostly first, second, or fourth dominants; in 1998, threeawns were the second dominant on most of the area where they occurred (figure 10-3b). Area dominated by dropseeds increased 10% from 1915–16 to 1998 with most of the increase occurring before 1928–29 (table 10-1). These increases in cover were probably due to population fluctuations in these widespread but relatively short-lived species. These species were primarily second and third dominants during all three time periods (figure 10-3c).

Areas dominated by burrograss declined 7% from 1915–16 to 1998, whereas

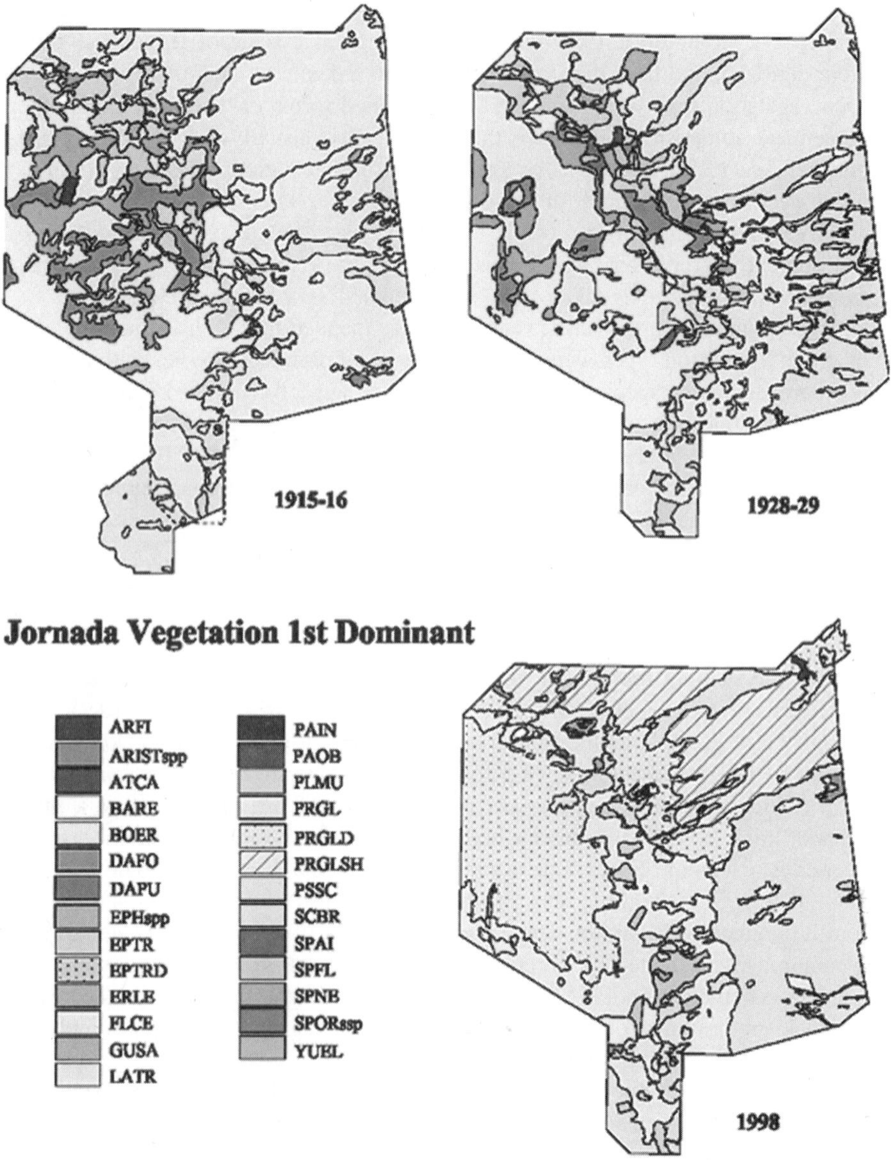

Figure 10-1. Vegetation maps of the first dominant species derived from reconnaissance surveys of the plains portion of the Jornada Experiment Range in 1915–16, 1928–29, and 1998. Species codes are defined in table 10-1. More area was fenced in 1915–16 than at present, but only the area shown by dotted lines and conforming to present boundaries was used in computing areas. The areas dominated by mesquite in 1998 were categorized as mesquite (soil accumulations at base of plants 20 cm or less), mesquite dunes (20 cm to about 3 m in height), or mesquite sandhills (greater than 3 m in height, sometimes 6 m or more). See color insert.

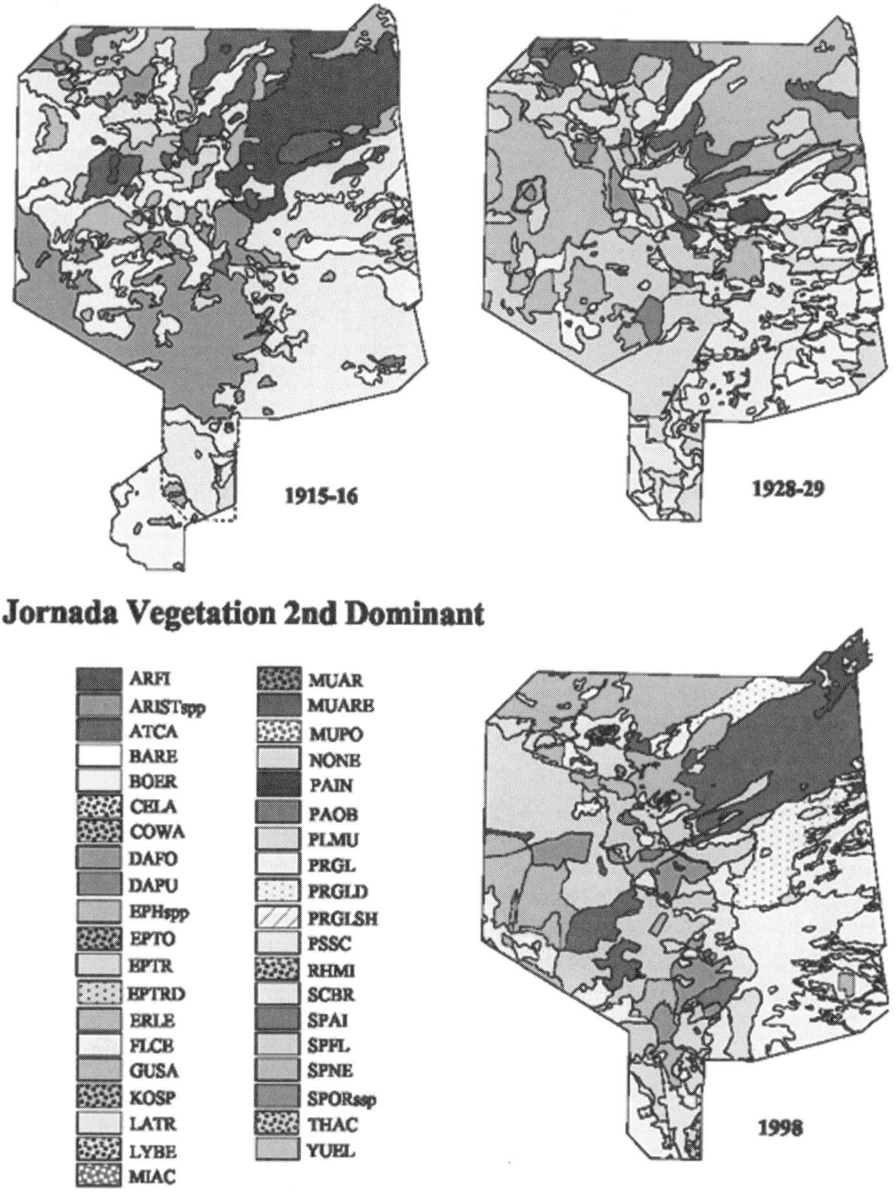

Figure 10-2. Vegetation maps of the second dominant species derived from reconnaissance surveys of the plains portion of the JER in 1915–16, 1928–29, and 1998. Species codes are defined in table 10-1. More area was fenced in 1915–16 than at present, but only the area shown by dotted lines and conforming to present boundaries was used in computing areas. The areas dominated by mesquite in 1998 were categorized as mesquite (soil accumulations at base of plants 20 cm or less), mesquite dunes (20 cm to about 3 m in height), or mesquite sandhills (greater than 3 m in height, sometimes 6 m or more). See color insert.

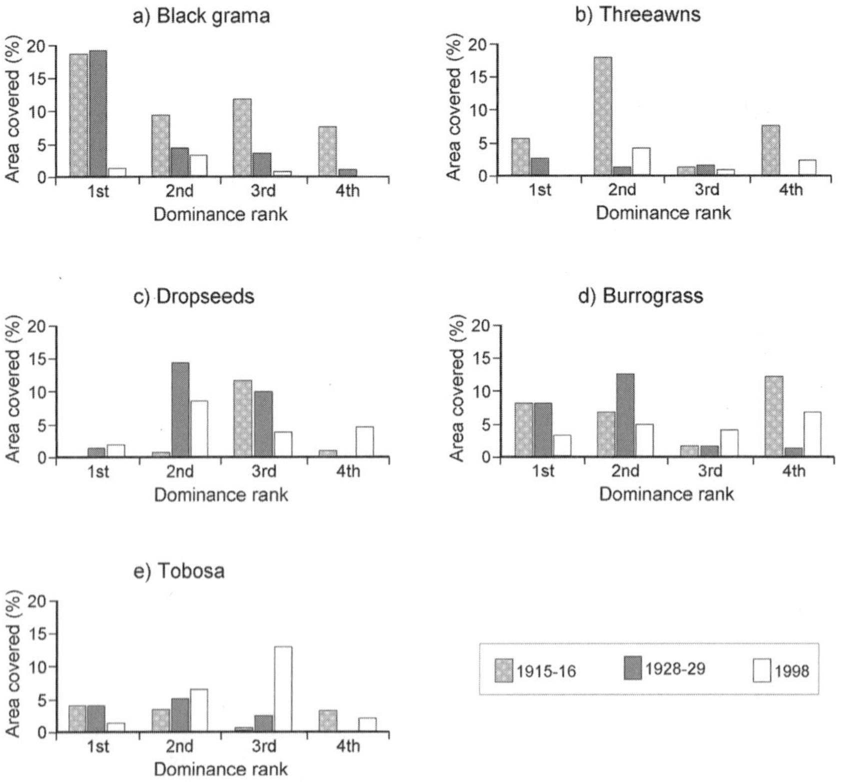

Figure 10-3. Percentage of area covered for five grass species or genera in three sample dates for four dominance classes: (a) black grama, (b) threeawns, (c) dropseeds, (d) burrograss, and (e) tobosa.

tobosa dominance did not change (table 10-1). Both species occurred frequently in most of the four dominance classes for all three time periods (figures 10-3d, e). Quadrat records indicate that these two species were less affected than black grama by the drought of the 1950s; reduction in dominance was generally due to encroaching shrubs and associated effects of aeolian deposits (Gibbens and Beck 1987). Several minor grass species showed no change in area dominated, including fluff grass, sand muhly (*Muhlenbergia arenicola*), and vine mesquite (*Panicum obtusum*). Bush muhly and the introduced Lehmann lovegrass (*Eragrostis lehmanniana*), on the other hand, have expanded.

Shrubs

Areas in which creosotebush was the first or second dominant increased 14% from 1915–16 to 1998 (table 10-1). In 1915–16, creosotebush was primarily a second dominant, whereas by 1998 this species was a primary dominant on most sites where it occurred (figure 10-4a). Much of this increase in creosotebush

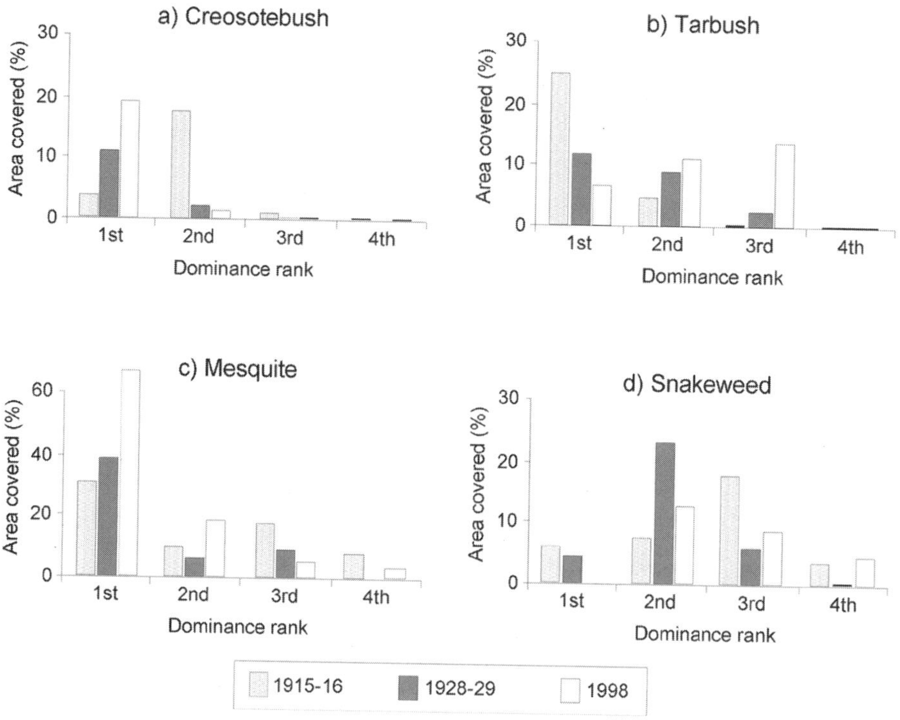

Figure 10-4. Percentage of area covered for four shrub species in three sample dates for four dominance classes: (a) creosotebush, (b) tarbush, (c) mesquite, and (d) snakeweed.

occurred at the expense of tarbush on the bajada slopes at the eastern side of the Jornada (figures 10-1, 10-2). Tarbush declined in area (12%) where it was the first and second dominant (table 10-1), although this species also increased on areas formerly dominated by burrograss and tobosa (figures 10-1, 10-2), and remains a second and third dominant throughout much of the area where it was a primary dominant in 1915–16 (figure 10-4b).

Of all shrubs, mesquite made the largest gain in area as a first or second dominant (40%) from 1915–16 to 1998 (table 10-1). Mesquite has been the first dominant on most areas where it occurred for all three time periods (figure 10-4c). If areas in which mesquite was estimated to occupy a subdominant role are included, this species was a major plant community component on approximately 95% of the Jornada in 1998. Although mesquite has been the primary object of shrub control measures and about 10,000 ha have been treated on the Jornada, this species continues to increase in density, cover, and importance (Gibbens et al. 1993). Areas occupied by mesquite include the sandhills, where this species was dominant in 1858 (Buffington and Herbel 1965) and in 1915–16 (figure 10-1). There is some correlation between size of mesquite dunes and age (Gadzia and Ludwig 1983). However, depth of material available for dune formation also plays an important role (chapter 4).

Table 10-1. Percentage of area covered by each species as a first and second dominant in each of three mapping years, and change in percentage of area from 1915–16 to 1998.

	Species Code	1915–16	1928–29	1998	Change in Area (%)
Grasses					
Threeawns (*Aristida* spp.)	ARIST spp	23	4	4	−19
Black grama (*Bouteloua eriopoda*)	BOER	28	24	4	−24
Fluffgrass (*Dasyochloa pulchella*)	DAPU		0.5		0
Lehmann lovegrass (*Eragrostis lehmanniana*)	ERLE			0.1	0.1
Sand muhly (*Muhlenbergia arenicola*)	MUAR, MUARE	0.3	0.3	0.3	0
Bush muhly (*Muhlenbergia porteri*)	MUPO			0.6	0.6
Vine mesquite (*Panicum obtusum*)	PAOB		0.3		0
Tobosa (*Pleuraphis mutica*)	PLMU	8	9	8	0
Burrograss (*Schleropogon brevifolius*)	SCBR	15	21	8	−7
Dropseeds (*Sporobolus* spp.)	SPAI, SPFL, SPNE SPORssp	1	16	11	10
Dominant shrubs					
Tarbush (*Flourensia cernua*)	FLCE	30	20	18	−12
Snakeweed (*Gutierrezia sarothrae*)	GUSA	15	27	13	−2
Creosotebush (*Larrea tridentata*)	LATR	12	16	26	14
Honey mesquite (*Prosopis glandulosa*)	PRGL, PRGLD, PRGLSH	34	38	74	40
Other shrubs					
Sand sage (*Artemisia filifolia*)	ARFI	11	1	1	−10
Four-wing saltbush (*Atriplex canescens*)	ATCA	6	7	14	8
Winterfat (*Krascheninnikovia lanata*)	KRLA			0.1	0.1
Feather-plume (*Dalea formosa*)	DAFO			0.3	0.3
Mormon tea (*Ephedra* spp.)	EPHspp, EPTO, EPTR, EPTRD	0.3	0.3	0.4	0.1
Crucifixion thorn (*Koeberlinia spinosa*)	KOSP			0.3	0.3
Silver wolfberry (*Lycium berlandieri*)	LYBE		0.1	0	0
Mariola (*Parthenium incanum*)	PAIN			0.1	0.1
Broom dalea (*Psorothamnus scoparius*)	PSSC	0.7	0.6	0.8	0.1
Prickle-leaf dogweed (*Thymophylla acerosa*)	THAC			0.2	0.2
Soaptree and banana yucca (*Yucca* spp.)	YUEL	1	3	5	4

Notes: Nomenclature follows Allred (2003). Similar species are combined in the table and denoted by multiple species codes. Areas are calculated for the current extent of the JER and do not include the southwestern extension marked by the dashed line in the 1915–16 map.

The suffrutescent broom snakeweed declined slightly (2%) in areas where it was the first and second dominant from 1915–16 to 1998 (table 10-1). However, total area where this species was a significant component of the plant community varied through time from 15% (1915–16) to 27% (1928–29) and 13% (1998). Similar to the dropseeds, broom snakeweed is relatively short-lived and populations fluctuate widely among years, often in response to winter precipitation. This species was a second or third dominant on most sites where it occurred (figure 10-4d). Of the remaining principal shrub species, only sand sage (*Artemisia filifolia*) showed a large decrease in area dominated from 1915–16 to 1998 (10%; figures 10-1 and 10-2, table 10-1). Fourwing saltbush and soaptree yucca increased in area dominated (8% and 4%, respectively). The remaining shrub species showed little or no change through time.

Drivers of Vegetation Change

At the Jornada, the widespread expansion of two shrubs (honey mesquite and creosotebush) into perennial grasslands has been particularly important. Much of this expansion has been accompanied by a decrease in cover and abundance of black grama, a key forage species that previously dominated many of the sandy loam uplands in the Chihuahuan Desert. The shrub species were formerly restricted to more localized sites in the area (Gardner 1951; Stein and Ludwig 1979). Because of the widespread expansion of these two shrub species at the expense of black grama, we focus on the key drivers influencing dynamics of these three species, although additional perennial grasses will be included when information is available. Next we summarize the major drivers involved in shrub invasion and the resulting effects on black grama and ecosystem processes. Two major aspects of vegetation dynamics are considered for each driver when possible: (1) the response of vegetation during the time that the driver was operative, and (2) the recovery of the vegetation following the cessation of the driver. The implications of these responses to shrub invasion are also considered.

Livestock

Livestock have often been implicated as important drivers in shrub invasion into desert grasslands. Livestock density in southern New Mexico increased markedly in the 1880s following the American Civil War (Fredrickson et al. 1998). At the Jornada, cattle densities have decreased over the past century (chapter 13). Although the peak in animal numbers in the late 1800s and early 1900s corresponds to a period of rapid invasion by shrubs in the region, other factors interacting with the effects of livestock likely contributed to the invasion process. A number of studies have examined the response of vegetation following livestock removal. In many cases (but not all) perennial grasses increased following exclusion of livestock. Most studies recognized the importance of other factors, including drought, fire, and small animals, as separate or confounding constraints to grass recovery (Glendening 1952; Branscomb 1958; Humphrey 1958; Schlesinger et al. 1990; Drewa et al. 2001).

Livestock have four major effects on vegetation that differ for grasses and shrubs: (1) herbivory on aboveground biomass (for grasses: tillers, inflorescences, seeds; for shrubs: leaves, stems, seeds), (2) dispersal of seed, (3) trampling of plants and soil, and (4) redistribution of nutrients, particularly nitrogen, through urine and fecal deposition. Herbivory can occur at the seedling or adult stage with different consequences for plant survival and vegetation dynamics. The importance of seed dispersal depends on the palatability of the flowering structures and seeds and the ability of germinable seeds to pass through the digestive tract of animals. Trampling has direct effects on plant mortality, as well as indirect effects on water availability through the modification of soil properties, such as bulk density and infiltration capacity, that influence plant growth. These effects are detailed in chapter 13. Here, we focus on responses of the vegetation to livestock herbivory and dispersal of seed with implications for interactions between grasses and shrubs.

Perennial Grasses

Black grama has been the most well-studied perennial grass in terms of response to grazing by cattle. Basal area and biomass of black grama can decrease as much as 50% in three to four years when utilization is $>$ 40% under seasonal or year-round grazing (Canfield 1939; Valentine 1970). In the absence of cattle or under light to moderate grazing ($<$ 40% utilization), black grama is primarily influenced by precipitation (Paulsen and Ares 1962). Overgrazing and excessive trampling have negative effects on inflorescence height, as well as the number and length of stolons (Valentine 1970; Miller and Donart 1979). Growth of black grama is further delayed when overgrazing is accompanied by drought: Regrowth of this species was negligible after 13 years of cattle exclusion (Nelson 1934; Canfield 1939). Reductions in biomass and basal area of dominant grasses can have important negative effects on annuals and perennial forbs, likely as a result of increased wind erosion and loss of soil particles (Drewa and Havstad 2001).

Shrubs

In addition to the indirect effects of grazing by cattle on grasses that reduce competition and positively influence shrub growth, grazing also has direct positive effects on shrubs. The most important positive effects of grazing are on shrub recruitment, particularly honey mesquite recruitment. Dissemination of seed by cattle has been invoked as a more important factor than the negative effects of grazing on grasses that influences mesquite invasion into black grama grasslands (Humphrey 1958; Wright 1982; Paulsen and Ares 1962). Cattle readily consume mesquite pods, and large numbers of seeds remain viable after passing through the digestive tracts of cattle (Paulsen and Ares 1962; Mooney et al. 1977). Seeds are often deposited at large distances from where they were consumed (Humphrey 1958; Paulsen and Ares 1962). A favorable microenvironment provided by cattle dung may facilitate germination, particularly if the dung is deposited in the rainy season (Paulsen and Ares 1962). Seedlings of honey mesquite readily become established; within four months following germination, seedlings develop roots to depths of 40 cm (Brown and Archer 1990). Seed germination increased seven to eight times in Texas prairies when the heights of perennial grasses were clipped to 25 cm compared with unclipped controls (Brown and Archer 1989). Germination, establishment, and survival of honey mesquite were not affected by level of perennial grass density or moderate and heavy grazing (Brown and Archer 1999). By contrast, seedling survival through time was dependent on the presence of large patches of soil without perennial grass cover (Whitford et al. 2001).

Small Animal Activity

Native rodents and lagomorphs are important components of Chihuahuan Desert ecosystems that affect grasslands and shrublands differently with the potential to influence shrub encroachment (chapter 12). These small animals primarily affect plants through their consumption of aboveground material and, in particular, seeds and seedlings. This consumption can have major impacts on the ability of grasses to recover following shrub invasion, as well as on the ability of shrub seedlings to become established in grasslands.

Perennial Grasses

When perennial grasses are abundant, small animals appear to have minimal impact on grassland dynamics (Norris 1950; Buffington and Herbel 1965; Gosz and Gosz 1996). However, in open areas following shrub invasion, jackrabbits (Lepus californicus) and rodents are generally more abundant (Vorhies and Taylor 1933), and small animal activity in overgrazed areas can prevent a recovery in perennial grasses (Norris 1950). In some cases, deteriorated rangeland is dominated through time by snakeweed (Parker 1938). Burrowing activities of bannertail kangaroo rats (Dipodomys spectabilis) occur more frequently in black grama grasslands compared with mesquite-dominated areas. These activities can shift species composition toward annuals instead of perennial grasses (Moroka et al. 1982). Granivory by ants may be more important than loss of seeds to rodents for various plant species in black grama grasslands (Kerley and Whitford 2000).

Grass recovery can occur rapidly following the removal of small animals. Basal area and biomass of perennial grasses increased four to five times following the exclusion of small animals for 8 years (Norris 1950). Lagomorphs were found to have differential effects on vegetation in shrub-dominated areas of the Jornada that were evident 50 years after their exclusion. Compared with controls, spike dropseed increased 30 times in plots where rabbits were excluded (Gibbens et al. 1993). Dropseeds are preferred forage for jackrabbits—up to 40% of their diet (Dabo et al. 1982). Fluff grass was more abundant in plots with rabbits, compared to controls, and may be more affected by water availability than grazing pressure (Gibbens et al. 1993). Positive effects of lagomorph exclusion on black grama were expected because this species can comprise up to 54% of the diet of jackrabbits in summer months (Fatehi 1986; Fatehi et al. 1988) and black grama seedlings are readily consumed in shrub-dominated areas (Bestelmeyer and Peters unpublished data). Black grama did not show a response until 51 years of lagomorph exclusion in areas where shrubs had been removed (Havstad et al. 1999). This delayed response is likely due to a limited seed source for recovery because very few black grama plants occurred on plots at the beginning of the study and a stand of black grama is not located near the study site (Havstad et al. 2003).

Shrubs

Small animals can have both negative and positive effects on shrubs, with most impacts occurring at the seed and seedling stage. The diet of jackrabbits is comprised of honey mesquite, creosotebush, and broom snakeweed stems that are used as sources of food and water (Dabo et al. 1982; Steinberger and Whitford 1983b; Fatehi 1986; Fatehi et al. 1988). High seedling mortality of creosotebush due to jackrabbits and rodents has been observed in several studies (Boyd and Brum 1983; Whitford et al. 2001). Canopy cover of honey mesquite and creosotebush were significantly higher on lagomorph exclusion plots within 10 years after the start of the treatments and persisted for > 40 years (Gibbens et al. 1993; Havstad et al. 1999). Evidence also exists for the positive effects of small animals in promoting shrub invasion. Rodents can aid in the dispersal of honey mesquite seeds into grassland areas in southern Arizona (Reynolds and Glendening 1949[a]) and presumably in the Chihuahuan Desert as well. Merriam kangaroo rats (Dipodomys merriami) collect seeds and store them in shallow caches; many of these seeds are not consumed and can germinate to contribute to mesquite expansion (Paulsen and Ares 1962). Burrowing activities of kangaroo rats can also have positive effects on the growth, flowering, and survival of creosotebush (Chew and Whitford 1992).

Drought

Periodic drought is a key characteristic of arid regions. Severe drought (Palmer drought index between -3 and -4) of 2–4 years' duration occurs on average every 20–25 years, whereas extreme drought (palmer index < -4) occurs every 50–60 years (Scurlock 1998). At the Jornada Basin site, extreme drought occurred in the early 1900s and again in the 1950s when 7 years out of 11 over the > 80-year record were classified as either severe or extreme drought based on annual precipitation. The drought in the 1950s was the most severe recorded over a 350-year period for the Southwestern United States (Fredrickson et al. 1998), although tree-ring reconstruction shows that similar droughts occurred prior to the seventeenth century (Woodhouse and Overpeck 1998). Studies conducted through time since the early 1900s have documented both small- and large-scale shifts in dominance from perennial grasses to shrubs with much of the change corresponding with the 1950s drought (Hennessy et al. 1983a; Gibbens and Beck 1987). These observations clearly show that drought is a contributing factor in shrub invasion (Herbel et al. 1972). Shifts in dominance from perennial grasses to shrubs occur due to differences in life history traits of species that determine their ability to tolerate drought and to recover following drought. Long-term data are critical in examining the effects of infrequent catastrophic events, such as drought, on ecosystem properties and dynamics (Conley et al. 1992).

Perennial Grasses

In general, basal area of black grama and other perennial grasses decrease during drought with black grama recovering more slowly than other grasses. The susceptibility of black grama to summer drought was noted early in the 1900s (Jardine and Forsling 1922; Nelson 1934). Similar responses were observed during the drought of 1934 when no new growth was reported for black grama and basal cover of this species decreased 77% between 1933 and 1935 (Campbell 1936; Canfield 1939). Most of our information comes from plant responses during and after the drought of the 1950s that are examined in the context of long-term responses beginning in 1915 and postdrought responses continuing to the present. During the drought from 1951 to 1956, growing season precipitation (July 1–October 1) decreased 43–47% from the long-term average; the broad-scale pattern in vegetation was a reduction in basal cover of black grama by 72–90% from predrought levels (Herbel et al. 1972). The largest reduction in black grama cover occurred on deep sandy soils, and the least reduction occurred on sandy soils with a shallow indurated caliche layer. Grazing intensity was found to have little effect on black grama basal area during extended drought, although recovery was faster for conservatively grazed plots compared with heavy grazing or protected plots (Paulsen and Ares 1962).

Detailed analyses of species responses during the drought of the 1950s have been conducted using a series of 1-m^2 permanent quadrats located throughout the Jornada starting in 1915. The quadrats were charted annually until 1947 and sporadically since that time. A total of 104 quadrats were located in black grama (57 quadrats), tobosa (22 quadrats), burrograss (12 quadrats), threeawn (6 quadrats), and communities dominated by other species (7 quadrats). Prior to 1950 black grama went extinct on 11 quadrats that were previously dominated by this species (Gibbens and Beck 1988). The most dramatic change in vegetation occurred during 1950–61 when black grama disappeared from an additional 37 quadrats. Recovery of black grama following drought is indicated because this species was present in 9 quadrats in 1979 and in 12 quadrats in 1995 (Yao et al. 2002a). The ability of black grama to persist on 16% of the quadrats and to recover on 5% is likely due to a number of factors. Recent analyses show that black grama persistence on these quadrats is positively related to the distance from the nearest shrub community in 1915 (Yao et al. 2002b). Other characteristics of each plot, including soil texture, annual precipitation, depth to caliche, and elevation, were less important than distance to shrubland. Management practices (e.g., herbicide treatment, grazing history, and distance to water sources) and small-scale redistribution of water that were not included in these analyses are also expected to be important to black grama persistence. Temporal variation in black grama basal area was found to be related to long-term precipitation (2–15 years following sampling) for quadrats with and without persistent black grama, although additional factors were more important than precipitation for quadrats where black grama went extinct (Yao et al. 2002a). By contrast, annual growth of black grama was related to short-term precipitation (< 15 months following sampling) for both types of quadrats.

The slow recovery of black grama following drought is supported by a clipping study conducted on the Chihuahuan Desert Rangeland Research Center (CDRRC; Lohmiller 1963). Postdrought measurements made in 1962 showed that yield of black grama had not recovered to predrought levels even though five of the six postdrought years had above-average precipitation. Recent simulation model analyses indicate that the slow recovery of black grama to drought in the 1950s and 1970s was consistent with an inability of seedlings to become established (Peters and Herrick 2001). Other perennial grasses and forbs responded more rapidly and dominated simulated plots until black grama became established and grew to dominate the plots.

Other important grass species were also negatively affected by drought in the 1950s, although generally their recovery was faster than that of black grama. On average, tobosa basal cover was lower in the 1950s compared with the 1930s, 1940s, and 1970s (Gibbens and Beck 1988). However, very few quadrats (< 10) were sampled in any given year. High basal cover of tobosa starting in 1965 indicates rapid recovery. Tobosa grasslands are typically located in low-lying areas with heavy soils that receive some run-in of water from overland flow (Herbel and Gile 1973) that may explain the fast recovery. Tobosa has also been reported to be more drought-tolerant than black grama, possibly due to its ability to become dormant under low soil-moisture conditions (Herbel et al. 1972). Burrograss in run-in locations showed similar responses to tobosa, both during the 1950s drought and in the following years as basal area recovered to predrought levels in 1960 (Gibbens and Beck 1988). Although only six quadrats were established in threeawn communities, all of them responded similarly to drought in that no threeawn occurred in these quadrats after 1952. Currently, all of these quadrats have converted to mesquite coppice dunes (Gibbens and Beck 1988). In a related study, threeawns as well as dropseeds were more susceptible to drought compared with black grama; the threeawns were eliminated by the drought, whereas the dropseeds made a partial recovery through seedlings (Herbel et al. 1972).

Shrubs

In general, shrubs are less severely impacted by summer drought compared with grasses. Most information is available from short-term experiments, observations on small plots, and aerial photography analyzed over time to indicate rates of invasion. However, long-term studies have not been conducted at a sufficiently large scale. Based on rainfall exclusion studies, both creosotebush and mesquite were found to be well adapted to withstand season-long droughts, in part because of their ability to use near-surface soil water during the summer (Reynolds et al. 1999b). In addition, creosotebush plants were able to shift growth and physiological activity to use temporally available moisture, carry out limited physiological activity during drought, and compensate for drought through enhanced growth in the season following drought. For mesquite, compensation occurred through increased water uptake associated with enhanced physiological activity in the season following drought. Both species have well-developed lateral and taproot systems that also may be important for survival under conditions of low precipitation (Gibbens and Lenz 2001).

Observations through time suggest that shrubs now dominating former black grama sites became established during or following the 1950s drought. An analysis of changes in dominant species showed that creosotebush, mesquite, and tarbush became dominants on former black grama sites by 1981 (Gibbens and Beck 1987). Although it is not always possible to determine when the shrubs became established, these species were clearly dominant by 1981. In some cases establishment date can be determined. Mesquite plants appeared on quadrats in 1916 and 1924 and persisted until 1981. One tarbush plant appeared on a quadrat in 1956 and persisted until at least 1981.

Several studies have shown the ability of mesquite to survive and even increase in cover and abundance during the 1950s drought. A black grama community that was hand-grubbed in 1939 to remove all mesquite plants had increasing numbers of mesquite, from 285 plants/ha in 1948 to 377 plants/ha in 1955 and 400 plants/ha in 1959 (Wright 1982). Size of mesquite plants also increased during the drought as evidenced by an increase in canopy cover from 0.6% to 2.3% from 1948 to 1959. An ecotone between mesquite coppice duneland and black grama grassland was studied using transects established in 1940 (Wright 1982). The mesquite dune type originally covered 25% of the first transect and 50% of the second transect. For both transects, density of mesquite increased from 1940 to 1948; density decreased but cover increased from 1953 to 1959. The decrease in density may have been due to the loss of small shrubs during the drought. Increases in mesquite cover occurred throughout both transects, whereas increases in numbers occurred primarily in the transition zone and grassland areas. Differences were also observed in the rate of encroachment along the two transects that were likely due to their different orientations with respect to the prevailing winds, as well as different soil textures. More recently, satellite images have been used to document short-term drought effects with grasslands being the most spectrally responsive to variation in precipitation compared with shrub-dominated areas (Peters et al. 1993).

Fire

Little is known about the role of fire in the Chihuahuan Desert, either in the United States or Mexico (Humphrey 1974; Drewa et al. 2001). Most information on fire in desert grasslands of the United States is from the Sonoran Desert in southeastern Arizona (Humphrey 1958; Wright 1980). In Chihuahuan Desert grasslands, natural fires most likely occurred in June, when the frequency of lightning strikes is high (Gosz et al. 1995), the vegetation is dry, and weather conditions exist to promote the spread of fire (low humidity, high temperatures, and high winds). A characteristic fire return interval is unknown, but a 9–10-year period has been estimated for desert grasslands in southeastern Arizona (Cable 1967; McPherson 1995).

Some have questioned if biomass and fuel load were sufficient prior to European settlement to carry extensive fire in the Chihuahuan Desert (Buffington and Herbel 1965; Dick-Peddie 1993). Prior to a period of intense livestock grazing and subsequent encroachment of shrubs during the late nineteenth century, fuel loads were likely higher than at present. Additionally, despite reduced fuel loads, natural fires are still observed every two to three years in ungrazed, black grama–dominated grasslands in central New Mexico at the Sevilleta LTER, and an extensive natural fire was observed in the early twentieth century on the Jornada. Thus, though fires continue to occur naturally in Chihuahuan Desert grasslands, it is likely that the size, frequency, and intensity of these fires have decreased over the past 150 years (Bahre 1995).

Perennial Grasses (Black Grama)

In general, effects of fire on perennial grasses are highly variable; grass response may be influenced by soil moisture conditions at the time of the fire, the amount of rainfall received during the growing season immediately following the fire, or the grazing intensity by cattle (Gosz and Gosz 1996). Most research conducted on black grama indicates high mortality following fire under drought conditions that persist into the immediate postfire environment. For example, only 10% of black grama ramets survived a fire under excessively dry conditions in the Sonoran Desert (Cable 1965). At the Jornada, cover of black grama decreased 13% in burned plots and increased 4% in unburned areas four years following a fire that occurred during a summer-long drought in 1995 (Drewa and Havstad 2001). Other recent studies suggest that black grama can recover rapidly from even year-round cattle grazing (utilization $< 40\%$) if growing season precipitation is at least equivalent to the long-term average immediately following a fire (Drewa et al. 2001).

At the Jornada, a June fire conducted in 1999 in a community codominated by black grama and honey mesquite resulted in low mortality of the former species (Drewa and Havstad 2001). Two years following fire, basal cover of black grama was 56% of unburned cover in grazed areas and 64% of unburned cover in ungrazed areas. Grass mortality was a function of the size of individual plants prior to fire. Small plants (basal area < 10 cm^2) had a higher probability of fire-induced mortality than larger plants (> 30 cm^2). These rapid responses are likely a result of above-average precipitation during the growing season immediately following the fire. Similarly, at the Sevilleta LTER site, a lightning-ignited fire in 1998 resulted in low mortality of black grama (Peters unpublished data). Precipitation was above average in summer 1998. Thus, black grama recovery is not only highly variable 4–8 years to > 50 years (Wright 1980; McPherson 1995; Gosz and Gosz 1996), but may depend more on soil moisture availability during the season of the fire than on grazing intensity when properly managed.

Shrubs

Prior to European settlement, natural fire may have been effective in deterring honey mesquite invasion by top-killing or completely killing plants. The degree of kill depends in part on shrub size, fire season, and intensity of burn. After one year following prescribed fires in June 1999, canopy area of honey mesquite decreased by 22% but increased 24% in unburned areas (Drewa et al. 2001). In the same study, shrub volume decreased 40% following fires and increased 30% after just one year in fire-excluded areas. In addition, only 3 small shrubs (< 65 cm height; 10 years old) of the 210 shrubs were killed completely.

Fire season and intensity were also found to influence the ability of honey mesquite to recover following fire (Drewa et al. 2001). Shrubs were 8% taller and resprouted 16% more after dormant season than growing-season fires. For shrubs clipped to simulate fire, resprouting was 35% greater than that after actual low-intensity fires (fueled by natural vegetation) and 60% greater than that after actual high-intensity fires (artificial fuel additions). In natural settings growing season fires may be more effective than those during the dormant season in reducing the stature of honey mesquite.

Fire may also be effective in limiting the recruitment of mesquite plants into grasslands. Fire can completely kill seeds and seedlings of mesquite (Cox et al. 1993; Brown and Archer 1999). Although complete kill of larger plants is rare, top kill of older life history stages delays reproductive maturity and setting of seed (Drewa et al. 2001). Less information is known about fire effects on other species of shrubs, including creosotebush or tarbush. Fire may be important in limiting recruitment of these species into grasslands, but the probability of adult mortality may be lower compared with mesquite, which is a more aggressive resprouter.

Climate Change

Broad-scale patterns in climate are largely responsible for large-scale patterns in vegetation. The presence of C_4-dominated grasslands in southern New Mexico is a result of high temperatures during the growing season combined with low and variable amounts of precipitation, as well as periodic drought (Schmutz et al. 1992). The region has experienced major changes in climate over geological time that have resulted in several major shifts between grasslands and shrublands over the past 10,000 years (VanDevender and Spaulding 1979; Monger et al. 1998). Historic climate has also been proposed as an explanation for the low ability of black grama to become established under current climatic conditions (Neilson 1986). The current shift from grass- to shrub-dominated ecosystems has likely occurred at a faster rate (50–100 years) compared with geologic changes over hundreds or thousands of years, and the influence of human activities has been particularly important in these rapid rates of change. However, shifts between grasslands and shrublands over geologic time suggest that the current shrub invasion episode may be reversed if climatic conditions once again become more favorable for the recruitment and growth of grasses.

Elevated concentrations of atmospheric CO_2 are another driver proposed to influence shrub invasion dynamics (Mayeux et al. 1991; Johnson et al. 1993; Polley et al. 1996). Woody plants typically possess the C_3 photosynthetic pathway that may confer an advantage under elevated CO_2 compared with C_4 grasses. Also, widespread encroachment of woody species into grasslands began shortly after atmospheric CO_2 concentration rose above its preindustrial level of 270–280 ppm. Thus the global nature of the increase in CO_2, the multiple benefits of higher CO_2 to C_3 plants, and the near synchrony of the increase in CO_2 and shift in vegetation from C_4 grasses to C_3 woody plants have led some researchers to propose that the historic increase in CO_2 is a major factor contributing to woody plant encroachment (Polley et al. 1992). However, others have questioned the validity of this explanation. Elevated CO_2 may have contributed to the general increase in woody plants globally, but local factors are more likely to be important to the rate, pattern, and extent of invasion at a particular site (Archer et al. 1995). Interactions among elevated CO_2 and other drivers are also likely to be important (Polley et al. 1996).

Directional changes in climate as a result of elevated concentrations of atmospheric CO_2 may have long-term effects on grass–shrub interactions with implications for invasion dynamics. Simulation models have been used to predict long-term responses of grasses and shrubs to directional changes in climate. Effects of climate on the probability of establishment of black grama were investigated along a climatic gradient in the southwestern United States using a daily time step soil water model, SOILWAT (Minnick and Coffin 1999). Sites were simulated using historical weather data and predicted weather from global climate models for sites spanning southern Colorado to southern New Mexico. Simulated black grama establishment decreased from north to south as historic annual precipitation decreased and temperature increased. Low simulated seedling establishment of black grama at the Jornada is similar to field results for this species (Neilson 1986). A directional increase in year-round temperature and increase in summer precipitation resulted in an increase in establishment of black grama at all sites. The large range in establishment probability along this gradient suggests that seedling establishment may not limit recovery of this species throughout its geographic distribution. Other recruitment processes, such as production of viable seeds and presence of germinable seeds in the soil, may become more important as temperatures decrease and precipitation increases (Peters 2002b). At the Sevilleta LTER, 250 km north of the Jornada, seed production of black grama occurs more frequently (viability $> 50\%$) than at the Jornada (Peters 2002b). Two periods of precipitation are predicted to be critical for black grama recruitment: Precipitation in June is important for seed germination followed by precipitation in July and August that is needed for seedling establishment (Peters 2000).

More recently, a simulation model (ECOTONE) was used to investigate the importance of multiple processes for shifts in species dominance and composition with changes in climate (Peters 2002a). Model results for a site at the northern boundary of the Chihuahuan Desert (Sevilleta LTER) show that black grama and creosotebush ecotones are stable under current climatic conditions without cattle grazing, small rodent activity, fire, or severe drought. However, a change in climate to an increase in summer precipitation and year-round increase in temperature resulted in a shift to black grama dominance. This shift in dominance was due to an increase in establishment of black grama seedlings followed by more favorable conditions for the growth of this species compared with the growth of creosotebush. These shifts in vegetation are not predicted to occur uniformly across the landscape (Peters and Herrick 1999b). Recovery of black grama is not expected on degraded sites currently dominated by creosotebush where the establishment of black grama has a very low probability of occurring. By contrast, sites that are currently codominated by black grama and creosotebush on soils with a moderate ability to hold water for plants are predicted to shift to black grama dominance. Seed availability is also predicted to influence the ability of black grama to recover on sites where it currently does not occur (Peters and Herrick 1999a; Rastetter et al. 2003).

Conclusions

Shrub invasion consists of a complex suite of processes and drivers that have occurred, in most cases, in combination for many sites worldwide. Within the Jornada Basin, a number of short- and long-term experiments combined with monitoring and simulation models provide one of the best opportunities to investigate shrub invasion processes through time and space. Our studies show that a suite of processes were important in transforming the JER from desert grassland to shrubland over the past 50–100 years. Large numbers of cattle consuming perennial grasses and dispersing mesquite seeds combined with severe drought in the early 1900s and again in the 1950s, along with an increase in small animal herbivory on grasses and a reduction in fire control of shrubs, apparently led to the landscapes that we see today. Elevated concentrations of atmospheric CO_2 may have conferred a growth advantage to shrubs; however, it is unlikely that this factor alone generated the shift from grasses to shrubs. Predicted directional changes in climate that increase summer precipitation and temperature may promote shifts in vegetation from grass–shrub savannas to grass dominance on certain sites in the future. In addition, changes in the size, timing, and intensity of rainfall events during the growing season may also shift the vegetation with grasses being favored by small, frequent rainfall and shrubs being favored by large rain events with deeper infiltration.

11

Patterns of Net Primary Production in Chihuahuan Desert Ecosystems

Laura F. Huenneke
William H. Schlesinger

The Jornada Basin of southern New Mexico has long been an important location for the study of productivity in desert ecosystems. Researchers have studied the magnitude and sustainability of plant production since the founding of the USDA Jornada Experimental Range (JER) in 1912. The consistent administration and research focus of the JER and of the Chihuahuan Desert Rangeland Research Center (CDRRC) have facilitated a number of long-term studies of vegetation dynamics and productivity. These long-term data sets are especially critical for understanding arid ecosystems, where interannual and decadal scale variation in climate is great and plant performance is strongly constrained by the physical environment. Long-term data, including the net primary productivity (NPP) data that are the focus of this chapter, are also essential for understanding the progression or rather, degradation of ecosystem structure that has been called desertification.

Through the years a variety of approaches have been used to evaluate plant production in the Jornada Basin. These approaches span the range from applied or management-oriented techniques, focused primarily on assessing patterns of palatable forage production, to more basic empirical studies based on dimension analysis or similar measurements of plant growth, to estimates based on photosynthetic measurements, to remote sensing and modeling approaches. NPP was a particular focus of the work performed during the International Biological Programme or IBP (1970s) and is still a major emphasis in the Long-Term Ecological Research (LTER) era. Thus, the Jornada provides a unique opportunity to compare the strengths and weaknesses of different approaches applied to a complex system.

Ecosystem science has provided a set of general hypotheses about the factors regulating NPP in arid and semiarid ecosystems (reviewed by Noy-Meir 1973;

Hadley and Szarek 1981; Ludwig 1986, 1987). These premises include the following:

1. Plant productivity is low relative to that of other ecosystems (Lieth 1975).
2. NPP is regulated primarily by localized water availability and hence should be correlated closely with precipitation (Le Houerou 1984; Le Houerou et al. 1988). This premise is related to Noy-Meir's (1973) definition of deserts as "water-controlled ecosystems with infrequent, discrete, and largely unpredictable water inputs." Water and energy flows are considered to be coincident because plant production cannot take place without water expenditure through transpiration.
3. A pulse-reserve pattern (Noy-Meir 1973) characterizes the behavior of populations (including producer populations) in deserts, such that the episodic availability of resources in excess of some threshold (such as the "pulse" of precipitation) stimulates growth and the production of a large reserve (e.g., of photosynthetic tissues, propagules, and organic matter).
4. Deserts, especially shrubland systems, are dominated by long-lived, stress-tolerant plants with slow growth and low population turnover. Low aboveground productivity and high biomass accumulation ratios, at least in woody plants, should reflect this low rate of turnover.
5. Soil texture has been proposed as an important determinant of aboveground NPP in arid and semiarid systems, with coarse sandy soils having greater infiltration of water to depth, lower rates of evaporative loss from the surface, and therefore greater water availability and NPP than fine-textured soils in the same climatic regime. This has been termed the inverse-texture hypothesis (Noy-Meir 1973; Sala et al. 1988), because the reverse is predicted to occur in more humid regions where water-holding capacity is presumed to be more important than infiltration in determining soil moisture availability.
6. Abiotic constraints and the physiological tolerances of organisms, rather than biological interactions among organisms, dictate productivity levels. In other words, deserts represent areas where plants are stress tolerators, rather than competitors or ruderals.

The history of vegetation conversion in the region and the compelling need to understand desertification processes and their implications for system productivity and stability suggest additional questions that can and should be addressed. Our objective in this chapter is to review the history of NPP studies in the Jornada Basin, examining the degree to which studies there have given insight into these general hypotheses and examining the following specific questions:

1. Is NPP lower in shrublands (desertified) than in grasslands?
2. Is there evidence that NPP has declined over time at Jornada sites with progressive loss of soil resources or degradation?
3. Is NPP related to the diversity or composition of the plant community in averages or temporal patterns; that is, do biological factors such as species or growth form differences play a role?

4. Is NPP in this arid environment constrained most strongly by current and local precipitation? Do shrublands differ from grasslands in temporal variability and response to current precipitation?
5. Do biomass accumulation ratios reflect the slow growth and low turnover predicted for woody desert species?
6. Are Jornada empirical data consistent with the inverse-texture hypothesis; that is, are there consistent differences in productivity between fine- and coarse-textured soils that support the idea of greater control of water availability by surface properties and processes (infiltration and evaporation) than by water-holding capacity of the profile?

The chapter concludes by identifying those areas in which future research should be focused to make the greatest advances in our understanding of ecosystem function.

History of NPP Work at the Jornada

Over the past century, a number of important studies in the Jornada Basin have contributed to our understanding of NPP in arid environments. Early work on the JER focused on observations of forage productivity and the basic patterns of growth of the primary forage (perennial grass) species (Nelson 1934; Canfield 1939). After dry periods in the 1930s and particularly after the onset of the 1950s drought, scientific work centered on attempts to understand forage production and its response to drought on different soil types. The chief long-term data set on plant production from the Jornada was a study maintained from 1957 to 1988, summarized by Herbel and Gibbens (1996). Forage (perennial grass) production was assessed by clipping at sites located on 12 different soils; production or yield was obtained in annual sampling during October or November of each year. Rain gauges provided measures of monthly precipitation at most sites. Sampling of most of these transects was maintained and extended from 1978 through 1998 using clipping approaches to estimate production of all species (not just forage grasses) (Gibbens unpublished data).

In the 1970s, ecosystem studies were carried out at two Jornada Basin sites (one creosotebush [*Larrea tridentata*] shrubland and one black grama [*Bouteloua eriopoda*] grassland) under the auspices of the IBP. An IBP report (Pieper et al. 1983) described plant biomass and productivity at the IBP grassland and drew comparisons with study sites at Fort Stanton, New Mexico (a more mesic site), and Santa Rita, Arizona. Three years of data were contrasted, one noted as being "dry." Standing crop in both grazed and ungrazed areas contained about 50% warm season grasses, but black grama was less common and less important in the grazed locations. Summer forbs were noted as an important component of summer/fall biomass. The general values of NPP observed were typical of published values for semiarid grasslands and desert scrub elsewhere, 80 g C/m^2/yr (or roughly 160 g/m^2/yr plant dry weight). Pieper et al. (1983) reported that aboveground NPP (ANPP) was usually higher in ungrazed than grazed locations at the IBP grassland. Values were on the order of 100–200 g/m^2 per year, excluding any estimate for yucca (*Yucca elata*) and mesquite (*Prosopis glandulosa*) (described as too variable and too poorly sampled). Warm season grasses were the most important component, but warm season forbs and shrubs also contributed significantly to total ANPP during these IBP studies.

During the initial phase of the Jornada LTER program, productivity studies were concentrated along transects from the base of Mount Summerford to the playa near the CDRRC headquarters. Permanent quadrats and line intercept transects located every 30 m along 3-km transects were used to assess plant composition and biomass. Annual fertilization of one of the two transects was carried out to investigate the degree to which nitrogen availability limited production and composition in this semiarid system. Reviews by Ludwig (1986) summarized the observations coming from this and other early LTER work. He emphasized high variability over both space and time, documenting localized spots of extremely high productivity and considerable variation among years. Measures of aboveground production varied from 2–15× among years (table 11-1).

Another Ludwig study, near Carlsbad, New Mexico, reported equally variable results for a creosotebush stand (243–416 g/m^2/yr) and a mesquite-invaded grassland (23–82 g/m^2/yr) over just two years. Ludwig's (1986) review of data from other deserts documented 3–10-fold increases in ANPP between dry and wet years. One of the major conclusions from the Jornada IBP work on desert shrublands (Ludwig 1987) was that localized production (in small parts of the landscape) can be extremely high, as table 11-1 suggests.

In the second phase of the Jornada LTER program, a network of 15 sites was established to facilitate formal comparisons of plant composition, biomass, and ANPP in five major vegetation types (three sites in each): black grama grassland, grass-dominated playas, creosotebush shrublands, mesquite dune systems, and tarbush (*Flourensia cernua*) shrublands. This LTER program has monitored ANPP using consistent methods since spring 1989; these measurements provide a core set of observations for comparison of ecosystem function and structure among

Table 11-1. Ranges of values for annual aboveground net primary production for various ecosystems at the Jornada, from Ludwig (1987).

	Lowest Annual Value (g/m^2/yr)	Highest Annual Value (g/m^2/yr)
Bajada alluvial fans	53	292
Bajada small arroyos	37	318
Bajada large arroyos	30	456
Basin slopes	48	179
Basin swales	292	592
Basin playa	52	258

Jornada plant communities. Each site contains a grid of 49 1-m^2 permanent quadrats arranged at 10-m intervals in a 7 × 7 square. Sampling occurs three times per year: February, to capture winter annuals; May, timed to measure shrub leafout and to reflect a late-season sample of the spring bloom of annuals; and September–October, representing the peak biomass for the greatest number of annuals and other fall-flowering species. Plants or plant parts within the rectangular volume above each quadrat are measured nondestructively; plant dimensions are used to estimate living biomass per species per quadrat based on regressions developed during the first few years of sampling from adjacent harvested samples.

Huenneke et al. (2001, 2002) summarized methods and reported spatial and temporal patterns of ANPP over the first decade of this study. Further details of sampling and analysis, including an assessment of the adequacy of sample size, are given by Huenneke et al. (2001). The resulting data are consistent in intensity and methodology across ecosystem types, as well as over time. The method also allows more complete characterization of community productivity, spatially explicit description of pattern, and greater temporal resolution of aboveground productivity than is typically provided by rangeland yield estimates. This study demonstrated the impact of the patchiness of arid vegetation and variability in composition on the adequacy of sampling. Comprehensive sampling of all vascular plants within 49 square-meter quadrats was only marginally adequate for describing community attributes (e.g., aboveground biomass) in the most heterogeneous of systems. Given this demonstration, we are reluctant to draw strong conclusions when comparing our estimates of production with those from other studies where sample sizes are markedly smaller.

Fourteen years of results (figure 11-1; Huenneke et al. 2002 with additional unpublished data more recently available) showed that grasslands exhibit the greatest variation in production values over time; at least some grasslands (and grass-dominated playas) in some seasons are capable of achieving high rates of ANPP, but these peaks do not appear consistently, nor are those peaks restricted necessarily to summer rather than spring. In contrast, creosotebush-dominated sites demonstrate extremely regular seasonality, with the peak of production nearly always occurring in spring. Mesquite shrublands were less markedly seasonal than creosotebush systems, and demonstrated considerable variability among the three sites. Tarbush-dominated systems were remarkably invariant in aboveground biomass and in ANPP over time despite a substantial grass component.

A primary focus for the NPP study was spatial heterogeneity; the Jornada desertification hypothesis (Schlesinger et al. 1990; see also chapter 1) predicts that shrub-dominated systems are patchier in structure than the more homogenous grasslands and that patchiness is exacerbated by self-reinforcing patterns of increased biotic function within shrub patches compared to decreased biotic function in interplant spaces. At the scale of our monitoring, aboveground biomass was patchier in shrub-dominated systems than in the grasslands. In most growing seasons, ANPP followed the same pattern (i.e., greater spatial heterogeneity in shrublands; Huenneke et al. 2002).

Variation in mean ANPP is considerable even within structurally similar sites in a single year, such that it is difficult to assess differences among ecosystem

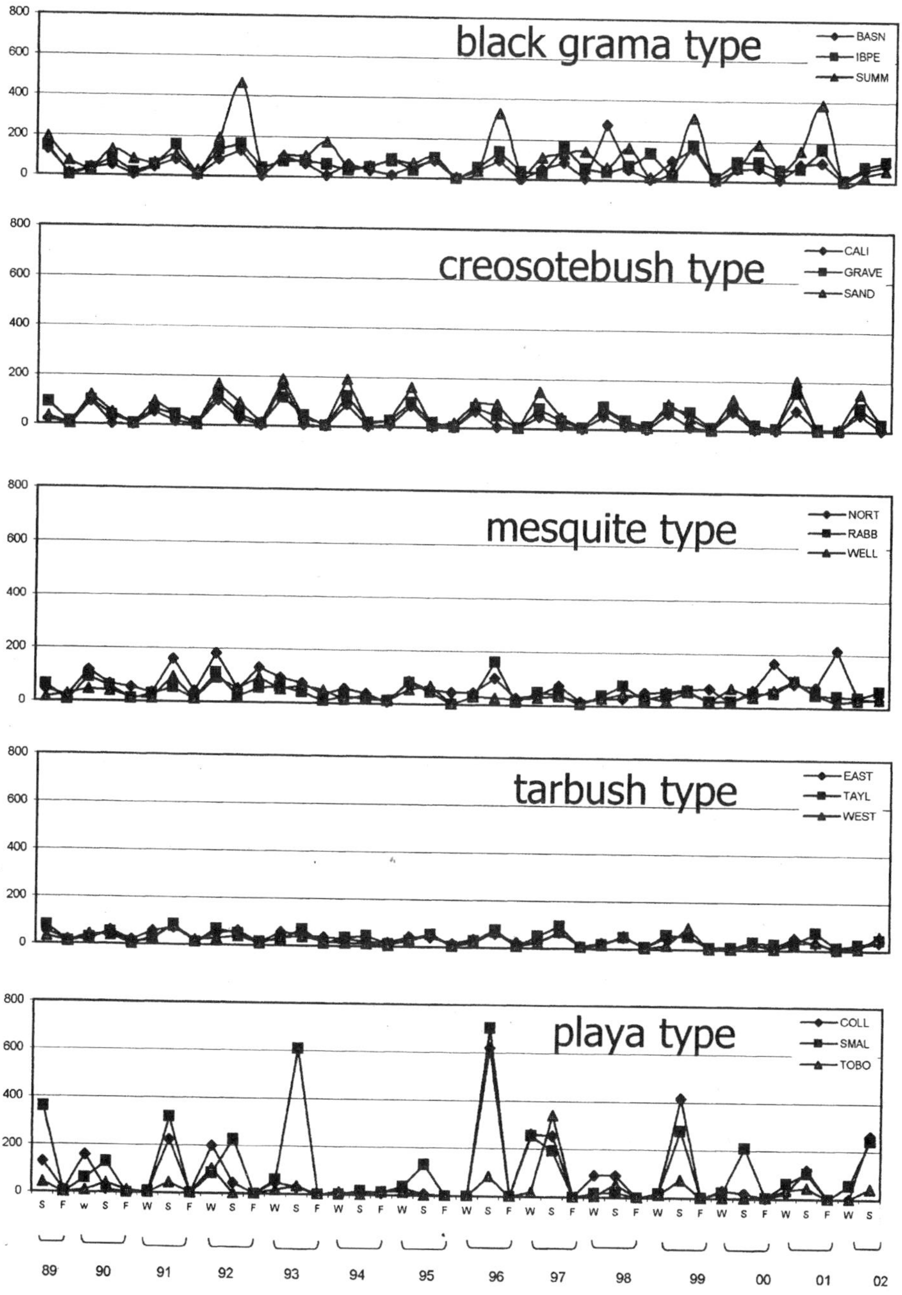

Figure 11-1. Annual aboveground net primary production (ANPP; g/m²) by season (Spring, Fall, and Winter) for 15 sites in 5 ecosystem types (3 sites per type with site acronyms for each type shown within figure) in the Jornada Basin. Figure reproduced from Huenneke et al. (2002) with additional unpublished data more recently available.

types. In any one season, ANPP values may have very similar ranges across all ecosystem types. However, during the 14-year (1989–2002) record, black grama grasslands were capable of supporting higher rates of production than shrub-dominated systems (Huenneke et al. 2002 with additional unpublished data more recently available). One of the next major challenges for the Jornada Basin LTER program is to better understand the drivers of variation in productivity among sites: What landscape-scale factors explain differences among the three sites sampled within each of the five ecosystem types? Based on differences in geomorphological setting or history, redistribution of water (e.g., from runoff to run-in relationships) and soil differences merit exploration to understand this variability.

Whether comparing different studies from the Jornada or attempting to compare rates of production at the Jornada to those measured at other locations, one quickly encounters difficulties. Most typically, studies present biomass and/or productivity of only the perennial grasses, the primary forage for livestock (e.g., Paulsen and Ares 1962; Valentine 1970; various studies summarized in Herbel and Gibbens 1996). It is clear, however, both from LTER studies (Huenneke et al. 2002; see also chapter 10) and from other work that woody plants, succulents, and other nonforage species can contribute substantial percentages of ANPP even in grasslands (e.g., Lane et al. 1998). In much of the arid and semiarid rangeland literature, reported production values greatly underestimate actual total values (and ecological potential). This makes direct comparison among studies difficult and has perhaps exacerbated the view that conversion to woody plant dominance represents extreme degradation. Our data (Huenneke et al. 2002) confirm that differences between grassland and shrubland productivity, though detectable over the long term, are not enormous when one considers the entire plant community. Another major difficulty, of course, is the great range of methods applied, from visual estimates to clipping to nondestructive dimension analysis. This history of diverse approaches greatly limits the power of comparison, whether among or between Jornada studies and those reported from other sites.

Applications of Jornada NPP Results to Hypotheses

Production Is Constrained by Water and Nutrient Availability

At least since the 1970s, the predominant model for arid ecosystems has posited that dryland ecosystems respond primarily to water with pulse response of both vegetative and reproductive growth to large precipitation events or episodes of water availability (e.g., Beatley 1974; Hadley and Szarek 1981). However, Jornada work during the IBP and early LTER periods tested the idea that primary productivity might be at least partially constrained by nitrogen (Ettershank et al. 1978). Ludwig (1987) reviewed fertilization results from that and other works, finding evidence of significant nitrogen limitation. In reality, water and nitrogen undoubtedly interact. Gutierrez and Whitford (1987a, b; Gutierrez et al. 1988) observed strong interactions between water and nitrogen amendments for annuals in the early LTER transect work. Herbel (1963) found a strong effect of nitrogen when water was available for *Hilaria* (=*Pleuromorpha*) *mutica*. For the herbaceous component at least, nitrogen availability appears to be an important constraint on primary production (chapter 6).

Temporal Variation in NPP: Correlation with Precipitation Patterns

The influence of water availability on production is typically investigated by examining the correlation of production with precipitation inputs. Herbel (1963) documented a positive correlation between growth in one important Jornada perennial grass (*Pleuromorpha mutica*) with a substantial lag effect (higher production in the year after water availability had been high, or the converse, drought in one year reducing response of the grass to next year's moisture). The year lag contrasted with studies of blue grama (*Bouteloua gracilis*) by Sala and Lauenroth (1982), who found that even very small rain events can stimulate short-term positive response. Preliminary inspections of correlations between more recent LTER data from the ANPP sites and precipitation records from those sites revealed very few positive correlations between rainfall and production at any one site with or without lag times. Local precipitation inputs may be a poor proxy for water availability due to the importance of runoff and run-in patterns in redistributing water. The aggregation of growth responses by species differing dramatically in life history, physiology, and growth form also probably obscures any simple correlation.

Noy-Meir (1985) suggested that efficiency ratios such as the slope of the relationship between NPP and rainfall or the slope dP/dr are best for comparing NPP among desert systems. Le Houerou et al. (1988) reviewed production studies from numerous semiarid ecosystems to estimate the efficiency with which precipitation is converted into plant production. The conclusion was that each mm of rain produces on average 4 kg/ha/yr (0.4 g/m^2). For Jornada Basin upland grasslands, the LTER productivity data are about 248 g/m^2 from a long-term average annual precipitation of 245 mm, or about twice the productivity predicted by Le Houerou et al. (1988).

There has also been substantial attention paid in the literature to the relationship between the variability in production and the variability in precipitation. Knapp and Smith (2001) analyzed long-term productivity data from 11 different LTER sites with particular attention to temporal variability in ANPP and the relationship to variation in precipitation inputs. Mean values for both ANPP and precipitation at the Jornada Basin were among the lowest among the 11 sites, and variability in precipitation was relatively high, as might be expected. However, the sites with the greatest interannual variation in precipitation did not, in fact, have the highest variability in ANPP. Jornada ecosystems (and the other LTER sites in their analysis that are dominated by grasses) showed the highest pulses of production in response to relative maxima in precipitation (maximum precipitation–mean/mean) (Knapp and Smith 2001, fig. 2B). Jornada ecosystems appear to have asymmetric responses to climate with greater pulses of ANPP in relatively wet years compared to the declines in production in relatively dry years (Knapp and Smith 2001; Huenneke et al. 2002). Pulses in ANPP may then cascade into pulses of consumer activity and/or reproduction, again with varying lag times depending on life histories (Ostfield and Keesing 2000; see also chapter 12).

One factor contributing to the poor fit of local ANPP values to climate data is the high degree of spatial variability in precipitation (chapter 3). Because a large proportion of the annual precipitation is delivered in the form of localized convective storms, rainfall inputs are extremely patchy, and no single location's rainfall record (e.g., the central LTER weather station) is an accurate measure of the precipitation received at any of the dispersed 15 ANPP (chapter 3) study sites. Most (13 of the 15) sites had simple collecting gauges at or near the production plots for much of the study period, so we were able to obtain crude local estimates of rainfall. Regressions of seasonal NPP values against local precipitation amounts for individual sites were nearly always insignificant. Various lag times have been proposed (or determined based on regression from individual species growth data) for different components of the vegetation; that is, growth appears to respond to precipitation received in a prior season or a prior year rather than the current season. Even when we incorporated different lag times into our regressions, the fit for total production values remained extremely poor. We suspect that different species have very different patterns of response to precipitation, and therefore no simple relationship exists between precipitation received in any one interval and total community aboveground production.

Finally, one additional source of variability likely serves to obscure the precipitation–ANPP relationship: landscape location and spatial context for a site. Local precipitation may not be the primary determinant of soil moisture availability; runoff and run-in can be substantial in certain topographic positions, and surface soil characteristics, animal disturbance, and other factors may constrain or enhance infiltration (chapter 5). Thus we expect that spatially explicit models may be required to explain variation among sites in NPP not adequately explained by local precipitation (chapter 18).

NPP Reduction in Shrublands versus Grasslands: Desertification as Degradation

Our study of 15 sites affords the first real opportunity to examine this question carefully, because it delves into patterns of productivity over many seasons and in multiple locations for each ecosystem type. Hence, we are able to assess differences among ecosystems relative to the tremendous spatial and temporal variation. For any given season, productivity rates in shrub-dominated systems may equal those in grasslands (Huenneke et al. 2002). Tarbush-dominated systems, however, are consistently low in productivity (despite a significant grass component). Grasslands demonstrate occasional higher peaks of production than those seen in any season in a creosotebush or mesquite system. Thus, although long-term means may not differ substantially, grasslands are at least capable of higher production. The Summerford grassland is unusual in supporting extremely high production rates in several years; the most likely explanation is its location at the base of Mount Summerford where runoff from the mountain is consistent and perhaps aspect or shelter from prevailing winds contributes to favorable site water balance.

Two provisos are necessary when discussing the higher ANPP values sustained in grassland relative to shrub-dominated systems. First, our current shrublands represent a mix of recently desertified former grassland sites and sites long occupied by shrubs. These diverse settings are being compared with sites that have resisted shrub invasion and the loss of grasses; sites that may be unusual in some respect (Gibbens et al. 2005; see also chapter 10). One should not rashly presume that comparing today's shrublands with today's grasslands is a comparison of equivalent sites differing only in the nature of the current vegetation.

Second, the tremendous range of variation (spatial and temporal) in ANPP values renders most comparisons with published values from arid ecosystems meaningless. Most studies fail to document the adequacy of sample numbers to capture a representative range of spatial variation (Huenneke et al. 2001). Similarly, most published studies are of relatively short duration and so do not portray production over long enough time periods to represent the range of variation possible at a given site. Without such assessment of the range of variation, attempts to compare the relative magnitude of production across locations are futile.

NPP in Relation to Soil Texture

The long-term data set described by Herbel and Gibbens (1996) furnishes an excellent opportunity to address the inverse texture hypothesis—the prediction that coarse, sandy soils will exhibit greater infiltration rates and therefore greater moisture availability and productivity than sites with clay-dominated soil textures. In fact, the Jornada Basin data show a clear and consistent pattern of coarser-textured soils supporting lower perennial grass yields than finer textured soils. Maximum, minimum, and average values of grass production all demonstrated the same pattern (table 11-2). ANPP appears to be greatest in sites on loamy soils, and the sandy or coarser soils do not generate ANPP values as high as (much less greater than) those from finer textured soils. Similar rejection of the hypothesis is supported by the LTER data: Mesquite and creosotebush sites have coarser soils than do grasslands but certainly do not have higher average productivity (Huenneke et al. 2002). Average ANPP (across three sites of each community type) for 13 years (1990–2002) for the clay-textured playa sites, the clay loam–textured tarbush sites, the loam-textured upland grassland sites, the gravel and sand–textured creosotebush sites, and the sand-textured mesquite sites were 205, 97, 248, 144, and 156 g/m^2, respectively (table 11-3). The highest ANPP values anywhere are on the heavy soils of playas and the loamy grassland soils. (However, it is true that ANPP is consistently low on the fine-textured soils of tarbush sites.) Results from Jornada productivity studies are generally not consistent with the inverse-texture hypothesis. The implication is that surface processes of water infiltration and evaporation from bare soil surface are not the primary factors

Table 11-2. Aboveground net primary production (kg/ha) of perennial grasses in relation to soil texture: a test of the inverse-texture hypothesis, which predicts that ANPP will be greater on coarser textured soils in arid and semi-arid ecosystems. Data from Herbel and Gibbens (1996).

Soil Texture, Soil[a]	Peak Yield	Mean Yield	Minimum Yield
Clay, A	1805	648	133
Clay loam, G	3718	1259	14
Clay loam, G	4061	2433	569
Silt loam, D	3017	785	61
Loam, B	6547	2126	20
Loamy sand, P	1031	401	7
Sand, O	918	188	9
Sand, O	453	207	11
Sand, T	800	354	38
Sand, T	1037	335	27

[a] See table 4-1 for description of A, G, D, B, P, O, and T soil map unit designations.

driving ANPP. Instead, perhaps the water-holding capacity of profile and the relative topographic position determining a sites runoff/run-in balance are more important as functional characteristics.

NPP in Relation to Diversity and Structure of Plant Community

The Jornada Basin long-term data sets represent good opportunities to study diversity–ecosystem function relationships with both observational and manipulative approaches. Chihuahuan Desert communities are less complex structurally than most forest types, though more challenging than more mesic grassland or herbaceous communities. The communities encompass a relatively wide range of richness values, from extremely simple mesquite sites to diverse grasslands. Our data are providing opportunities for others to test hypotheses about the shape of the productivity–species richness relationship. Fertilization-induced increases in productivity in the first years of the Jornada LTER program were accompanied by modest decreases in species richness (Gough et al. 2000).

In the long-term LTER study of the 15 ANPP sites, the low degree of evenness is striking across virtually all ecosystem types. The single most dominant species in each site contributes the largest proportion of ANPP, and only a small number (one or two) of subdominants add any substantive proportion to total community production (chapter 10). There has been no change in the identity of those dominants at any site over 14 years of study. Our pattern of the single dominant (either

Table 11-3. Mean biomass (g/m²), mean ANNP (g/m²), and biomass accumulation ratios or mean residence time of aboveground plant biomass (the ratio of biomass to ANNP) for the five main community types in the Jornada Basin.

Community Type	Mean Biomass	Mean ANPP	Ratio
Creosotebush	222	144	1.5
Upland grassland	319	248	1.3
Mesquite	277	156	1.8
Lowland playa	192	205	0.9
Tarbush	159	97	1.6

Notes: Mean biomass was calculated as the average (across three sites of each ecosystem type) of the annual peak aboveground biomass for 14 years (1989–2002). ANPP was calculated as the average (across three sites of each community type) of the 12-month aboveground production for 13 years (1990–2002) (Huenneke at al. 2002 with additional unpublished data).

a shrub or a perennial grass, depending on the ecosystem type) contributing most of the biomass and production contrasts with the suggestion of Hadley and Szarek (1981) of a general pattern of annuals contributing about half the annual NPP in hot desert systems. Certainly annuals can constitute high numbers and high surface cover in some sites in some seasons, but the larger perennial plants with long-lasting aboveground structures dominate ecosystem function in the sense of ANPP in the Jornada Basin. Chew and Chew (1965) estimated that 70% of the production of a creosotebush community near Portal, Arizona, was contributed by the dominant shrub, a somewhat smaller percentage of the total than at the Jornada.

We observe no compensatory behavior among species (in terms of contributing to productivity) between dry and wet years; rather, there are additive effects of species richness in grassland with across-the-board increases for all species in favorable years. Nor is there any simple correlation of richness with ANPP, either within a season or over the length of the study.

An ongoing LTER experiment addresses the diversity–function relationship through plant removals that have reduced species and functional group richness. In 1995 we initiated a long-term plant diversity experiment in which groups of species have been removed from a mixed creosotebush-mesquite shrub site. In the short term, this experiment tests the hypothesis that growth and resource use of one species group (e.g., perennial grasses) are constrained by interactions (such as competition for soil water) with other species groups (e.g., shrubs). However, in the first five years of this removal experiment, we observed no strong positive responses, questioning that the removal of any one species group in fact releases any other group from water limitation.

NPP Reduction Observed within a Site or Community over Time: Desertification as Degradation

Even with the long history of study at the Jornada, there has been little consistency of method over any reasonable time period, so we have had no direct way of comparing results or providing a true long-term perspective of basin-wide changes in ANPP. Year-to-year (and also decadal scale) variation is so great that it is not immediately possible to say that within a given site there has been significant degradation or loss of productive capacity. Given that shrublands have ranges of production largely overlapping with the range of values of grasslands (except for peaks), it seems that primary change has been alteration in the identity of the producers and a loss in the capacity to respond to favorable years, rather than an overall directional decrease in production rates.

Biomass Turnover and the Dynamic Nature of Arid Plant Communities

We calculated the biomass accumulation ratio for each of the five ecosystem types as an indicator of the mean residence time of aboveground plant biomass (table 11-3). The biomass accumulation ratio or mean peak biomass divided by annual productivity (Whittaker 1975) is 0.9 for the playas, which contain primarily short-lived herbaceous vegetation whose aboveground portions are replaced completely in each wet season. Surprisingly, even the mesquite sites, so strongly dominated by the woody shrub component, had a biomass accumulation ratio of only 1.8 (suggesting complete turnover of aboveground biomass on average within two years). One mesquite site (M-NORT) showed a multiyear directional trend in biomass accumulation with aboveground biomass increasing during the 1990–94 period. The magnitude of the increase was substantial. However, the 1994–95 dry conditions coincided with a sharp decline in biomass to former levels, and in the subsequent years there has been no directional change, merely fluctuation in NPP that is typical of the other sites being studied.

Belowground Productivity

There have been few formal studies of belowground plant biomass and productivity in Jornada ecosystems, although there has been considerable descriptive work. Belowground biomass (and allocation belowground or root:shoot ratio) is generally assumed to be high in arid and semiarid systems. However, Chew and Chew (1965) found that creosotebush near Portal had a rather low proportion of root biomass. Ludwig (1977) presented data for eight species of woody plants at the Jornada showing root:shoot ratios of mean 0.9; he described these as generalized (not specialized) root systems. On the other hand, Gibbens and Lenz (2001) have documented very extensive root systems for a number of Jornada plants, demonstrating impressive proliferation in specific soil volumes and suggesting highly plastic and responsive allocation. Brisson and Reynolds (1994) similarly described patterns of root distribution in creosotebush at the Jornada that suggested intricate and plastic relationships among neighboring individuals.

Pieper et al. (1983) found root:shoot ratios greater than 1 (1.4–2) for vegetation sampled in the long-term forage production study with higher values in grazed than in ungrazed communities. Grasses constitute much of this belowground biomass. In absolute terms, belowground biomass was less in grazed areas (due largely to the difference in species composition with fewer grasses and thus fewer shallow fibrous roots).

In the productivity study carried out across the 15 LTER sites, we were not able to study the belowground portions of these ecosystems and so cannot extrapolate from our measurements of aboveground biomass and productivity to total values. Ludwig (1977) reviewed a number of studies and concluded that the often stated generalization that desert shrubs have high root:shoot ratios is too simplistic; values for creosotebush at the Jornada, for example, varied from 0.23 to 2.7. Mean values for eight Jornada species averaged 0.925. If we apply this ratio to our average aboveground biomass values (table 11-3), we would estimate that total plant biomass (above- and belowground) ranges from about 300 g/m^2 in the tarbush ecosystem type to more than 500 g/m^2 in the grassland and mesquite ecosystems.

Remaining Challenges

Our challenge at the Jornada Basin is to better understand the sources of variation in NPP—the most basic measure of ecosystem function. To what degree is variation a simple reflection of variable site characteristics or local precipitation? Spatially explicit models may be necessary to understand local variation: topographic position, fluxes of materials, or organisms from one landscape position to another, and so on (chapter 18).

Are remote sensing methodologies suitable for monitoring NPP on the spatial and temporal scales appropriate for these landscapes? There is little evidence suggesting that remotely sensed data are adequate for assessing plant biomass and ANPP in this arid region (poor correlations). Variety of indices have been tried and different indices fit different ecosystems best.

Which mechanisms of nutrient losses are of a magnitude sufficient to cause degradation of productive capacity within a site? Mesquite sites are areas of greatest sand and dust generation; however, net flux may be different due to landscape position relative to prevailing winds, characteristics of vegetation structure (roughness) that would trap dust or waterborne sediments, for example. Thus both eolian and hydrologic fluxes are crucial to study at landscape as well as any localized scale (chapter 9).

Conclusions

What have we learned from the Jornada Basin about NPP in arid ecosystems? First, these ecosystems are more complex in time than a simple pulse-reserve model would suggest: There are no simple relationships between precipitation input and system function (not even threshold relationships). Instead, the varying life span and multiple modes of response (from vegetative growth to belowground storage to seedling recruitment) may contribute to the potential for complex multiyear dynamics. More significant as an immediate problem is the tremendous spatial variation in rates of aboveground production within ecosystems of a given type. This variation makes it difficult to extrapolate from point-based measurements to larger, management-relevant scales. We have found it necessary and desirable to move beyond local determinants (such as soil features and local precipitation) to spatially explicit parameters that reflect a site's landscape location in order to better understand temporal and spatial patterns of production (chapter 18).

A significant contribution of our studies of plant production is that total community composition includes significant contributions from species other than perennial grasses. Hence, traditional estimates of ANPP (based on forage alone or on clipping of grasses) are far lower than our more dynamic and comprehensive estimates. In the short term, this makes it frustrating or impossible to compare our estimates with those from previous work or other sites; in the long run, this more comprehensive view provides a sounder basis for applications, such as remediation efforts, or for understanding organic matter or carbon dynamics and sequestration.

12

Chihuahuan Desert Fauna: Effects on Ecosystem Properties and Processes

Walter G. Whitford
Brandon T. Bestelmeyer

This chapter focuses on the direct and indirect effects of animals on ecosystem processes and/or their effects on ecosystem properties. This set of effects has been the primary focus of animal studies on the Jornada Experimental Range (JER) and the Chihuahuan Desert Rangeland Research Center (CDRRC) during the twentieth century. Early studies dealt with animal species that were thought to reduce the amount of primary production that was available to support livestock. With the establishment of the International Biological Programme (IBP) in the late 1960s and its premise that ecosystems could be modeled based on energy flow, animal studies were designed to measure energy flow through consumer populations. Those studies yielded estimates of consumption of live plant biomass between 1% and 10% of the annual net primary production (NPP) (Turner and Chew 1981). From these studies Chew (1974) concluded that in most ecosystems consumers process only a small fraction of the NPP as live plant material but play important roles in ecosystems as regulators of ecosystem processes rather than energy flow. Chew's hypothesis was then the focus of animal studies in the Jornada Basin for nearly 30 years. Studies of animals as regulators of ecosystem processes led to the expansion of Chew's hypothesis to include the effects of animals on ecosystem properties, such as patchiness.

Many of the studies examined in this chapter support the hypothesis that animals affect spatiotemporal heterogeneity and in turn are affected by it. Because this research focused on the role of animals in ecosystems, studies of animal populations were conducted simultaneously with functional studies. Population and behavioral studies were considered an integral part of the central theme because they supported an understanding of the spatial and temporal variation of desert ecosystem properties.

The Distribution and Abundance of Animals and Their Effects

We review animal studies that focused on spatial patterns in the distribution and ecosystem effects of several taxa and guilds. Large-scale ecosystem degradation and vegetation changes in the Jornada Basin occurred prior to studies of animal populations (Buffington and Herbel 1965). Therefore, it is important to bear in mind that the published data on animal populations reflect vegetation and ecosystem conditions that are very different from the conditions in which many Chihuahuan Desert species existed only a century before (see chapter 10).

Factors affecting the distribution of vegetation types have probably had strong effects on small mammal diversity. Overall, the most abundant and widespread rodents on the Jornada belong to the family heteromyidae (kangaroo rats [*Dipodomys* spp.], silky pocket mice [*Perognathus* spp.], and coarse-haired pocket mice [*Chaetodipus* spp.]). Merriam's kangaroo rat (*Dipodomys merriami*) is most abundant in the shrub-dominated habitats, and Ord's kangaroo rat (*Dipodomys ordii*) is most abundant in the grassland habitats. The banner-tailed kangaroo rat (*Dipodomys spectabilis*), a grassland specialist that plays a keystone role in these ecosystems (Mun and Whitford 1990), is absent in the desertified mesquite (*Prosopis glandulosa*) coppice dunes and creosotebush (*Larrea tridentata*) and tarbush (*Flourensia cernua*) shrublands. Nonetheless, both the abundance and species richness of rodents were higher in shrub-dominated areas than in desert grassland (Wood 1969; Whitford 1976; Whitford et al. 1978b). The subdominant species in desert grasslands included grasshopper mice (*Onychomys* spp.), spotted ground squirrels (*Spermophilus spilosoma*), and silky pocket mice (*P. flavus*). Dry lake basin grasslands and tobosa (*Pleuraphis mutica*) grass swales are thought to support cotton rats (*Sigmodon hispidus*) (Wood 1969), whereas pocket gophers (*Thomomys bottae*) are limited to the piedmont grassland at the base of Mount Summerford of the Doña Ana Mountains (see figure 2-1 in chapter 2). Studies in other regions of the Chihuahuan Desert suggest that vegetation growth form, vegetation cover, landscape position, and soil texture determine the spatial distribution patterns of rodents. Black-tailed prairie dogs (*Cynomys ludovicianus*) occurred in scattered colonies in the basin prior to 1917. During World War I these populations were exterminated by government programs to increase forage area for livestock to promote red meat production during the war period. These populations have not returned (Oakes 2000).

Black-tailed jackrabbits (*Lepus californicus*) and desert cottontails (*Sylvilagus auduboni*) are important midsize herbivores. Their abundance fluctuates greatly over time in response to rainfall patterns, desertification status, and productivity of the landscape units on the Jornada. Mean annual abundance of black-tailed jackrabbits was 36/km^2 in mesquite shrublands, 30/km^2 in mesquite coppice dunes, and approximately 8/km^2 in creosotebush and tarbush shrublands. Mean annual abundance in grassland was 5.7/km^2. Desert cottontail abundance varied from 1.0 to 7.2/km^2 in shrublands but only 0.25/km^2 in grasslands.

Rodents and other small mammals may create spatial heterogeneity through their digging activities. Foraging pits serve to trap windblown seeds and plant litter (Steinberger and Whitford 1983a). When the pits are filled in by eolian or water-transported sediments, the seeds in the pit escape collection by harvester ants and probably escape other seed feeders. More than half of tagged foraging pits in black grama (*Bouteloua eriopoda*) grassland produced threeawn (*Aristida* spp.) seedlings and/or seedlings of globe mallow (*Sphaeralcea subhastata*) (Jackson and Whitford unpublished data). Furthermore, rodent digging activities may accelerate erosion rates when loosened sediment is washed away (Neave and Abrahams 2001, see chapter 7).

Birds and rodents exhibit similar patterns of abundance and species richness. Breeding bird densities in black grama grasslands (9.8 breeding pairs/km^2) were considerably lower than in the creosotebush shrublands (28.8 pairs/km^2) (Raitt and Pimm 1978). Intact grasslands supported fewer species and lower abundances of breeding birds than the most degraded areas (mesquite coppice dunes) (Whitford 1997). The breeding birds in desert grassland were insectivores. Breeding/nesting birds were completely absent from the desert grassland site in a year with below-average growing season rainfall. The grassland breeding birds nested in soapweed (*Yucca elata*) and in mesquite $>$ 3 m tall, providing evidence for the importance of vegetation height diversity for breeding bird abundance and species richness. It also documented dependence of desert grassland breeding bird abundance and species richness on rainfall.

Breeding bird densities on a creosotebush-dominated bajada averaged 28.8 breeding pairs/km^2 during the two years of study (Raitt and Maze 1968). Nests of the most abundant species, black-throated sparrows (*Amphispiza bilineata*), were primarily in creosotebushes at the margins of small drainage channels (86% of the nests). Nests of verdins (*Auriparus flaviceps*), the second most abundant species, were predominantly in whitethorn (*Acacia constricta*) growing along the margins of large and small drainage channels (79% of the nests). The nests of other species recorded in this study were located in large, riparian shrubs growing in the channels or margins of large arroyos (cactus wren [*Campylorhynchus brunneicapillus*], crissal thrasher [*Toxostoma dorsale*], black-tailed gnatcatcher [*Pilioptila melanura*], and loggerhead shrike [*Lanius ludiovicianus*]). Other species that were recorded on the study area included scaled quail (*Callipepla squamata*), mourning dove (*Zenaida macroura*), and roadrunner (*Geococcyx californianus*).

Dissemination of seeds and production of safe germination sites are among the important processes contributed by birds. Nest construction by cactus wrens concentrates many viable seeds of grasses and forbs in the materials collected. Common plants used in cactus wren nests were bush muhly (*Muhlenbergia porteri*) and the exotic Lehmann's lovegrass (*Eragrostis lehmanniana*) (Milton et al. 1998). Nest material that was subjected to germination trials produced 375 seedlings, mostly monocots, from 20-g samples from 12 nests. Seed in the nest material is dispersed when the nest is abandoned and disintegrates. The inclusion of an exotic lovegrass in the nest construction material may contribute to the dispersal of this species, which is a competitor with native grasses.

The most abundant lizard species on the Summerford watershed was the western whiptail (*Cnemidophorus tigris*). Densities of this species ranged from 30–50/ha on the playa fringe and creosotebush bajada. There were 12 species of lizards recorded in the various habitats on the watershed. Several species were transients or immigrants from nearby source habitats, including the Chihuahuan spotted whiptail (*Cnemidophorus exsanguis*), Great Plains skink (*Eumeces obsoletus*), and lesser earless lizard (*Holbrookia maculata*). Only four species were permanent residents of the creosotebush bajada: western whiptail, checkered whiptail (*C. tesselatus*), round-tailed horned lizard (*Phrynosoma modestum*), and long-nosed leopard lizard (*Gambelia wislizeni*). The greater earless lizard (*Cophosaurus* [*Holbrookia*] *texana*) was limited to the large arroyo habitats on the bajada. The side-blotched lizard (*Uta stansburiana*) was associated with the dense vegetation of the feeder arroyos and the large arroyos. Seven lizard species were residents in the mesquite–Mormon tea (*Ephedra* spp.) area fringing the playa lake basin: western whiptail, checkered whiptail, Texas horned lizard (*Phrynosoma cornutum*), round-tailed horned lizard, side-blotched lizard, desert spiny lizard (*Sceloporus magister*), and long-nosed leopard lizard (Whitford and Creusere 1977). The abundance and diversity of lizards was higher in mesquite coppice dunes than in adjacent grasslands. The higher species richness in desertified habitats is the result of the addition of shrubland species to the grassland lizard assemblage.

Five species of anurans inhabit several areas of the Jornada Basin around ephemeral lakes. Estimated densities of adult anurans (based on mark and recapture) at a playa lake were: western spadefoot (*Scaphiopus hammondi*), 238/ha; Plains spadefoot (*Scaphiopus bombifrons*), 206.3/ha; Couch's spadefoot (*Scaphiopus couchi*), 79.4/ha; green toad (*Bufo debilis*), 39.7/ha; and Great Plains toad (*Bufo cognatus*), 11.9/ha (Creusere and Whitford 1977). As soon as the playa lake flooded, species of breeding frogs began to occupy different parts of the playa. Western spadefoots were concentrated in open water with depth greater than 12 cm, Plains spadefoots and green toads were concentrated in areas of sparse vegetation with green toads in the shallower water, and Couch's spadefoots were concentrated in shallow water areas with dense vegetation.

Anurans may play a role in moving nutrients from areas of nutrient concentration, that is, ephemeral lakes, which serve as collection points for runoff and transported sediments. Juvenile toads move to areas of sandy soil around the margins of the playa, where they burrow into the soil to estivate until the next growing season. Overwintering juvenile toads suffered high mortality (70–80%) during the first winter and 50–58% during the second winter (Creusere and Whitford 1977). Survival of juvenile toads is dependent on the quantity of fat the toads can accumulate prior to burrowing into the soil for winter and on the moisture of the burrow sites (Whitford and Meltzer 1976). Juvenile toads that die in their overwinter burrows return nutrients concentrated in their natal ponds to the surrounding area. Based on the conservative estimate of juvenile toads of 18,333/ha, this can be an important mechanism of spatial redistribution of nutrients from a nutrient sink to the surrounding landscape.

Studies of arthropods have focused on those taxonomic groups that have a role in seed dispersal and the fate of seeds, decomposition and nutrient cycling processes, and/or the formation of patch heterogeneity of Chihuahuan Desert landscapes. Because of their effects on decomposition, nutrient cycling, water infiltration, and spatial distribution patterns of organic matter, subterranean termites are considered keystone species in the Chihuahuan Desert (Whitford 2000). The abundance and spatial distribution of subterranean termites (primarily Isoptera: Termitidae, *Gnathamitermes tubiformans*) was studied by use of bait rolls (Johnson and Whitford 1975; Nash et al. 1999). These studies found that subterranean termites were equally abundant in all desertified and undegraded habitats, except for those areas inundated for periods of a month or more (ephemeral lake basins). These keystone insects have not been affected by vegetation change resulting from desertification. The average live biomass of termites, estimated from numbers of termites removed from bait rolls, on the Desert Biome watershed was 3.6 kg/ha. The ratio of termite biomass to livestock biomass based on average stocking rates was 4.4, indicating the potential importance of termites in energy flow in Jornada ecosystems (Johnson and Whitford 1975).

Abiotic processes (heat and ultraviolet light) decompose detritus located on the soil surface (MacKay et al. 1987a; Moorhead and Reynolds 1989a) or the detritus is consumed by invertebrate detritivores and decomposed in their guts by the symbiotic microflora and microfauna (Crawford 1988). There were large differences in ground-dwelling arthropod communities at the base and on the piedmont slopes of the IBP Desert Biome (Summerford) watershed. The most abundant ground-dwelling arthropods were tenebrionid beetles and orb spiders (mostly black widow [*Lactodectus mactans*]). Most of the taxa reported for the creosotebush bajada habitat were also reported for desert grasslands in the Jornada Basin. However, in the grasslands, small tenebrionid beetles (*Araeoschizus decipiens*) were several times more abundant than any of the other ground-dwelling arthropods. Another difference between the bajada shrubland and the basin grassland is the abundance of sand roaches, Polyphagidae (mostly *Arenivaga* spp.) (Whitford et al. 1995). The abundance of dung beetles in the traps in the grassland sites represents dispersing individuals that were trapped en route to adjacent areas that were grazed by livestock. Overall, the most abundant taxa in both shrubland and desert grasslands are detritivorous (scarabids, tenebrionids, polyphagids, and gryllacridids). The gut symbionts of these detritivores include bacteria, fungi, protozoans, and nematodes (Crawford 1988). Gut symbionts allow relatively high assimilation efficiencies (30–70%) even when the animals consume dead plant materials that are primarily cellulose and lignin. Several species of stink beetles (*Eleodes* [Coleoptera: Tenebrionidae]) have been observed feeding on the chaff accumulations around the nest disks of seed-harvesting ants (Whitford 1974). Desert cockroaches (*Arenivaga* spp.) feed on decaying leaves and roots of desert shrubs (Hawke and Farley 1973). Because most of the annual, aboveground NPP enters the dead plant material or detritus pool, detritivores are the most abundant Chihuahuan Desert animals and account for the highest biomass of primary consumers (Ludwig and Whitford 1981).

The spatial distribution of the most abundant seed-harvester ant species appears to be primarily related to soil texture characteristics (Whitford et al. 1976, 1999). Although there is overlap in the distribution of the two widespread species (rough harvester ant [*Pogonomyrmex rugosus*] and desert harvester ant [*Pogonomyrmex desertorum*]), *P. rugosus* is absent from mobile and stabilized sand dune areas, and *P. desertorum* is absent from the clay and silt soils of the tobosa grass swales. In a broom dalea (*Dalea scoparia*) sand dune area, Maricopa harvester ant (*Po-*

gonomyrmex maricopa) replaces rough harvester ant as the large harvester ant in the system. The large nests of these harvester ants in turn contribute to patchiness in soil properties due to the to the concentration of nutrients to nests, bioturbation, and chemical alterations of soil via the vertical redistribution of calcium carbonate (Whitford and DiMarco 1995; Wagner et al. 1997; Whitford 2002).

Soil texture appears to be the most important factor limiting the distribution of most of the other ant species in Chihuahuan Desert ant communities (Whitford et al. 1999; Bestelmeyer and Wiens 2001a). Creosotebush shrublands on the gravelly soils of the Summerford bajada had the highest site species richness (36 species) recorded in the Jornada Basin. The Summerford bajada is a valuable environment from a biodiversity perspective because it is a source of newly discovered (and potentially rare) ant species (MacKay and MacKay 2002) and possesses a higher richness than similar environments elsewhere (e.g., the Sevilleta LTER; Bestelmeyer and Wiens 2001b). Within sandy soils, however, mesquite cover may be a key factor governing species abundance (Bestelmeyer 2005). For example, an attine ant species (*Trachymyrmex smithii neomexicanus*) that is absent in grasslands occurs in high abundance in adjacent mesquite coppice dunes and has also been recorded in creosotebush communities on deep sandy soils (Wisdom and Whitford 1981). The attine ant species collect senesced leaves and senesced floral parts (petals and sepals) (Schumacher and Whitford 1975). The leaves and floral parts are broken down by fungi cultured by the ants in fungal gardens (Gamboa 1975).

The earliest studies of microarthropods found that their densities were directly correlated with the amount of surface litter (Santos et al. 1978). They reported that nanorchestid mites were found in all habitats sampled on a creosotebush-dominated bajada. Prostigmatid mites were the most numerous acari in all but one area: litter under dense shrubs on margins of an arroyo. In the arroyo margin area, cryptostigmatid (oribatid) mites were the most abundant order of acari. There were 18 orders of soil microarthropods recorded from 12 sites on the Summerford watershed (Wallwork et al. 1985). Mites and collembolans dominated all sites and, except for a honey mesquite site at the edge of the ephemeral lake, mites were more abundant than collembolans. The collembolan density at the honey mesquite site was estimated at 24,460/m^2. Genera of several families of prostigmatid mites dominated most of the sites in the Jornada Basin (Nanorchestidae: *Speleorchestes* sp., *Nanorchestes* sp.; Tarsonemidae: *Tarsonemus* sp.; Tydeidae: *Tydeus* sp., *Tydaeolus* sp.). These genera were the most abundant mites in black grama grassland, tobosa grass swales, mesquite coppice dunes, and tarbush shrublands (Steinberger and Whitford 1984, 1985; Silva et al. 1989a; Kay et al. 1999).

Cryptostigmatid mites were dominant only where the organic matter content of the soil was > 30%. In leaf litter, the highest density of mites was recorded from juniper (*Juniperus* spp.) litter (29,486/m^2), and the lowest density was recorded in creosotebush litter (8,274/m^2). Among the most abundant cryptostigmatid mites reported from the LTER watershed was a previously undescribed species, *Jornadia larreae* (Wallwork and Weems 1984).

Microarthropods were isolated from honey mesquite rhizosphere soils from depths up to 13 m (Silva et al. 1989b). Many of the microarthropods that characterized the rhizosphere fauna at depths > 1 m were species that were abundant in surface soils, that is, prostigmatids (*Speleorchestes* sp., *Tarsonemus* sp., *Nanorchestes* sp., and *Tydaeolus* sp.), cryptostigmatids (*Bankisonoma ovata* and *Passalozetes neomexicanus*), and the collembolan (*Brachystomella arida*).

The initial stage of decomposition of belowground litter is primarily via soil bacteria. The bacteria are grazed by protozoans, primarily naked amoebae, and by bacteriophagous nematodes. The numbers of protozoans and nematodes are regulated by several species of omnivorous microarthropods (Acarina) that prey on the nematodes (Santos et al. 1981). When microarthropods were eliminated from buried litter by a broad-spectrum insecticide, bacteriophagus nematode numbers increased dramatically. The large numbers of nematodes overgrazed the bacteria, thus reducing the rate of decomposition. In mesic ecosystems, microarthropods affect decomposition and mineralization processes by masticating the litter and passing it through their guts. This increases the surface area and inoculates the litter with microflora from the gut of the arthropods. The dominant soil acari in mesic systems are cryptostigmatid (oribatid mites). Oribatids constitute a small fraction of the soil acari community in arid and semiarid ecosystems (Wallwork 1982; Wallwork et al. 1985). Thus, in arid and semiarid regions, the role of microarthropods as regulators of the rate of decomposition is indirect via predation on nematodes and/or fungi, rather than directly by consumption of dead plant material.

The later stages of decomposition and mineralization in dry soils are regulated by some of the same species of omnivorous mites feeding on fungi. Fungi replace bacteria as the primary microfloral decomposers in dry soils. Experiments in which microarthropods were removed showed that rates of nitrogen mineralization were significantly reduced compared with the rates measured when microarthropods were present (Parker et al. 1984b). Elimination of fungivorous and omnivorous mites resulted in a large increase in fungal biomass. Mineral nitrogen from soil surrounding dead roots or buried litter is incorporated into fungal biomass. The fungi use the carbon in litter or roots as energy sources but scavenge nitrogen from the surrounding soil to produce fungal biomass. The nitrogen incorporated into fungal biomass is considered immobilized, that is, not available to be absorbed by plant roots. Soil microarthropods that graze on fungal hyphae release immobilized nitrogen as mineral nitrogen in the form of excretory products. These experiments demonstrate that mineralization of nitrogen in desert ecosystems requires the activities of soil microarthropods.

Arthropods can indirectly affect spatial patterns of plant growth by stimulating the activity of the soil microflora or by supplying soluble nitrogen directly to the soil beneath the plant canopies. The most abundant insects on shrubs in deserts are sucking insects (*Homoptera* and *Hemiptera*) (Lightfoot and Whitford 1987; Schowalter 1996; Schowalter et al. 1999). The frass and honeydew production from these insects fertilizes the litter under the shrub canopies with soluble carbohydrates and nitrogen. This readily available form of carbon and nitrogen stimulates the growth of microflora on the litter (Lightfoot and Whitford 1987). Rapid growth of soil microflora as a result of inputs of high-carbon, low-nitrogen substrates results in the immobilization of soil nitrogen in the rapidly growing microbial biomass (Parker et al. 1984b). Nitrogen immobilization imposes severe nitrogen limitations on the biomass production of shrubs and of ephemeral and perennial herbaceous species.

Synthesis Topics

In this section we review additional animal studies with respect to seven key topics in ecology that have guided Jornada research over the last three decades.

Pulse-Reserve and Source-Sink Models

In deserts, where many animal species live close to their limit of physiological tolerance for one or more abiotic factors, the responses of species populations to fluctuations in the abiotic environment must be understood before questions concerning the role of animals in ecosystems can be addressed. The pulse-reserve paradigm has been the primary conceptual model for responses of desert organisms to the abiotic environment (Noy-Meir 1974). Following this conceptual model, a rainfall pulse stimulates reproduction and growth of animal populations. Population reserves may be in the form of desiccation-resistant eggs, specialized "replete" workers that store food in honey pot ants (*Myrmecocystus*), or stored energy as fat reserves.

Whereas the pulse-reserve model addresses temporal heterogeneity, the source-sink conceptual model addresses the consequences of spatial variation in the transfer of individuals (and resources) between elements of landscape mosaics. This model considers landscapes to be composed of three types of habitats: (1) source habitat, in which reproduction exceeds mortality and the expected per capita growth rate is greater than one; (2) sink habitat, in which limited reproduction is possible but will not (on average) compensate for mortality, and the per capita growth is between zero and one; and (3) unusable habitat through which animals disperse, which comprises the matrix of all habitats that are never exploited by the species in question and in which patches of source and sink habitats are embedded (Danielson 1992). Results from several vertebrate studies (e.g., the cotton rat and amphibian cases reviewed earlier) can be interpreted according to an integration of the pulse-reserve and source-sink conceptual models.

Bottom-Up or Top-Down Population Regulation

Temporal and spatial variation in terrestrial herbivore populations has generated controversial hypotheses concerning the regulation of herbivore numbers (Strong 1988; Matson and Hunter 1992). The top-down hypothesis focuses on the impacts of predators and/or parasitoids on animal numbers, and the bottom-up hypothesis focuses on the consequences of resource quality. Top-down impacts on creosotebush canopy phytophages were studied by experimentally excluding avian and arthropod predators from creosotebushes. In the two years of the study, exclusion of predators resulted in increases in densities of phytophagous insects (Floyd 1996). The effects of bird and arthropod predation on phytophage populations was additive one year but were compensatory the following year. In the second year of the study, predatory arthropod numbers were lower on creosotebushes from which birds had been excluded than on the shrubs from which birds had not been excluded. The relative effects of predators on herbivore populations varied among seasons and among sites in both years. The impacts of predators on the herbivorous insects were not correlated with known gradients of climatic or of resources quality heterogeneity. The results of this study "confirm the important direct and cumulative effects of multiple predator guilds, even against a complex background of temporal and spatial heterogeneity" (Floyd 1996).

Two studies of creosotebush arthropods provided evidence of bottom-up regulation of herbivorous insect abundance. Herbivorous arthropods were more abundant on creosotebushes on nitrogen-fertilized plots than on creosotebushes on irrigated or control plots (Lightfoot and Whitford 1987). The morphology of creosotebush affects the nutrient status of soils under the shrubs. Insect abundance was higher on high-nutrient shrubs than on low-nutrient shrubs, confirming in part the bottom-up regulation (Lightfoot and Whitford 1989, 1991). A study of the effects of plant stress using "rain-out" shelters to impose drought stress revealed that only 2 of the 44 insect species studied increased in abundance on the stressed creosotebushes. Eight species of phytophagous insects exhibited increased abundance on creosotebushes in irrigated plots. The abundances of the remaining species did not change significantly among treatments (Schowalter et al. 1999). These studies show that phytophagous insects on creosotebush are regulated by both top-down and bottom-up processes.

The Roles of Social Insects

By virtue of their collective activities and ecological success, eusocial insects exert a strong effect on the functioning of desert ecosystems (MacKay 1991). Ants, for example, are among the most abundant arthropods in most of the world's deserts. The high forager abundance and flexible foraging habits of ants are two reasons for their success. Seed-harvester ants of the genus *Pogonomyrmex* are among the most abundant and widely distributed ants in the Chihuahuan Desert. The numbers of foragers in colonies varied among species: 1,000–6,000 in rough harvester ants (*P. rugosus*), approximately 1,000 in California harvester ants (*P. californicus*), and 200–600 in desert harvester ants (*P. desertorum*) (Whitford and Ettershank 1975). Peak foraging activity of rough harvester ants measured as percent of colonies foraging and rate of foragers returning to colonies, occurred following periods of drought. Rough harvester ants foraged at night during midsummer and colonies ceased foraging when granaries were filled. Desert harvester ants and California harvester ants exhibited only diurnal foraging behavior and did not exhibit larder-hoarding satiation. Soil surface temperature and saturation deficit accounted for 10–40% of the variation in foraging activity in harvester ants. Foraging in harvester ants was primarily affected by forage availability and secondarily by microclimate (Whitford and Ettershank 1975).

Scavenging and honey dew collection are other important activities performed by ants (Van Zee et al. 1997). Nearly all small arthropod cadavers that reach the ground are scavenged by ants, particularly the piss ant (*Forelius* or *Iridomyrmex* spp.) and the bicolored crazy ant (*Dorymyrmex* or *Conomyrma bicolor*). Larger ants and ants adapted to high soil surface temperatures tend to remove these materials over larger distances to their nests. The abundance of these two kinds of species at the Jornada leads to scales of redistribution that are nearly seven times that found in the shortgrass steppe biome (Bestelmeyer and Wiens 2003).

Chihuahuan Desert ant species vary widely in their daily activity patterns, including species that have specialized to nearly lethal diurnal temperatures (the piss ant and honey-pot ants *Myrmecocystus* [Endiodioctes subgenus]) to nocturnality (*Myrmecocystus* [Myrmecocystus subgenus]). The numbers of active colonies of seed-harvesting ants (*Pheidole* spp.), long-legged ants (*Aphaenogaster* [*Novomessor*] *cockerelli*), crazy ants (*Dorymyrmex* [*Conomyrma*] spp.), piss ants, fire ants (*Solenopsis xyloni*), perpilosa formica ants (*Formica perpilosa*), honey-pot ants, and New Mexico leaf-cutter ants (*Trachymyrmex smithii*) also vary seasonally, generally with peak numbers of active colonies in midsummer but with some important exceptions among years (Whitford 1978). As with harvester ants, the foraging activity patterns of these species appear to primarily reflect variation in forage availability and secondarily microclimate. Experimental studies of foraging ecology of long-legged ants demonstrated the importance of forage quality as a factor affecting foraging activity. When long-legged ant colonies were provided both grass seeds and tuna fish, those colonies provided with tuna fish extended their foraging time and remained active until soil surface temperatures reached lethal levels (Whitford et al. 1980a). Colonies provided with seeds ceased foraging at midmorning, the same time that colonies provided with no supplemental forage ceased foraging. It was concluded that foraging activity of ant species is only partly a function of microclimate, in other words, soil surface temperatures and air relative humidity. Most of the ant species responded to availability of preferred forage and to quantities of food stored in the nests. Ant species exhibited satiation when luxury amounts of preferred forage was available, and colonies ceased foraging when satiated. Because many ant species store food in their nests (i.e., harvester ants, honey-pot ants, fire ants, long-legged ants, and desert leaf-cutter ants) and behave as larder hoarders, stored food is an important regulator of colony foraging activity. Overall, the suite of foraging behaviors used by ants ensures that their populations and foraging activities are sustained even through the most stressful periods.

Although less diverse and apparent than ants, the collective activities of termites may drive nutrient cycling to a greater degree. Where studies have been conducted on quantities of materials consumed by termites, it was estimated that they processed between 3% and 50% of the annual input of detritus and herbivore dung (Whitford et al. 1982) (table 12-1). The large variation in the fraction of dung consumed by termites (15.1–95.6%) was attributed to the rate at which foragers located individual dung pats. The variation in the fraction of creosotebush leaf and stem litter consumed by termites was found to be related to the availability of other preferred plant species, that is, some grasses and annuals. On the

Table 12-1. Estimated annual percent of total input of dead plant material and dung consumed by subterranean termites in the Chihuahuan Desert.

		Reference
Shrubs		
Larrea tridentata (dead stems)	3.0%	MacKay et al. (1989)
L. tridentata (leaves and twigs)	40.0%	Whitford et al. (1982)
L. tridentata (leaves and twigs)	0.0%	Fowler & Whitford (1980); MacKay et al. (1987b)
Yucca elata (flowering stalks)	12.0%	Whitford et al. (1982)
Grasses		
Erioneuron (Tridens) pulchellus (roots)	50.0%	Whitford et al. (1988a)
E. pulchellus (standing dead)	30.0%	Silva et al. (1985)
E. pulchellus (standing dead)	43.0%	Whitford et al. (1982)
Aristida purpurea (standing dead)	40.0%	Schaefer & Whitford (1981)
Annual plants, aboveground		
Crypthantha spp.	45.0%	Schaefer & Whitford (1981)
Lepidium lasiocarpum	50.0%	Whitford et al. (1982)
L. lasiocarpum	30.0%	Schaefer & Whitford (1981)
Eriastrum diffusum	2.6%	Whitford et al. (1982)
E. diffusum	4.0%	Schaefer & Whitford (1981)
Eriogonum trichopes	4.2%	Whitford et al. (1982)
Eriogonum rotundifolium	5.0%	Schaefer & Whitford (1981)
Astragalus spp.	0.0%	Schaefer & Whitford (1981)
Baileya multiradiata	45.0%	Whitford et al. (1982)
B. multiradiata (roots)	50.0%	Whitford et al. (1988a)
Cattle dung	46.5%	Whitford et al. (1982)

Jornada, there was no annual plant production and very low abundance of fluff grass (*Dasyochloa pulchella*) on the creosotebush bajada in the only year in which creosotebush leaf and stem litter was consumed by termites (Fowler and Whitford 1980; Whitford et al. 1982; MacKay et al. 1987b). In the Chihuahuan Desert, subterranean termites also consumed large quantities of dead roots of grasses and annuals (Whitford et al. 1988a). There is indirect evidence that termites consume a large fraction of dead roots of shrub species, if termites locate the roots (Mun and Whitford 1998). Although there are no quantitative data on the fraction of litter and dung consumed by termites in desert grassland, mesquite coppice dunes, mesquite-grass mosaic, and tarbush habitats in the Jornada Basin, recent studies on termite galleries and sheeting in these habitats suggest that termites may be consuming a larger fraction of the detritus than has been reported for the creosotebush-dominated bajada.

The processing of dead plant material by the gut symbionts of termites has important implications for the organic matter content of soils and the cycling of nutrients. The gut symbionts of some species of termites have the capacity to decompose lignins and other recalcitrant organic molecules (Butler and Buckerfield 1979). Thus, termites produce only small quantities of feces. Termite feces contain very little recalcitrant carbon to contribute to the soil organic matter pool. The soil organic matter content of soil patches on a Chihuahuan Desert watershed was found to be strongly negatively correlated with the abundance/activity of subterranean termites (Nash and Whitford 1995). Since the rates of processes such as nitrogen mineralization are directly related to the soil organic matter content (Whitford et al. 1986), termites can indirectly affect the availability of essential nutrients for plant growth.

There are other characteristics of termites that contribute directly to nutrient cycling processes. Many species of termites have been shown to fix atmospheric nitrogen via hindgut symbionts (Beneman 1973; Schaefer and Whitford 1981; Bentley 1984), which allows termites to use foods with high carbon:nitrogen ratios. This nitrogen enters the nitrogen cycle in desert ecosystems primarily through the many predators that feed on termites (Schaefer and Whitford 1981). Termites also contribute significantly to cycling of other soil nutrients, such as phosphorus and sulfur, via this same pathway (Schaefer and Whitford 1981). The materials used to construct sheeting over potential food materials are also enriched with such nutrients as calcium and potassium (Bagine 1984).

All ecosystem processes and properties that are modified by the activities of termites taken together make a strong case for considering subterranean termites keystone organisms in Chihuahuan Desert ecosystems. Termite consumption of a large fraction of the input of dead plant material and dung affects spatial patterns of soil organic matter and spatial variability in macroporosity resulting from foraging galleries. Soil turnover and nutrient turnover resulting from construction of surface galleries and sheeting further contribute to variation in water infiltration, water storage, and soil nutrient concentrations. This spatial variation in soil water content and nutrients affects the species composition and productivity of the plant community.

The Roles of Soil Microfauna in Decomposition

Although termites and other macroinvertebrates are the primary processors of dead plant materials that remain on the soil surface, the decomposition of roots and litter that is buried occurs through the interactions of a complex of soil micro- and mesofauna and the microflora. Plant materials trapped in animal-produced pits may be buried by windblown soil or by runoff water sediment (Steinberger and Whitford 1983a). The decomposition and nutrient mineralization of buried materials and dead roots differs greatly from the processing of plant materials retained on the soil surface. Soil microarthropods, especially prostigmatid mites

(Tydeidae and Tarsonemidae), contribute significantly to the decomposition of buried litter (Santos and Whitford 1981). When microarthropods were eliminated from decomposing litter by an insecticide, decomposition rates were significantly reduced.

Because of their importance in decomposition and mineralization processes, a number of studies examined temporal variation in community composition and abundance of soil microarthropods. Experimental studies of timing and characteristics of rainfall, the characteristics of leaf litter accumulations on composition, and abundance of soil microarthropods and other soil invertebrates were conducted by irrigation studies and by use of rain-out shelters. Microarthropod abundance on decomposing roots was relatively independent of rainfall (including supplemental irrigation) (Whitford et al. 1988b). The prostigmatid mite genera that were most abundant in the vicinity of shrub roots and herbaceous plant roots were the same genera associated with the live roots of mesquite. The abundance of microarthropods associated with decomposing roots peaked in the warm-wet season (July–September) and decreased dramatically as soils cooled to minimum annual temperatures. Breeding activity in Chihuahuan Desert mites coincides with the summer rainfall season (Wallwork et al. 1986). The seasonal breeding pattern was not affected by irrigation during other seasons of the year. The strict seasonality of reproduction in mites was interpreted as an outcome of strong selection pressure to recruit only when food quality and quantity is maximal and when soil microclimate is most favorable. One species, Joshua's orabitid (*Joshuella striata*), which is widely distributed species found in summer and winter rainfall deserts of North America, produced eggs in response to winter rainfall and during the summer wet season (Wallwork et al. 1986).

Because soils dry rapidly after small rain events, protozoans and nematodes that are active only in water films on soil particles encyst or enter anhydrobiosis. In desert soils, protozoans and nematodes are in an inactive state most of the time because soils are at soil water potentials of approximately −6.0 MPa much of the year (Whitford 1989). Approximately 50% of the protozoan population is encysted at a soil water potential of −0.1 MPa, and virtually the entire protozoan population is inactive at −0.4 MPa (Whitford 1989). Ninety-nine percent of the nematode population is anhydrobiotic at water potentials between −3.0 MPa and −5.0 MPa (Freckman et al. 1987). Therefore, in Jornada desert soils, taxa of soil microarthropods are the only active microfaunal component of soil food webs during much of the year.

The Roles of Granivores and Native Herbivores

The removal of seeds by vertebrate and invertebrate granivores was recognized as a potentially important ecosystem process early in the Desert IBP program. Studies of seed consumption, effects of granivory on the seed bank, and the impacts of various groups of granivores on vegetation composition have produced clear answers regarding the significance of granivory on desert ecosystems.

Seed consumption by ants (*Pheidole* spp. and *Pogonomyrmex* spp.) was estimated by collecting the booty of returning foragers and counting the rate of return of foragers to the nests (Whitford 1979; Whitford et al. 1981a). The numbers of seeds transported to ant nests were compared to the estimated seed production by the plants in the area. The three species of *Pogonomyrmex* harvester ants harvested varying percentages of the monthly seed production of the dominant forbs and annual grasses. For example, the harvester ants harvested approximately 100% of the seed production by the annual grass six weeks grama (*Bouteloua barbata*) in August but only 32% of the seeds of this annual grass in September (Whitford 1979). The seeds of six weeks grama and four species of forbs (desert marigold [*Baileya multiradiata*], mealy goosefoot [*Chenopodium incanum*], Abert's buckwheat [*Eriogonum abertianum*], and little desert trumpet [*Eriogonum trichopes*]) accounted for most of the seeds harvested by harvester ants. Large *Pogonomyrmex* harvester ants concentrated their seed harvesting activity in the summer months (June–September), and small *Pheidole* harvesters concentrated their seed-harvesting activity in the late summer and early autumn (August–November). *Pheidole* harvested nearly 10 times more seeds than were harvested by *Pogonomyrmex*. It was estimated that *Pheidole* harvested 3.44×10^8 seeds/ha in creosotebush communities on a bajada, 3.11×10^8 seeds/ha in a mesquite Mormon tea community at the base of a watershed, and 9.7×10^8 seeds/ha in a black grama grassland community. *Pheidole* collected large quantities of seeds from fluff grass. However, there were large differences in the percentages of grass seed collected by the two most abundant species. More than 50% of the seeds collected by small militant harvesters (*Pheidole militicida*) were seeds of annual forbs, whereas 75% of the seeds collected by small arid harvesters (*Pheidole xerophila*) were grass seeds (Whitford et al. 1981a). Based on these studies, it was concluded that small seed-harvesting ants had a larger effect on the seed reserves than did the larger harvester ants and that ants consumed a significant fraction of the seed production of some species of grasses and annual forbs.

The impacts of herbivory by native animals differ from those of livestock, especially with regard to shrubs. Observations on the large numbers of terminal stems that were killed by girdlers or node borers led to a study that examined the effects of stem girdlers and node borers on the growth of mesquite (Whitford et al. 1978a). Stems killed by Bostrichids represented approximately 1% of the total stem biomass and between 1.4% and 53.4% of the leaf biomass of the shrubs sampled. Twig girdlers (*Oncideres rhodisticta*) killed stems on 45% of the mesquite shrubs on the site. The girdled mesquite stems provide oviposition sites and larval development sites for a number of other insects: buprestid, cerambycid, clerid, and scolytid beetles, as well as some butterflies and moths (*Lepidoptera* spp.). Scolytid beetle larvae were the most abundant insect larvae found in one- and two-year-old girdled stems of mesquite (Whitford et al. 1978a). Simulated girdling of 40–80% of the appropriate size branches of mesquite demonstrated that there was no reduction in shoot and leaf growth in either natural or simulated girdled plants in comparison to ungirdled controls. Girdling has the effect of pruning mesquite plants and stimulating growth of new stems from lateral nodes below the girdle. The removal of terminal stems of creosotebush by rabbits results in compensatory growth with several stems originating from below the severed stem (Whitford 1993). Creosotebushes that are pruned by rabbits on a regular basis develop a dense canopy and a hemispherical morphology.

Predator Ecology

Autecological studies of predators can provide information about the regulation of herbivore, granivore, and detritivore guilds and the structure of food webs (Polis 1994). In this regard, studies of coyotes (*Canis latrans*) would have been extremely useful, but such studies were not advisable at the Jornada because coyotes were subjected to control practices until the late 1980s. Recent studies of coyote behaviours have been limited and of short duration, though still insightful (Windberg et al. 1977).

Among other predators, the most abundant avian insectivore in the Jornada shrublands is the black-throated sparrow (*Amphispiza bilineata*). Black-throated sparrows nest from early April through the summer. The adult birds forage intensively to feed the chicks. Zimmer (1993) reported that when creosotebush and tarbush were flowering, there was an increase in abundance of foliage arthropods. Black-throated sparrows are opportunistic predators. Following summer rains that stimulated the emergence of termite alates (winged reproductives) the sparrows shifted to termites and brought loads of 3–10 alates per trip to the nest. When there was an unusual emergence of mydas flies (Mydidae), for about one week the mydas flies became the second most frequent prey item. In black-throated sparrows, the clutch size was regulated by prey availability. Clutch sizes were larger in years when grasshoppers were abundant. Approximately one-third of the clutches were lost to predators (Zimmer 1993).

The only large raptor that breeds in the Jornada Basin is the Swainson's hawk (*Buteo swainsoni*). Average density of nesting pairs during the summers of 1974 and 1975 was one pair per 9.4 km^2 (Pilz 1983). Forty-eight percent of the hatchling hawks fledged in 1974, and 72% fledged in 1975. The average number of chicks produced per nest was 2.4–2.5. Prey items brought to the chicks were 55% reptiles (22% of the biomass). Small mammals accounted for 42% of the prey items but accounted for 79% of the prey biomass. The most frequent prey were horned lizards, which were 14% of total prey items but only 7% of the biomass, and western whiptails, which were 33% of the total but only 8% of the biomass. Rabbits (jackrabbits and desert cottontails) accounted for 36%; spotted ground squirrels, 19%; banner-tailed kangaroo rats, 15%; and lizards, 14% of the prey biomass brought to the nestlings. Other mammals taken by Swainson's hawks included Ord's kangaroo rats, packrats and woodrats (*Neotoma* spp.), and hispid cotton rats. Other reptiles included the lizards (round-tailed horned lizard, long-nosed leopard lizard, and desert spiny lizard), and the snakes (glossy snake [*Arizona elegans*] and coachwhip [*Masticophis flagellum*]). The variety of prey taken by these hawks suggests that predation by Swainson's hawks has little effect on the abundance of the prey species.

The abundance of ants in the Chihuahuan Desert supports specialized predators: horned lizards of the genus *Phrynosoma*. Texas horned lizards were reported to feed mostly on two species of harvester ants, *Pogonomyrmex rugosus* and *P. desertorum* (Whitford and Bryant 1979). Although the average number of ants taken per feeding stop (15) was higher when lizards were positioned near nest disks or near columns of foragers (feeding stops, 14), horned lizards made more feeding stops (46) in areas not associated with nests or columns of foragers (average per feeding stop = 4.7). Individual Texas horned lizards consumed between 30 and 100 ants per day. Simulated predation on rough harvester ant and desert harvester ant colonies revealed that colonies losing approximately 25% of the estimated forager population ceased foraging for up to five days. It was concluded that horned-lizard densities are regulated by the abundance and productivity of *Pogonomyrmex* ants. Round-tailed horned lizards are considerably smaller than Texas horned lizards and select much smaller ants as their primary prey. The most dependable prey for round-tailed horned lizards were honey-pot ants (Shaffer and Whitford 1981). Hairless honey-pot ants and mimicking honey-pot ants (*Myrmecocystus depilis/mimicus*) collected honey dew and leaf exudates from mesquite. Round-tailed horned lizards consumed foragers returning from the shrub canopy. The ant species composition of *P. modestum* changed its diet following summer rainfall and increase in activity of ants other than honey-pots. Species of large harvester ants, crazy ants, small seed harvesters, and long-legged ants contributed a significant proportion of the round-tailed horned lizard's diet during the warm-wet season. Other Chihuahuan Desert lizards that exhibit extreme prey specialization are the western whiptails and other whiptail (*Cnemidophorus*) species. On the Jornada, western whiptails and checkered whiptails are very effective in finding termites by rooting through leaf litter under shrubs (personal observation). At the Mapimi Biosphere Reserve in the southern Chihuahuan Desert, termites accounted for 79% of all of the prey items in the stomachs of whiptails (Barbault et al. 1978).

Possibly the most important predator–prey interactions are those in the detrital food webs in the Chihuahuan Desert. One of the most unexpected findings in studies of the soil microfauna was that soil mites of the family Tydeidae fed on nematodes and depressed nematode numbers (Santos and Whitford 1981). Nematophagy by soil microarthropods has since been reported as common in shortgrass steppe and in the Rocky Mountains (Walter 1988). Many microarthropod species that were thought to be mycophagous were found to be omnivorous. Omnivorous and predaceous mites that prey on bacteriophagous, fungivorous, and omnivorous nematodes are key elements in detrital food webs (Elliot et al. 1988).

The Relationship of Native Animals to Desertification

In the Jornada Basin, the transformation of desert grasslands to honey mesquite coppice dunes, mesquite-grass mosaics, tarbush shrubland, and creosotebush shrubland (chapter 10) has had a number of effects on animal populations and on the processes and properties that they affect. Studies of rodent and rabbit populations in the Jornada Basin have consistently documented low abundance and species diversity in desert grasslands and higher abundance and diversity in the desertified shrublands (Wood 1969; Whitford 1997). This consistency is remarkable considering that the studies cover a span of 40 years with considerably different rainfall and productivity patterns in the years preceding the trapping studies. Studies of rodent populations were initiated in the 1960s because "their populations can represent a large portion of the vertebrate weight, or biomass, of an area and often impose a greater impact on the community than the more conspicuous game or livestock species" (Wood 1969). Wood's study suggested that feedbacks between vegetation change and rodent community structure could contribute to maintaining desertified ecosystems in a stable, altered state. Wood (1969) reported the rodent biomass in mesquite coppice dune areas (0.72 kg/ha) was double that of the black grama grassland (0.35 kg/ha). Although species populations fluctuated throughout the study (1960–63), the mean rodent biomass in the grassland and other communities sampled remained stable. The rodent biomass in the mesquite coppice dune site, however, fluctuated from a high of 0.94 kg/ha to a low of 0.60 kg/ha.

The spatial and temporal variation in rodent populations may exacerbate the desertification processes and contribute to the irreversibility of the desertified state (Whitford 1993). In places and times of high rodent abundance, rates of herbivory and graminivory may increasingly constrain grass seed production even as grass cover declines (Dabo 1980; Kerley et al. 1997). Rodent cache pits and soil disturbances, on the other hand, may increase the germination rates of some grass species. Thus it is possible that the activities of animals may produce either positive or negative feedbacks on the ecosystem structure, but it is not yet clear which of these effects is most important.

Conclusions

Taken together, the body of research on animals at the Jornada reveals three intriguing patterns: (1) patterns of shrub cover and water redistribution are dominant elements structuring the environments of Chihuahuan Desert animals, (2) feedbacks from animals influence nutrient availability and plant demography via several direct and indirect pathways, and (3) the contributions of native animals to desertification remains unclear.

The idea that variation in habitat complexity and differentiation is important for animal diversity is well established in ecology, but this idea has seldom been connected to desertification. Although certain species are associated with grasslands, data for several taxa indicate that shrubs play a positive role for animal diversity (table 12-2) despite the suggestion that shrubs do the opposite (Muldavin et al. 2001). The role of water redistribution patterns in creating habitat differentiation is less well understood but has important contributions to the development of arroyo vegetation and ephemeral water bodies used by distinct animal groups. Because run-off is increased as grass cover declines, it is possible that grassland degradation has accentuated habitat differentiation in some cases. The relationship of conventional notions of degradation to biodiversity involves several mechanisms and is not always clear cut (Bestelmeyer et al. 2003b).

The importance of animals for nutrient flux and other feedbacks to ecosystem

Table 12-2. A summary of the key relationships described in this chapter, including the effects of variation in ecosystem structure on different taxa or functional groups as well as the feedbacks exhibited by taxa on ecosystem properties.

Taxon/Functional Group	Effect of Ecosystem Structure	Feedbacks to Ecosystem
Rodents and lagomorphs	Increased density in shrublands	Graminivory, herbivory reduces grass reproduction
		Foraging pits favor seed germination for some grasses
Birds	Increased density/richness in shrublands	Redistribute grass seeds (natives and exotics)
	Arroyo vegetation used by some species	
Lizards	Increased density/richness in shrublands	Consumption of ants and termites
Anurans	Positively affected by water redistribution	Redistribute nutrients to surrounding watershed in their bodies
Ants	Additional species in shrublands	Granivory effects on plant reproduction
		Nutrient concentration in nests and soil patchiness
		Bioturbation and vertical redistribution in soils
		Food for specialist predators (*Phrynosoma*)
Termites	Ubiquitous except in inundated areas	Rapid breakdown of roots, litter, and dung
		Reduce soil carbon and N mineralization rates
		N fixation via hindgut symbionts
		Increase macroporosity and water infiltration
Phytophagous insects	Specialized to shrubs	Frass locally alters nutrient availability
Macro-detritivores	Species sort among grasslands/shrublands	Decomposition of litter
Microarthropods	Track litter amounts, but ubiquitous	Regulate fungi and N availability to plants
		Control nematode predation on bacteria, decomposition rates

properties lends support to Chew's hypothesis. This is especially true of termites and the detrital pathway, as has been found in other desert systems (Stafford Smith and Morton 1990). Perhaps more remarkable is the diversity of indirect pathways that has been uncovered (table 12-2). Only through detailed studies of natural history could such diversity be revealed. These observations also suggest a further modification to Chew's hypothesis: A given taxon may have more than one important effect on ecosystem properties (e.g., limiting N availability but increasing infiltration). The consequences of these effects for plants and soils may reinforce or counteract one another to varying degrees. Such considerations are critical in establishing the true functional roles and redundancies of species in ecosystems (Rosenfeld 2002).

The multiple effects of different animal taxa for different ecosystem properties preclude simple statements about the role of animals in desertification or other landscape change. We do not have enough information to gauge the relative importance of various animal effects for plant recruitment and mortality, especially against a background of livestock grazing, historical legacies, and soil and climate variability. Nonetheless, the research summarized here allows us to frame the next generation of questions much more effectively.

13

Grazing Livestock Management in an Arid Ecosystem

Kris M. Havstad
Ed L. Fredrickson
Laura F. Huenneke

The history of livestock grazing in the Jornada Basin of southern New Mexico is a relatively recent story, but one of profound implications. For four centuries this region has supported a rangeland livestock industry—initially sheep (*Ovis aries*), goats (*Capra aegagrus hircus*), and cattle (*Bos taurus* and *Bos indicus*), but primarily beef cattle for the past 130 years. Throughout this brief history of a domesticated ruminant in an ecosystem without a significant presence of large hoofed mammals as part of its evolutionary development, the livestock industry has continually grappled with high degrees of temporal and spatial variation in forage production. Management of this consumptive use, whether during Spanish, Mexican, U.S. territorial, U.S. federal, or New Mexican governments, has constantly reaffirmed the need for grazing management to be flexible and responsive to the stress of droughts. The history of anecdotal experiences has been more recently augmented by scientific investigations first initiated in 1915. This chapter outlines the general history of livestock in this region, defining characteristics of herbivory in arid lands, and principles of grazing management derived from nearly a century of studies on grazing by large domesticated herbivores.

General History

Seventeen ships carried 1,200 people and enough cattle, horses, sheep, and pigs to colonize northern Hispaniola during Columbus's second voyage in 1493. Livestock originating from the Andalusian Plain of southern Spain were loaded aboard ship at the southern port of Cádiz and the Canary Islands before making the 22-day voyage (Rouse 1977). It was not until 1521 that Gregorio Villalobos unloaded livestock in New Spain (Mexico) near Tampico; the actual number of cattle and their origin are disputed. Rouse (1977) claimed that 50 calves were transported to the mainland from either Cuba or Hispaniola, whereas Peplow (1958) and Wellman (1954) claimed 6 animals arrived from Hispaniola. Irrespective of the initial numbers, livestock were soon moved north from the Mexico City area during the early sixteenth century with both missionaries and resource extraction industries as retired military officers and Spanish nobility built a mining- and grazing-based economy throughout the region of present-day northern Mexico. By 1539 livestock had reached the present-day United States–Mexico border with the greatest concentrations being along the coasts and the central plateau. This northern expansion of a ranching frontier in North America was to development of Hispanic America what western expansion of farming across the continent in the nineteenth century was to Anglo America (Morrissey 1951). There were a million cattle in New Spain by 1600 with grazing associations formed under formal Spanish law (Bowling 1942). By 1609, the city of Santa Fe was a northern distribution point for livestock in the Americas.

Livestock were given to colonists by the Spanish government as an enticement for settlement (Bowling 1942). Scurlock (1998) reported estimates of livestock numbers in New Mexico from 1598 to 1830 (table 13-1). Sheep were the principal species in the New Mexico region of New Spain during this Spanish settlement period. Individual herds of 4,000–5,000 were common throughout the region (Hastings and Turner 1965). A transhumant grazing system was common, as flocks of sheep were annually driven from present-day northern New Mexico south through the Rio Grande Valley into Mexico to service livestock markets in Chihuahua and Durango (Scurlock 1998). The first reports of localized overgrazing by livestock appeared in the 1630s (Ford 1987). When Mexico gained independence from Spain in 1821, many of the Spanish settlements were abandoned, and livestock numbers declined. For example, there were fewer than 5,000 cattle in the Arizona territory during the mid-1800s.

A grazing-based economy was reestablished following the 1848 Treaty of Guadalupe-Hidalgo and the conclusion of the American Civil War in 1865. By 1891, there were 1.5 million cattle in the Arizona and New Mexico territories, a region covering the current states of Arizona and New Mexico. This expansion in numbers was accompanied by an expansion onto rangelands not previously grazed by livestock (Hastings and Turner 1965). This regional exploitation was driven by speculation by Eastern and European investors capitalizing on new technologies for pumping water for livestock and fencing lands (McNaughton 1993). Aggressive programs to control predators and concurrent establishment of railroad networks that moved cattle to growing Eastern markets undoubtedly aided this expansion. Escalation of livestock numbers and their expansion into areas not previously grazed had serious ramifications (Buffington and Herbel 1965). By the early 1900s, reports on the widespread destruction of Southwestern rangelands by livestock overgrazing were common (Smith 1899; Wooton 1908). Historical

Table 13-1. Livestock numbers in New Mexico, 1598–1830 (from Scurlock 1998).

Year	Sheep	Cattle	Goats	Horses	Mules	Totals
1598	4,000	1,000	1,000	150	—	6,150
1694	3,100	—	—	—	—	3,000
1697	4,000	650	170	—	—	4,820
1757	112,182[a]	16,157	[b]	7,356	—	135,695
1777	69,000	—	—	—	—	69,000
1820s	1,000,000	5,000	—	850	2,150	1,008,000

[a] Includes Hopi flocks.
[b] Included with sheep.

details of the livestock industry's beginnings in the Jornada Basin are presented in chapter 1.

Cattle numbers in the Southwest peaked at over 1 million head in 1890, during World War I, and again in 1920 but by 1990 had declined to 900,000 head in Arizona and New Mexico (Fredrickson et al. 1998). Currently, forage demand in New Mexico and Arizona is approximately 10 million annual unit months, of which 37% are supplied from federally managed rangelands in these two states (Torell et al. 1992). The regional economy includes a grazing-based component that is predominately comprised of cattle. In New Mexico, 9,000+ ranching operations, totaling 600,000 head of beef cattle, generated approximately $800 million in cash receipts from livestock sales in 2001 (USDA 2001). The industry is an unconsolidated amalgamation of small businesses with highly variable economic viabilities (Fowler and Torell 1985). Most ranching enterprises have fewer than 250 cattle, employ fewer than 5 people, have been in operation for an average of 19 years, and annually spend $18,000 for community services and $19,000 on structural land improvements (Fowler 1993).

Herbivory

There are numerous general theories on the role of herbivores in shaping grassland and shrubland ecosystems. These theories include the autogenic hypotheses (Noy-Meir 1979/80), optimization theory (McNaughton 1979), evolutionary gradients of grazing history (Milchunas et al. 1988), plant traits adapted to large mammalian grazers (Mack and Thompson 1982), keystone guilds (Brown and Heske 1990), and plant chemical–mediated defoliation (Bryant et al. 1991). None of these theories easily accommodate the inclusion of an exotic large herbivore within an arid ecosystem such as the northern Chihuahuan Desert.

Though it is likely that domestication of cattle has altered some behaviors that were characteristic of their predecessors, the aurochs (*Bos primigenius*), particularly a lessening of their gregarious nature (Hemmer 1990), inherent foraging patterns of cattle are similar to other wild generalist ungulates. Diurnal behaviors are sensitive to environmental conditions such as day length and ambient temperatures (Arnold and Dudzinski 1978), vegetative conditions such as species composition and available biomass (Holloway et al. 1979), physiological states such as lactation (Wagner et al. 1986), and the history of prior grazing experiences (Burritt and Provenza 1989). Forage preferences can be extremely plastic, as diet selection is mediated by the central nervous system and mitigated by intrinsic feedbacks and external stimuli (Provenza et al. 1998).

Native Herbivores

Desert grasslands have historically supported low chronic levels of herbivory by native vertebrates (chapter 12). In the Jornada Basin, native ungulate densities are low, and herbage consumption by small mammals has been estimated at < 5 g/m^2/yr (Pieper et al. 1983). These intake levels are typically $< 10\%$ of aboveground net primary production (ANPP) (Pieper et al. 1983). In this environment of erratic and low productivity, herbivory by native species has been a historically chronic and minimal feature where most of the energy within this ecosystem is traditionally channeled through decomposers rather than herbivores. The black-tailed prairie dog (*Cynomys ludovicianus*) was an endemic species often found on heavy-textured playa soils common throughout the Jornada Basin (Oakes 2000). This species, possibly a keystone herbivore on these playa sites (Miller et al. 2000, and accompanying citations), was poisoned and eradicated prior to and during World War I to reduce forage competition with cattle. This action was justified by the performing federal agency as a means to increase meat production in support of the U.S. war effort. The prairie dog has remained extirpated from much of its former habitat in the Jornada Basin. Prior to extermination efforts, presence of this animal may have prevented woody plant dominance within more productive desert grassland sites receiving external surface and subsurface water flows (Weltzin et al. 1997). Other mammals, particularly kangaroo rats (*Dipodomys* spp.), are extremely important as both granivores and gramivores within this ecosystem (Heske et al. 1993; Kerley et al. 1997; see also chapter 12). Kangaroo rat presence or absence can be more influential on plant community dynamics than the presence or absence of livestock (Brown and Heske 1990). Kangaroo rats also were the targets of private and federal poisoning campaigns during the 1920s to improve forage conditions for livestock. These campaigns were quickly abandoned when the extent of the task was fully realized (Jornada Experimental Range Annual Reports 1925–26 unpublished) and likely resulted in large alterations in the demographics of native mammalian herbivores and their predators for short periods. As a consequence, grass–shrub interactions and other aspects of vegetarian dynamics were likely altered to some unknown degree as well. Competition for forage among cattle and native mammalian herbivores is relatively slight in desert environments. Dietary overlap is most pronounced between cattle and black-tailed jackrabbits (Wansi et al. 1992).

Though jackrabbits can influence numerous processes, their population densities are highly variable and independent of cattle presence. Although data on the amount of standing crop consumed and dietary overlap between herbivore species are useful, they do not account for degree of selectivity or the possible effects one herbivore may have on the diets of more selective herbivores. For example, although newly emergent plants and plant parts may constitute a relatively small percentage of the overall standing crop, their removal by selective herbivores, such as jackrabbits, may greatly alter vegetation dynamics. Removal of decadent plant material by generalist herbivores like livestock may facilitate greater selectivity for meristematic tissue by more selective native herbivores, ultimately affecting plant survival and native herbivore fecundity. In this case, the effect on vegetation dynamics of herbivores when viewed independently may not be as great as when the interactive effects of two or more herbivorous species are combined.

Livestock

For many arid and semiarid ecosystems, the amount of biomass supported per unit of primary production is about an order of magnitude greater under rangeland livestock production than under natural, nonagricultural systems (Oesterheld et al. 1992). This observation appears valid for the Jornada Basin. Biomass of native consumers present in upland grassland communities in the Jornada Basin is 0.03 g/m^2 (Pieper et al. 1983). Under conservative stocking rates of nine animal units per 259 ha (640 acres or one section of land) during years of average forage production, the biomass of cattle supported on these grasslands would be about 1.7 g/m^2. Only under extremely low stocking rates or for grazing seasons of just a few months' duration per year would livestock biomass be lowered to levels equivalent to the native herbivore biomass supported by these grasslands.

Mature cattle consume 5–15 kg (dry matter basis; NRC 1996) of forage daily. A classic recommendation for stocking desert grassland is 1 cow/260 ha/25 mm precipitation/yr. This stocking level would result in a harvest rate of 7–21 g/m^2/yr from an area receiving 245 mm of precipitation. Reported values for forage consumption by cattle under conservative stocking of desert grasslands have been 8–14 g/m^2/yr (Pieper et al. 1983). Annual forage consumption during the widespread overstocked periods of the late nineteenth and early twentieth centuries may have ranged from 30 to 60 g/m^2.

In an unpublished report, Cassady and Valentine (1938) summarized results from one of the earliest studies of forage intake by cattle grazing black grama (*Bouteloua eriopoda*) grasslands. During the winter dormant seasons of 1936–38, mature cows (average body weight of 328 kg) consumed 6.9 kg (dry matter basis) per day of perennial grasses, of which 82% was black grama. This is an intake rate of 2.1% of body weight per day and a rate that nearly meets the nutritional requirements of a range beef cow in the last trimester of gestation. Available forage averaged 319 kg/ha (32 g/m^2), and the stocking rate during the study period resulted in a utilization of 54% (17.3 g/m^2) of the perennial grass forage.

Basal cover of perennial grasses on this study area in 1937–38 was estimated at 7%. Based on the forage intake results in this study, these authors estimated that the winter carrying capacity for this range would be 15.5 cows per section per year. This figure would have to be adjusted for the increased body weight (500 kg) of today's animal unit (AU) and a vastly increased milk production potential, resulting in a 52% greater daily forage intake. The general carrying capacity would be adjusted to 10.8 AU per section per year. This figure would also have to be adjusted for a utilization rate of 35% instead of 50%, reducing the general grazing capacity to 7.6 AU per section. Also, basal cover of black grama is highly variable across the Jornada Basin. Areas of desert grassland on the Jornada Experimental Range (JER) today average 3–5% basal cover, which would imply a general grazing capacity of 3.5 AU per section for dormant season use when forage intake would be about 2.1% (or less) of body weight. Thus, Cassady and Valentine's study in the 1930s helped define the relatively low grazing capacities that are inherent to the desert grasslands in the Jornada Basin.

Jornada desert grassland ANPP during a 3-year period of near average total annual precipitation and protected from cattle grazing ranged from 125–186 g/m^2 (Sims and Singh 1978). Production with conservative stocking was estimated at 58 ($\pm$ 20) g/m^2 over a 15-year period, which included years of severe drought (Paulsen and Ares 1962). During some seasons, even conservative stocking can result in acute harvest rates within pastures or across ranches. Distribution of use is uneven due to physical, biological, and structural features of the environment (Holechek et al. 1999). Generally, a high proportion of tillers will be ungrazed, defoliated tillers will usually be grazed only once, and biomass removal from grazed tillers will be high (Senock et al. 1993). Sims and Singh (1978) reported maximum growth rates of warm season grasses at the Jornada were 1.5–3.4 g/m^2/day under nongrazing by livestock and 0.6–3.3 g/m^2/day with livestock grazing.

Long-Term Effects of Open-Range Cattle

There is an extensive body of literature on plant responses to herbivory (e.g., Detling 1988; Heitschmidt and Stuth 1991; Huntley 1991). Given that grassland ecosystems are governed by numerous direct and indirect biotic interactions (Lockwood and Lockwood 1993), of which grazing is an integral process (McNaughton 1991), the effects of herbivory cascade throughout these ecosystems. Its effects can be neutral, adverse, or beneficial (Sims and Singh 1978; Lacey and Van Poollen 1981), but interpretations are greatly influenced by dynamics of scale (Turner 1989).

Prior to the introduction of cattle, large (10^2–10^3 kg) native ungulates had been rare within the northern Chihuahuan Desert since the Pleistocene (McDonald 1981). Given this uneven history of megafauna presence, the Chihuahuan Desert hosts a range of plants with different degrees of adaptation to large herbivores. With the introduction of livestock, plants better adapted to this presence of large herbivores flourished. A prominent example is honey mesquite (*Prosopis glandulosa)* with both chemical and morphological traits that deter herbivory, and with seed characteristics which encourage ingestion and exploit ungulate dispersal. Cattle directly and indirectly affect numerous ecological processes in similar fashion to native large herbivores and other types of disturbances (Pykala 2000). Their effects include alterations of NPP and plant–water relations, seed dispersal, species composition and life form, nutrient cycling and retention, energy flow efficiencies, food web interactions, and factors such as fire frequency (Sims and Singh 1978; Detling 1988; Archer and Smiens 1991; Hobbs et al. 1991).

The effects of herbivory on ecosystems are best understood in regard to long-term dynamics (Huntley 1991). We view this perspective as particularly appropriate for subtropical grasslands for two primary reasons. First, this region is undergoing continual transitions between vegetation types, albeit in discontinuous fashions (Grover and Musick 1990). Ecotones (at several scales) are key study areas for elucidating dynamics of these systems (Gosz 1993; Neilson 1993). We know very little about the ecological dynamics of transitional states and their responses to disturbances such as defoliation. For example, competition for soil resources between perennial grasses and mesquite may be minimal early in the mesquite life cycle and at low mesquite density (Brown and Archer 1989). However, the dynamics of this competition are significantly altered under conditions of resource redistribution and increased mesquite density. Effects of herbivory, even examined at similar temporal and spatial scales, would be substantially different across this gradient of vegetation transition.

Second, we are dealing with a situation in which the primary large ungulate is an exotic domesticated ruminant whose density is directly regulated by humans. The fundamental question is not grazing as an optimization process within the ecosystem but of the sustainability and long term consequences of grazing by livestock. At its core, grazing is a behavioral process, and the key aspect of grazing behavior is the expressed forage preferences of livestock. The primary effects of livestock grazing in the Chihuahuan Desert are a function of diet selection. Pieper (1994) correctly stated that it could be extremely difficult to predict how livestock will affect rangeland resources because their effects will be highly dependent on the diversities and activities of the grazing animals. Different species and kinds of livestock have different forage preferences and those preferences are related to the array of choices, that is, available plant species. Studies of dietary selection by livestock in the Jornada Basin are summarized in table 13-2. Basically, grazing is species specific (Hobbs and Huenneke 1992). For example, annual species are not typically found as major components in cattle diets. Kelt and Valone (1995) reported that only 2 of 79 annual species responded (increased) significantly following livestock removal. As with other deserts, understanding individual species responses to defoliation can serve as a good approximation to the understanding of many ecological phenomena in deserts (Noy-Meir 1979/80).

Grazing Management

The primary initial research objectives of the Jornada Range Reserve in 1915 were to quantify the carrying capacity of desert rangelands, establish a system of forage utilization consistent with plant growth requirements, and develop a range management plan to minimize stock loss during droughts (Havstad and Schles-

Table 13-2. Livestock dietary preferences by forage class in the Jornada Basin.

Reference		% of Dietary Composition			Comments
		Grasses	Forbs	Shrubs	
Herbel & Nelson (1969)	Hereford and Santa Gertrudis cows	58	30	12	5–7 species comprised 54–77% of diets; averaged across 4 seasons; 4 year study; upland and lowland sites; little difference between cattle breeds
Rosiere et al. (1975)	Hereford steers	43	32	19	Between 2–8 species comprised 72% of diets across 4 seasons; unknown species comprised 6% of diets
Anderson & Holechek (1983)	Hereford x Angus heifers and steers	35	51	19	4 plant species comprised 55–60% of diets; summer grazing season; primarily a tobosa grass lowland; heifer and steer diets similar
Hakkila (1986)	Hereford, Brangus, and H x B steers	55	20	25	6 species comprised 83% of diets; averaged across 4 seasons; 1 species (mesa dropseed) used year round, and 2 other species (soapweed yucca and red threeawn) used> 80% of the time
Smith (1993)	Hereford and H x Santa Gertrudis cows	65	29	6	2 grass species comprised 33–55% of cattle diets; one forb species comprised 41–66% of sheep diets; averaged across seasons and years; sheep diets relatively constant across seasons and years
	Rambouillet ewes	8	87	5	

inger 1996). A key problem for range management was the inaccurate judgment of carrying capacity (Wooton 1915). Jardine and Forsling (1922), Canfield (1939), and Paulsen and Ares (1962) established guidelines for carrying capacities of black grama rangelands. These classic studies used three different experimental designs to evaluate perennial grass responses to different livestock grazing strategies. Jardine and Forsling (1922) evaluated large-scale pasture responses on the Jornada Reserve and adjacent rangeland from 1915–19, a drought period. They measured basal cover responses of black grama to three coarsely applied management practices: (1) heavily grazed yearlong until 1918 and lightly grazed during the 1918 and 1919 growing seasons, (2) grazed yearlong 1915–19, and (3) reduced grazing during the growing season but fully utilized during the dormant seasons, 1915–19. Basal cover responses of black grama, compared to an area protected from livestock grazing, clearly favored treatment 3, and the authors concluded that light grazing during the growing season was the appropriate grazing strategy for black grama dominated rangelands. Canfield (1939) conducted a small plot evaluation of black grama responses to different intensities and frequencies of clipping over an 11-year study. In evaluating black grama responses to clipping to either a 2.5 cm or 5 cm residue height at 2-, 4-, or 6-week intervals or once at the end of the growing season, by the end of the study all 1 m^2 plots clipped during the growing season were denuded. The obvious conclusion was that moderate or heavy use of black grama over an extended period was inappropriate. Paulsen and Ares (1962) summarized observations from small 1 m^2 plots arrayed across the JER where basal area of perennial grasses was recorded annually from 1916 to 1953. Plots were stratified to reflect nonuse and light, moderate, and heavy utilization by livestock. Results from this extensive long-term study clearly reflected the need to conservatively ($<$ 40% of current year's growth removed) graze black grama and severely reduce or eliminate use during extensive drought periods. Subsequent research has reinforced the consistency of these guidelines, as Campbell and Crafts (1938), Paulsen and Ares (1962), and Holechek et al. (1994) have reported. These authors all concluded that proper utilization of black grama should be less than 40% of current year's growth.

The original philosophy was that proper utilization of the leaves and stems of the main forage plants was the basic principle of range management (Canfield 1939). General management guidelines published in the 1910s and 1920s are very similar to those promoted today. For example, nearly 80 years ago Jardine and Forsling (1922) recommended the following drought strategies: (1) limit breeding stock to carrying capacities during drought, (2) add surplus stock during good forage years depending on market conditions, (3) adjust range use seasonally depending on growth characteristics of key species, (4) establish permanent watering points no more than 5 miles apart, and (5) establish both herding and salting practices that achieve optimal stock distribution. Similar recommendations for drought conditions are outlined in one of the most current textbooks on range management (Holechek et al. 1998a). Interestingly, strategy #5 may have accelerated shrub expansion into areas formerly desert grasslands. Though livestock dispersal of mesquite seed was seen very early as a reason for mesquite encroachment (Campbell 1929), management practices were not employed to limit further dispersal. Enhancing livestock distribution with stock water and salt placements may have actually promoted mesquite seed dispersal.

Initial research on livestock production also emphasized strategies for drought. Most of the original efforts focused on supplemental feeding programs, especially those that used locally available foodstuffs, such as cottonseed products. For example, general recommendations were to feed 450–900 g per cow per day of supplemental protein to augment range forage for maintenance (Forsling 1924) with slightly higher quantities suggested for growth of stockers (Jardine and Hurtt 1917). These general recommendations have persisted over ensuing decades. Supplementation research has now typically narrowed its focus to mechanisms of and animal responses to protein and energy supplements to trigger specific physiological activities for specific animal production stages (Gambill et al. 1994).

More novel research has emphasized specialized practices for emergency feed conditions and management of poisonous plants. Soapweed (*Yucca elata*) was found to be a palatable emergency feed when fed chopped and fresh (Forsling 1919). Ensiling was not determined to be necessary. Other plant species were either deemed not suitable as emergency feeds (i.e., *Dasylirion wheeleri* and *Yucca macrocarpa*) or required spine removal (*Opuntia* spp.). Interestingly, burning spines from prickly pear cactus (in 1924 Forsling estimated that one person could prepare cactus feed for 200–400 had of cattle in a day) was employed during the 1994–95 drought in the Southwestern United States, though not in the Jornada Basin. However, even in the 1910s and 1920s the use of emergency feed practices was not viewed as responsible management.

As in other Western rangeland regions, studies of poisonous plants provided both initial guidelines for livestock management and insight into the difficulties of plant control in a desert environment. For southern New Mexico, drymaria (*Drymaria pachyphylla*) became a problem in response to overgrazing in the late 1800s and early 1900s (Little 1937). Drymaria is highly toxic and causes death within hours of consumption of a lethal dose. Though generally unpalatable, losses can occur for all classes of livestock especially in summer months if other forage is unavailable. For clay soils, drymaria was viewed as an early seral species with infestations characteristic of degraded areas (Campbell 1931). Avoidance of grazing in drymaria-infested areas was the recommended management strategy. Various measures of control (fencing, burning, spraying, and revegetation) were determined to be either too expensive or ineffective. The recommended control practice was hoeing, but eradication was not viewed as a viable possibility. These general characteristics relative to management and control recommendations for poisonous plants persist today (James et al. 1993).

Though most complex grazing systems have not been shown to improve rangeland conditions in the desert Southwest (Martin 1975), specialized grazing systems have demonstrated some merit. In Arizona, rotation grazing did not improve ranges that were in good condition, but a rotational seasonal rest and grazing system may accelerate recovery of ranges in poor condition (Martin and Severson 1988). The benefits of more intensive grazing systems, such as short-duration grazing, are generally negative (Bryant et al. 1989). A few examples of good rangeland conditions under intensive grazing management in the arid zone exist, but these examples are undocumented in the scientific literature. The success of these specific situations is probably due to a unique combination of progressive management and a thorough understanding by the ranchers of the ecological characteristics of their specific rangeland. In the Jornada Basin, Beck and McNeely (1997) reported results of a long-term study comparing continuous, year-long grazing with seasonal use by cattle. Herbage production varied 100-fold over the course of this study, irrespective of the two grazing strategies. Forage quantity and quality in this environment limited the average calf production to 0.32 g/m^2. These data illustrate the overriding restricting effect of annual variation in primary production on the options for creative and intensive management in this environment. Other studies comparing continuous use to a rotational or seasonal use system in tobosa-dominated (*Pleuraphis mutica*) grasslands have demonstrated some differences in grazing effects (Senock et al. 1993) but no differences in animal performance (Tadingar 1982).

The grazing system developed at the Jornada in recognition of the dynamic nature of forage production in this region was the "best pasture" (Herbel and Nelson 1969). This pasture-scale system is highly flexible in terms of grazing season, and it exploits ephemeral growth of forage. The best pasture system does not involve rotation of livestock at a predetermined calendar date. The only use of a grazing capacity concept is the estimation of an average stocking rate (animal units per section) for each pasture. However, this capacity is recognized to be quite variable depending on actual forage production. The best pasture grazing system requires flexible herd management where livestock numbers and class are adjusted to forage production and location. The latter is an extremely important consideration in this environment because ANPP can be spatially highly variable. In an unpublished study, livestock production from 1940 to 1951 under flexible herd management was compared to the period of 1927 to 1934 under constant stocking. Though more cattle were stocked from 1927 to 1934, the annual production per cow under flexible management was 31% higher. The best pasture system is not routinely discussed as a specific method for grazing management in the Southwestern United States. However, the general principles of flexible herd management and adjustment of stocking in response to variation in forage production are widely used in many range livestock operations.

There have been a few long-term studies of the effects on arid rangelands of extended rest periods with no grazing by livestock. In southeastern Arizona, Bock and Bock (1993) reported that exclusion of livestock for 22 years increased the total cover of perennial grasses on a site with an average annual precipitation of 430 mm. In the drier Jornada Basin, Atwood (1987) examined four exclosures in black grama–dominated grasslands after 17, 22, 32, and 48 years of rest. Basal cover of black grama was greater in the 32- and 48-year exclosures compared to adjacent grazed areas. However, no differences between grazed and rested areas were noted after 17 years of rest, and basal cover of black grama was actually greater in the grazed area compared to the exclosure receiving 22 years of rest. Obviously, black grama is slow to respond to protection, and responses can be highly variable depending on ecological conditions at the time of rest initiation.

Various techniques (such as esophageal fistulation) for animal nutrition research allow investigations of the interactions between plants and livestock. Cattle genotypes with relatively modest performance traits, such as milk production, might be more successful in this nutrient-sparse environment. It is possible that some desired characteristics would mirror those inherent in the original cattle breeds introduced to North America in the sixteenth century.

Research on plant–animal interactions now reflects the widespread diversity of shrubs in the Chihuahuan Desert. Foraging behaviors are strongly mediated by secondary plant chemistry (Estell et al. 1994), and chronic ingestion may have postingestive consequences that further shape preferences (Fredrickson et al. 1994). The use of livestock as biocontrol agents for remediation will require detailed knowledge of this chemically mediated interaction to be an effective technology.

Conclusions

In summarizing 45 years of grazing research in the arid region of south-central New Mexico, Paulsen and Ares (1961) wrote: "Sustained grazing capacity does not exist on the semi-desert ranges . . . stocking may be high in some periods (meaning that primary production is high and high livestock numbers would be appropriate) and in others there is virtually no capacity."

Our knowledge of various effects of livestock grazing in arid environments has been well synthesized (Pieper 1994). We have a general understanding of the importance of controlling timing, intensity, and frequency of grazing (Holechek et al. 1998b). It is also well recognized that livestock grazing under poor management or excessive use can have various negative effects, some of which are severe and long-lasting. Proper utilization of forage species has long been recognized as a key component of livestock grazing management (Canfield 1939). Jardine and Forsling (1922) established early guidelines for carrying capacities of desert grasslands. These authors and others have repeatedly concluded that proper utilization of arid grasslands should be less than 40% of current year's growth (Campbell and Crafts 1938; Paulsen and Ares 1962; Holechek et al. 1994, 1999).

The primary problems related to management of livestock grazing in arid and semiarid rangelands are those faced by producers since the seventeenth century: (1) coping with temporal variations in forage production, (2) manipulating an animal behavioral process (grazing) that is plant species–specific, (3) managing grazing across landscapes with limited (if any) ability to monitor or assess impacts, and (4) controlling dispersal of seeds. The most persistent problems are the annual and seasonal deficits in available forage due to the natural recurrent disturbance of drought in this environment. Forage production on upland desert rangelands can average between 150 and 250 g/m^2 (see chapter 11) during years of normal precipitation but may be $<$ 100 g/m^2 during drought years (Herbel and Gibbens 1996; see table 11-2 in chapter 11). Almost any grazing during severe drought years would exceed proper utilization. Conservative stocking at 10–30% below capacity has also been recommended as both a strategy to cope with drought and as a means to improve vegetation conditions on some ranges (Holechek et al. 1999). Though Paulsen and Ares (1961) concluded that grazing could not be viewed as sustainable, to some extent this depends on the spatial scale of livestock management. Conservative stocking is probably the most important practice to improve conditions and approach sustained livestock use of New Mexico's arid rangelands. Based on our knowledge of the role of native consumers in this system, this recommendation reflects intent to minimize the affects of livestock on energy flows and appropriately manage their effects on ecosystem processes.

14

Remediation Research in the Jornada Basin: Past and Future

Jeffrey E. Herrick
Kris M. Havstad
Albert Rango

It seems as if the entire set of changes in the soil environment . . . ensures that mesquite will occupy those sites for long periods of time.

—Wright and Honea (1986)

Land degradation in most of the Chihuahuan Desert is characterized by a shift from grass- to shrub-dominated plant communities (Ballín Cortés 1987; Grover and Musick 1990; Fredrickson et al. 1998; see also chapter 10). This shift is associated with increased soil resource redistribution and spatial variability at the plant-interspace scale (Schlesinger et al. 1990; see also chapter 6). Earlier descriptions focused more specifically on the loss of plant species, such as black grama (*Bouteloua eriopoda*), which were palatable to livestock (Nelson 1934). In 1958, it was estimated that one section (3.2 km^2) of black grama grassland could support 18 animal units yearlong, while a similar area dominated by mesquite (*Prosopis glandulosa*) dunes could support just three animal units (Jornada Experimental Range Staff 1958; see also chapter 13). It was recognized that overgrazing facilitated the increase of less palatable species, including shrubs. Consequently, the objectives of the first organized rangeland research in the Southwest were to identify proper techniques to restore grasslands that had been overgrazed (Jardine and Hurtt 1917; Havstad 1996). Today, we recognize the importance of multiple, interacting factors in addition to overgrazing, and research is more broadly focused on the recovery of ecosystem functions necessary to support multiple ecosystem services. This chapter details this extensive history of research to identify and develop technologies to revegetate, restore, reclaim, rehabilitate, or more generally remediate degraded rangelands.

The Society for Ecological Restoration considers that "an ecosystem has recovered when it contains sufficient biotic and abiotic resources to continue its development without assistance or subsidy. It will demonstrate resilience to normal ranges of environmental stress and disturbance. It will interact with contiguous ecosystems in terms of biotic and abiotic flows and cultural interactions"

(Society for Ecological Restoration Science and Policy Working Group 2002). Although restoration of perennial grasslands is often cited as the ultimate objective of management intervention in the Southwest, we recognize that in many if not most cases complete restoration of a preexisting plant and animal community is impossible, even if we had perfect knowledge of all of the elements they contained. We also recognize that many of the historic management interventions discussed herein had more limited objectives. Revegetation is a primary objective where perennial plants have been completely lost from a site, whereas the term reclamation is used to refer to management designed to address a narrowly defined objective or set of objectives, such as erosion control. The word remediation is used throughout this chapter to include management designed to support revegetation, reclamation, and restoration objectives.

This chapter is organized into three sections. The first is designed to provide an overview of approaches developed and tested in the Jornada Basin and elsewhere in the region. The second places this work in the context of existing knowledge about ecological processes. The third section looks toward developing dynamic, ecologically based, landscape-level approaches that will benefit from improved knowledge of long-term processes and patterns.

Historic Approaches

The history of remediation research in the Jornada Basin can be conveniently divided into three periods. The first began with the creation of the Jornada Experimental Range (JER) and emphasized improved livestock management. The second was associated with availability of inexpensive labor during the Great Depression of the 1930s, coinciding with the recognition that livestock management alone might be insufficient to reverse shrub encroachment into grassland. The third period was dominated by increased reliance on herbicides and mechanized shrub control.

Livestock Management (1912–1930s)

A number of state and federal experiment stations were established at the end of the nineteenth and beginning of the twentieth century. Research was designed to estimate the carrying capacity of rangelands during drought and nondrought years, to evaluate and demonstrate the use of water, mineral feeding stations, and fencing to control livestock distribution and increase the quantity and quality of meat production. A number of livestock exclosures were established, and in some cases, clipping trials were initiated to determine sustainable levels of plant utilization. On the JER, four large (250 ha each) exclosures were established during the 1930s: the natural revegetation, the mesquite sand hills artificial revegetation, the gravelly ridges (creosotebush [*Larrea tridentata*]) artificial revegetation exclosure, and the Doña Ana moisture conservation plots exclosure. Changes in vegetation composition inside the exclosures generally followed changes in the surrounding landscape, a pattern repeated in many areas in the Southwestern United States.

While perennial grass cover fluctuated both inside and outside exclosures in response to rainfall variability, the general trend was increased shrub dominance. The natural revegetation exclosure was located on a mesquite–black grama ecotone. It is now completely covered by mesquite duneland. The plant communities in the other three exclosures were already shrub dominated when established (Buffington and Herbel 1965) and continue to be shrub dominated today.

Livestock management research continues today with the development of improved technologies to control livestock distribution through the use of GPS and GIS technologies. These technologies will allow site-specific management of relatively small areas in extensive rangeland systems without fencing (Anderson 2001; see also chapter 13).

Labor-Intensive and Mechanical Approaches (1930s)

Campbell (1929) suggested that in the absence of grazing, grasslands would eventually replace mesquite dunelands through natural succession. Although livestock exclusion continues to be discussed even today as a key to the restoration of Southwestern landscapes (Donahue 1999), the failure of livestock management alone to reverse shrub invasion was beginning to be recognized by the 1950s (JER Staff 1958). It is now clear that some degraded sites will not be remediated simply by exclusion of domestic livestock (Bestelmeyer et al. 2003a; see also chapter 10).

Much of the early experimentation in the Jornada Basin was completed by individuals employed through government programs designed to stimulate the economy by increasing employment. Despite the fact that few of them had formal scientific training, their experiments and other projects have generated extremely valuable data due to their long-term nature and because treatments were often replicated across the landscape. The Civilian Conservation Corps (CCC) provided a large supply of labor during the Depression years of the 1930s. There were approximately 50,000 CCC workers in New Mexico between 1933 and 1942 (Melzer 2000). There were three permanent camps and one temporary CCC camp located in the Jornada Basin. The workers built roads, established many of the experiments listed in table 14-1, and developed a recreational facility at Rope Springs on the eastern side of the basin. Few detailed records of the manipulations completed by the CCC have been preserved, but many of the structures and patterns created can be detected both on the ground and in aerial photographs dating to 1935 (figure 14-1; Rango et al. 2002), and the objectives and results of some of the experiments are summarized in internal reports. Individuals hired with funding from the National Industrial Recovery Act in 1933 were responsible for much of the project planning, supervision, and documentation (Valentine 1942). Many of these individuals continued with careers in rangeland management after World War II (Ares 1974). Economic Recovery Act funding was also used during this period for supervisory personnel.

There was a clear understanding in the 1930s that grass establishment in shrublands, especially mesquite dunelands, was probably limited by multiple factors, including soil instability (which leads to seed burial and root exposure), prefer-

Table 14-1. Nonherbicide remediation trials in the Jornada Basin (USDA-ARS Jornada Experimental Range and NMSU College Ranch, now the Chihuahuan Desert Rangeland Research Center). Original units and plant names from the reports are retained. For a summary of herbicide trial results, see text and Herbel and Gould (1995).

Treatment	Livestock Excluded?	Year(s)	Plant Community	Soil	Location (Pasture #)	Results	Source
Rodent exclusion + mesquite brush piles + denude dune tops (to facilitate wind erosion) + trenches between dunes (to catch topsoil and seed) + diesel oil applied to mesquite	Yes, in 1934	1934 (rodent exclude 1936)	Mesquite dune	Onite (coarse-loamy, mixed Typic Haplargid)	Mesquite Sandhills Artificial Revegetation Exclosure (4)	Unsuccessful	JER Staff (1958); Valentine (1942)
Rabbit and rodent control attempted throughout 640-acre exclosure and 1-mile buffer strip with poisoned grain in Jan, Feb, and Nov 1934 for rodents and poisoned salt blocks for rabbits	Yes, in 1934	1934		Onite (coarse-loamy, mixed Typic Haplargid)	Mesquite Sandhills Artificial Revegetation Exclosure (4)	Unsuccessful. No data on kill % encountered	Valentine (1942)
Livestock exclusion	Yes, in 1934	1934	Mesquite/grassland transition	Onite (coarse-loamy, mixed Typic Haplargid)	Mesquite Sandhills Natural Revegetation Exclosure (4)	By 1955, area occupied by mesquite increased 89% and area occupied by black grama decreased 91%. By 1980, there was a net loss of 4.6 cm of soil and the entire 640 acres was covered by mesquite dunes	JER Staff (1958); Valentine (1942); Gibbens et al. (1983)
Trenching to 26 inches around dune interspaces + transplanting grasses and *Atriplex canescens*	Yes, in 1934	1934–	Mesquite dune	Onite (coarse-loamy, mixed Typic Haplargid)	Mesquite Sandhills Artificial Revegetation Exclosure (4)	Unsuccessful	JER Staff (1958); Valentine (1942)

(continued)

Table 14-1. *(continued)*

Treatment	Livestock Excluded?	Year(s)	Plant Community	Soil	Location (Pasture #)	Results	Source
Contour terraces constructed with a road grader. Twenty-one terraces 8 feet wide × 16–20 inches high leaving 6–8 foot wide shallow pit on upslope side. Spacing 100–300 = with closer spacing at top of slope. Added rock "weeps" to allow water to percolate through after several years to minimize breaks. Some terraces seeded to "various native and cultivated plants"	Yes (for at least 11 years)	1935	Black grama (upper slope), creosote (mid-slope) and Mormon tea/ short-lived perennial grasses (footslope)	A compact gravelly clay loam underlaid by a heavier compact clay loam and in places caliche at 24–36 inches. Higher permeability higher on slope.	210 acres on north-facing slope of Doña Ana Mountains	"Little improvement that can be attributed to the terraces has taken place over the area in general." Temporary snakeweed increase on upslope side of terraces. Black grama increased in the one inter-terrace quadrat in which it occurred, but was less vigorous than in other areas. Good grass establishment behind some terraces seeded to semidesert-adapted grasses. Based on six m^2 quadrats in bottoms of basins above terraces and five on intervening undisturbed areas between terraces	Valentine (1947)
Brush dams placed across slope between mesquite dunes. 12–16 inches high, tied down with wire ("brush water spreaders")	Light grazing	1937	Mesquite dunes with fourwing saltbush	Loose sandy loam over compacted sandy loam, sometimes exposed at surface. Caliche at > 30 inches.		"Structures have brought about only the slightest improvement and this is restricted entirely to the areas immediately beneath the brush dams"	Valentine (1947)

Brush water spreaders. Brush held down at 2-foot intervals by wire ties anchored by driving knotted ends 10–12 inches into soil. Spreaders across slope, but trending downslope (1/2pc). Water supplied from small dams across gullies. Seed scattered in brush of some seeders	Yes (for at least 11 years) in one pasture; Light stocking in other pasture	1937	Variable. Black grama. Black grama/tobosa. Creosote. Sparse fluffgrass with annual grasses and forbs	Variable. Sandy loam (6–10 inches) over coarse sandy loam with one small area of compact clay loam	N and E facing slopes 2–4% slope at foot of Doña Ana Mountains	"In general it is impossible to identify any area either above or below the spreaders that have been benefited from them." Spreaders effective until dams broke after several years. Within 9 years, most water not diverted. Effective in creating microsites for seedling establishment. Noted patches of perennial grass establishment 8 years after construction. *Based on observations and five pairs of square meter quadrats located above and below spreader	Valentine (1947)
Crescent-shaped dams, contour furrows (at 10–20-foot intervals), check dams in gullies using grain sacks filled with soil and, in some cases, manure. Check dams were also constructed using	Yes	August 1937	Creosote		North side of Doña Ana exclosure around 1915 black grama clipping	Heavy rains in September 1937 and summer, 1938 "greatly damaged furrows and check dams," but by the end of 1938, grasses had increased to 30% of total cover	JER Staff (1939)
Handgrubbing	N/A	1937–?	Tarbush	N/A	N/A	Substantial increase in forage yield by 1945	JER Staff (1958)
Lagomorph exclusion + shrub removal + seeding + furrowing in a factorial design	Yes	1938	Creosote	Canutillo gravelly sandy loam Doña Ana) Upton gravelly loam (Ragged and Parker tanks)	Doña Ana Exclosure (6) Gravelly Ridges Exclosure (20) Parker Tank (20)	1956: No effects of furrowing and seeding. Grass density increased in shrub removal plots and increased most in shrub removal + lagomorph exclusion plots	JER Staff (1958); Korzdorfer (1968); Gravelly Ridges only: Gibbens et al. (1993); Havstad et al. (1999)

(continued)

Table 14-1. (*continued*)

Treatment	Livestock Excluded?	Year(s)	Plant Community	Soil	Location (Pasture #)	Results	Source
						1995: Increased shrub and grass cover in lagomorph exclusion plots. Increased black grama cover in shrub removal plots, but minimal no significant increase in total grass cover due to higher *Muhlenbergia porteri* in shrub-intact plots	
Intensive contour structures and rabbit exclusion. 6-inch deep furrows located 4–6 feet apart, 8-inch deep furrows located 25–35 feet apart, 8-inch deep ridge contour furrows located 30–45 feet apart, 6-inch deep × 24–30 inch wide ripper furrows located 10–12 feet apart	N/A	1939	Creosote, yucca, snakeweed, mesquite, fluffgrass and *Croton corymbolus*	Coarse sandy loam over coarse sand at 24 inches	NE facing slope extending from foot of Doña Ana Mountains	No effect of contour structures on either perennial vegetation or soil moisture (measured to 24 inches), and no effect of rabbit exclusion	Valentine (1947)
Contour channels. Flat-bottomed trench along contour 24–30 inches wide × 6 inches deep with soil formed into ridges on each side by a road ripper with a piece of steel fastened to teeth. Intervals of 25–75 feet on 2–3% slope. 50–60 acres	Yes	1939	Creosote, mesquite, snakeweed with scattered patches of bunchgrasses	Loose sandy loam over caliche at 0–30 inch depth	In Doña Ana exclosure near lagomorph exclusion study (?)	As a whole, the treated area is little if any different from the surrounding untreated area and it may be fairly concluded that the treatment has been ineffective in bringing about any improvement of the site	Valentine (1947)
Mesquite brush piles to capture *Atriplex canescens* seeds and improve miscrosite conditions for establishment in interspaces	Yes, in 1934	1939–42	Mesquite dune	Onite (coarse-loamy, mixed Typic Haplargid)	Mesquite Sandhills Artificial Revegetation Exclosure (4)	Successful (1939); unsuccessful (1942)	JER Staff (1958); Valentine (1942); Jornada Staff (1939)

Mowing	N/A	1939–?	Snakeweed	N/A	N/A	"Gave some promise"	JER Staff (1958)
Burning with a flame gun	N/A	1939–?	Snakeweed	N/A	N/A	Effective except when dormant	JER Staff (1958)
Grubbing	N/A	1939–?	Snakeweed	N/A	N/A	Effective at all times of year	JER Staff (1958)
Mowing	N/A	1939–?	Snakeweed	N/A	N/A	"Gave some promise"	JER Staff (1958)
Trenching to 26 inches around dune interspaces + cutting mesquite roots + seeding grass, shrub and subshrub species	Yes, in 1934	1930s	Mesquite dune	Onite (coarse-loamy, mixed Typic Haplargid)	Mesquite Sandhills Artificial Revegetation Exclosure (4)	Only *Sporobolus* spp. and *Paspalum stramineum* emerged. None survived 5 years	JER Staff (1958); Valentine (1942)
Rodent, rabbit and livestock exclosures (various combinations). Design-limited: nonreplicated and one treatment had 2.5X more grass at study initiation as other three	See treatments	1940	Mesquite–snakeweed			1948 yield of "desirable grasses" dramatically higher in rabbit and rodent excluded plots in mesquite-snakeweed. Variable results in other 2 plant communities	NM Ag Experiment Station 1949
Rodent, rabbit, and livestock exclosures (various combinations)	See treatments	1940	*Erioneuron pulchellum, Aristida* spp				NM Ag Experiment Station 1949
Rodent, rabbit, and livestock exclosures (various combinations)	See treatments	1940	*Bouteloua eriopoda*				NM Ag Experiment Station 1949
Seeding. Various attempts to plant native and exotic, including *Eragrostis* spp., *Bouteloua eriopoda, Sporobolus flexuosus* and *Atriplex canescens* using flat planting, furrow planting, loose seedbed and compacted seedbed		1947–49	Snakeweed and creosote (2 sites)	"Good soil" and "poor soil," respectively	Various on CDRRC (College Ranch)	Poor to no establishment 1947, 1948, and 1949. In 1948, concluded that "success with about 50% of plantings is the best that can be expected"	NM Ag Experiment Station 1949; 1950

(continued)

Table 14-1. *(continued)*

Treatment	Livestock Excluded?	Year(s)	Plant Community	Soil	Location (Pasture #)	Results	Source
Handgrubbing areas averaging 30 mesquite plants/acre	Yes	1948	Black grama/ mesquite[a]		Artificial revegetation exclosures (?)	13.3% reinvasion after 7 years	JER Staff (1958)
Handgrubbing areas averaging 805 mesquite plants/acre	Yes	1948	Mesquite/ bunchgrass[a]		Artificial revegetation exclosures(?)	64.5% reinvasion after 7 years	JER Staff (1958)
Handgrubbing areas averaging 30 mesquite plants/acre	Yes	1948	Black grama/ mesquite[a]		Artificial revegetation exclosures(?)	13.3% reinvasion after 7 years	JER Staff (1958)
Railing	N/A	1952	Tarbush	N/A	N/A	Fair control in May	JER Staff (1958)
Contour terraces constructed with a road grader. Five, concentric, 3 inches high, on contour on a <1% slope. Seeded and biosolids applied	Yes, initially	1975–79	None			Experiment abandoned 8/79 due to repeated structural failure of dikes during heavy rain events and apparent lack of plant establishment. Resurvey in 1997 showed highly successful establishment	Walton et al. (2001)

[a] Grass component not specified; inferred from mesquite density.

ential grazing on grasses by native herbivores, and competition by existing plants. Various approaches were tried to enhance grass establishment. The treatments can be classified into five basic groups: seeding and transplanting, microsite manipulations, shrub removal, water redistribution, and small mammal control. Livestock were excluded from most of the experimental areas. In some cases, livestock exclusion was included as an experimental treatment. Although many of the manipulations appeared to increase grass establishment temporarily, few had significant lasting effects on plant community composition. Some of the treatments applied in the 1930s could no longer be detected in aerial photographs by 1968, whereas others were still visible in 1996 (table 14-2).

Seeding and Transplanting

Seeding and transplanting were generally done in association with one of the other treatments. A large variety of species were tried (Valentine 1942). Four-wing saltbush (*Atriplex canescens*) was one of the most popular species on sandy basin soils because of its value as a forage crop and its observed potential to compete with or at least coexist with mesquite. However, few of the treatments were successful. Valentine (1942) concluded that success depended on rainfall distribution and amount during the establishment period and during subsequent years. Many of the seedings in the artificial revegetation exclosure were planted in 1934, a year with barely half the long-term average rainfall.

Microsite Manipulations

Manipulations designed to improve conditions for seedling establishment at the microsite scale (Fowler 1986) included piling brush (often cut from the top of mesquite dunes), digging trenches (to trap soil and seeds), and cutting roots (to reduce root competition from mesquite). Limited establishment was observed in response to these treatments and there was virtually no survival after five years (Valentine 1942). Seed burial was cited as a problem in many cases, particularly in the interdune areas.

Shrub Removal

Grubbing (shrub removal using hand tools) was frequently cited as one of the most effective approaches for maintaining or increasing the productivity of perennial grasslands. Attempts to combat shrub invasion by hand continued well into the second part of the twentieth century. As late as 1958, Herbel et al. concluded "grubbing light stands of young mesquite plants is the most economical means of controlling mesquite" (Herbel et al. 1958). At that time it cost $2.05/ha to remove an average of 200 plants/ha. Careful examination of belt transects later showed that the grubbers missed 7% of the plants. Grubbing becomes more difficult as plant size increases because the root crown must be removed to prevent

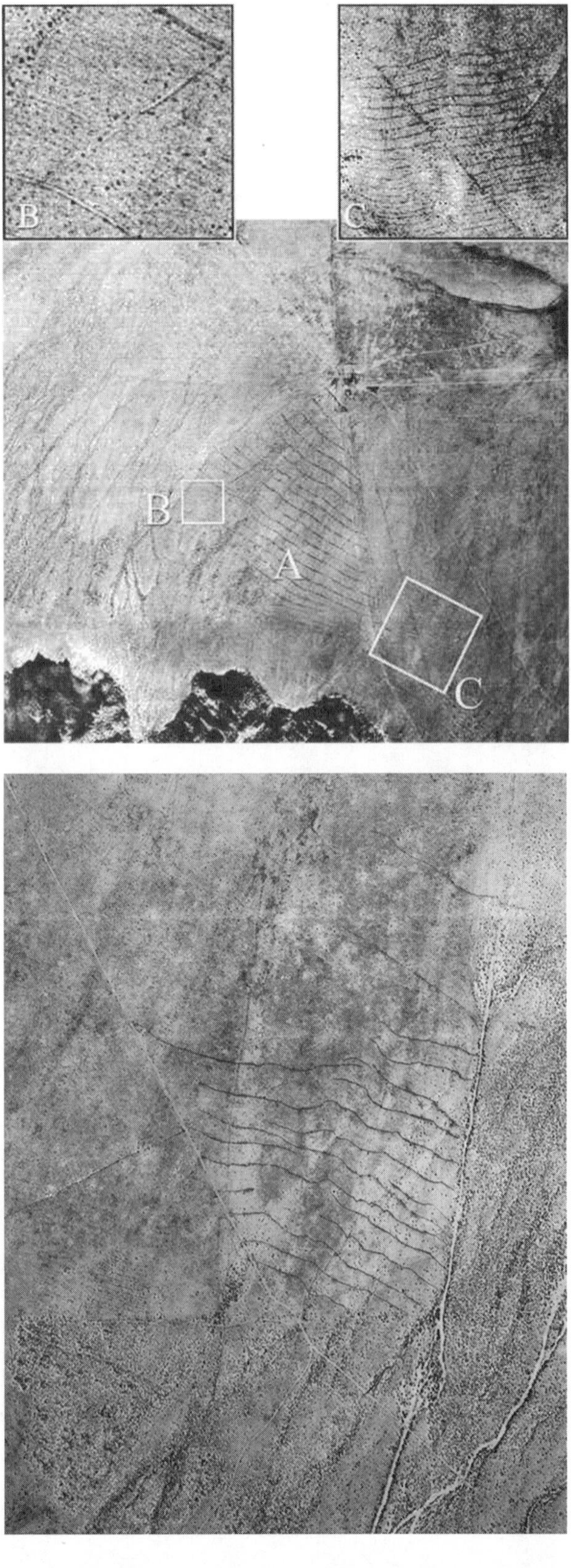

Figure 14-1. Early remediation experiments visible in aerial photographs (from Rango et al. 2002; see table 14-2).

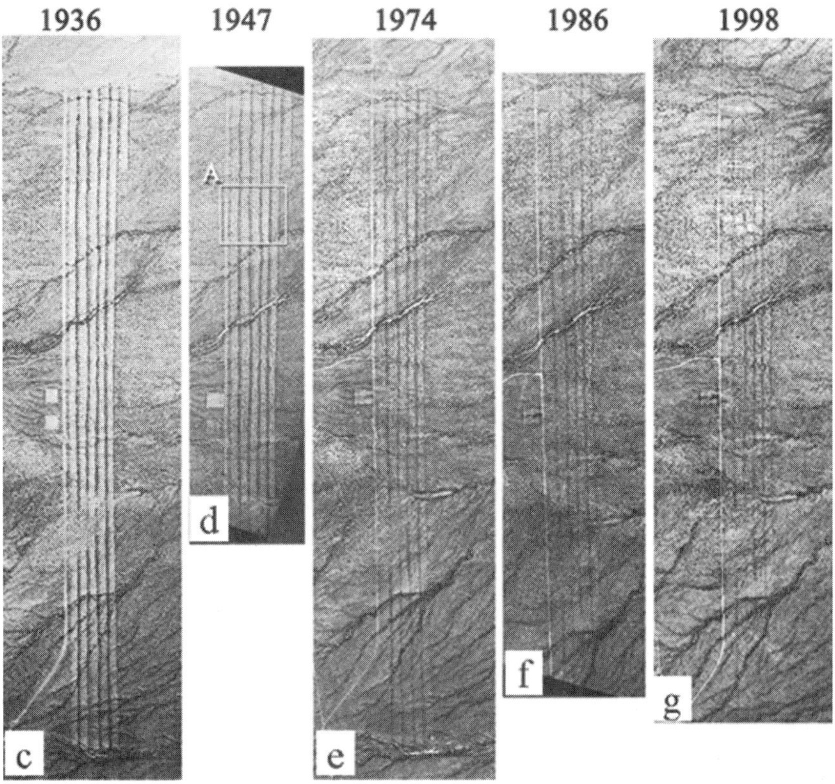
1936
1947
1974
1986
1998
A
c
d
e
f
g

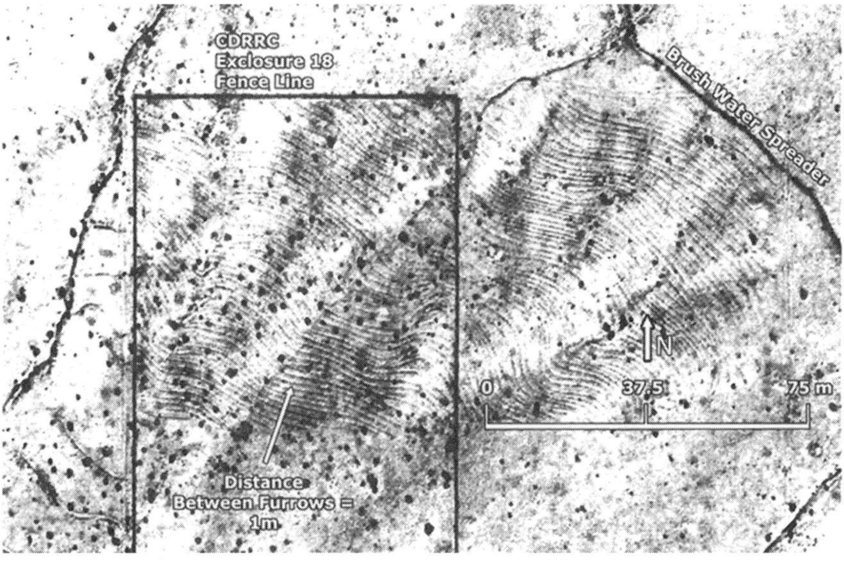
CDRRC
Exclosure 18
Fence Line
Brush Water Spreader
N
0
37.5
75 m
Distance
Between Furrows =
1m

Table 14-2. Longevity of remediation treatments visible in aerial (1936, 1947, and 1996) and satellite (1968, 1972, and 1983) photos in the Jornada Basin (modified from Rango et al. 2002). Y, P, and N indicate that the treatments were visible, partially visible, and not visible, respectively.

Treatment	Year Constructed	1936	1947	1968	1972	1983	1996
Contour terraces (A in fig. 14.1a)	1935	Y	Y	Y	Y	P	P
Contour terraces (B in fig. 14.1a)	1935	Y	P	N	N	N	N
Contour terraces (C in fig. 14.1a)	1935	Y	P	N	N	N	N
Brush spreader (fig. 14.1b)	1937	N/A	Y	Y	Y	P	P
Grubbed strips (fig. 14.1c–g)	1936	Y	Y	Y	Y	Y	Y
Exclosure 18 (fig. 14.1h)	1939	N/A	Y	P	P	P	N

resprouting. Tarbush (*Flourensia cernua*) and creosotebush, relatively weak sprouters compared to mesquite, were sometimes simply cut off at the soil surface.

Water Redistribution

Water redistribution was attempted at a variety of scales using both hand tools and tractor-mounted implements (table 14-1). In most cases, the objective was to slow the movement of water across the landscape through the creation of soil dikes, terraces, or furrows (figure 14-2). In at least one case, brush dams were constructed between mesquite dunes in areas with a gentle slope. Linear brush piles were also used to spread water from rock dams in gullies across creosotebush-dominated slopes.

Limited data available from a few plots showed only partial, if any, grass response to these structures by the mid-1940s (Valentine 1947). In a review of the water redistribution projects of the 1930s, Valentine (1947) reported that "In general it is impossible to identify any area either above or below the spreaders that have been benefited from them. This is true even of the area above the spreaders where the marks left by standing and running water give evidence that they were instrumental, at least occasionally, in bringing water to and holding it on these limited areas." Interestingly, however, many of these treatment areas are visible six decades or more after their establishment due to higher grass or shrub cover and/or changes in species composition.

These long-term changes are often associated with a persistent shrub response to many of the water redistribution treatments. The concentration of even relatively limited quantities of water and nutrients on the contour terraces and behind the dikes could explain the apparently higher shrub biomass visible in figures 14-1 and 14-2. Valentine (1947) also observed that the brush water spreaders created more favorable microsites for seedling establishment. This observation is supported by Gutierrez-Luna (2000) who showed that water-dispersed seeds are pref-

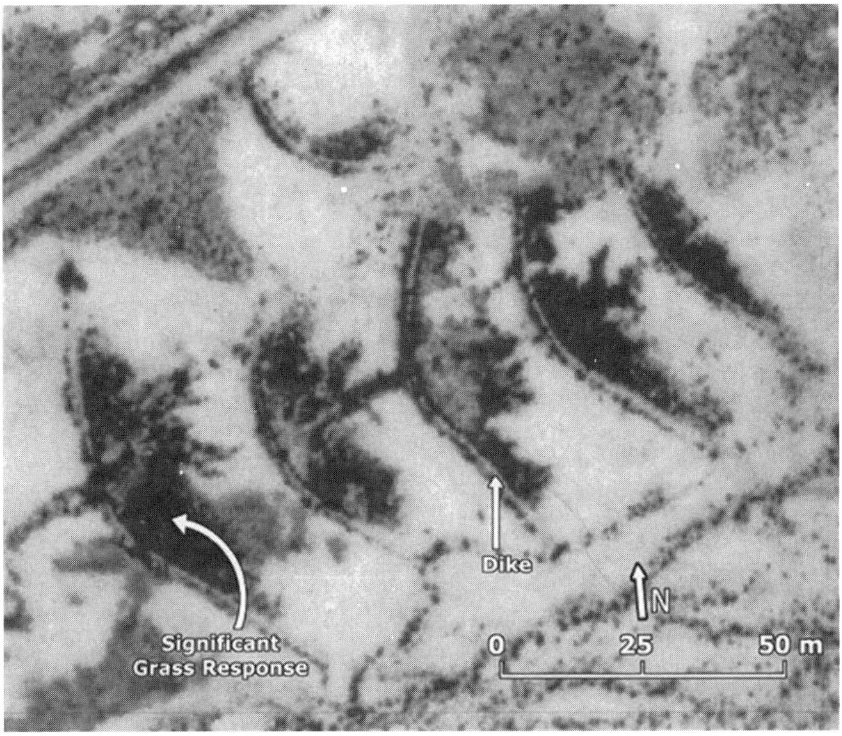

Figure 14-2. Aerial photo (1998) of contour dikes constructed in the 1970s showing vegetation establishment.

erentially deposited in naturally formed litter dams, and moisture and temperature conditions in these microsites tend to be more favorable for seedling establishment. Seedlings survive longer where there is litter on the surface.

Most of the water-retaining structures constructed during the 1930s were built on relatively coarse-textured soils with slopes $> 2\%$. In at least one case in which soil moisture content was measured, the water-holding capacity of the soil was so low that no increase in moisture availability was detected behind the ridges (Valentine 1947). Additional water-retention structures were built in the 1970s on heavier soils with lesser slopes. Unlike the water-retention structures built in the 1930s, which were in shrub-dominated areas, the 1970s contour dikes were established in areas that were devoid of vegetation. The 1970s structures were abandoned within four years due to high maintenance costs and limited grass establishment in spite of seeding and sewage sludge applications. Twenty-two years later, however, native species had revived (Walton et al. 2001; see also figure 14-2).

These data, together with the fact that none of the water-retention structures constructed during the 1930s was maintained for more than five years, suggest that these strategies might be better viewed as being only partially tested rather than rejected as failures. In fact, some of the most vigorous grass patches on the Jornada are located upslope of the most carefully maintained water redistribution structure in the Basin: the access road that connects all of the CCC camps and the JER headquarters to the city of Las Cruces. These patches extend up to 20 m upslope from the road, and there are a number of areas in which production is correspondingly reduced downslope.

Small Mammal Control

Work on small mammals was divided between attempts to control them using poisoned grain (for rodents) and poisoned salt blocks (for rabbits) and experiments designed to quantify their effects on grass production and survival. Valentine (1947) suggested that rodent and rabbit damage was one reason for the failure of brush dams to increase grass establishment between mesquite dunes. In 1939, jackrabbits (*Lepus*) and kangaroo rats (*Dipodomys*) were described as the "principal range-destroying rodents" which were "controlled by shooting and poisoning" (JER Staff 1939). Data from three replicated lagomorph exclusion experiments show that over a 28-year period, perennial grass cover was on average higher at two locations and that there was no significant effect of exclusion at the third location (data from Korzdorfer 1968). Data collection continued at one of the locations where in 1967 basal cover was close to zero in nearly all plots. By 1995, cover at this one location had rebounded to 3.5% in lagomorph-excluded plots and 2.9% in the plots to which rabbits had free access (Havstad et al. 1999). This experiment is unique in that it is one of the oldest manipulative experiments (other than livestock exclosures) replicated at the landscape scale.

The successful efforts to remove prairie dogs (*Cynomys ludovicianus*) from the basin in 1916–17 and the unsuccessful attempts to eradicate kangaroo rats were part of early remediation attempts and were apparently based on the assumption that these animals competed extensively with livestock for forage (chapter 12). Confidence in the efficacy of the eradication treatments was low, as illustrated by the fact that rodent and rabbit exclosures designed to quantify the impacts of small mammal herbivory in an artificial (seeded and transplanted) revegetation experiment were established near the center of a 9-square-mile area in which poison baits were repeatedly applied (table 14-1). While small mammals reduce perennial grass density near their burrows and increase the wind- and water-erodible soil and in unvegetated patches (chapter 12), they also remove shrub seedlings and are important seed dispersal agents that may actually contribute to future remediation strategies (Havstad et al. 1999). For example, Gibbens et al. (1992) reported that 81% of mesquite seedlings emerging in response to a July 31, 1989, rainstorm were dead by the following May. Of these, all but 2% had been bitten off "at or slightly below the cotyledonary node, which causes death." All of the surviving seedlings also showed signs of herbivory.

The Promise of Technology (1941–1980s)

The apparent failure of grazing management alone to reverse the shrub invasion of grasslands, together with the uncertain effectiveness of the Depression-era manipulations led to an intensified search for viable alternatives that could be applied with minimal labor. During the boom years following the war, fossil fuels replaced human labor as the most cost-effective input and "shrub control" became the new mantra. The authors of the 1958 annual report for the JER observed that "there is no evidence of recovery [of mesquite sand dunes to grassland] after 25 years, even in areas completely protected. To the contrary, severe duning has been spreading even with conservative grazing . . . the absence of grazing use will not retard that spread . . . The suggested control method is grubbing" (JER Staff 1958).

The reduced emphasis on water redistribution-based approaches attempted during the 1930s was probably driven in part by the absence of any evidence of positive impacts of these approaches during the drought of the 1950s, when annual precipitation (from 1950–56) averaged just 150 mm/year. Although there are few quantitative records of the effects of the Depression-era treatments, the quadrat studies (Gibbens and Beck 1987) and shrub removal/lagomorph exclusion experiments suggest that grass recovery in response to management was virtually erased during the accelerated loss of grassland during the 1950s drought. Attempts to control water distribution did continue, however (table 14-1), and the development of new equipment allowed for more labor-efficient if not more effective resource redistribution.

Herbicides

Herbicides offered the opportunity to reduce shrub competition without disturbing the soil surface. With the virtual disappearance of inexpensive labor in the boom years of the 1950s, shrub control methods that could be applied to large areas with minimal human effort were in high demand.

Early Trials

Herbicide studies were initiated in the 1930s. Plants were sprayed with sulfuric acid, kerosene, sodium chlorate, and diesel oil and dusted with mixtures of borate and sodium chlorate. Only sodium chlorate and atlacide (a chlorate-based material) were effective in trials with snakeweed (*Gutierrezia sarothrae*) and only at concentrations in excess of 10% (JER Staff 1958). In the 1940s a number of materials were injected into the soil at the base of mesquite plants. The only material that was consistently successful was a mixture of diesel oil and 10% ethylene dibromide (New Mexico Agricultural Experiment Station 1949).

Synthetic Herbicides

The development of phenoxy herbicides, specifically 2,4,5–trichlorophenoxyacetic acid (2,4,5-T), after World War II opened a new era in shrub control. These materials could be applied aerially over large areas and, with timely application, could result in significant reductions in shrub productivity and density. Although 2,4,5-T can no longer be legally applied to rangeland, there are a number of other materials, including clopyralid, tebuthiuron, and monuron, that are still available. Shrub density has been successfully reduced on relatively large areas in the Chihuahuan Desert in both the United States and Mexico, and herbicide applications are frequently included in brush management plans. Much of the remaining grassland on the JER and the Chihuahuan Desert Rangeland Research Center (CDRRC) has been treated at least once with herbicide.

Management recommendations for the use of these herbicides based on several decades of research are summarized in a New Mexico State University Agricultural Experiment Station Bulletin (Herbel and Gould 1995). Even this relatively optimistic publication, however, concludes with the cautionary note that "it is possible to renovate brush-infested rangelands with herbicides, but some of the practices are costly." The practices, including both seeding and the herbicide applications themselves, frequently must be reapplied several times (Ethridge et al. 1997). Economic limitations often limit their use to more productive sites or to sites that have multiple values, such as watershed protection and wildlife habitat conservation (Bovey 2001). Although there are many herbicides available, those that were most toxic to animals or caused other environmental problems have been taken off the market. The environmental costs of rangeland herbicide applications continue to be of concern in some areas (Bovey 2001).

Mechanical Treatments

Numerous implements were designed by engineers to exploit the ever increasing power of agricultural and engineering machinery to remove shrubs, prepare a seedbed, create small pits where water could accumulate, and plant seeds. The Arid Land Seeder accomplished all four operations in a single pass. Developed at the JER and tested during the 1960s (Abernathy and Herbel 1973; Herbel et al. 1973), this implement consisted of a standard root plow mounted on a bulldozer trailed by a conveyor assembly and rangeland drill. Brush was removed by the root plow, carried by the conveyor assemble over the top of the drill, which planted the seed, and replaced the brush over seeded rows to form a mulch over the seeds. Various modifications of the machine were used to clear and seed 23 plots dominated by creosotebush and/or tarbush between 1966 and 1970. The plots were each approximately 2 ha in area and were scattered throughout southern New Mexico. Approximately 50% of the seedings were considered successful based on qualitative evaluations made six years or sooner after treatment (Abernathy and Herbel 1973). Drought was cited as the primary reason for failure, with overgrazing of recently established grasses also contributing (Herbel et al. 1973). Other devices used to remediate Southwestern rangelands, especially in southern Arizona, are summarized in Jordan (1981) and Roundy and Biedenbender (1995). Some of these approaches were also tested in the Jornada Basin (JER Staff Annual Reports unpublished).

All of these treatments result in high levels of soil surface disturbance, increasing erosion susceptibility (Wood et al. 1991). They also require significant energy inputs and the availability and maintenance of expensive machinery. They can, however, be effective in reducing shrub density and cover. Following an evaluation of 92 years of reseeding efforts in the Southwestern United States and Mexico, Cox et al. (1984) concluded that although most of the approaches evaluated were at least occasionally successful, "no single seedbed treatment has been shown to be superior to any other over time." It was also concluded that reseeding studies needed to be conducted more systematically.

Comparisons of the relative cost and effectiveness of mechanical versus chemical shrub control have yielded variable results. Tebuthiuron was generally more effective at reducing creosotebush canopy cover and increasing forage production at four sites in the Chihuahuan Desert of southeastern Arizona and northern Chihuahua, Mexico (Morton and Melgoza 1991). Holechek and Hess (1994) estimated that burning cost \$2.50–\$12.50/ha (where sufficient fuel exists), herbicide cost \$30–50/ha, and mechanical control cost \$60–125/ha. Estimates include the cost of materials, machinery depreciation, and labor.

Ethridge et al. (1997) found that only 1 in 10 rangeland seedings in southern New Mexico was profitable when the costs were compared to income from increased livestock production. Jones and Johnson (1998) pointed out that some of the failures may have been unnecessary as, "sophisticated analyses of ecological adaptation and genetic variation were rarely considered in early trials." In other words, at least some of the failures may be partially attributed to the seeding of species and varieties that were not adapted to the local edaphic and climatic conditions. However, many of the failures were clearly due to inadequate soil moisture and the rapid reestablishment of shrub species.

Other Technologies

A number of other technologies have been tried, ranging from applying hot wax to increase runoff from mesquite dunes (Gibbens personal communication) to using polyacrylamide to reduce soil crust resistance to seedling emergence and increase infiltration capacity. Greenhouse trials indicated that polyacrylamides may effectively increase seedling emergence (Rubio et al. 1992); however, the results of the infiltration studies were inconclusive (Rubio Arias 1988).

Data Quality and Reliability of Conclusions

Experimental designs varied widely. Most of the studies were not replicated and baseline data were frequently not recorded or were recorded on such small plots as to be of little value. Many studies for which there is some baseline information have been unreplicated, and in many cases, the most useful information is qualitative.

In spite of these caveats, these studies in aggregate represent a tremendous resource. They addressed, either directly or indirectly, virtually every factor potentially affecting grass establishment and survival. Many of the treatments were applied in logical combinations in an attempt to remove multiple constraints (brush mulches combined with water redistribution or reduction in rodent or rabbit populations). The treatments were often replicated across the landscape and applied during different years, providing a nonstatistical but potentially more relevant form of replication. The discussion summarizes a relatively small fraction of the work completed. Although many records no longer exist, others persist in archives scattered throughout the country. One of the reports cited (JER Staff 1939) was found at the National Agricultural Library in Washington, D.C., but had obviously been rescued from the USFS Allegheny Forest Experiment Station library, which stamped it "March 13, 1939, RECEIVED." Our ability to interpret the long-term impacts of the historic treatments will continue to increase as we relocate and resample them using a combination of historic and contemporary aerial photographs, ground-based rephotography, and on-ground surveys, together with the archival records cited here.

Summary

Four generalizations can be drawn from the historic literature. The first is that most past attempts to manipulate the system failed. The second is that there are enough successes to convince some, at least, that the system can be controlled (Cassady and Glendening 1940). The third is that it often takes decades to determine the success or failure of a particular manipulation. It may take decades for positive effects to appear as plant-soil feedbacks gradually change soil water-holding and infiltration capacities. Meteorological conditions that facilitate establishment occur relatively rarely and may be poorly timed relative to other factors affecting establishment, such as the number of germinable seeds in the soil and herbivore population dynamics. Conversely, short-term successes can rapidly turn into failures as landscape-level processes gradually overwhelm patch-scale treatments and droughts that cause widespread mortality among long-lived perennial plants occur just once every several decades. The fourth, and perhaps most important generalization, is that the success or failure of a manipulation depends on multiple interacting factors. The relative importance of each factor varies across space and time.

Understanding Ecosystem Processes

Until the start of the International Biology Programme (IBP) in 1970, most research in the Jornada Basin was specifically designed to determine which treatments could be used to improve management. Over the past three decades, emphasis on improving understanding of basic ecosystem processes has increased. The objectives of the research vary and, in some cases, continue to be management-driven. Increasingly, however, it is recognized that the success of future management systems will depend on a more thorough understanding of these processes, and it is not always obvious which need to be studied. In many cases, the most obvious process is not necessarily the most important one. Chesson and Huntly (1997) argue that species coexistence is favored by harsh but fluctuating conditions that create opportunities for establishment that vary through space and time.

In 1986, Wright and Honea concluded the abstract of an article on soil properties in mesquite dunelands with the statement, "It seems as if the entire set of changes in the soil environment . . . ensure that mesquite will occupy those sites for long periods of time." The changes include soil organic matter, nitrogen, cations, phosphorous availability, and soil texture, particularly where there is an argillic or other horizon with a texture different from that at the soil surface (figure 14-3). It has also been shown that the effects can persist long after shrub removal. Thirteen years after velvet mesquite (*Prosopis juliflora*) removal, canopy–interspace differences in soil carbon were virtually identical to those where mesquite had been left intact at a site in southeastern Arizona, which has slightly higher temperatures and winter rainfall than the Jornada Basin (Tiedemann and Klemmedson 1986).

Schlesinger et al. (1990) describe a framework for understanding how various processes combine to reinforce the grass–shrubland transition. This framework, together with the landscape-level dynamics currently being addressed by the research program in the Jornada Basin, can serve as a starting point for defining the types of interventions that are most likely to be successful and identifying the parts of the landscape and times (relative to drought, extreme precipitation events, and fluctuations in herbivore populations and seed banks) when grassland recovery is most likely to occur.

Schlesinger et al. (1990) also highlighted the importance of the formation, maintenance, and deterioration of resource islands in deserts throughout the world. Studies in Israel (Boeken and Shachak 1994), and Australia (Tongway and Lud-

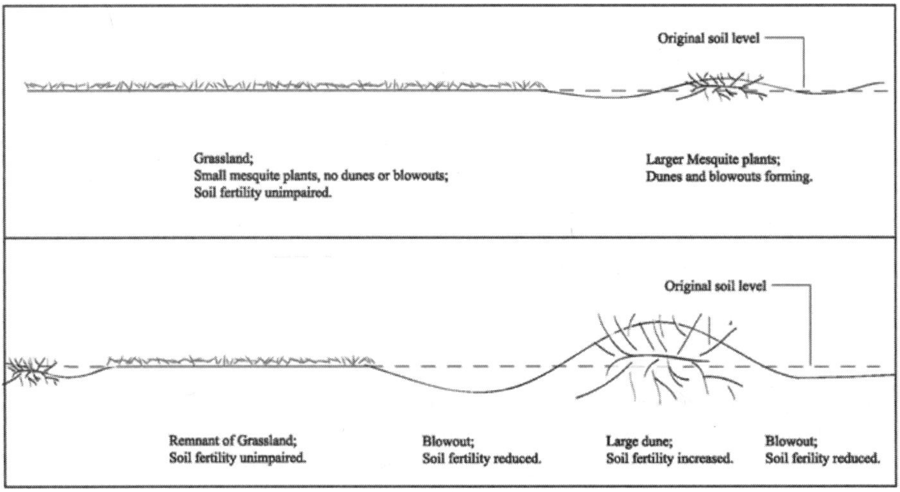

Figure 14-3. An early representation of the fertility changes associated with mesquite invasion (Valentine 1941).

wig 1997), as well as the United States (Valentine 1941; Schlesinger et al. 1996; Wainright et al. 1999b), have addressed the mechanisms by which human- and vegetation-formed patches affect plant production and resource availability and redistribution at multiple spatial scales. This trend toward understanding the processes responsible for the patterns should help target those processes that limit the success of remediation attempts.

Understanding the importance of resource redistribution at the plant-interspace level generated by the Jornada Basin research has led to a renewed interest in the possibility of manipulating resource availability to trigger changes in vegetation composition and structure. Most now agree that one of the keys to the persistence of shrublands in spite of diverse efforts to remove them is their ability to acquire resources from both greater depths and larger areas than grasses and to concentrate and retain those resources in self-reinforcing islands of fertility (figure 14-3; Valentine 1941; Wright and Honea 1986; Schlesinger et al. 1990). The key to maintaining production during drought years is the ability to tap deep water, while extensive shallow-root systems allow shrubs to compete with grass for water from brief or low-intensity rainstorms (Gibbens and Lenz 2001). The effect of reduced nutrient availability on grass production in mesquite dune interspaces was documented over 60 years ago (Valentine 1941).

Landscape-level controls on recovery are also clearly important, though less well understood. The importance of long-term geomorphic stability and inherent edaphic characteristics (McAuliffe 1994) has been clearly demonstrated and must be considered together with resource redistribution patterns (chapter 7) in predicting which parts of the landscape are most likely to respond to management inputs.

The Landscape Context: Edaphic Controls on Resource Availability

McAuliffe (1994) documented strong relationships between soil development and vegetation in southeast Arizona. Similar relationships have been reported for the Jornada (chapters 2 and 4). Perennial grasslands tend to persist on soils with an argillic horizon near but not at the soil surface. Argillic horizons are rich in clay and tend to retain more water at a depth that is accessible to grass roots. These soils tend to be located on older, more stable surfaces (as argillic horizons take thousands of years to develop). Similar patterns have been documented in the Chihuahuan Desert. Where black grama grasslands persist on coarse-textured soils in the Jornada Basin, there is often a calcic or petrocalcic horizon near the soil surface (Teaschner 2001). Highly developed calcic horizons appear to be relatively impervious to both water and roots. Most of these horizons, however, are heavily invaded by roots and have higher water-holding capacity than the loamy sands typical of many Chihuahuan Desert basin soils now dominated by mesquite. The failure of at least one intensive effort to increase grass establishment by concentrating water was attributed to the fact that the water-holding capacity of the soil was too low to support grassland (Valentine 1947).

The Landscape Context: Geomorphic and Climatic Stability

Geomorphically stable surfaces are necessary for the development of many of the soils that appear to have the greatest resistance to shrub invasion, such as those with an argillic horizon or a near-surface calcic or petrocalcic horizon (McAuliffe 1994). The amount of time required for recovery of some plant communities may be longer than past periods of geomorphic (and climatic) stability (Webb et al. 1987). The importance of lag time between climate and plant community changes associated with soil-vegetation feedbacks is poorly understood. Continuing soil erosion in shrub-invaded areas may effectively reduce the proportion of the landscape within which recovery of pre-1850 plant communities is possible or at least dramatically increase recovery time (Coffin Peters and Herrick 1999). The ultimate effects of soil degradation and loss on establishment probabilities for different species also depend on climate change and, particularly, precipitation amount and seasonal distribution.

The Landscape Context: Resource Redistribution

Identification of the scale at which resource redistribution affects plant community dynamics is critical to developing ecologically based approaches to remediation. This issue, which is now being addressed by Jornada researchers, is very poorly understood and is defined as one of our research objectives. We know that during most precipitation events, the majority of water redistribution occurs at the plant-interspace scale. We also know that some plant communities, such as those in playas, rely on larger, rarer events that result in water redistribution at the landscape scale. We do not understand how changes in the intensity and frequency of these redistribution events affect plant establishment, production, and survival at the two scales. We know even less about the effects of nutrient redistribution at multiple scales.

Similar issues apply to the dissemination of seeds, a key resource that can be quickly depleted through granivory and germination (Peters 2003). The maintenance of many plant communities depends on local seed production, whereas species invasions can be promoted through both short-distance dispersal along ecotones and long-range dispersal from distant populations. Seed banks represent yet another form of dispersal, but in time instead of space.

The relative importance of these three dispersal mechanisms for shrub persistence and grass reestablishment needs to be defined as part of any attempt to reestablish grassland in currently shrub-dominated systems. W. G. Whitford (personal communication) estimated that honey mesquite produced over 100 seeds per square meter in a mesquite-dominated community based on seed and pod counts. Tschirley and Martin (1960) reported that 9% of originally germinable seeds of the congeneric velvet mesquite were still germinable 10 years after burial at the Santa Rita Experimental Range in southeast Arizona. However, mortality can also be quite high. One study showed that less than 1% of seed produced remains in the seedbank one year after production with insect damage accounting for most of the mortality (Owens unpublished data). Mesquite seeds are dispersed over relatively short distances by rabbits and other small mammals and over longer distances by livestock (Tschirley and Martin 1960).

Fire: A Missing Link

Our understanding of one of the most fundamental agents of change in many ecosystems—fire—is extremely rudimentary, particularly when compared to our understanding of its role in forested and more mesic grassland systems (Knapp et al. 1998). Fire has been used in more mesic areas to both manage the growth form of mesquite (Teague et al. 1997) and prevent mesquite encroachment. There is some evidence to suggest that fire may be used in the Chihuahuan Desert to limit shrub expansion if it is applied during a relatively wet year when grasses can recover. The effects of a burn in most systems depend on careful timing relative to current weather, soil moisture, fuel load, and the size and growth stage of both the herbaceous and woody components (Drewa and Havstad 2001). These factors can all be measured or predicted with a relatively high level of confidence. In systems in which precipitation is low and unpredictable, however, one of the most important factors is also among the least predictable: the weather conditions following the fire. Long periods without precipitation following a fire can result in increased erosion and mortality of the less deeply rooted grasses and perennial forbs. In addition, seldom do fuel loads on the northern Chihuahuan Desert meet minimum levels of 600 kg/ha that have been recommended by Wright (1980) for desert grasslands.

Results of two sets of experiments in mesquite-invaded, black grama grasslands showed that fire reduces mesquite shrub volume but does not result in shrub mortality. Four years following fire in 1995, perennial grass cover was 13% lower in burned 8 × 12 m plots and 5% higher in unburned plots, and frequency decreased 30% in the burned areas compared to a 10% increase in unburned areas (Drewa and Havstad 2001). Black grama mortality was 45%, 27%, and 19% for small, medium, and large clumps one year after a 400-ha burn in 1999, compared with 11%, 4%, and 3% in 4-ha unburned controls. Fire conditions were nearly ideal with high winds during the May fire immediately followed by unseasonably early rains in June. Shrub volumes declined 40% in the burned areas and increased 30% in the unburned controls, but there was little evidence of shrub mortality (Drewa et al. 2001). The results of both studies together with the relatively high costs of propagating fire in these patchy arid environments imply that although fire may help slow shrub invasion, benefits, costs, and potential risks should be carefully considered before application to large areas.

Future Scenarios

Restoration of native grasslands and other plant communities may be limited by one or more of the following factors: (1) invasion of highly competitive, persistent species; (2) loss (both documented and undocumented) of plant, animal, and microbial species from the system; (3) loss of soil and/or modification of dynamic soil properties; (4) infrequency of suitable establishment periods with adequate soil moisture; (5) landscape level processes that overwhelm small-scale manipulations; and (6) those sites that are the most resilient will not necessarily be the most resistant to future degradation. Successful strategies in the future will need to address each of these limitations. We also need to find indicators that can be used to predict future trends in systems in which persistent changes in perennial vegetation may not appear for decades. To increase the probability of success, we need to focus more on restoring the resistance and resilience of soils and plant and animal communities rather than on short-term similarities to a particular community structure or composition.

Future success will also depend on recognizing that neither the soils nor the climate nor the faunal community are the same today as they were during the latter part of the nineteenth century when shrub invasion into grasslands began to accelerate. They are likely even more different than they were when the perennial grasslands became established. Loss of soil and animal species and additions of new species to the system, together with climate change, may mean it is no longer possible to reestablish some plant communities or that it may be possible to reestablish them only in selected parts of the landscape.

Figure 14-4 illustrates one possible approach to future remediation in which limited inputs are targeted to parts of the landscape with a high potential for change, triggering change at that site and, ultimately, change in the surrounding area (Herrick et al. 1997). A first step to identifying these high-potential areas is

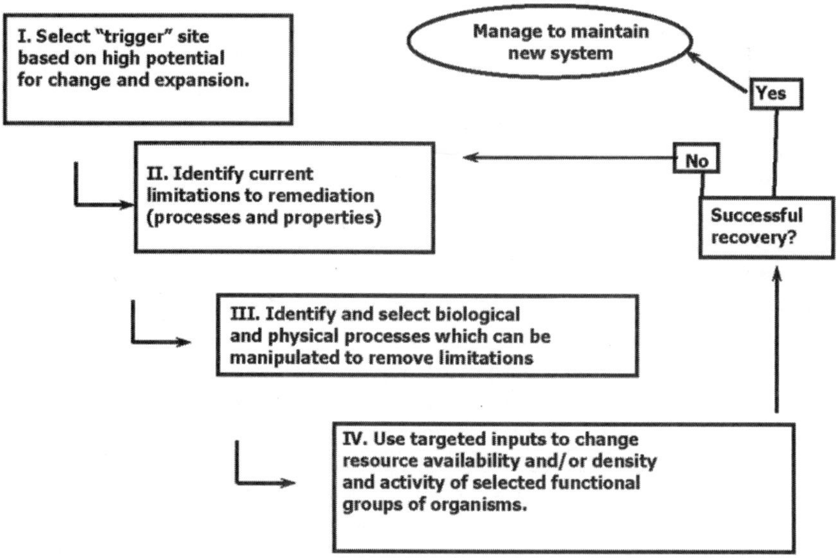

Figure 14-4. Low-input remediation system focusing on removing limitations in parts of the landscape with high potential for recovery (modified from Herrick et al. 1997).

to eliminate areas from consideration that have clearly crossed a threshold beyond which recovery is unlikely (Chambers 2000). Similar ideas are described in greater detail in Whisenant (1999). Based on historic documents, the concept was familiar to Jornada scientists in the 1930s. The authors of a 1939 report argued that although many of the intensive revegetation approaches "would be uneconomical for general application . . . certain key or strip areas could be so treated on the basis that the entire area would in time be improved by natural spread from the treated portions" (JER Staff 1939). The concept of targeting areas with naturally higher resource availability was resurrected by Herbel (1982) in a proposal to focus herbicide-based brush control on run-in areas, while leaving shrub-dominated upland areas to produce runoff. The argument was based both on the low benefit-cost ratio for treating the upland, runoff-producing areas, the higher returns for the run-in areas, and the recognition that maintaining runoff from the upland areas could be used to increase production on the lower areas.

It would appear from the location and design of some of the historic treatments, however, that the trigger site approach was tested numerous times during the 1930s. And it failed. The selection of potential trigger sites may be improved with technology and an enhanced understanding of processes operating at multiple scales. Nevertheless, comments in the historic records, supported by our observations and recent landscape-level analyses (see chapters 2 and 4) suggest that application of the trigger site concept is limited by loss of native species, historic soil degradation, potential threats of invasive species, and the infrequency of suitable establishment periods. It is also potentially limited by the tendency for both native and domesticated herbivores to converge on areas with the highest quality forage, as illustrated by rabbit pellet and cattle dung patch counts in small patches (less than 5 ha) in which perennial grasses have been successfully established (Fuhlendorf and Smeins 1997).

The relative importance of resistance and resilience also needs to be considered. Sites that are most resilient are not necessarily most resistant to future degradation (Seybold et al. 1999). Degraded riparian zones in the Southwestern United States, for example, recover relatively quickly compared to upland systems when livestock, and sometimes elk (*Cervus*), are excluded except when invaded by salt cedar (*Tamarix*). However, remediated systems are often not very resistant to subsequent overgrazing until the woody species have grown beyond the reach of livestock, though much of the research in this area is anecdotal (Sarr 2002). For example, the location of riparian sites in lower parts of the landscape also makes them susceptible to more intense flooding caused by upslope watershed degradation than they would have experienced before they were degraded. In this case, their resistance may have recovered; but a change in the disturbance regime (flooding) may affect their ability to function as they did before the uplands were degraded.

All of these limitations can potentially be overcome through a combination of: (1) careful analysis to identify the factors and processes that are most likely to limit establishment and survival at a particular time and location; (2) patience and a flexibility to initiate interventions when they are most likely to be successful, rather than when funding and logistical support are available; and (3) attention to landscape-level controls. Creative integration of multiple approaches, including short- and long-term experiments and monitoring, gradient analyses and descriptive studies, and conceptual, empirical, and theoretical modeling, will be necessary to develop effective remediation strategies based on an understanding of key ecological factors and processes (Archer and Bowman 2002).

Relevance to the Southern Chihuahuan Desert and Other Deserts

The basic patterns and processes described here are similar to those described for many other parts of the world. Roundy and Biedenbender (1995) draw similar conclusions based on their review of literature primarily focused on Arizona. Ballín Cortés (1987) identified a similar suite of limitations to the recovery of ecosystems in his analysis of desertification in the southern Chihuahuan Desert at approximately 21° N latitude. Lovich and Bainbridge (1999) concluded their assessment of the potential for restoration of southern California deserts by stating that "restorative intervention can be used to enhance the success and rate of recovery, but the costs are high and the probability for long-term success is low to moderate." The recommended strategy for the future is similar to that proposed by Whisenant (1996) and Tongway and Ludwig (1997), based on their experiences in central Texas and Australia, respectively.

Conclusions

Early investigators had an intuitive and practical understanding of the system in which they worked. They saw the individual limitations to grassland recovery and attempted to address them. To be more successful than they were, we must begin to work at spatial and temporal scales that are relevant to the processes we hope to affect and target interventions to those locations during those periods when the processes are most susceptible to change. We must also, as the earlier workers did, simultaneously target multiple processes with the objective of increasing the resistance and resilience of the modified ecosystems.

The role of societal values in defining remediation objectives and of economics in defining restoration success has been referred to several times in this chapter: remediation efforts through the 1980s were primarily designed to produce more forage for livestock (a societal benefit) and to reduce soil erosion (a cost). The success of remediation efforts, however, has been generally defined usually in terms of the net economic benefit to livestock producers. Evolution of societal values, human population growth, and the associated redistribution of financial resources will inevitably lead to shifts in the ways that costs and benefits of remediation efforts are evaluated.

Social and political values help define how ecologists view ecosystems, what we decide to emphasize in our research, and how we describe the results of our research. In some cases, the relationships are direct: Societal interest in restoring or maintaining grasslands to reduce soil erosion to improve air and water quality or to produce more livestock forage explain why we know much more about how to establish exotic grasses than native sagebrush in the Great Basin (Young et al. 1981). In other cases, there is a more subtle effect on where ecologists look first for answers to explain the success or failure of an intervention. For example, there is an ongoing debate on the relative importance of changes in root versus fungal biomass (Hodge 2001). Both are difficult to measure accurately, and biases can easily result in the selection of measurements, techniques, or levels of replication that favor one variable over another.

The successful future application of basic and applied ecological research to remediation depends on the ability and willingness of ecologists to include a discussion of the assumptions and potential biases that drove the selection of the research questions, experimental design, and interpretation of the results. Failure to do so may result in a repetition of the past in which single- or dual-factor treatments applied at a single location and single point in time were overwhelmed by complex interactions, many of which remain unidentified or at least unquantified.

15

Applications of Remotely Sensed Data from the Jornada Basin

Albert Rango
Jerry Ritchie
Tom Schmugge
William Kustas
Mark J. Chopping

Like other rangelands, little application of remote sensing data for measurement and monitoring has taken place within the Jornada Basin. Although remote sensing data in the form of aerial photographs were acquired as far back as 1935 over portions of the Jornada Basin, little reliance was placed on these data. With the launch of Earth resources satellites in 1972, a variety of sensors have been available to collect remote sensing data. These sensors are typically satellite-based but can be used from other platforms including ground-based towers and hand-held apparatus, low-altitude aircraft, and high-altitude aircraft with various resolutions (now as good as 0.61 m) and spectral capabilities. A multispectral, multispatial, and multitemporal remote sensing approach would be ideal for extrapolating ground-based point and plot knowledge to large areas or landscape units viewed from satellite-based platforms. This chapter details development and applications of long-term remotely sensed data sets that are used in concert with other long-term data to provide more comprehensive knowledge for management of rangeland across this basin and as a template for their use for rangeland management in other regions.

In concert with the ongoing Jornada Basin research program of ground measurements, in 1995 we began to collect remotely sensed data from ground, airborne, and satellite platforms to provide spatial and temporal data on the physical and biological state of basin rangeland. Data on distribution and reflectance of vegetation were measured on the ground along preestablished transects with detailed vegetation surveys (cover, composition, and height); with hand-held and yoke-mounted spectral and thermal radiometers; from aircraft flown at different elevations with spectral and thermal radiometers, infrared thermal radiometers, multispectral video, digital imagers, and laser altimeters; and from space with

Landsat Thematic Mapper (TM), IKONOS, QuickBird, Terra/Aqua, and other satellite-based sensors. These different platforms (ground, aircraft, and satellite) allow evaluation of landscape patterns and states at different scales. One general use of these measurements will be to quantify the hydrologic budget and plant response to changes in components in the water and energy balance at different scales and to evaluate techniques of scaling data.

Our concept involves overlaying various remote sensing capabilities at the same time in an area that has been studied for a long time using conventional approaches and where ground measurements can be taken at the time of the remote sensing observations. Jornada Basin study sites are ideal for this approach. The same types of data are acquired from different platforms with varying detail of measurement and with the objective of developing a way to spatially scale up or down and provide increased information content. Data have covered parts of both the Jornada Experimental Range (JER) and the Chihuahuan Desert Rangeland Research Center (CDRRC).

Conventional and Remote-Sensing Ground Measurements

These data were collected for comparison with aircraft- and satellite-measured radiance and reflectance. Sites (figure 15-1) were established in each of the five major vegetation types: black grama (*Bouteloua eriopoda*) upland grasslands, tobosa (*Plueraphis mutica*)-dominated playas, creosotebush (*Larrea tridentata*) shrublands, mesquite (*Prosopis glandulosa*) dune systems, and tarbush (*Flourensia cernua*) shrublands (as used for aboveground net primary production estimates as described in chapter 11). An additional site was located in an area of transition between the black grama grassland and the mesquite dune system. At these six sites both a 150-m permanently marked linear transect and a 30− × 30-m, 49-point square grid with grid points 5 m apart were established with remote sensing overflights in mind. Linear transects were used for vegetation surveys, leaf area index observations, and hand-held, full-range spectroradiometer measurements. Vegetation measurements were made of species cover, maximum plant height, and standing litter using vertical line-point intercept techniques (Canfield 1941; Eberhart 1978). In general, as ground cover decreased from grass to transition to mesquite communities, reflectance measured from all instruments increased, indicating a potential for change in heat and water balance of these ecosystems if shrubs continue to expand into the grass communities. These data suggest that changes of vegetation from grass-dominated to mesquite-dominated communities in the Jornada Basin could have significant effects on the albedo and the surface temperatures measured at the different sites and on the overall water and energy budget within the basin.

Leaf area index (LAI) measurements were made with a LICOR LAI-2000 instrument. Measured LAI has varied greatly over the time period (table 15-1) and trends are difficult to determine. In general, LAI is higher at the grass site in the spring (May) than in the fall (September/October). At the transition site the reverse is generally true; namely, the LAI is higher in fall than in spring. The

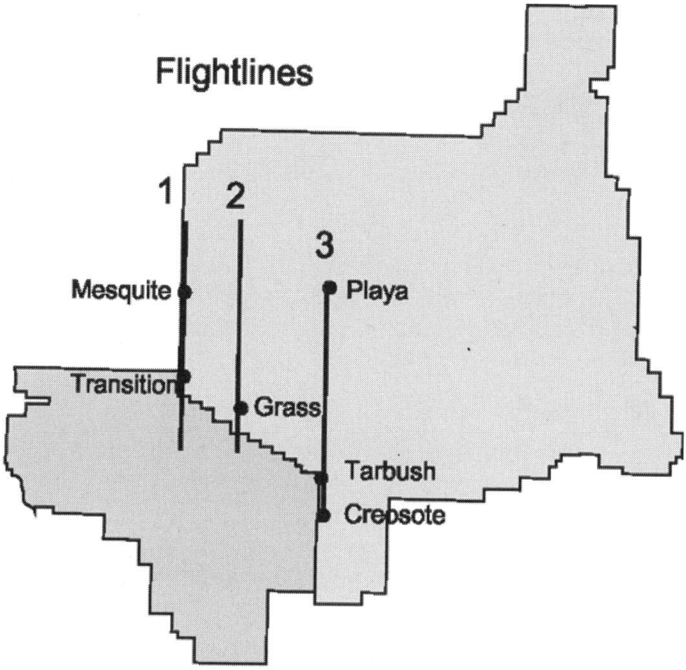

Figure 15-1. JER and CDRRC map with locations of six sites and the primary low-altitude aircraft flight lines.

mesquite site LAI has no consistent pattern, whereas the creosote site, over a more limited number of years, generally has a greater LAI in spring. LAI was low at all sites in September 2001. Variability in the range of LAI has increased since 1995 with great fluctuations in measurements in recent years.

Radiometric plant canopy and soil reflectance measurements were made with a Barnes modular multispectral radiometer at the grass, transition, and mesquite sites in September 1995 and September 1996 (Everitt et al. 1997). Spectral radiance measurements were made between 1100 and 1400 hours under sunny conditions. Radiometric measurements were converted to reflectance for a common solar irradiance reference condition (Richardson 1982). Additionally, multispectral measurements were made using an Exotech four-band radiometer, corresponding to the first four bands of Landsat TM and mounted on a shoulder-configured yoke system to measure the 30- × 30-m grid points and along the transect. A thermal radiometer was also mounted on the yoke.

Spectral radiance of vegetation communities were measured at 1 m above ground level using an Analytical Spectral Device (ASD) spectrometer at the 30- × 30-m grids at 5-m intervals and along the vegetation transects. The ASD makes measurements from 0.350–2.500 μm at 0.001-μm intervals. At each study site, separate radiometric measurements were made for each frequently encountered plant species, litter, and bare soil (10 measurements of each selected cover type).

Table 15-1. Average leaf area index measure along 150-m transects within 6 vegetation types (see text) at the JER with an LAI-2000.

Date	Grass	Transition	Mesquite	Creosote	Tarbush	Playa
September 1995	0.77	0.93	1.03			
February 1996	0.99	0.80	0.52			
May 1996	0.96	0.68	0.91			
September 1996	1.32	1.58	0.95			
May 1997	0.71	0.69	1.17			
September 1997	0.68	0.70	0.72			
April 1998	0.54	0.41	0.38			
September 1998	0.30	1.15	1.15			
May 1999	0.83	0.22	0.25	2.18		
September 1999	0.83	1.06	1.54	1.69		
May 2000	0.47	0.66	1.66	1.27		
September 2000	0.31	0.97	1.25	0.75		
May 2001	1.36	1.37	1.89	2.13		
September 2001	0.51	0.71	0.37	0.66	1.54	
May 2002	1.27	0.90	1.91	1.89	1.62	
October 2002	0.85	0.35	0.50	0.80	0.82	
May 2003	0.68	0.49	1.30	0.93	1.29	1.96

ASD spectral radiance measurements made on the 30 × 30 m^2 grids at 5-m intervals for the grass, transition, mesquite, and creosote sites for May and September 2001 were averaged to simulate a 30-m^2 pixel. All sites have similar overall radiance patterns over the entire spectrum (0.350–2.500 μm).

Though the spectra patterns are similar, there are differences in the magnitudes of radiance measured over the different vegetative communities. Mesquite communities have the highest reflectance with the exposed soil cover fraction up to 0.75. The transition communities have the next highest radiance, followed by the grass communities. The creosote and tarbush communities are usually between the grass and transition communities. Both of these communities have significant grass cover between and under the creosote and tarbush shrubs.

Moderate Resolution Imaging Spectrometer (MODIS) and Advanced Spaceborne Thermal Emission and Reflection (ASTER) are multispectral sensors on Terra, NASA's Earth Observation System (EOS) satellite. MASTER is the MODIS-ASTER simulator and has 50 channels from the visible to the thermal infrared (TIR), again with 2.5 mrad IFOV and a swath of about 45°. A comparison of reflectance measurements from the ASD, MASTER, and ASTER instruments showed that reflectance increased from grass to transition to mesquite communities from all three platforms (Figure 15-2). Patterns of reflectance were similar from the three platforms with MASTER and ASTER having similar absolute values. Comparison of absolute values of ASD ground measurements with MASTER and ASTER measurements differed with different vegetation communities, but in general, ASD reflectance measurements were slightly lower than MASTER or ASTER measurements. These differences are assumed to be related to the different footprint sizes of the instruments and probably the lesser percent of base soil cover in the ASD footprint.

Ground-based multiangular reflectance measurements have been acquired as part of the NASA EOS Grassland PROVE (prototype validation exercise) (Privette et al. 2000) field campaign and from 1999–2002 as part of our program (Rango et al. 1996; Ritchie et al. 1996). The PROVE data sets were acquired by a team from the University of Nebraska-Lincoln using a short tilting mast to acquire off-nadir samples over the semi-natural grassland, soil, and shrubland surfaces. Our data sets were acquired primarily in support of aircraft multiangle campaigns using a similar apparatus. At times a boom truck with a 30-m-high extension capability was used to measure sky and surface radiance.

From 1995 to 2000, surface landscape (soil and vegetation) temperatures were

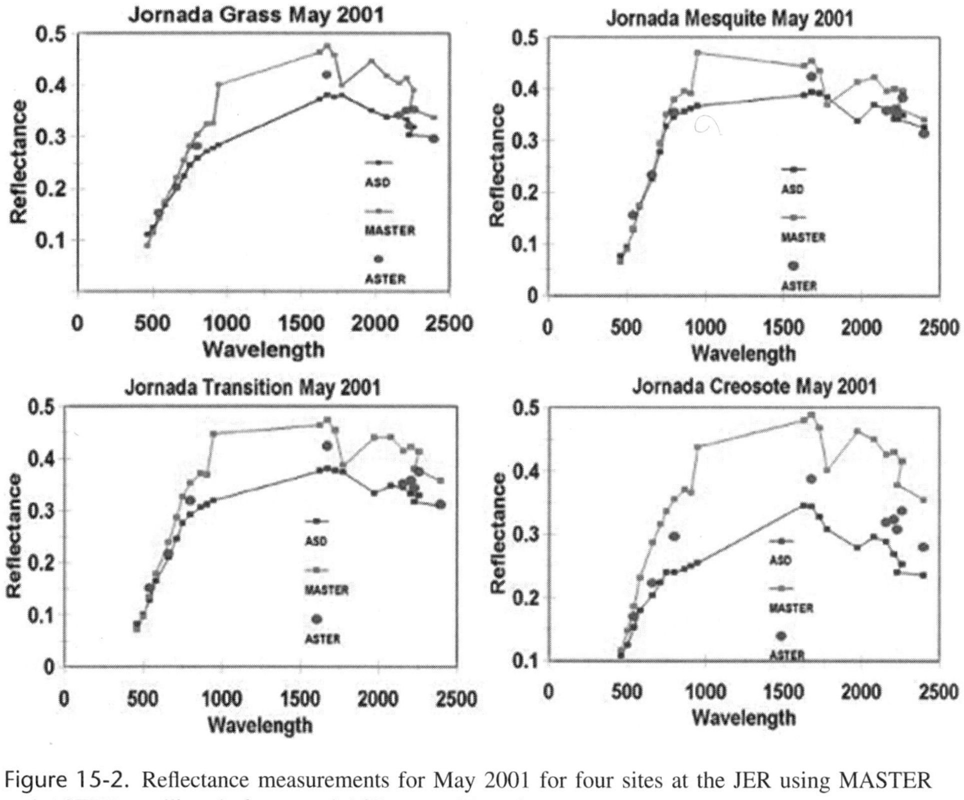

Figure 15-2. Reflectance measurements for May 2001 for four sites at the JER using MASTER and ASTER satellite platforms and ASD ground based measurements.

measured with an Everest thermal infrared radiometer (IRT) with a band pass of approximately 8–13 μm. Since 2000, temperature measurement has been made with an Apogee radiometer recorded on a Fluke thermometer.

Surface energy fluxes have been monitored on a nearly continuous basis from 1995 to 1997 using the Bowen ratio energy balance method (BREB). During the episodic field campaigns from 1997 through 2002 and continuously since May 2002, they have been monitored with eddy correlation (EC) and Bowen ratio systems primarily at grass and mesquite locations.

The Radiation and Energy Balance System (REBS) surface energy balance system (SEBS) was installed primarily at grassland and mesquite dune sites to measure energy flux, weather, and supplementary data. The SEBS is an integrated system of sensors designed for the U.S. Department of Energy's Atmospheric Radiation Measurement program. The net radiation (R_n) was measured using a REBS Q7 net radiometer positioned between 2 and 3 m above the surface. Air temperature and vapor pressure differences were measured with modified Vaisala HMP35A temperature-humidity probes with a 2-m separation.

Wind speed and direction were measured using a Met One anemometer and wind vane located at a nominal height of 3 m above the local topography. Atmospheric pressure was measured using a Met One barometric pressure sensor at one of the sites. Soil moisture was measured at over 5 cm depth using three soil moisture resistance sensors (REBS SMP-2). This value of soil moisture was used with an estimate of bulk density to compute the soil heat capacity, which is then used with a time rate of change in soil temperatures.

EC measurements were made using a one-dimensional Campbell Scientific CA27 with fine-wire thermocouple and, in later years, using a three-dimensional Campbell Scientific CSAT3 sonic anemometer to measure vertical wind speed (w) and air temperature (T) and a KH20 krypton hygrometer to measure vapor density (ρ_v) (Tanner et al. 1985; Tanner 1988).

Preliminary analyses of the BREB and EC data indicate significant discrepancies in flux estimates between the two techniques, particularly for the mesquite dune site where heterogeneity (distances between dune and mesquite clumps) exists at scales of tens of meters (Hipps et al. 1977; Ramalingam 1999). In general, H estimated by the BREB method tended to be significantly higher than when measured by the EC method; interestingly, the λE-values were somewhat smaller. This results in the BREB method estimating a larger value of the Bowen ratio β (= $H/\lambda E$) than given by the EC system.

The BREB method assumes that available energy ($R_n - G$) can be measured without error. For uniform surfaces with large values of vertical gradients, the Bowen ratio technique works well. However, there are serious problems for heterogeneous surfaces because some of the assumptions are simply not valid (Hipps and Kustas 2001). In fact, a recent study by De Bruin et al. (1999) suggests that many environmental and landscape conditions existing in desert environments are likely to cause a breakdown in temperature-humidity similarity and, hence, in the equality of the eddy diffusivities. However, differences in flux estimates between EC and BREB methods over heterogeneous terrain are usual. For example, similar discrepancies were observed over more uniform grasslands (Fritschen et al. 1992).

Efforts continue to evaluate and understand the factors causing discrepancies between the turbulent heat fluxes estimated by BREB and EC methods, but there have also been detailed studies of the uncertainty in measuring R_n and G for these desert communities. Not only is the available energy a necessary quantity for computing the heat fluxes using the BREB method, it is also a common approach for gaining a level of confidence in the turbulence flux measurements by assessing the closure ratio (CR) defined as $(H + \lambda E)/(R_n - G)$. The value of CR is generally less than 1 and may be due to a number of factors, including sensor measurement errors, horizontal flux divergence, and a significant mismatch in source areas of the measured fluxes (Stannard et al. 1994). For these sparsely vegetated surfaces, the mismatch in source areas is likely to be extreme between the measurement of available energy and the heat fluxes.

For net radiation, Kustas et al. (1998) showed that different models of radiometers from the same manufacturer required recalibration to achieve an uncertainty of less than 5% for this particular environment. Net radiometers from different manufacturers could also be made to agree under a dry condition by using regression or autoregression techniques. However, the resulting equations would induce bias for the wet surface condition. Thus, it is not possible to cross-calibrate two different makes of radiometers over the range of environmental conditions observed. This result indicates that determination of spatial distribution of net radiation over a variable surface should be made with identical instruments that have been cross-calibrated. However, the need still exists for development of a radiometer and calibration procedure that will produce accurate and consistent measurements over a range of surface conditions (Kustas et al. 1998).

Kustas et al. (2000) used an array of 20 soil heat flux plates and soil temperature sensors to characterize the spatial and temporal variability in soil heat flux as affected by vegetation and microtopographic effects of mesquite dune communities. Maximum differences in soil heat flux among sensors were nearly 300 W/m^2. Maximum differences among individual sensors under similar cover conditions (i.e., no cover or interdune, partial or open canopy cover, and full canopy cover) were significant, reaching values of 200–250 W/m^2. The "area average" soil heat flux from the array was compared to an estimate using three sensors from the nearby BREB station. These sensors were positioned to obtain soil heat flux estimates representative of the three main cover conditions, namely, no cover or interdune, partial or open canopy cover, and full canopy cover. Comparisons between the "array average" soil heat flux and the three-sensor system indicate that maximum differences on the order of 50 to nearly 100 W/m^2 are obtained in the early morning and midafternoon periods, respectively. These discrepancies are caused by shading from the vegetation and microtopography. The array-derived soil heat flux also produced a significantly higher temporal-varying, soil heat flux/net radiation ratio than that observed in other studies under more uniform cover conditions. Results from this study suggest that microtopography, as well as clustering of vegetation, needs to be accounted for when determining the number and location of sensors needed for estimating area average soil heat flux in this type of landscape.

Future investigations will include source-area footprint analyses (e.g., Schuepp et al. 1990; Schmid and Oke 1990; Horst and Weil 1992) to assess surface properties of the upwind fetch affecting the measurements. High-resolution remote sensing data will be critical for evaluating the surface properties within each of the source-area footprints. In addition, the 20-Hz turbulence data collected from recent experiments in 2000 and 2002 will be used to explore the applicability of the Monin-Obukhov similarity theory (MOST), relating mean and variance to covariance statistics (i.e., flux-gradient and flux-variance similarity) of temperature, humidity and momentum (wind) for the different desert communities. This work will ultimately prove useful for assessing the utility of MOST because land–atmosphere transfer schemes used in atmospheric models and remote-sensing energy balance approaches rely on MOST parameterizations (Hipps and Kustas 2001).

Low-Altitude Aircraft Missions

Laser profile altimetry measurements were made on four north–south and four east–west flight lines designed to cross the three study sites. Laser altimetry flights were made in May 1995, September 1995, February 1996, and May 1997. Flights were made at an altitude of approximately 200 m AGL. The altimeter was a pulsed gallium-arsenide diode laser transmitting and receiving 4,000 pulses per second at a wavelength of 0.904 μm. The field of view of the laser was 0.6 milliradians, which gives a footprint on the ground that is approximately 0.06% of the altitude. The timing electronics of the laser received allowed a vertical resolution of 5 cm for each measurement.

Laser altimeter–measured transects at the black grama grass, mesquite, and transition sites show differences in surface topography and roughness at the sites. The grass site (figure 15-3) is relatively uniform in surface roughness with an occasional shrub or taller vegetation on the surface present on the underlying topography. The mesquite site has evidence of dunes present on the underlying landscape with vegetation on top of the dunes. The transition site shows similarities to both the grass and mesquite sites with the beginning of what appear to be small dunes. Fractional vegetation cover and vegetation height measured from the laser altimeter data were comparable to measurements made on the ground (Ritchie 1996).

Fractal analysis of laser altimetry data from the grass, mesquite, and transition sites (four transects at each site for February, May, and September) supports the possibility of distinguishing between these landscapes using fractal properties of the laser data. A specific range of scales of fractal dimension can be used for distinguishing between grass, mesquite, and transition landscapes. The fractal dimensions tend to increase in a sequence from grass to transitional to mesquite sites in this range of scales. A fractal technique for estimating cover type represents a new technical opportunity to quantify landscape roughness (Pachepsky et al. 1997; Pachepsky and Ritchie 1998).

Studies have also shown (Menenti and Ritchie 1994; Menenti et al. 1996) that

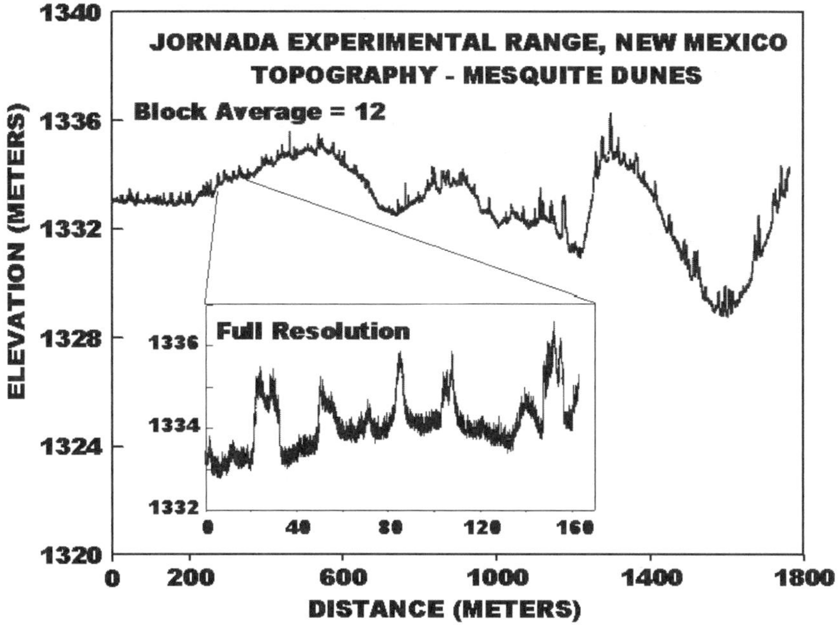

Figure 15-3. Laser altimeter measurement of topography and surface roughness at the grass site on May 19, 1995, at the JER.

the effective aerodynamic roughness at the shrub site can be estimated using the high-resolution laser altimeter measurements of land surface roughness. Studies at the mesquite site on the Jornada show that estimations of aerodynamic roughness and displacement height of a complex terrain, consisting of coppice dunes with bare interdunal areas, are plausible using simple terrain features computed from high-resolution laser altimeter data (De Vries et al. 1997, 2003).

A scanning laser altimeter was flown over the sites in February 1998. The laser instrument was the Swedish Saab/TopEye system flown on a helicopter platform using an across-track scanning system with a Z-shaped ground target path and with the along-track sampling rate determined by the speed of the aircraft. The wavelength of the laser is in the near infrared at 1.064 μm. The instrument operates in a number of modes, providing a sampling density adjustable from 1–15 samples per m2 and with swath widths from 21–168 m, corresponding to minimum and maximum footprints of 0.06 and 3.84 m. The laser pulse has a frequency of 7 Hz, and up to five returns are recorded per sounding. The maximum scan angle is 20°, reducing but not entirely eliminating the impact of relief displacement, shadowing, and layover.

As a recent improvement over the profiling laser, namely, Rango et al. (2000) found the potential of moderate sampling density (1–2 m) scanning laser technology for providing accurate models of shrub-coppice-dune morphology to be viable, with excellent matches between a DEM from simple three-point interpolation and transit data from engineering surveys.

A twin-engine research aircraft has been used to collect airborne spectral (Exotech), thermal, three-band multispectral video, digital imagery, laser altimeter, and bidirectional reflectance data. Airborne campaigns have been scheduled to make measurements centered on the date of an overpass of Landsat or EOS-Terra. GPS navigation (Trimble Transpack II) was integrated with the systems on the airplane to measure the flight direction (bearing), altitude, time, ground speed, latitude, and longitude coordinates. Video imagery was obtained with a three-camera, multispectral, digital video imaging system (Everitt et al. 1995). The three cameras are visible near infrared (0.4–1.1 μm) equipped with yellow-green (0.555–0.565 μm), red (0.623–0.635 μm), and NIR (0.845–0.857 μm) filters. Imagery was acquired at altitudes between 300 and 3,000 m AGL, between 1000 and 1400 local time, under clear conditions during each campaign.

Since 2000, digital images have also been captured with a high-speed, multiband, digital capture camera designed and built by DuncanTech. It consists of a single F mount with three CCD arrays filtered by filters centered on 0.550, 0.650, and 0.800 μm with band widths of 0.070, 0.040, and 0.065 μm (Schiebe et al. 2001).

Airborne measurements of surface temperature were made with an Everest thermal sensor having a 15° field of view. An Exotech four-band radiometer was used to make radiance measurements corresponding to the first four bands of the Landsat Thematic Mapper. The IRT and Exotech were mounted looking nadir. A second Exotech was placed on the ground looking upward to measure irradiance. A color video camera, borehole-sighted with the IRT and Exotech, recorded color images of the flight line. Each video frame was annotated with GPS data. Airborne data from these instruments were collected for the three flight lines at the approximate time of the Landsat or ASTER overpass. Each flight line was approximately 10 km long. Flights were made at altitudes of approximately 125 m and 300 m AGL with passes in opposite directions at each altitude on each flight line.

The aircraft flew an inexpensive hyperspectral imaging system (Yang et al. 2001) during fall 2002. The effective spectral range of this system is 0.467–0.932μm in 32–1,024 bands. This low-altitude hyperspectral capability should provide some valuable data for comparison with high-altitude, AVIRIS hyperspectral data.

High-Altitude Aircraft Data

An improved camera for producing vertical aerial photographs with minimal distortion was developed by Sherman Fairchild in 1917 for the U.S. military (Thompson and Gruner 1980). By the early 1930s, the USDA started to systematically photograph agricultural lands in all states. Black-and-white aerial photographic coverage over parts of the Jornada Basin began in 1935. Aerial coverage has continued to the present in association with nationwide mapping projects or various remote-sensing research projects where color infrared photography is flown for comparison with new types of remote-sensing instrumentation.

To cover areas of the Jornada Basin other than the grids and transects, data have been acquired by aircraft since 1997. The flight lines were chosen to cover mesquite, transition, grass, creosote, tarbush, and playa test sites as shown in figure 15-1. In 1997 the airplane flew both the Thematic Mapper Simulator (TMS) and the Thermal Infrared Multispectral Scanner (TIMS) instruments. TMS measures radiances in 12 bands ranging from the visible to thermal infrared; 7 bands are identical to the 7 bands of the Landsat TM. TIMS has six channels in the TIR, 8–12-μm region of the electromagnetic spectrum.

The MASTER VIS, NIR, and TIR data were corrected for atmospheric effects using the MODTRAN path radiance model with atmospheric profiles appropriate for the conditions. All the profiles were adjusted for the surface temperature and humidity conditions observed by weather station data from the Jornada.

Even higher altitude aircraft coverage is periodically provided by the NASA ER-2 aircraft at an altitude of 20 km flying both the Airborne Visible/Infrared Imaging Spectrometer (AVIRIS) and the RC-10 mapping camera. The AVIRIS instrument has 224 detectors covering the entire spectral range from 0.380 to 2.500 μm. These hyperspectral data are useful for vegetation mapping and can be compared with the ground-based ASD measurements. The RC-10 camera can also provide color infrared photography over the same area imaged by AVIRIS at 3–8-m resolution to compare with the multispectral 210-m resolution AVIRIS data.

Satellite Remote Sensing

Landsat was the pioneering Earth resources technology satellite, and it has been a very reliable platform. At present, Landsat 7 with the Enhanced TM + is operable. It provides coverage in six visible, near infrared, and mid-infrared channels with 30-m resolution; one thermal channel with 60-m resolution; and one panchromatic band with 15-m resolution. Because Landsat has been in existence since 1972 (in various instrument and wavelength configurations), it is treated as a standard against which other sensor products are compared.

With a ground-projected field of view between NOAA-AVHRR and Landsat TM, the MODIS instrument on the Terra satellite (and now the Aqua satellite also) is now providing coverage of earth every one to two days, acquiring data in 36 spectral bands. Two visible/near infrared bands nominally acquire data at 250-m resolution at nadir, 5 bands acquire data at 500-m resolution, and 29 bands (including the thermal bands) possess a 1-km resolution capability. The MODIS data are freely available through a number of Web sites, but some important data corrections are necessary for analysis.

ASTER radiometer (Yamaguchi et al. 1998) also flies on Terra and has three subsystems to collect data: the visible and near infrared (four bands, 15-m resolution), the shortwave infrared (six bands, 30-m resolution), and the TIR (five bands, 60-m resolution). A stereo capability is available from ASTER in the visible and near infrared. The thermal coverage, however, is the most interesting because of its multiple bands and high resolutionm which allows estimation of surface emissivity.

The Ikonos satellite system (named from the Greek word for *image*), which was launched in 1999, operates in a multispectral mode (visible and near infrared bands) with a resolution of 4 m and in a panchromatic mode with a resolution of 1 m. The swath width of data acquisition is limited to 11 km. It has great flexibility by being able to point off-nadir to acquire data.

More recently, QuickBird was launched in 2001. It is similar to Ikonos but has an improved multispectral spatial resolution of 2.44 m and a panchromatic resolution of 0.61 m. It has a swath width of 16.5 km and an off-nadir pointing capability. It is usually possible to get synchronous coverage over the Jornada test sites on the same day. NOAA-AVHRR, MODIS, and Landsat overpasses follow in quick succession, and ASTER can be scheduled to acquire data because Terra is already overhead at the same time. Although their revisit times are variable, data from Ikonos and QuickBird are possible at the same time (plus or minus a day) using their pointing capability.

Multiangular Reflectance Measurements and Thermal Infrared Analysis

The bidirectional reflectance distribution function describes the angular distribution of scattered light from a surface with respect to that of solar irradiance on the surface. Thus, it is a potential source of both noise and information on the surface. It is accessed by sampling the bidirectional reflectance using sensors that view off-nadir and/or at varying sun zenith angles, these data often being called "multiangle," "multiple-view angle," or "multiangular."

One of the objectives for the multispectral, thermal, infrared data from ASTER is to obtain both the surface temperature and multispectral emissivity from the five bands. The approach was demonstrated using aircraft data (Schmugge et al. 2001). The emissivity results using ASTER data from four Jornada sites for dates from May 9, 2000, to May 31, 2002, are given in figure 15-4. There is good agreement amongst the ASTER results for all five channels for both the mesquite and transition sites. At the mesquite site, we made field measurements with the CIMEL 312 (Legrand et al. 2000) radiometer and found there was good agreement between the ground measurements for the dominant dark soil and the ASTER results. As seen in figure 15-4, there was reasonable consistency among the ASTER results for the five different days.

The results from the grass and creosotebush sites show reasonably good agreement for the three dates in 2000 and 2001. However, the two dates in 2002 show consistently lower emissivity values, especially for the short wavelength (8–9 μm) channels. We suspect that this difference could be due to lower vegetation cover resulting from the drier than normal conditions during 2002.

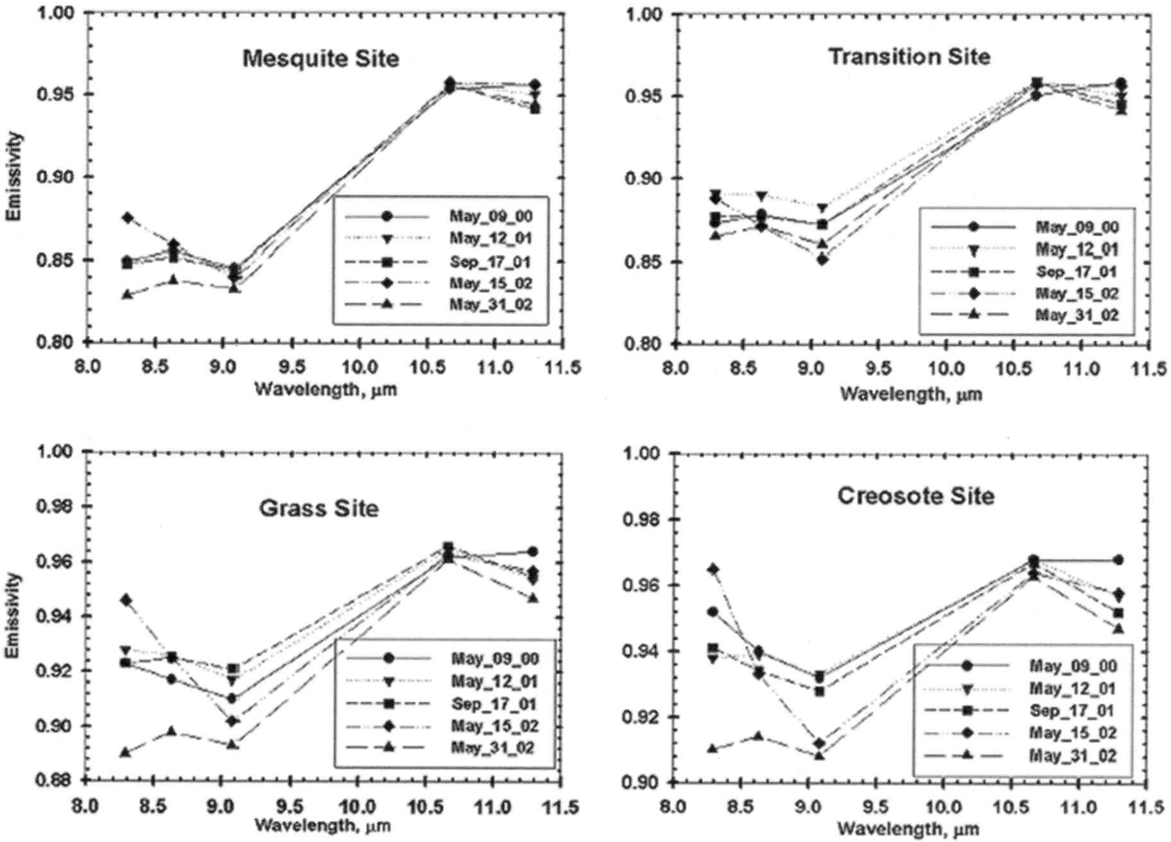

Figure 15-4. ASTER emissivity results for four sites at the JER from five observation dates between May 9, 2000 and May 31, 2002.

Ground measurements of surface brightness temperature, T_B, are summarized in figure 15-5 for the four dates on which we have nearly coincident ground measurements and an ASTER overpass. The ground data are presented in box plots at the times bracketing the overpass (i.e., 1130 and 1230 MDT) or coincident with the overpass (about 1200 MDT) for the four sites. The box shows the range for 50% of the observations, and the bright line in the box indicates the median for the data. The lines represent outliers in the measurements beyond that expected for a normal distribution (indicated by the brackets). The range of T_B for the five ASTER channels is presented by the up-and-down triangles. It was expected that the broadband emissivity for the ground measurements would be within the range of those for the five ASTER channels and, thus, the average ground T_B would be within the ASTER T_B range. It can be seen that there is reasonable agreement, but the ASTER data are a bit cooler. For the May 12, 2001, case, data from the aircraft instrument, MASTER, were also acquired. The range of MASTER T_Bs is indicated by the triangles to the left of the box. The channels represented by this range cover approximately the same wavelength range as that covered by the ASTER channels. The agreement of MASTER and ASTER is quite good.

These results indicate that a temperature emissivity separation algorithm developed for use of ASTER data appears to work as well with the data from space as it did with the aircraft data presented earlier (Schmugge et al. 2001). This is encouraging for the application of the technique for mapping emissivity over large areas.

Large Area Coverage

Two important capabilities are desirable for rangeland research in the Jornada Basin. The first is high-resolution, remote-sensing coverage over the entire basin or over large parts of the basin. Aircraft aerial photography is, of course, a possibility. Unfortunately, to acquire high-resolution data (< 1 m), the airplane needs to fly at low to moderate altitude. As a result, many frames of aerial photography would be required to cover both CDRRC and JER and even more to cover the entire Jornada Basin. It is probably more efficient to use 0.61-m, panchromatic (or 2.44-m multispectral) resolution of the commercial QuickBird satellite to get the high-resolution coverage desired. It takes at least three swaths of the QuickBird multispectral sensor to get complete coverage of the Jornada Basin. The excellent resolution, however, may outweigh any difficulties in working with several different swaths, especially when rangelands are being assessed. Because this is a commercial satellite product that would require numerous swaths to cover the entire Jornada Basin, the total cost of the image acquisition may become too expensive. With the drawbacks of getting both aerial photography or QuickBird data when needed and at a reasonable price, alternatives like the use of Unmanned Aerial Vehicles (UAVs) are currently being explored. Acquisition of high resolution images by UAVs has certain advantages over piloted aircraft missions including lower cost, increased safety, flexibility in mission planning, and closer proximity to target areas. UAV images at Jornada have been used to produce

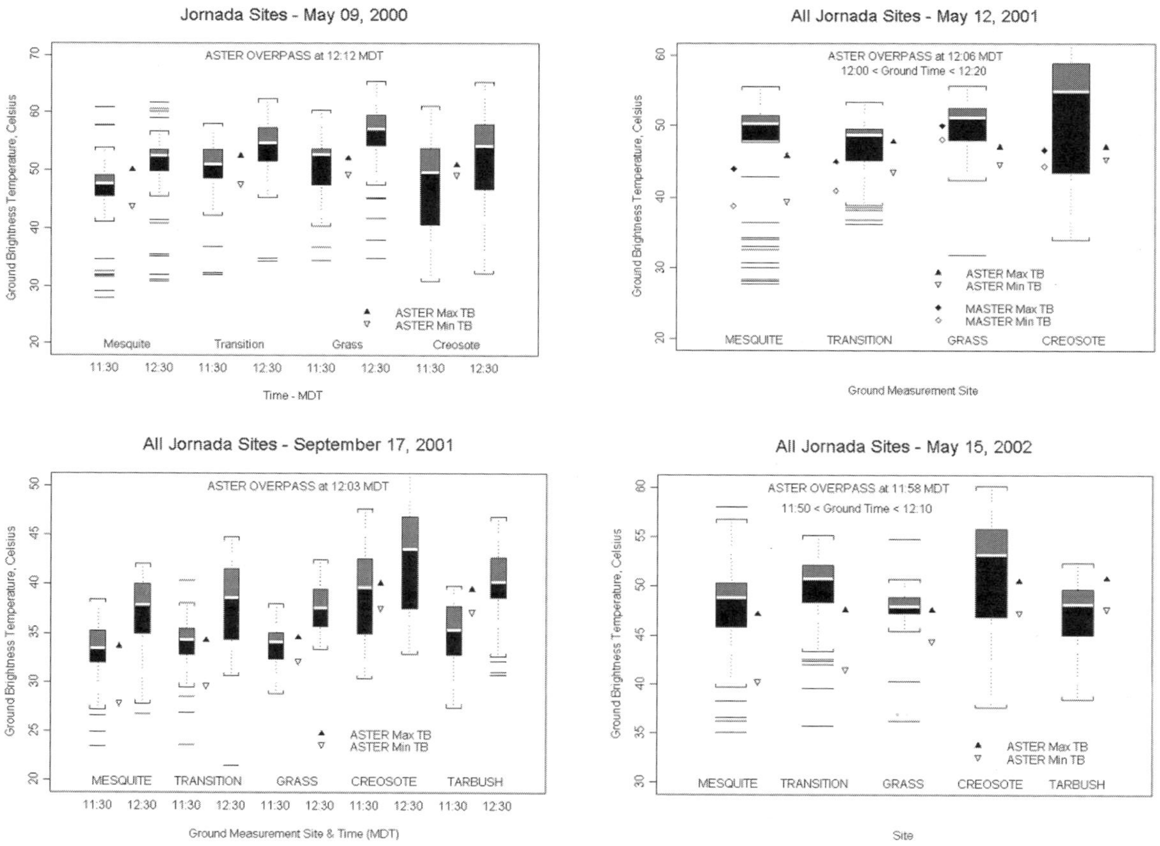

Figure 15-5. ASTER brightness temperatures at the surface compared with nearly coincident ground observations. Note that the May 12, 2001 case shows nearly coincident MASTER brightness temperatures.

measurements used in rangeland health determinations like vegetated and bare soil cover percentages and gap and patch sizes at 5 cm resolution. Eventually, resource management agencies, rangeland consultants, and private land managers should be able to use small and lightweight UAVs to acquire improved data at a reasonable cost that can be used to enhance management decisions.

To get a landscape perspective at reasonable cost without mosaicing, Landsat TM data seem to be ideal. The entire Jornada Basin is covered in one frame. The 30-m, multispectral resolution is sufficient to see major dirt roads, stock tanks, a variety of rangeland treatments, playa lake basins, ecotones, and the aforementioned southwest to northeast depositional patterns. On later Landsat platforms, the ETM+ sensor additionally provides 15-m panchromatic data. Repetitive coverage is possible every 16 days and is usually acquired successfully because of frequent clear skies outside the heart of the monsoon rainy reason. When very high-resolution imagery is not the primary goal, Landsat TM data are a very useful product for rangelands. The current Landsat platforms are old and becoming unreliable, so it is critical that follow-on Landsats be designed and launched very soon.

Conclusions

Using a multifaceted remote-sensing approach allows a number of applications in rangeland situations. First, it permits extrapolation of point-and-plot knowledge to large areas of rangeland that would never be observed normally because of inaccessibility, cost, lack of manpower and a variety of other problems. Second, it could be used as part of a system of related measurements to monitor rangeland resources across vast areas. When the cost can be afforded, the presently available 0.61–2.44-m (or similar) resolution, multispectral data are so useful that they can take the place of aircraft photography. With growing competition, the cost of this high-resolution satellite data is expected to decline. We also see a real immediate opportunity for incorporating UAV acquired high resolution imagery into this mix of data and their applications. Our approach for remotely sensed data has provided a framework for testing new, experimental technologies and techniques, such as scanning LIDAR and multiangle remote sensing from air and space. The fusion of different types of remote sensing data, such as multispectral, hyperspectral, radar, laser, and thermal spectral regions, is also expected to greatly increase information content in the future.

16

Modeling the Unique Attributes of Arid Ecosystems: Lessons from the Jornada Basin

James F. Reynolds
Paul R. Kemp
Kiona Ogle
Roberto J. Fernández
Qiong Gao
Jianguo Wu

The Jornada Basin is typical of arid ecosystems of the Southwestern United States and many other regions of the globe: It is water-limited with low annual net primary production (ANPP) and low-standing crop (Szarek 1979; Ludwig 1987). Yet paradoxically, arid ecosystems are structurally and functionally quite complex, exhibiting a remarkable range of species compositions and system behaviors. This can be attributed in part to the presence of complex topography and landscape physiography (Mabbutt 1997; see also chapter 2) which, when combined with extreme variability in precipitation (Cavazos et al. 2002; Weltzin et al. 2003; see also chapter 3), produces striking spatial and temporal heterogeneity in the availability of essential limiting resources, such as water and mineral nutrients (MacMahon and Wagner 1985; see also chapters 5 and 6).

In view of these complexities, one of the long-term objectives of the research in the Jornada Basin is to develop a synthetic understanding of the mechanisms and processes governing the complex patterns of arid land structure and functioning. It is clear that understanding and predicting potential cause–effect relationships will require considerable insights at multiple spatial and temporal scales (chapter 18). Models are expected to play an important role in this synthesis because most experiments and observations tend to take place at small spatial (e.g., 1–100 m^2) and brief temporal scales (e.g., days, months, one to five years) (Levin 1992), whereas many ecosystem responses are the result of interacting factors and feedbacks operating over larger spatial and longer time periods (O'Neill et al. 1989; Levin 1992).

In this chapter, we present a summary of some of the mechanistic models we developed as part of the Jornada Basin research program. Although our initial goal was largely focused on the relationship between precipitation and ecosystem functioning in the Jornada Basin, our work is sufficiently general that it should be applicable to other arid land regions of the world. Simulation modeling has a key role to play because it is difficult to experimentally examine even a partial spectrum of ecosystem-level responses that could result from abrupt perturbations, such as overgrazing and especially longer term external forcings, such as shifts in precipitation. Models permit us to distill data we have gathered to date, explore the consequences of various assumptions regarding specific cause–effect relationships, and examine responses of ecosystems to potential changes in forcing functions.

The majority of our efforts have been devoted to analyses of ecosystem carbon, nutrient, and water dynamics using the patch arid lands simulator-functional types model (PALS-FT) (table 16-1). PALS-FT is a physiologically based ecosystem model that simulates one-dimensional fluxes of carbon (C), water, and nitrogen (N) in a representative patch of desert vegetation of approximately 1–10 m^2 (figure 16-1). For landscape-level phenomena, a spatially explicit version of PALS-FT (MALS) is used (table 16-1). PALS-FT consists of four principal modules: (1) soil water distribution and extraction via evaporation and transpiration; (2) soil, surface, and canopy energy budgets; (3) plant growth, including phenological and physiological responses of key principal plant functional types; and, (4) nutrient cycling, including soil organic matter, decomposition, availability of inorganic N. We examine the progress that has been made with respect to the potentials of these models to strengthen understanding of various phenomena—from physiological responses of plants to highly variable pulses of water and nutrients, to long-term processes, such as nutrient cycling and landscape dynamics—as well as the critical limitations of models to accurately represent some of the unique attributes of arid ecosystems in the Jornada Basin.

Conceptual Framework

Aridland Paradigms

As described in several chapters in this volume, heterogeneity in arid lands is exemplified by the spatial distribution of soil nutrients (e.g., islands of fertility; Charley and West 1975; Virginia and Jarrell 1983), the vertical partitioning of soil water availability (e.g., the two-layer hypothesis of Walter 1971), and temporal pulses in nutrient and water availabilities (e.g., mineralization rains; Charley 1975). Consequently, many unique nonlinear behaviors emerge, such as nutrient cycles that are out of phase with abiotic driving variables (Charley 1972), non-equilibrium relationships among key abiotic and biotic variables (Ellis and Swift 1988; Illius and O'Connor 1999), and uncertain regeneration and growth pulses of many plant species, which involve complex survival mechanisms linked to episodic precipitation events and plant phenology (West et al. 1979; Kemp 1983; Reynolds et al. 2000).

Traditionally, arid land ecologists have sought to explain the complex and variable responses of arid land systems through simple paradigms. Two widely

Table 16-1. Versions of PALS showing specific modules. All patch-level simulations presented in this chapter were conducted using PALS-FT (shaded column).

		Patch Ecosystem Vers.			Flowpath Ecosystem Vers.	Key: ŏ = implemented; i = in progress
Principal Modules	Descriptions	PALS-*SWB*	PALS-*FT*	PALS-*CO2*	MALS[a]	Key Reference(s)
Soil water	Infiltration; flux between soil horizons; soil water potentials, etc.	ŏ	ŏ	ŏ	ŏ	Moorhead et al. (1989); Kemp et al. (1997); Reynolds et al. (2000); Ogle et al. (2004); Reynolds et al. (2004)
	Plant-soil relations: water uptake by roots; plant water potentials: rooting distributions etc.	ŏ	ŏ	ŏ	ŏ	
	Hydrologic surface run-in/run-off: 2-dimensional flow fields between contiguous patches (grid cells)	i	i	i	ŏ	Gao and Reynolds (2003)
Energy budget	Radiation interception; PET; bulk canopy conductance; etc.	ŏ	ŏ	ŏ	ŏ	Kemp et al. (1997)
	Soil temperature	ŏ	ŏ	ŏ	ŏ	Kemp et al. (1992)
Plant growth	Allocation; respiration; photosynthesis; growth rates; etc.	i	ŏ	ŏ	ŏ	Reynolds and Cunningham (1981); Reynolds (1986); Reynolds et al. (1997, 1999a, 2000, 2004); Shen et al. (2004)
	Seasonal changes in stomatal behavior; photosynthetic temperature acclimation	i	ŏ	i	i	Ogle and Reynolds (2002)
	Phenology: seed germination; reproductive/vegetative growth phases; etc.	i	ŏ	ŏ	ŏ	Bachelet et al. (1988); Kemp and Reynolds (2000)
	Seed dispersal	i	i	i	ŏ	Gao and Reynolds (2003)
	CO_2 effects: down-regulation; max photosynthesis rates; etc.	i	i	i	i	Modules described in Reynolds et al. (1992, 1996a); Chen and Reynolds (1997)
	Plant FT competition (other than water)	i	i	i	i	Modules described in Brisson and Reynolds (1994, 1997)
Nutrient cycling	Decomposition; PALS-CENTURY[b] (nutrient cycling)	i	ŏ	ŏ	ŏ	Moorhead and Reynolds (1989a); Kemp et al. (2003)

[a] Mosaic Arid Land Simulator (includes PALS-FT)
[b] CENTURY model as modified for use as a module in PALS-FT (see *Decomposition and Nutrient Cycling*)

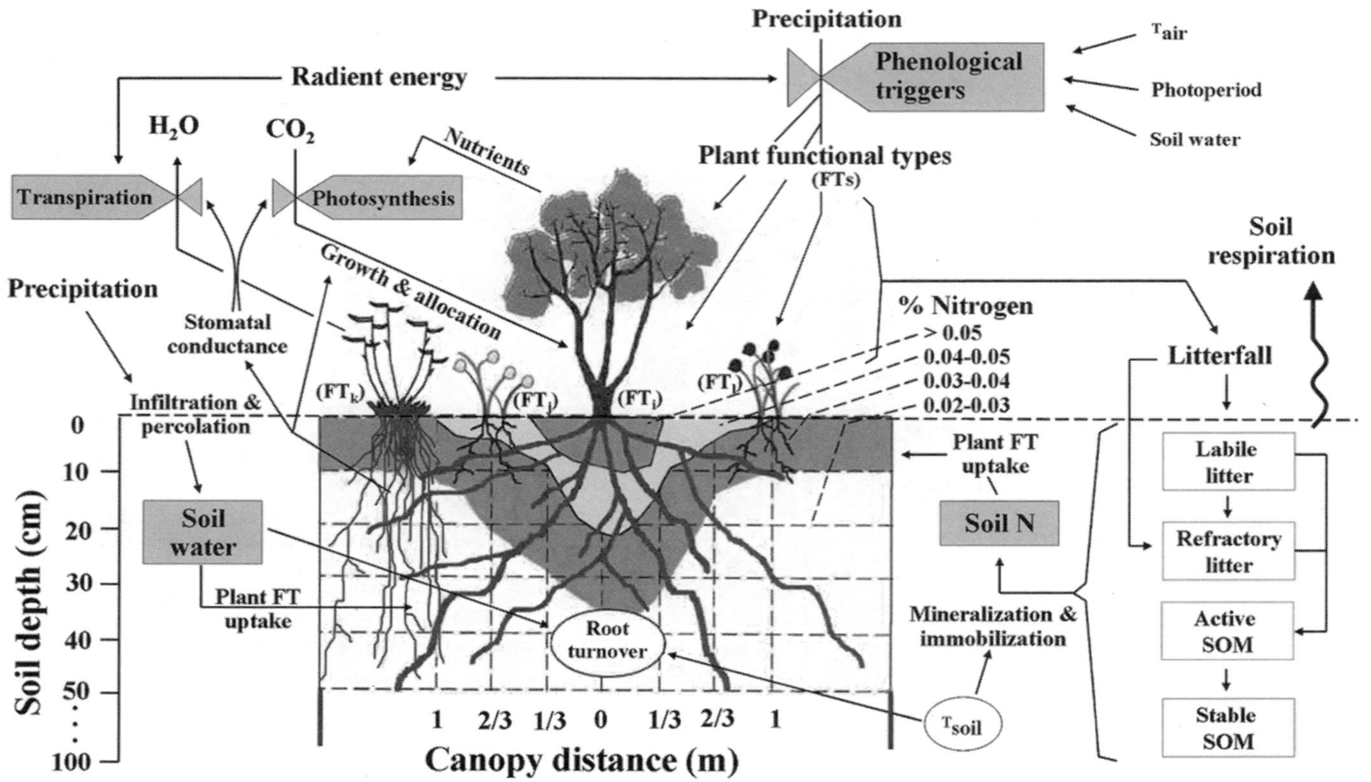

Figure 16-1. Schematic depicting the four principal modules of the PALS: (1) vertical soil water stratification and its extraction via evaporation and plant transpiration, (2) energy-budget/atmospheric environment, (3) carbon-nitrogen cycling in soil organic matter pools and resulting availability of inorganic N, and, (4) the phenology, physiology, and growth of key principal plant functional types found in the warm deserts of the Southwestern United States.

cited examples—the pulse-reserve (Noy-Meir 1973) and two-layer (Walter 1971) conceptual models—are often evoked to explain the translation of episodic rainfall into soil water availability and resultant plant production. In its simplest form, the pulse-reserve paradigm (figure 16-2a) describes how a rain event "triggers" a pulse of production (i.e., germination, growth, or reproduction), some of which is diverted to "reserve" (seed for annuals and perennials; storage organs for perennials). This paradigm suggests a simple, direct link between discontinuous and unpredictable rainfall and the long-term function and survival of arid land plant species (see review in Ogle and Reynolds 2004). The two-layer hypothesis, a complementary paradigm, suggests that seasonal pulses of moisture become vertically separated into shallow and deep soil water pools that are differentially utilized by plants, notably shallow-rooted grasses and deeply rooted woody plants.

Though heuristically compelling, these paradigms are not as simple or as straightforward as they appear. Studies at the Jornada Basin suggest that both paradigms are limited in terms of explaining productivity responses to rainfall.

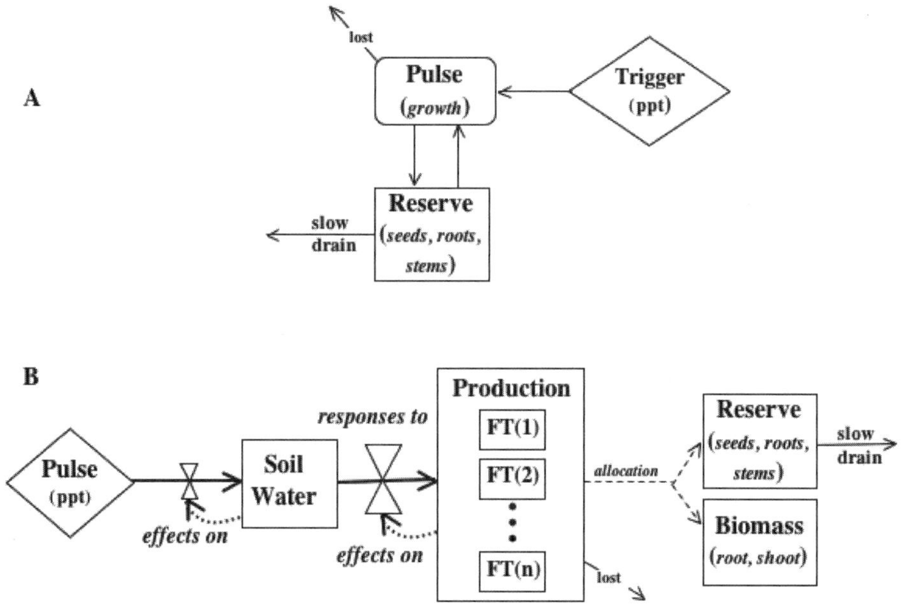

Figure 16-2. (a) Pulse-reserve paradigm (as presented in Noy-Meir 1973; based on Bridges and Westoby unpublished). (b) Modified pulse-reserve model, which explicitly identifies three components of the relationship between precipitation and plant production: (1) pulses of precipitation, (2) the role of soil water (e.g., antecedent conditions, soil type), and (3) plant functional types (FTs). Integrating plant water use with soil water availability makes it possible to distinguish "plant responses to" from "plant effects on" soil water. For example, when plants deplete soil water ("−" effect), this contributes to plant growth ("+" response) and various plant FTs can be good competitors either by being good extractors (effect pathway) or good tolerators (response pathway) (redrawn and modified from Reynolds et al. 2004).

Rather, productivity is a manifestation of interactions among numerous soil, plant, and atmospheric variables that result in complex patterns of soil water storage and water use by plants (chapter 11). In an effort to achieve a more explanatory conceptual model, we recently invoked several modifications to the classic pulse-reserve paradigm (Reynolds et al. 2004). As illustrated in figure 16-2b, the most important of these include (1) the translation of precipitation into usable "soil moisture pulses," i.e., soil water pools (storage), which allow for antecedent soil moisture and may dampen or amplify the effect of individual rainfall events; and, (2) water use by different plant species.

Modeling Focus Areas

Our proposed revision of the pulse-reserve model (compare figure 16-2a and 16-2b) is not meant to be all-encompassing. Rather, it serves as a general guide for identifying those key processes that engender observed patterns of structure and functioning in arid ecosystems. Our conceptual model suggests three indispensable topics of focus, which we discuss next.

Plant Functional Types

Plant functional types, which have different phenological, physiological, and morphological characteristics, are important components of arid ecosystems. As detailed in chapter 10, in the late 1800s the Jornada Basin consisted largely of warm-season, C_4, perennial grasses; a century later, these communities have largely been replaced C_3 shrub-dominated communities. In an effort to account for the different combinations of plant functional types representative of these historical grass–shrub transitions, most of the model development and validation reported in this chapter is based on data collected from the 2,700-m transect (figure 16-3), which encompasses a variety of plant assemblages and soil types (Wierenga et al. 1987: Cornelius et al. 1999).

In PALS-FT, we group the principal species that occur along the transect into eight plant functional types based on similarity of rooting patterns, seasonal activity, growth forms, and physiological responses to soil water deficits (table 16-2). Within a specific plant functional type, there may be some variation in the phenology and growth of individual species or in species composition from year to year. Nevertheless, the functional type concept is a powerful tool for modeling collections of relatively similar species (Smith et al. 1997).

Distinct Patch Types

Conceptually, the Jornada Basin can be characterized as a two-phase mosaic, which is typical of many arid ecosystems; that is, there are scattered patches of plants with relatively high within-patch cover interspersed within a matrix of relatively bare soil (Aguiar and Sala 1999). The two-phase mosaic constitutes the conceptual framework underlying the development of PALS-FT. To account for

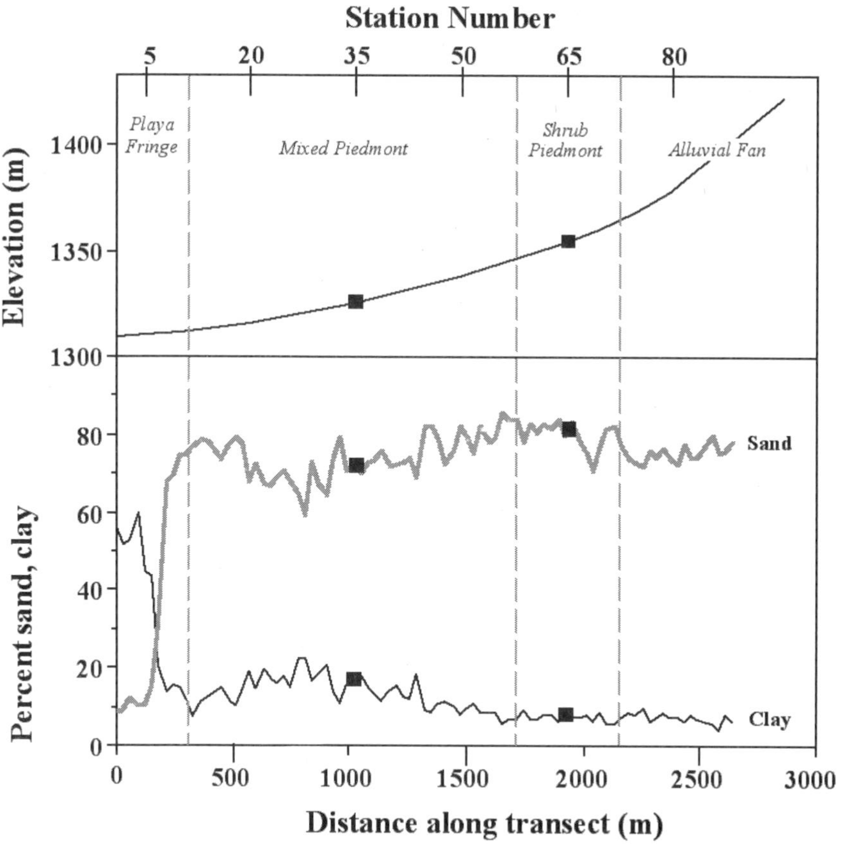

Figure 16-3. Schematic of Jornada Basin transect, established in 1982 as part of the Jornada LTER I study. Along the 2,700-m transect, 90 sampling stations were established at 30-m intervals, extending from the floor of a small, closed-basin watershed (containing a small ephemeral lake bed or playa), up a broad alluvial slope to the footslopes of an isolated, granitic mountain. Soils along the transect are relatively coarse textured (typic haplargids or torriorthentic haplustolls) with the exception of the playa, which contains fine-textured vertisols. Small squares show location of two stations (#35, #65) used in the simulations of grass- versus shrub-dominated patches.

the hierarchical nature of arid lands, this conceptual framework is depicted as a spatially nested hierarchy in table 16-3.

The smallest spatial unit in the hierarchy is an individual plant. Multiple cooccurring individual plants plus their immediate soil and atmospheric environment constitute a patch (figure 16-1), which is the basic unit for our ecosystem-level simulations with PALS-FT. We assume that a patch is internally homogenous, that is, located in a particular soil type where size is a function of the size of various contiguous plant functional types. Different patch types (e.g., grass- or shrub-dominated ones) have different functional (e.g., photosynthesis, reproductive rate,

Table 16-2. Plant functional types (FTs) used in the PALS-FT simulation model. Complete list of species in the Jornada Basin can be found online at http://usda-ars.nmsu.edu/jer/plantlist.htm.

Plant FT	Brief Description	Examples	Occurrence of Highest Cover along LTER I Transect (see fig. 16-3)[a]
Perennial forbs	Active from spring through autumn	Leatherweed (*Croton pottsii*) Hairyseed bahia (*Bahia absinthifolia*)	*Playa* (stations 1–7) and *mixed piedmont* (broad zone of relatively open, mixed vegetation; stations 11–57)
Deciduous shrubs	Winter deciduous	Honey mesquite (*Prosopis glandulosa*)	*Playa fringe* (stations 8–10)
Deciduous subshrubs	Winter-dormant	Desert zinnia (*Zinnia acerosa*), threadleaf snakeweed (*Gutierrezia microcephalum*)	*Mixed piedmont* (stations 11–57)
Spring annuals	Germinate in winter or spring and flower anytime from late spring to late summer	Abert's buckwheat (*Eriogonum abertianum*), Steve's pincushion (*Chaenactis stevioides*)	
Evergreen shrubs	Drought-tolerant	Creosotebush (*Larrea tridentata*)	*Shrub piedmont* (station 58–72)
Winter annuals	Germinate in autumn or winter and flower in winter or spring	Tansy mustard (*Descurainia pinnata*), Gordon's bladderpod (*Lesquerella gordonii*)	*Mixed piedmont* (stations 11–57) and *alluvial fan* (stations 73–90)
Summer annuals	Germinate and flower in summer	Sixweeks grama grass (*Bouteloua barbata*), lemonscent (*Pectis angustifolia*)	
Perennial grasses	C_4, summer-active	Black grama grass (*Bouteloua eriopoda*), bush muhly (*Muhlenbergia porteri*), fluffgrass (*Dasyochloa pulchella*)	Sporadically codominant in *mixed piedmont* (stations 11–57) and dominant on *alluvial fan* (stations 73–90)

[a] Some plant FTs can periodically be found at all locations along the 3 km transect.

Table 16-3. Hierarchical view of ecological systems showing the spatial simulation units used in developing PALS-FT (described in Reynolds et al. 1993; Reynolds and Acock 1997).

Simulation Unit	Spatial Scale (m^2)[a]	Structural Components	Typical Coupling Variables: Water	Typical Coupling Variables: Energy/Nutrients	Systems[c]: Plant	Systems[c]: Ecosystem	Systems[c]: Ecosystem
Plant FT	10^{-4}–10^{-1}	• Leaves • Stems • Roots • Soil volume	• Soil water • Transpiration • Plant water potential • Water uptake	• Photosynthesis • Respiration • Nutrient uptake • Herbivory	**L**	L − 1	L − 2
Patch ecosystem	10^{0}–10^{4}	• **Plant FTs** • Soil properties • Microbial populations • Surface litter • Soil organic matter	• Precipitation • Infiltration • Runoff/run-on • Evapotranspiration • Soil water balance	• Net carbon balance • Net ecosystem carbon flux • Nutrient cycling • Decomposition • Trace gas flux	L + 1	**L**	L − 1
Flowpath ecosystems[b]	10^{2}–10^{5}	• **Patch ecosystems** • Toposequences or soil catenas • Connectivity	• Soil water flux and discharge [Hillslope Darcian flow dominant]	• Mass flux of sediment • Mass flux of dissolved nutrients • Aeolian transport • Dispersal • Trace gas flux	L + 2	L + 1	**L**
Landscape	10^{6}–4×10^{6}	• **Flowpath ecosystems** • Groundwater • Channel storage	Channel flow [Turbulent flow dominant]		L + 3	L + 2	L + 1
Region[b]	10^{7}–10^{10}	• **Landscapes** (=Integrative flow systems) • Lakes; rivers		• Hydrologic transport of sediments and nutrients • Aeolian transport • Migration	L + 4	L + 3	L + 2

Notes: The level of concern (system **L**) is itself a component of a higher-level system (L + 1), the latter of which may influence (limit, bound, etc.) the behavior of **L**; in turn, **L** can be subdivided into components of the next lower level (L − 1), which serve as state variables in models of **L** and are studied to explain the mechanisms operating at **L** (e.g., leaf-level photosynthesis [L − 1] as a component of whole-plant growth [**L**]). We can generally ignore levels higher than L + 3 and lower than L − 3 when trying to understand **L** behavior (O'Neill et al. 1989). While this scheme is arbitrary, it is biased towards plants as the central unit of study, and the distinction between levels is somewhat vague, it serves to illustrate the complexity of multiscaled systems and provides a useful framework for discussion. Furthermore, terms such as *ecosystem* and *landscape* may connote very different meanings unless their spatial and temporal limits are explicitly defined. Based on Reynolds et al. (1993, 1996b). Shaded row/column indicates patch-level simulations presented in this chapter.

[a] Typical values from Osmond et al. (1980), Woodmansee (1988), and Walker and Walker (1991).

[b] Equivalent to mesoscale in Walker and Walker (1991), which includes second-order watersheds.

[c] **L** is the level at which one is interested.

phenology) and structural (e.g., canopy volume, rooting depth, leaf area) characteristics. Patch behavior is a function of the various interactions and feedbacks of the patch with environmental drivers.

A series of connected patches form flowpaths (schematics provided in Reynolds and Wu 1999). Distinct geomorphic surfaces (alluvial fans, piedmonts, etc.) or topographic features (e.g., watersheds) often form natural boundaries for flowpaths (chapters 2, 4, and 7). As illustrated in figure 16-3, the LTER transect consists of a number of contiguous patches arranged linearly along an elevation and soil gradient. Thus, variable seasonal rainfall, downslope redistribution of water and organic matter, and soil texture–related variation in infiltration and water-holding capacity all interact to generate a complex spatial and temporal gradient of patch types with differing soil water and nitrogen availabilities.

Soil Water Availability

One of the main difficulties in understanding and predicting responses of arid ecosystems to variable precipitation stems from the fact that plants and most other ecosystem components (e.g., decomposers) respond primarily to soil water, not precipitation per se (see chapter 5). Arid environments present a unique suite of concerns with respect to soil water, including low and episodic rainfall, patchy distribution of plant cover, poorly developed soils with variable textures, vertical redistribution of soil water due to water potential or thermal gradients, and discontinuous and depth-dependent distributions of soil water (often due to relatively impermeable layers of calcium carbonate or clay). Accordingly, our modeling activities and supporting experiments have focused on the extent to which precipitation is translated into soil water and how this, in turn, affects plant-, patch-, and flowpath-level processes and dynamics (table 16-3).

Mechanistic Modeling in a Hierarchical Framework

Our models are hierarchical in that behavior of the system at any specific level (L) (e.g., organ, whole plant, community, ecosystem, etc.) is modeled by representing the interactions of the component systems at one (L-1) or two (L-2) levels smaller in the hierarchy (table 16-3). For example, whole plant dynamics are modeled via descriptions of organs (e.g., allocation between roots and shoots, leaf photosynthesis), whereas community dynamics is modeled based on whole plant dynamics (for examples, see Reynolds and Leadley 1992).

The high spatial and temporal variability in arid ecosystems presents a challenge for achieving the appropriate level of detail in models. Linear methods of scaling, such as averaging or summing over spatial units or through time, are likely to be unsatisfactory because of extremes and nonlinear relationships. Hence, we must consider both the spatial and temporal hierarchies in which processes occur. Rainfall, for example, epitomizes this difficulty: There is a multifarious relationship between the magnitude of a rainfall event, the degree of ecological responses, and the time scale over which these responses unfold (Schwinning and

Sala 2004). Small pulses of rainfall may trigger germination, but larger (or cumulative) amounts are necessary to sustain growth, flowering, fruiting, and seed production (Beatley 1974; Bachelet et al. 1988). A single precipitation event may affect plant survival and growth, but the effect is dependent on its timing relative to other events (Reynolds et al. 2000).

In summary, addressing questions of how plant functional types may respond to as well as affect soil water and nutrient availability involves a careful match of scales of observation (patterns), theory (explanations), and models (mechanistic descriptions involving key processes, interactions, and feedbacks) (Reynolds et al. 1993). Our goal is to develop models capable of prediction at the level of interest while avoiding scaling errors.

Modeling Arid Ecosystems

In the following sections, we summarize results from our hierarchical modeling approach, including plant-, patch-, and flowpath-level processes. We highlight important and often surprising outcomes of our modeling efforts with the goal of illustrating how model development and results can provide insight into the complex interactions within these arid ecosystems. We also discuss significant uncertainties and identify some of the many challenges that remain.

Plant Level

Leaf-Level Acclimation in Creosotebush

In most arid ecosystems, the physiological and growth behaviors of individual plants are highly variable with respect to time of day, season, and landscape position (Smith and Nowak 1990). Many plants tend to grow in seasons or time periods with the most reliable resource availability (e.g., the rainy season) although a few, such as creosotebush (*Larrea tridentata*) are physiologically active throughout most of the year. Although much dogma exists concerning the ability of creosotebush to cope with extremes in temperature and water availability, surprisingly few supporting data actually exist. Because of its ecological dominance on bajadas within the Jornada Basin and its distinct phenology and physiology, creosotebush (evergreen shrub, table 16-2) is an important component of PALS-FT. Thus we were motivated to reassess the physiological and growth behavior of creosotebush, especially with regard to its magnitude and mode of temperature acclimation and water-stress response.

We developed A-Season, a semi-mechanistic mathematical model that explicitly links CO_2 assimilation rate (A), stomatal conductance to water vapor (g), leaf internal CO_2 concentration (C_i), and plant water potential (Ψ_p) and incorporates the effects of temperature (T), atmospheric vapor pressure deficit (VPD), light intensity (photon flux density, PFD), and soil water availability (as reflected in values of Ψ_p). See Ogle and Reynolds (2002) for derivation of models and other details. A-Season contains a standard Fickian diffusion equation for photosynthesis: $A = 0.625 \cdot g \cdot (C_a - C_i)$, where C_a is atmospheric CO_2 concentration and 0.625 is the ratio of the diffusivity of water vapor to CO_2 in air. We derived an expression for C_i based on a linearization of a typical $A - C_i$ curve (A is a saturating function of C_i; see for example Lambers et al. 1998) making C_i dependent on g (Katul et al. 2000). The A-Season model assumes that VPD and Ψ_p affect g, and the threshold function is consistent with a biophysical model of stomatal regulation of leaf water potential (figure 16-4a; Oren et al. 1999). We also allow for a temperature acclimation response by expressing maximum daily g (g_{max}) as a function of T_{gro} (average 24-hour, 7-day air temperature) and Ψ_p (figure 16.4b). We analyzed A-Season by fitting it to field data from the Jornada Basin collected under a broad range of environmental conditions and by testing it against data gathered under dissimilar conditions.

What did we learn from this modeling exercise? In general, we found that temperature acclimation of stomatal behavior may be much more important to seasonal changes in photosynthesis than previously recognized. For example, the inclusion of growing-season temperature (T_{gro}) is necessary to reproduce the seasonal variability in A and g (figure 16-5). Moreover, our study sheds light on the current understanding of CO_2 demand in creosotebush. Many researchers assume that plants maintain nearly constant C_i (Wong et al. 1979; Yoshie 1986), yet our field measurements over a 2-year period found diurnal and seasonal variation in C_i (range $\cong$ 55–510 ppm). The inclusion of variable C_i in the photosynthetic response functions enables us to capture the hysteresis in instantaneous C_i that results from a coupling between C_i and g and large diurnal fluctuations in PFD, T, and VPD (figure 16-6). Models that neglect stomatal temperature acclimation and/or assume constant C_i may yield unrealistic predictions of plant and ecosystem carbon and water dynamics.

Root–Soil Interactions

Our work on the aboveground physiology of creosotebush underscores the importance of root–soil linkages and controls of water uptake within the belowground environment. Root structure, including density, depth of penetration, and horizontal distribution within the soil matrix, can have significant effects on transpiration, competition, and growth of plants. However, the belowground environment represents a challenge in developing models for the Jornada Basin because (1) quantitative, accurate assessments of root biomass are difficult to obtain; (2) rooting depths vary greatly across life forms and species (e.g., Gibbens and Lenz 2001; Schenk and Jackson 2002); (3) even within a species, rooting patterns may be quite variable from site to site and year to year (Brisson and Reynolds 1994; Gibbens and Lenz 2001); (4) root structural properties may not be good indicators of root functional activity, such as water and nutrient uptake (Plamboeck et al. 1999).

The majority of our modeling of the belowground environment has been somewhat empirical. For example, Kemp et al. (1997) used a simple water uptake formulation and empirical estimates of root distributions in PALS-FT to elucidate the sensitivity of various plant functional types to changes in root distributions.

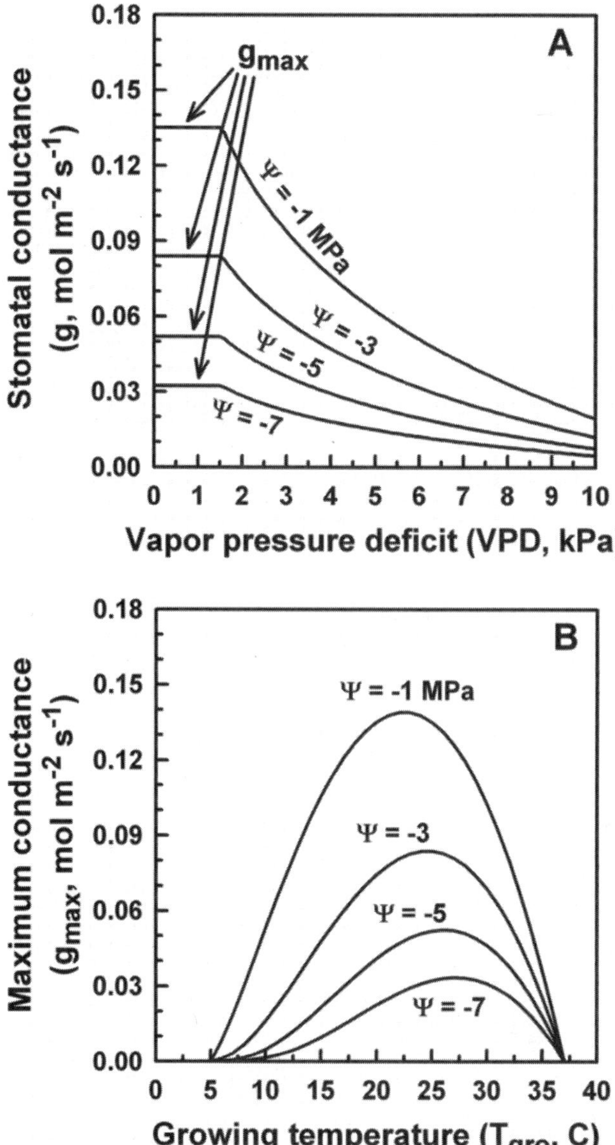

Figure 16-4. "A-Season model" predictions of stomatal conductance (g) in creosotebush as a function of vapor pressure deficit (VPD), plant predawn water potential (Ψ_p), and average growing temperature (Tgro). (A) Plots of g versus VPD, where each curve is for a specific value of Ψ_p. The curves show: (1) stomatal sensitivity to VPD decreases (slope becomes more shallow) as Ψ_p becomes more negative (increasing water stress), and (2) g exhibits a threshold-type response such that stomata are operating at their maximum ($g$max) for VPD < 1.54 kPa. (B) Plots of $g$max versus Tgro (same Ψ_p values as in panel a). The curves illustrate: (1) $g$max is reduced by water stress, (2) stomata acclimate to temperature, whereby $g$max depends on Tgro, exhibiting highest values for Tgro between 22°C to 27°C, and (3) Tgro and Ψ_p interact to affect conductance (i.e., the shape and optimum of the $g$max versus Tgro curves vary with Ψ_p). Reproduced from Ogle and Reynolds (2004).

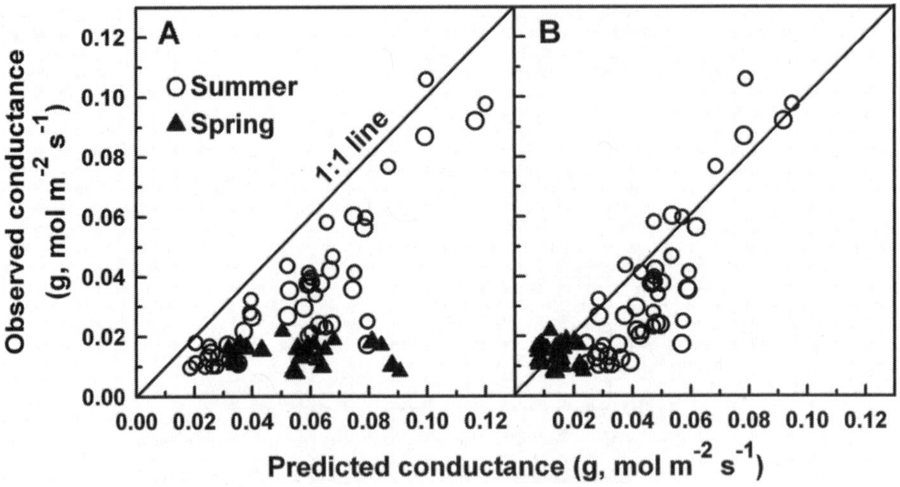

Figure 16-5. Plots of observed versus predicted stomatal conductance (g) for different assumptions regarding the effects of growing temperature (Tgro) on maximum conductance ($g$max) in the A-Season model. (A) No Tgro effect, thus g is independent of Tgro. (B) Tgro affects g as shown in Fig. 16-4b. Data collected during a cool, wet period in spring 1999 (open triangles, mean g-values) and from warm summer 1998 (solid circles). The thick diagonal line is the 1:1 line. The incorporation of a Tgro effect substantially improves the ability of the A-Season model to capture seasonal dynamics in creosotebush's stomatal and photosynthetic (not shown) behavior. Reproduced from Ogle and Reynolds (2004).

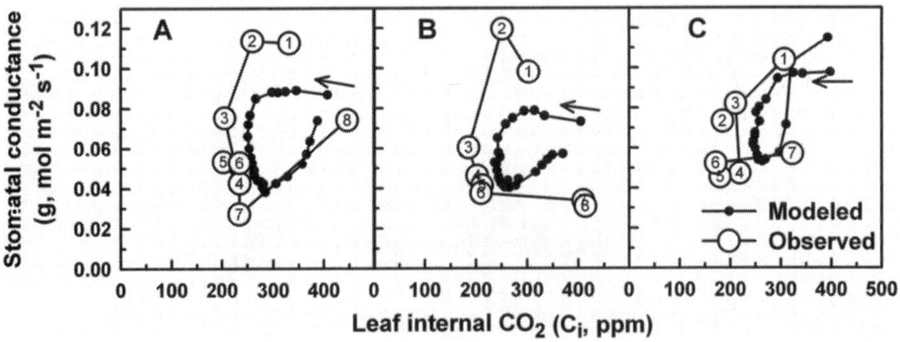

Figure 16-6. Predicted and observed diurnal courses of stomatal conductance (g) versus leaf-internal CO_2 concentration (C_i). Open circles are observed means and numbers (1–8) correspond to time (e.g., 1 = first measurement of the day). Predictions of the A-Season model (solid line, dots) are computed from 30-minute averages of climatic variables obtained from a weather station located in the study site; arrows are placed near early morning and point in the direction of increasing time. Plots are for three days for which modeled and observed C_i are in close agreement: (A) 7/28/98, $\Psi_p = -1.9$ MPa, (B) 7/31/98, $\Psi_p = -2.6$ MPa, and (C) 8/4/98, $\Psi_p = -2.1$ MPa. The model successfully captures the observed hysteresis patterns, whereby the g versus C_i trajectory depends on time of day. Reproduced from Ogle and Reynolds (2004).

Recently, Ogle et al. (2004) took a more mechanistic approach, linking stable isotope technology and a biophysical water flux model to estimate the vertical distribution of roots for creosotebush. We sought to answer several key questions, such as where do active roots occur within the soil profile, and how does the spatial or vertical distribution of active roots interact with soil water availability to affect plant water sources and soil water dynamics? The emergence of stable isotope technology has proved a powerful tool to unravel some of the uncertainties with respect to such questions (Brunel et al. 1995). For example, the relative abundance of deuterium and ^{18}O (heavy oxygen) in plant stem water reflects the specific soil depths from which roots obtain water (Ehleringer and Dawson 1992).

We addressed these questions by developing the root area profile and isotope deconvolution (RAPID) algorithm, which explicitly couples measurements of plant water potential, stable isotopes, soil water, and soil physical properties. Details of the RAPID algorithm, including an outline of the logical steps, a list of assumptions, description of the root uptake model, fits to observed data, and numerous examples, are given in Ogle (2003) and Ogle et al. (2004). RAPID incorporates a model for root water uptake within a Bayesian statistical framework with biologically realistic constraints placed on root area profiles. We used a simple biophysical water flux model (modified from Campbell 1991) that captures key elements of root–soil–water interactions. The model describes the fraction of water acquired, p_i, from soil layer i as a function of the fraction of active root area in that layer, $f_{r,i}$. The p_i values are then used to determine the isotopic signature of the stem water that results from a particular root distribution. The objective is to estimate p_i and $f_{r,i}$. Because the actual root distribution for most species is unknown, we assume the distribution of active root area can be described by a mixture of gamma density functions that capture a wide range of unimodal and bimodal root densities with depth (for equations see Ogle 2003). We then used RAPID to estimate the vertical distribution of active roots for creosotebush plants at a site in the Jornada Basin and explored the implications for water uptake during a dry-down period following a summer rainfall event.

At first glance the reconstructed, active root area profile for creosotebush appears unimodal with about 96% of the active root area between 20 and 45 cm (figure 16-7a). However, the RAPID algorithm predicts that there is a small but statistically significant fraction of active root area in the top 2–5 cm (figure 16-7a, inset).

What did we learn from this modeling exercise? Although the small fraction of active roots in the surface layers normally contribute little to total plant water uptake, their role may be to acquire water immediately following a rain event (BassiriRad et al. 1999). For example, nearly 30–60% of the water taken up during the first and second days following a 17-mm rainfall came from the top 10 cm (figure 16-7b, c). This water is probably critical to creosotebush growth and carbon dynamics because photosynthesis is often enhanced immediately following rain events (Reynolds et al. 1999b). Essential nutrients are also concentrated in the topsoil (Jobbágy and Jackson 2001), and the small fraction of active roots near the surface may allow creosotebush to capitalize on improved nutrient availability following rain (BassiriRad et al. 1999). As the shallow soil dries, the

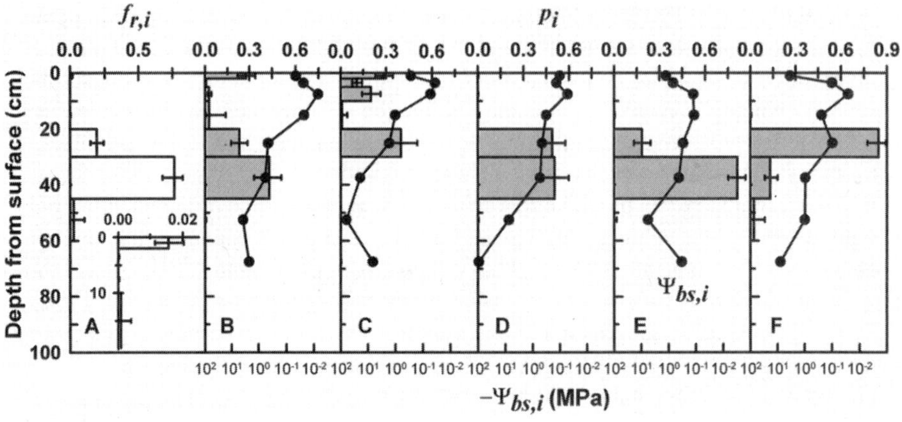

Figure 16-7. Results using the RAPID algorithm for the posterior estimates of (A) active root area and (B–F) water uptake profiles for creosotebush. (A) White bars depict posterior means for the fraction of active root area in each soil layer (i.e., $f_{r,i}$) and the whiskers are the 95% posterior credible intervals (CrI; i.e., the 2.5th and the 97.5th quantiles posterior samples). The inset is a magnification of the fractions in the top 20 cm. (B–F) The filled bars are the posterior means for the fraction of water taken up from each soil layer (i.e., p_i) for five sampling dates; the whiskers are the 95% CrIs. The filled circles are the mean bulk soil water potentials ($\Psi bs,i$) for each layer, estimated from soil water content and texture data. The samples were collected after a rain event and during the following dry-down period. Dates associated with each panel are: (B) 7/25/98, 1 day after moderate rainfall (17.2 mm); (C) 7/26/98, 2 days after rain; (D) 7/29/98, 5 days after rain; (E) 7/31/98, 7 days after rain; and (F) 8/2/98, 9 days after rain. Reproduced from Ogle et al. (2004).

middle layers are the primary water source, and no water is acquired from the topsoil (figure 16-7d and 16-7f). The large fraction of active roots in the 20–45-cm range may provide a stable water source essential for maintaining everyday function throughout the year.

Patch Scale: Soil Water and Nutrient Cycling

Soil Water and Evapotranspiration

The spatial and temporal distribution of soil water in arid lands governs many biotic and abiotic processes (Wainwright et al. 1999b). Thus it is vital to have a robust, accurate model of the availability of water in desert soils. Given that many soil water models have been developed over the years, our approach for developing a relevant submodel for PALS-FT involved evaluation of three existing soil water models (2DSOIL, SWB, SW)(table 16-4). Our goal was to develop a robust model of minimal complexity. By comparing results from different models for a wide range of soil properties, vegetative cover, precipitation, and microclimate, we could identify the minimum information necessary to describe soil water dynamics in the Jornada Basin.

Table 16-4. Assumptions and functional relationships of the three soil water models (see text) used in Kemp et al. (1997) to examine soil water dynamics in the Jornada Basin.

	Model Intercomparison			
Attributes	2DSOIL	SWB	SW	PALS-SWB
Time step	Variable	1 day	Variable	1 day
Soil layers • *Number (total)* • *Thickness of individual layer*[a]	24 Variable (1–5 cm)	6 Variable (10 or 20 cm)	20 Variable (1–10 cm)	3–20 Variable (5–20 cm)
Soil water flux • *Moisture retention* • *Hydraulic conductivity*	Darcy-Richard's equation; finite element; van Genuchten (1980) (to calculate matrix potential) Gardner (1958)	None Campbell et al. (1993) (to calculate WHC) None	Darcy-Richard's equation; predictor-corrector; Campbell et al. (1993) (to calculate matrix potential) Gardner (1958)	None Campbell et al. (1993) (to calculate WHC) None
Root distribution	Uniform	Optimized for SWB	Estimated from literature	Estimated from literature
Evapotranspiration • *Transpiration* • *Soil evaporation* • *Water uptake*	f (Canopy E budget, LA, average Ψ_{soil}) f (Surface E budget) f (Ψ_{soil} in layer)	f (Canopy E budget average Ψ_{soil}, VPD) f (Surface E budget coupled with model of Linacre (1973)) f (Average Ψ_{soil})	f (VPD, σ_{stom}, LA, Ψ_{soil} in each layer) f (Surface resistance, vapor gradient) f (Ψ_{soil} in layer)	f (VPD, σ_{stom}, LA, average root-weighted Ψ_{soil} across layers) f (Surface E budget coupled with model of Linacre (1973)) f (Average root-weighted Ψ_{soil} across layers)

Notes: Based on this intercomparison, PALS-SWB was developed, which is a modified version of SWB and SW. Key to abbreviations: Ψ_{soil} = soil water potential; VPD = vapor pressure deficit; LA = leaf area; WHC = water holding capacity; E = energy; σ_{stom} = stomatal conductance. Details and rationale for model formulations provided in Kemp et al. (1997) and Reynolds et al. (2000). Right column indicates module used in patch-level simulations presented in this chapter (see table 16-1).

[a] Layers increase in thickness with increasing depth

The 2DSOIL model (Pachepsky et al. 1993) is a mechanistic model that emphasizes physical aspects of soil water fluxes; SWB (Baier and Robertson 1966) is a simple water budget model that does not consider soil water redistribution and includes simplified schemes for soil evaporation and transpiration by different plant functional types; and SW (see Kemp et al. 1997) is a semi-mechanistic model (intermediate in complexity between 2DSOIL and SWB) that includes soil water fluxes and the physiological control of water loss by different plant life forms. Transect data (figure 16-3) were used to investigate the comparative behavior of the models: the models were parameterized using data from 1986 and validated using data from 1987.

Each model predicted soil water distributions reasonably well, though there were differences in the way each partitioned water loss between evaporation and transpiration (details given in Kemp et al. 1997; Reynolds et al. 2000). The SW model, which accounts for transpiration by each plant functional type, predicted the greatest differences in transpiration along the transect, reflecting differences in plant community types; 2DSOIL, which emphasizes soil physical properties, tended to be more sensitive to changes in soil texture; and SWB was most sensitive to shifts in rooting distributions.

The results of these simulations provide insight into spatial and temporal variation in evapotranspiration (ET) in the Jornada Basin. Each model predicted variable amounts of transpiration versus evaporation as a function of location along this transect, which can be attributed to both vegetation and soil differences (table 16-5). All three models predicted lowest transpiration (about 40% of total ET) for the creosotebush community (Station 65, figure 16-3) with the lowest plant cover (30% peak cover) and highest transpiration (58–70% of ET) for the mixed vegetation. However, these differences in transpiration were also a function of soil texture differences between the communities, which can affect water-holding capacity of surface soils and thus E (i.e., the reverse-texture hypothesis of Noy-Meir 1973). The models were not conclusive regarding the effect of soil texture on soil evaporation. In comparing two transect locations with similar plant functional type cover but different soil texture (Station 20 versus 50), 2DSOIL indicated that evaporation was marginally higher on the coarse-textured soil (Station 50), whereas SW predicted greater evaporation on the fine-textured soil (Station 20).

What did we learn from this modeling exercise? Experimental studies of ET have shown that the percentage of total ET attributable to transpiration varies from 7% to 80% in various arid and semiarid ecosystems in North America (reviewed in Reynolds et al. 2000). Given the complexity of atmospheric, plant, and soil relationships, it is difficult for experimental studies to quantify the importance of these interactions, especially given the short time periods of most studies. An important "take-home lesson" is the importance of the interdepen-

Table 16-5. Predicted values from three soil water models (see text) of evaporation (E), transpiration (T), and T as a percent of total annual evapotranspiration (ET = E + T) at various locations along the LTER I transect during 1986. E and T are cumulative values for the year. Based on Kemp et al. (1997).

	2DSOIL			SWB			SW		
Station[a]	T	E	T/ET	T	E	T/ET	T	E	T/ET
20	18.0	12.9	58%	16.7	16.4	50%	22.6	12.8	64%
35	16.0	13.4	54%	12.5	16.5	43%	15.4	20.8	43%
50	14.7	13.1	53%	16.8	15.8	52%	25.2	11.0	70%
65	9.2	15.5	37%	12.7	17.8	42%	10.5	17.3	38%
80	14.1	14.1	50%	16.4	15.8	51%	19.8	12.4	61%

[a] See figure 16-3.

dency between transpiration and evaporation as a result of competition for soil water between the atmosphere and the plants. Our modeling research suggests that studies attempting to study transpiration or evaporation in isolation from one another are likely to reach erroneous conclusions regarding the relative contributions of these processes (see chapter 5). Furthermore, our models suggest that plant cover and rooting distributions are more important in determining soil water distribution than physical processes. These findings helped us develop a version of a soil water module (PALS-SWB, table 16-4) that represents the minimum degree of complexity necessary to capture these key dynamics.

Decomposition and Nutrient Cycling

Arid ecosystems have soils that are, for the most part, poorly developed and low in organic matter and nutrients. Litter and/or nutrient inputs are sporadic (Crawford and Gosz 1982), and both litter and nutrients are usually spatially heterogeneous, reflecting heterogeneity in the vegetation (Schlesinger et al. 1996). Although our understanding of nutrient cycling in arid lands is rudimentary, studies at the Jornada Basin have provided insight into the unique aspects of litter decay and soil nutrient dynamics under arid conditions (see chapter 6).

Our patch-level modeling has addressed a number of these unique attributes by focusing on two aspects of nutrient cycling: (1) microbial-mediated decay of litter as a function of its chemical composition; and (2) prediction of litter decay and carbon and nitrogen mineralization over time, including the effects of extended drought. Model development with respect to the first focus area emphasized decomposition and mineralization processes associated with different chemical pools within the litter as well as the pool of microbial decomposers acting on these pools, all of which were incorporated into the mechanistic decomposition module GENDEC (Moorhead and Reynolds 1989b). Using GENDEC to simulate decay of leaf and root litter of creosotebush revealed that much of the observed dynamics of C and N in both litter and soil could be accounted for by microbial processes and that differences between surface and belowground litter losses were largely accounted for by physical weathering of surface litter (Moorhead and Reynolds 1989a). These results also suggested that decomposition of litter was strongly limited by N availability to microbes during the early phases of decay, providing an explanation for N immobilization in litter pools reported by Skujins (1981). However, when we attempted to use GENDEC for long-term simulations, we encountered surprising inaccuracy: The predicted mass loss and N dynamics of litter were too rapid and overly sensitive to soil environmental conditions. These results motivated us to compare GENDEC to two well-known models of decomposition and nutrient cycling: the Andrén and Paustian scheme (A&P, Andrén and Paustian 1987) and CENTURY (Parton et al. 1987).

Characteristics of the three models are given in table 16-6. Each model has unique assumptions for describing N dynamics and each represents a different degree of complexity. Each model simulates decay from pools of litter and soil organic matter, governed by decomposition rates modified by soil temperature and moisture, and by the chemical composition of the C and N pools. In general,

Table 16-6. Comparison of three decomposition models tested against observed decomposition data from the Jornada Basin. The best fit to data was obtained with a modified version of CENTURY, a modified version of which was subsequently incorporated as a module in PALS (i.e., PALS-CENTURY).

	Model intercomparison			
Attributes	GENDEC	A & P	CENTURY	PALS-CENTURY
Time-step	1 day	1 day	1 month	1 day
Partition of litter into chemical fractions	Observed (data or literature)	Observed (data or literature)	Function of lignin/ N ratio of litter	Observed (data or literature)
OM pools / decay Constants (k)[a] Pool 1/ k (d^{-1}) Pool 2/ k (d^{-1}) Pool 3/ k (d^{-1}) Pool 4/ k (d^{-1}) Pool 5/ k (d^{-1})	Solubles / 0.20 Cellulose / 0.08 Lignin / 0.01 Live microbial / 0.20 Dead microbial / 0.30	Labile / 0.367 Refractory / 0.0067 Active OM / 0.029 Stabilized OM / 0.0005	Metabolic / 0.05 Structural / (0.013)[b] Active OM / (0.02)[c] Slow OM / 0.00054 Passive OM / 0.000019	Metabolic / 0.04 Structural/ (0.010) Active OM / (0.016) Slow OM / 0.00043 Passive OM / 0.000015
Abiotic decay-rate controls	Soil temperature Water Nitrogen	Soil temperature Water	Monthly PPT/PET Air temperature	*Root litter* Soil temperature Water *Leaf litter* PPT/PET Air temperature
Efficiency of decay of litter and OM pools (1 − respiratory loss)	0.60 for all pools	0.36 for all pools	0.40–0.70, pool dependent	(same as CENTURY)
N pools and conversions	Explicit N pools / N:C is fixed or N dependent	Implicit pools / N:C is function of C decay	Explicit pools / N:C is fixed except for Metabolic	(same as CENTURY)

Notes: Details of modifications to CENTURY provided in Kemp et al. (2003). Right column indicates module used in patch-level simulations presented in this chapter (see table 16-1).

[a] Decay constants converted to daily values.
[b] k modified by lignin fraction (L_s); $k' = k.\ e^{(-3.0.\ L_s)}$ (with 20% lignin, $k' \approx 0.007$).
[c] k modified by the fraction of silt + clay in the soil (T): $k' = k.\ (1 - 0.75.\ T)$ (with 25% clay + silt, $k' \approx 0.016$).

GENDEC focuses on microbial metabolism and litter chemistry with a high degree of mechanism, A&P is the simplest formulation with the fewest pools, and CENTURY includes potentially important pools of recalcitrant soil organic matter and is intermediate in its complexity. We compared simulated mass loss, N content, and N mineralization of leaf and root litter using data obtained during a three-year field study at the LTER site, which examined the decomposition and N release from leaf and root litter of creosotebush and honey mesquite (*Prosopis glandulosa*) (Kemp et al. 2003).

For brevity, we show results only for creosotebush leaf litter, which demonstrate how the models tended to differ with respect to one another. Under both ambient and drought conditions, the A&P model underestimated rates of surface mass loss for litter (figure 16-8a and 16-8b) and gave the poorest estimates of N content of leaf litter (figure 16-8c and 16-8d), suggesting that a simple model that was originally developed for temperate, mesic agroecosystems is not sufficient to describe nutrient cycling in arid lands. GENDEC was the only model to predict N immobilization in leaf litter, although amounts were excessive (figure 16-8c and 16-8d); in addition, rates of mass loss in the first 10 months were too high (figure 16-8a and 16-8b). Last, although CENTURY underestimated leaf mass loss rates and did not predict N immobilization in litter, it provided the best overall description of decay and N mineralization across the spectrum of litter (root and leaf), species (creosotebush and mesquite), and treatments (ambient and drought).

This model comparison was instrumental in developing a decomposition and nutrient cycling module for PALS-FT. Given that CENTURY had the least error overall and that this model has been successfully used for describing long-term nutrient cycling in many semiarid grasslands (see Hall et al. 2000), we opted to use it in PALS-FT. To improve on its shortcomings for arid lands, we made several significant modifications: (1) We changed the time step from monthly to daily to

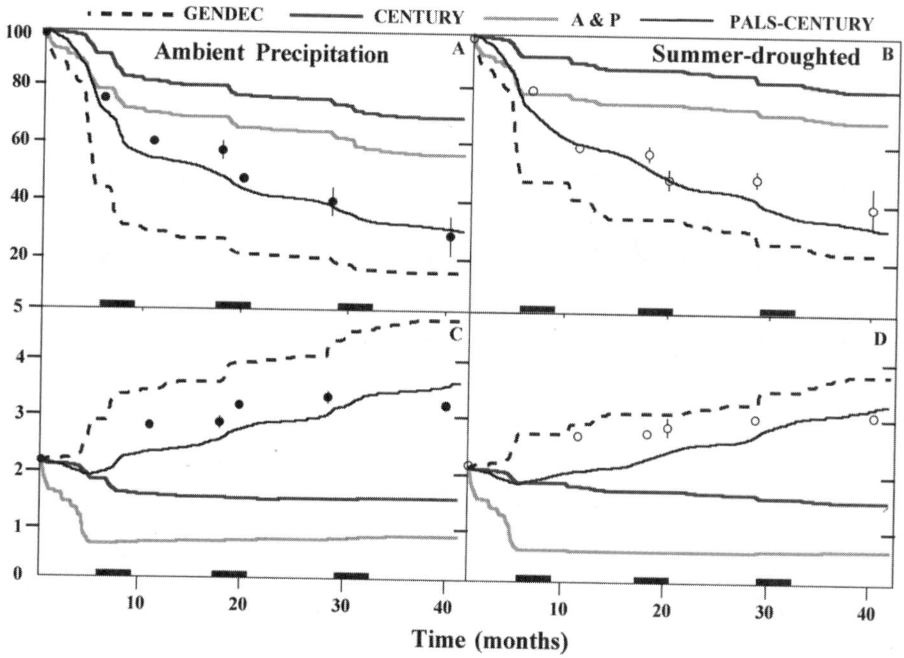

Figure 16-8. Comparisons of four decomposition models (GENDEC, CENTURY, A&P, and PALS-CENTURY) to predict decay of creosotebush leaf litter (A, B) and N content of decaying litter mass (C, D). Observed data (points) are means (+/− ISE) of 6–8 litterbags obtained from a field study (December 1991–June 1995) during which litterbag samples received either ambient precipitation (A, C) or had summer rainfall withheld (B, D–summer period is indicated by thick bars along the abscissa).

account for short-term variability in soil moisture and droughts of varying duration; (2) we added a term to account for the effect of soil moisture on the rate of root litter decay; and (3) we changed the method of partitioning litter into metabolic and structural fractions. In table 16-6, this submodel of PALS-FT (called PALS-CENTURY) is compared to the three original decomposition models (for further details, see Kemp et al. 2003).

What did we learn from these modeling exercises? We found that soil water availability plays less of a role during the initial phases of decomposition of surface litter than during latter stages, because surface litter is partly degraded by physical processes. Our studies with PALS-CENTURY suggest that nutrient cycling in the Jornada Basin may be poised between C limitation and N limitation, because small changes in the model led to relatively large qualitative changes in predictions of N behavior (figure 16-8c and 16-8d). Whether N from decaying litter is immobilized by microbial decomposers or released as mineralized N depends on whether microbial decomposers are primarily C- or N-limited (MacKay et al. 1987a; Montaña et al. 1988). Our modeling analysis suggests that decomposers are balanced between these two limiting nutrients and that any natural or anthropogenic factors (such as climate change and elevated CO_2) that affect litter amount or quality could change this balance.

Patch Scale: Unraveling NPP–Precipitation Relationships

A central objective of our modeling efforts in the Jornada Basin has been to elucidate cause–effect relationships underlying the tremendous variability of net primary production (NPP) over time and space. Although we intuitively expect that NPP in arid lands should be directly related to rainfall, we find that interannual variation in NPP for a given arid land site is, in fact, only weakly correlated with precipitation (Paruelo et al. 1999; Oesterheld et al. 2001; Wiegand et al. 2004). Le Houerou et al. (1988), for example, found that the variability in annual NPP measured at 77 rangeland sites was 50% greater than the corresponding variability in annual rainfall (figure 16-9a), indicating a complex relationship that defies simple conceptual models (e.g., figure 16-2a). Identifying cause–effect relationships between rainfall and NPP is crucial because the arid and semiarid vegetation zones of these regions appear to be particularly vulnerable to changes in precipitation (MacMahon and Schimpf 1981; McClaran and Van Devender 1995; Weltzin and McPherson 2003). A shift in seasonal precipitation and/or changes in the frequency and magnitude of extreme rain events could potentially lead to significant ecological and biogeochemical impacts (Higgins et al. 1998; Reynolds et al. 1999a, 2004; Sheppard et al. 2002)

One of our main challenges has been to address what constitutes "biologically significant" rainfall. Scores of factors could be responsible for poor correlations between precipitation and plant productivity (see chapter 11). For the Jornada Basin region, these include the timing of precipitation (Prince et al. 1988; Reynolds et al. 2000), composition of plant species and their moisture requirements (Kemp et al. 1997), preconditioning effects of drought (e.g., Rockström and Falkenmark 2000), compensatory responses following drought (Reynolds et al.

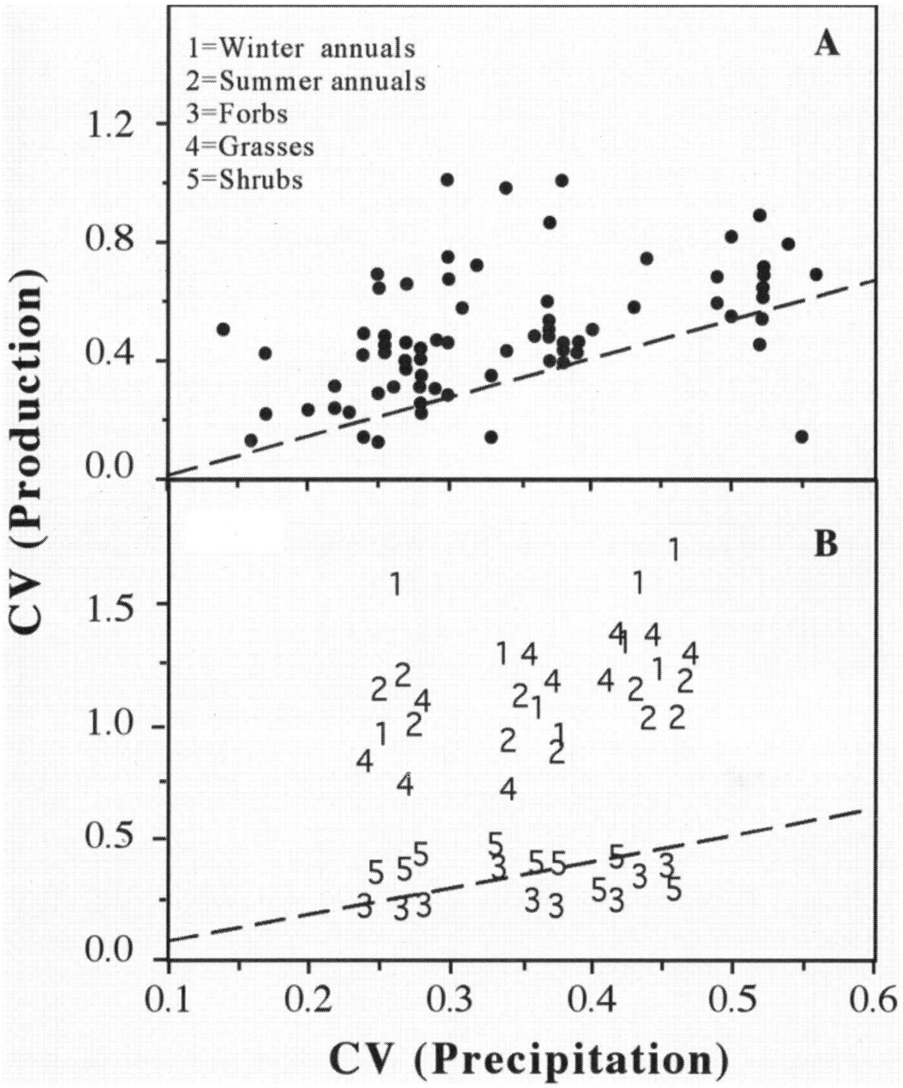

Figure 16-9. (A) Coefficient of variation (CV) in NPP shown as a function of the CV in precipitation. Based on data in Le Houerou et al. (1988) obtained from 77 rangeland sites (< 600 mm annual precipitation) scattered throughout the world, and (B) CV in NPP of various plant FTs obtained from simulations of PALS-FT as a function of CV in precipitation for decadal periods using rainfall from the Jornada Basin (1915–95). Numbers correspond to specific CV values for each FT from each decade. Modified from Reynolds et al. (1999a).

1999b), and hydrological factors, including variation in the infiltration capacity of soils (chapters 5 and 7), spatial variability in run-in and runoff (Gao and Reynolds 1993), and antecedent soil moisture levels (Reynolds et al. 2004). To ascertain the biological significance of precipitation, numerous factors must be considered, including the length of time between rain events, the size of an event and its intensity, the current status of soil water content, and the physiological activities of the plants. For example, the response of well-hydrated plants to a given moisture input will be different than drought-stressed plants (e.g., BassiriRad et al. 1999), and some arid land species require a minimal precipitation event to trigger a transition from a state of lower to higher physiological activity (Schwinning and Sala 2004).

In sum, elucidating cause–effect relationships between plant production and precipitation necessitates an integrated, multifactor approach. Using PALS-FT, we have explored a multitude of potential ways in which certain abiotic and biotic factors may directly or indirectly affect the dynamics between precipitation and plant productivity. In the following sections, we describe a series of simulations in which we scrutinized the role of individual precipitation events or precipitation pulses (Schwinning and Sala 2004), antecedent soil moisture, and long-term (decadal) period drought cycles on NPP dynamics.

Storms, Antecedent Moisture, and NPP

In the Jornada Basin, the majority (60%) of precipitation events are less than 5 mm (see chapter 3; Reynolds et al. 2004, fig. 3). Such relatively small rainfall events generally recharge only the uppermost portion of the soil profile, whereas large rain events are necessary to recharge deeper layers (see chapter 5). Given the ecological significance of how seasonal pulses of moisture could be vertically separated into shallow and deep soil water pools—which can be differentially utilized by shallow-rooted grasses and deep-rooted woody plants (i.e., the two-layer hypothesis)—our interest was piqued to assess the biological effectiveness of small versus large rainfall events.

First, returning to our conceptual model (figure 16-2b), the significance of any individual rain event must be appraised in terms of its impact on soil moisture recharge. Because the majority of all individual precipitation events in the Jornada Basin are < 5 mm, we hypothesized that small events may be "amplified" to some extent if they were to occur on sequential days, which we refer to as *storms*. In analyses of the long-term weather records for the Jornada Basin (see Reynolds et al. 2004; see also chapter 3), on average, about half of all rain events each year occur as individual rain events (with no rainfall on either the previous or following day), and the other half occur as storm events. Furthermore, the majority of storms produced < 5 mm of total precipitation. Second, analyses of length of the intervening period between storms in the Jornada Basin have revealed that a majority of rains and storms (56%) were separated by five days or less. This clustering occurs more frequently in summer, with about 70% of the summer storms occurring within five days of each other.

We used PALS-FT to explore the extent to which precipitation events and storms elicited plant responses, especially production. We analyzed differences among plant functional types regarding their growth responses to an 85-year Jornada Basin rainfall record as a function of (1) the physiological status of the plant (a function of prior abiotic conditions, as indicated by antecedent plant water potential); and (2) specific rainfall inputs, for example, storm size. In general, our simulations revealed many nonlinear responses. Large events and storms generally elicited much greater growth responses than small ones, especially in the evergreen creosotebush when plants were severely drought-stressed. Small storms elicited large growth responses in annual plants and grasses when plants were moderately drought-stressed. The aboveground biomass of severely drought-stressed grasses tended to decline following any size rain event, reflecting the reallocation of biomass from crowns and shoots to new root growth (see Reynolds et al. 2004, for a complete discussion of simulation results).

In relation to the two-layer hypothesis of water partitioning, we explored the extent to which water use by different plant functional types was preferentially derived from upper versus lower soil layers. Based on long-term simulations using the Jornada Basin rainfall record, we found that there was a difference in depth of soil water recharge depending on soil texture. For medium to fine-textured soils (> 18% clay) there was no consistent recharge below 60 cm, whereas there was consistent (⅔ of years) recharge below 60 cm for coarse-textured soils (< 12% clay). Next, we calculated the amount of water transpired from each soil layer by the various plant functional types for two types of ecosystem patches: one with a coarse (9% clay) soil and another with a fine-textured (21% clay) soil. These results indicated that there was little vertical water partitioning among the plant functional types (figure 16-10). For both soil types, the largest water use was from the top 20 cm, which provided 35% of the water transpired for the coarse soil and 46% of the water transpired for the fine soil.

Plant Functional Types Are "Drinking from Same Cup"

Indices of water use efficiency of individual plant species or entire communities have been a useful way to understand the connection between rainfall and production in arid lands (Fischer and Turner 1978). In principle, plants that use or transpire more of the available soil water relative to the amount lost as evaporation will have greater production. Hence, external factors—natural or anthropogenic—that influence the relative partitioning of soil moisture between transpiration and evaporation can potentially alter community productivity. For example, if climate change leads to increased winter rainfall in the Jornada Basin, we could speculate that this will lead to an increase in transpiration relative to total ET, because of decreased evaporation and thus increased soil water storage in winter. Similarly, we could hypothesize that if overgrazing caused a shift from grass- to shrub-dominated plant communities, the ratio of transpiration to ET (T/ET) would be reduced because of greater evaporation associated with increased bare soil. Beyond such speculations, to make future assessments of the extent to which external factors will result in changes in T/ET in arid land communities, we must first

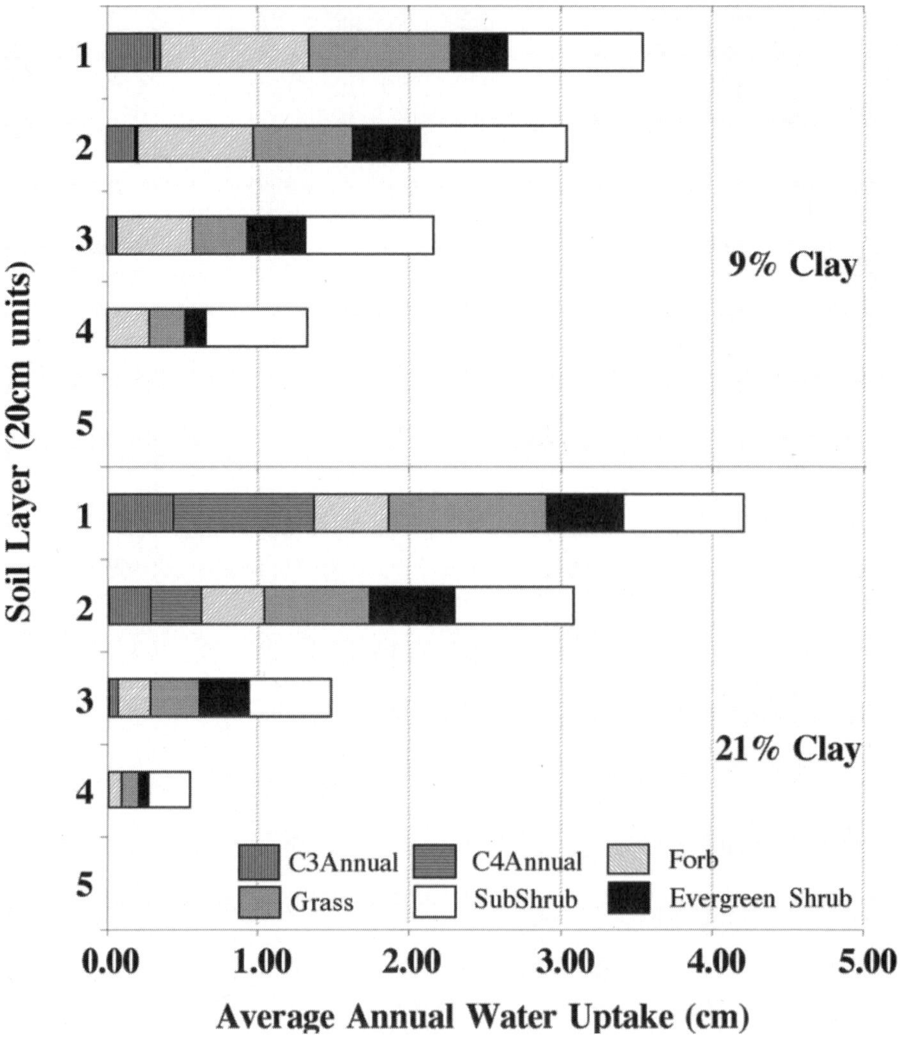

Figure 16-10. Average annual water uptake by various plant FTs shown for 20-cm increments in soil depth. Average is based on simulation covering the period 1915–2000. Redrawn from Reynolds et al. (2004).

have a good understanding of the natural variability in T/ET, as well as the expected divergence in T/ET between different community types (e.g., grass- versus shrub-dominated).

To address these latter two questions, we used PALS-FT to examine long-term variability in plant water use by communities dominated by different plant functional types. We examined differences in the timing of transpiration and depth of soil water distribution and water uptake within two ecosystem patch types characteristic of the Jornada Basin: a shrub-dominated patch (shrub piedmont zone) and a grass-dominated patch (mixed piedmont zone) (see figure 16-3). Contrary to our expectation of large differences in community water use, the simulation showed that over a 100-year period, the average T/ET was 34% for *both* the grass- and shrub-dominated communities (using Las Cruces, NM, weather data; see Reynolds et al. 2000). On the other hand, there were large annual differences in total and seasonal transpiration for a particular patch type and for a given year. This is illustrated in figure 16-11, where for two successive years, 1965 and 1966—with nearly identical total rainfall (177 mm and 167 mm, respectively)—the T/ET values predicted by PALS-FT for the grass-dominated patch ecosystem were 22% and 32%, respectively, compared to 33% and 43% for the shrub-dominated patch. These differences are a function of plant functional type responses to the different distributions of the total rainfall throughout the two years. Note that, for example, total T/ET increased in the shrub community in 1966 as compared to 1965 (by 10%) whereas total rainfall actually decreased 10 mm!

The availability of soil water to plants depends largely on the inter- and intra-seasonal distributions of the annual precipitation, interacting with the phenological behavior of the plants. Hence, the species composition of a community can result in relatively different patterns of soil moisture use. For example, a year with a large amount of precipitation in midsummer would favor growth (transpiration) of C_4 grasses and summer annuals, whereas a year with a large amount of precipitation in winter and early spring would favor evergreen shrubs and winter annuals. However, over a number of years total water use is apt to be relatively similar in different communities because differences that favor one or the other tend to balance out. An important corollary is that we found little evidence of vertical separation of soil water use between the grass- versus shrub-dominated communities, supporting Hunter's (1991) analogy for warm desert systems that most of the plants are "drinking from same cup with different straws," but as our simulations show, not always at the same time.

Drought Cycles and Water Use by Plant Functional Types

Because the Jornada Basin and other desert regions may be characterized by roughly decadal-length periods of drought or above average rainfall (Conley et al. 1992; Reynolds et al. 1999a), we used PALS-FT to further examine the variability in NPP as a function of such natural climate cycles. From the long-term rainfall records in the Jornada Basin (average annual rainfall = 245 mm), we selected three periods to examine: (1) a normal decade (1968–77, 250 mm average rainfall), (2) a dry decade (1947–56, 33% below normal or 166 mm), and (3) a wet decade (1984–93, 32% above average rainfall or 325 mm). In table 16-7 the details of the simulation schemes employed for each decade are given, as well as a summary of the results (for complete details, see Reynolds et al. 1999a). The results are also presented in terms of the coefficient of variation in rainfall and NPP, which facilitates a direct comparison to the study of Le Houerou et al. (1988).

The dry decade is characterized by somewhat greater reduction in summer

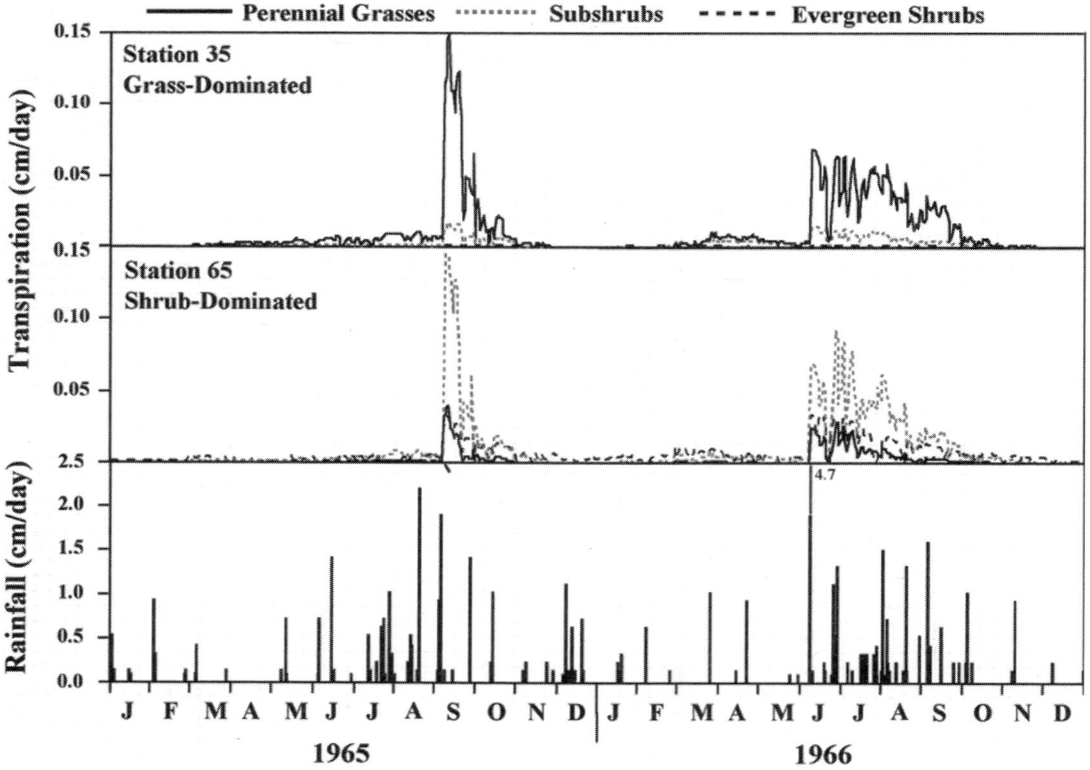

Figure 16-11. Simulation of plant FT transpiration using PALS-FT for a grass- and shrub-dominated patch (Stations 35 and 65, respectively; see figure 16–3). These simulations used daily rainfall for 1965 (177 mm) and 1966 (167 mm), which are years with nearly identical total precipitation. For clarity, only three plant FTs occurring in both patches are shown. Modified from Reynolds et al. (2000).

Table 16-7. Simulated annual NPP of various plant FTs of the Jornada Basin using PALS-FT in response to decadal variations in rainfall

	Plant FTs[a]	NPP (g m^{-2})	CV (NPP)	% Change in NPP vs. Normal
Normal decade (1968–77)	C_3 winter annuals	5.19	1.20	—
Average rainfall = 250 mm	C_4 summer annuals	14.57	1.16	—
(CV rain = 0.29)	Perennial forbs	15.48	0.29	—
	Perennial grass	22.39	1.12	—
	Evergreen shrubs	34.92	0.45	—
	Total	***92.54***	***0.56***	—
Dry decade (1947–56)	C_3 winter annuals	4.18	1.65	–20
Average rainfall = 166 mm	C_4 summer annuals	7.37	0.97	–50
(33% below normal)	Perennial forbs	10.79	0.59	–30
(CV rain = 0.35)	Perennial grass	8.98	0.62	–60
	Evergreen shrubs	26.48	0.73	–25
	Total	***57.81***	***0.60***	***–38***
Wet decade (1984–93)	C_3 winter annuals	11.27	1.75	+ 120
Average rainfall = 325 mm	C_4 summer annuals	18.29	1.58	+ 25
(32% above normal)	Perennial forbs	18.01	0.45	+ 15
(CV rain = 0.26)	Perennial grass	138.17	0.89	+ 500
	Evergreen shrubs	89.57	0.33	+ 150
	Total	***275.30***	***0.52***	***+ 300***
Le Houerou et al. (1988) (fig. 9) (CV rain = 0.34)	77 arid rangelands	**137**	**0.49**	

Note: for each decade, 10 simulations of 10 years each were run, randomly mixing the chronological sequence of years for simulation runs 2–9. Modified from Reynolds et al. (1999a).

[a] See table 16-2 for representative species.

rainfall compared to winter, and simulated NPP of C_4 perennial grasses was found to decline by 60%, whereas NPP of the shrubs declined by only 25% (table 16-7). These results are consistent with the findings of Gibbens and Beck (1988), who reported that aboveground cover of the principal range grasses of the Jornada Basin were severely impacted by this dry decade. The wet decade was a period of slightly increased summer moisture (10%) and greatly increased winter and spring rainfall (50% and 85% above normal, respectively). The PALS-FT simulations predict that the C_4 perennial grasses would again be most impacted, having a 500% increase in productivity over this decade (table 16-7). We attribute this to the increased spring rainfall, which would benefit the simulated production of C_4 perennial grasses by providing carryover moisture from early spring (when they break dormancy) until midsummer when the normal monsoon rains usually begin.

Although these results demonstrate significant shifts in plant functional type productivity and standing biomass, they reveal that the variability in total annual NPP is considerably greater than the variability in rainfall both within and between decades (table 16-7). The 33% decline in rainfall during the dry decade resulted in a nearly 40% decline in NPP, whereas the 32% increase in rainfall during the wet decade caused a 300% increase in NPP. These findings are consistent with observed productivity measurements reported by Le Houerou et al. (1988), which suggest that annual productivity is not linearly related to rainfall (see figure 16-9a). Unlike Le Houerou et al., by using a simulation model we are able to identify the responses of individual plant functional types from the total variation; hence, we are able to ascertain the *cause* of the variation in total production: namely, some plant functional types have extreme variation in NPP (e.g., annuals species and grasses), whereas others (e.g., shrubs) are less variable (see discussion in Reynolds et al. 1999a; figure 16-9b).

What Did We Learn from These Simulations?

First, it is apparent from our simulation studies that a great diversity of plant productivity responses can be produced by year-to-year variations in precipitation. However, it is inappropriate to make sweeping statements regarding specific growth responses to rainfall per se because of pronounced interactive effects between precipitation, soil water, and plant functional type responses (figure 16-2). Our studies further suggest that for most soil types and in most seasons, there is little separation of soil water with depth. Thus the coexistence of plant functional types is most directly a function of the divergent abilities of plant functional types to capitalize on temporal differences in soil moisture availability.

Second, in spite of greatly differing productivity responses of plant functional types to variable annual rainfall amounts and distribution, our simulations of evapotranspiration over long-term periods indicated that there was a relatively consistent pattern of total soil water use by different community types (grass- versus shrub-dominated). This is somewhat unexpected given our conclusion of differences in specific patterns of productivity (and associated water use) by the different plant functional types. However, although there are certainly differences in the specific daily, weekly, or even seasonal patterns of water use associated with different plant functional type compositions, we conclude that the bulk of the seasonal rainfall will be lost as evaporation and transpiration via whichever plant functional types are active at the time soil moisture is available.

Last, our simulations of decadal shifts in total amount of precipitation were, of course, expected to have relatively strong effects on plant productivity. Indeed, our simulations demonstrate the expected: that the decade of summer drought impacted summer-active C_4 perennial grasses much more than C_3 shrubs, or other winter/spring-active plant functional types. But these simulations also produced a somewhat counterintuitive result: that C_4 grasses would also be most responsive to shifts in late winter or spring rainfall. These findings contrast with a study at a nearby site in the Chihuahuan Desert, in which Brown et al. (1997) reported that C_3 shrubs had the greatest response to increased winter/spring rainfall over nearly the same period (1979–92) as for our simulation. The shifts in precipitation over this period were apparently different across the northern Chihuahuan Desert (Brown et al. 1997), suggesting that relatively subtle shifts in seasonal precipitation patterns could elicit relatively large differences in ecosystem responses across these arid land systems—thus lending support to the hypothesis that a relatively delicate balance exists between grass- versus shrub-dominated ecosystems, which can be tipped in part by seasonal shifts in precipitation (Reynolds et al. 1997).

Flowpath Scale

The PALS-FT model has been tested under a range of arid and semiarid environmental conditions, although it is restricted to small-scale, homogenous patches. To overcome this limitation, we developed the mosaic arid land simulator (MALS, table 16-1), which is a spatially explicit adaptation of PALS. We implemented MALS along the LTER transect (figure 16-3) by dividing the transect into 270 contiguous grid cells, each 10 m long × 30 m wide, where each cell receives run-in from adjacent uphill cells and loses runoff to adjacent downslope cells. Complete details of MALS, including a description of our spatially explicit hydrological model, are provided in Gao and Reynolds (2003).

We used MALS to examine two climate change scenarios considered plausible for the arid and semiarid regions of the United States (Sheppard et al. 2002): (1) a shift in rainfall seasonality, to wetter winters and drier summers; and (2) an increase in the number of large precipitation events. Both of these scenarios have been touted as potentially facilitating shrub invasion into grasslands; for example, an increase in the number of large precipitation events increases soil water recharge at deeper layers, which favors deep-rooted shrubs. Details of how we implemented these scenarios are given in Gao and Reynolds (2003). Here, we focus only on the importance of runoff and run-in on the impacts of these scenarios on plant production by implementing two versions of MALS: one with and the other without runoff/run-in redistribution.

We found the effects of runoff/run-in redistribution on plant responses and soil water dynamics to be generally important and, in several instances, dramatic. In the shifting rainfall seasonality scenarios (figure 16-12a–c), the presence of runoff/run-in flows partially ameliorated the decrease in grass and herb biomass with shifts to greater winter precipitation, although the relative amount of this effect diminishes with a greater shift in rainfall. In the adjusted rainfall event sizes scenarios (figure 16-12d–f), the presence of runoff/run-in redistribution is generally much less important. The patterns are similar to shifts in rainfall seasonality but tend to disappear as larger rainfall events are shifted.

What did we learn from this modeling exercise? Our simulations generally support the hypothesis that an increase in the number of large precipitation events may favor shrub establishment and growth although these results are equivocal, depending on what constitutes a large event and the timing of such events. We found complex interactions among the amount/seasonality of rainfall and its redistribution in the landscape via run-in and runoff, and subsequent water availability for growth and reproduction of shrubs versus herbaceous plants at various landscape positions. These results suggest that a mechanistic understanding of the linkages between biotic and abiotic factors over the landscape is crucial to predict large-scale changes in arid lands.

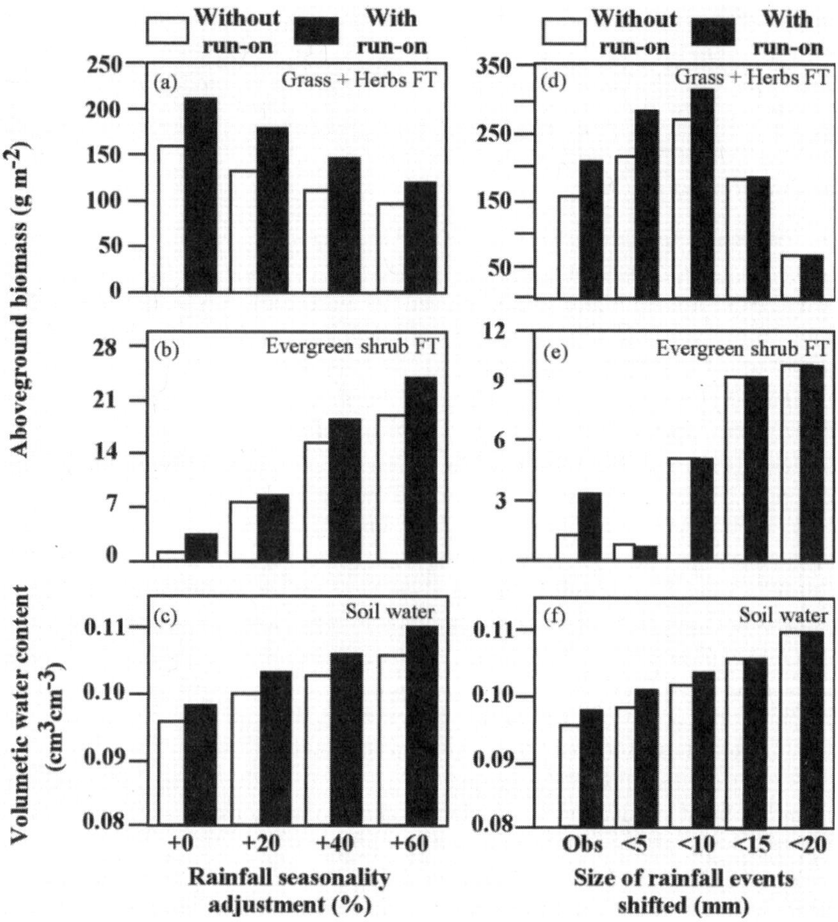

Figure 16-12. Aboveground biomass for grass and herbs (a, d), evergreen shrubs (b, e), and volumetric soil water (c, f). For panels a–c the amount of daily rainfall from October 1 to May 31 each year was increased by 0% (observed), 20%, 40%, and 60%, and rainfall from June 1 to September 30 was adjusted accordingly to preserve the total observed annual rainfall. For panels d–f, four scenarios that progressively redistribute rainfall into larger size events while maintaining the total observed rainfall each year (and the ratio of summer to winter/spring precipitation). Results are shown for two versions of the model: with runoff/run-on redistribution (black bars) and without runoff/run-on redistribution (white bars). In the without runoff/run-on redistribution, runoff produced at an individual grid cell (i.e., when precipitation > infiltration) was removed from the system. Redrawn from Gao and Reynolds (2003).

Conclusions

Throughout this chapter we have used models to explore the manifestation of discontinuous and unpredictable rainfall as available soil water, and its effects—at various hierarchical levels—on the short- and long-term functioning of ecosystems of the Jornada Basin. We now turn to a consideration of some important caveats.

Natural and human disturbances will continue to affect the Jornada Basin and other arid ecosystems of the globe in unknown and complex ways. In spite of much progress, significant gaps remain in our knowledge, and as a result, simulation modeling will continue to play a major role in global change research in arid lands. Although we must rely heavily on models, they are based on an uneven mixture of facts, assumptions, and conceptualizations. There is an omnipresent danger that in the absence of full (or even partial) understanding, untested hypotheses will become incorporated into models and over time be forgotten (or ignored) with unknown, yet potentially severe consequences.

Such concerns are especially germane for arid ecosystems, which consist of copious slow processes that may abruptly switch their rates of change in response to changing environmental drivers. Responding to these and other concerns, we have previously noted that this leads to a somewhat troubling paradox: in the absence of data and understanding, there tends to be a heavy reliance on models, the quality of which are in turn highly dependent on the quality and availability of data and understanding (Reynolds et al. 1996b, 2001).

In spite of these limitations and concerns, we must continue to develop models at local, regional and global scales to address the questions being posed by resource managers and policy makers. It is important to recognize that although we should not necessarily trust models to accurately predict arid land responses to climatic change or other human perturbations, model simulations can help us understand ecosystem functioning and indicate how sensitive arid lands may be to projected human impacts. As our knowledge improves, our models will improve. Nevertheless, attempts to predict the future behavior of arid ecosystems are operating at the frontier of our science and its methods.

17

A Holistic View of an Arid Ecosystem: A Synthesis of Research and Its Applications

Brandon T. Bestelmeyer
Joel R. Brown
Kris M. Havstad
Ed L. Fredrickson

A primary objective of the Jornada Basin research program has been to provide a broad view of arid land ecology. Architects of the program, more recently scientists with the Jornada Basin Long-Term Ecological Research (LTER) program, felt that existing ecological data sets were usually of too short a duration and represented too few ecosystem components to provide a foundation for predicting dynamics in response to disturbances (NSF 1979). This recognition gave rise to the approach of using long-term and multidisciplinary research at particular places to advance a holistic and broad-scale but also mechanistic view of ecological dynamics. Such a view is essential to applying ecological research to natural resources management (Golley 1993; Li 2000). In this synthesis chapter we ask: What has this approach taught us about the structure and function of an arid ecosystem? How should this knowledge change the way we manage arid ecosystems? What gaps in our knowledge still exist and why?

The Jornada Basin LTER was established in 1981 with the primary aim of using ecological science to understand the progressive loss of semiarid grasslands and their replacement with shrublands. This motivation echoed that which initiated the Jornada Experimental Range (JER) 69 years earlier. The combined, century-long body of research offers a unique perspective on several core ideas in ecology, including the existence of equilibria in ecosystems, the role of scale, landscape heterogeneity and historic events in ecosystem processes and trajectories, and the linkage between ecosystem processes and biodiversity. From this perspective, we examine key assumptions of this research tradition, including the value of the ecosystem concept and the ability to extrapolate site-based conclusions across a biome. The Jornada Basin research program is also uncommon in its close ties to long-term, management-oriented research. The research questions first asked by the U.S. Forest Service and later by the Agricultural Research Service (ARS), such as how to manage livestock operations, frame much of the Jornada Basin research. This allows us to consider the contributions of this research and synthesis toward answering management questions.

The Jornada Basin Ecosystem

The research presented in this volume suggests that the Jornada Basin (rather than the individual watersheds within it) provides a reasonable delineation of an "ecosystem object" (in a narrow sense, as in Golley 1993), or perhaps a metaecosystem (Loreau et al. 2003), in which internal connections are relatively strong across several compartments (e.g., hydrologic and eolian fluxes and animal movements). Of course, there are fluxes into and out of the basin (chapter 9) and the basin is also part of a greater whole.

Within the basin, the patch has served as the fundamental unit of organization. The patch includes plants and their associated interspaces (Schlesinger et al. 1990). Feedbacks between plants and soil comprising a patch (chapter 5) lead to patch persistence in the face of abiotic forces (e.g., erosion) that may otherwise tend toward patch disintegration. Disturbances and regeneration of vegetation lead to changes in patch identity (e.g., a grass or shrub patch) and location over time (White and Pickett 1985).

Past and current Jornada research suggests there are general rules by which patch mosaics (and their effects) are organized within landscapes via geomorphic patterning (Ludwig and Cornelius 1987; McAuliffe 1994; Wondzell et al. 1996; see also chapter 16). Because research conducted in different parts of the landscape can be compared and connected via these rules, we review past results following a multiscale landscape-geomorphic framework (figure 17-1). Although the elements of the framework are specific to parts of the Jornada Basin and the Chihuahuan Desert, the processes it represents are observed throughout the Basin and Range Physiographic Province of North America and in other topographically diverse, arid systems of the world (Gile et al. 1981; McAuliffe 2003). These patterns are the foundation for understanding the spatially interactive mechanisms of ecosystem change described in chapter 18. In the sections that follow, we summarize four key insights derived from Jornada research that contribute to this framework and elaborate on the questions asked at the inception of this research program.

Plant–Soil–Animal Feedbacks Govern Patch Transitions

Perhaps one of the most significant contributions of the Jornada Basin program to date is the recognition that several parallel feedbacks govern changes in the characteristics of patches (Schlesinger et al. 1990). Nowhere are these feedbacks more evident within the Jornada Basin than in the progression of grass-to-shrub transitions. Schlesinger et al. (chapter 5) succinctly describe such transi-

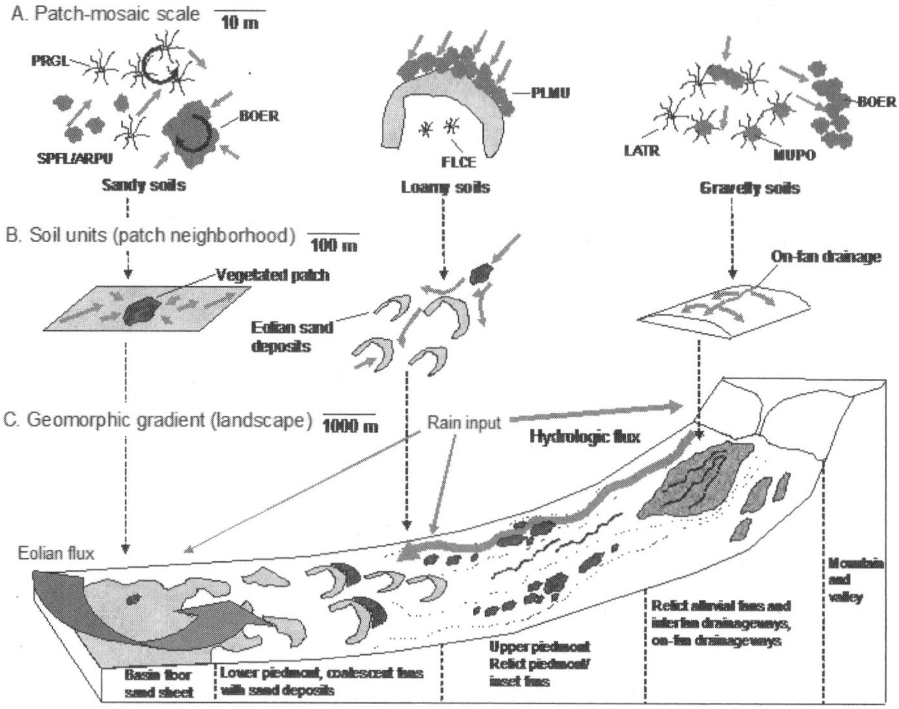

Figure 17-1. A graphical framework describing the relationships of plants and material fluxes occurring on common geomorphic surfaces at the Jornada Basin. The diagram is based on the west slope of the San Andres Mountains. (a) The spatial organization of plant patch mosaics and small-scale fluxes within each landform; ARPU = *Arisitida purpurea* (purple threeawn), BOER = *Bouteloua eriopoda* (black grama), FLCE = *Flourensia cernua* (tarbush), LATR = *Larrea tridentata* (creosotebush), PLMU = *Pleuraphis mutica* (tobosa), PRGL = *Prosopis glandulosa* (honey mesquite), SPFL = *Sporobolus flexuosus* (mesa dropseed). The directionality of various fluxes occurring in each of the dominant soil types are represented by arrows. Gray arrows are water, and dotted arrows are eroded soil material. (b) The soil/topographic context of patch mosaics. Dark patches represent dense vegetation, light areas represent exposed sandy surfaces. Sandy soils occur on mostly level surfaces, loamy areas occur on gentle (1–3%) slopes, and gravelly soils may have variable morphometry (e.g., convex). (c) The geomorphic gradient receives variable rain inputs with elevation depending on the interaction of weather systems with mountain ranges. Runoff from mountain footslopes is concentrated into interfan drainageways during high-intensity storms. The basin floor is dominated by sandy soils, and southwesterly/westerly winds deposit materials from the basin floor to positions higher upslope. Hydrologic and eolian fluxes interact in the central zone characterized by arcuate sand deposits.

tions as a reconfiguration of biotic activity toward shrubland. A host of interactions have been identified that regulate the rate and nature of transitions.

There is historical evidence that variability in the magnitude and coincidence of multiple stressors, particularly extended drought periods cooccurring with instances of overgrazing by livestock, have led to episodic losses of grass patches. These periods include years during the early 1890s, 1910s, 1930s, and 1950s (chapters 10 and 13). Climate records reveal that these droughts varied with respect to the combination of rainfall and temperatures (chapter 3), and this produced varying effects on vegetation. Summer droughts associated with increased frequencies of El Niño periods (featuring high winter rainfall) over the past century may have favored shrub establishment and survival at the expense of perennial grasses (Brown et al. 1997; see also chapter 3). It is unclear why particular patches of certain species (e.g., black grama, *Bouteloua eriopoda*) are lost while others of the same species survived during a given drought episode (Gibbens and Beck 1988). Consequently, current approaches to explaining patterns in the Jornada Basin's long-term quadrat data have emphasized landscape context in addition to patch characteristics (chapter 10).

Unlike grasses, locally invading shrubs are little affected by livestock and can access deeper, more reliable sources of soil water than can grasses (Burgess 1995; Gibbens and Lenz 2001; see also chapter 6). Thus, shrubs are more likely to survive periodic droughts and capitalize on the space and resources made available as grasses decline. Once shrubs, particularly honey mesquite (*Prosopis glandulosa*) and creosotebush (*Larrea tridentata*), become established in a patch, a number of characteristics favor their persistence. Mesquite, for example, is less reliant on N mineralization (which declines by half as grasslands change to shrublands) due to its ability to fix nitrogen. Additionally, shrub physiognomy slows rainfall impact at the soil surface, thus reducing local erosional losses under shrubs when compared to interspaces (Whitford et al. 1997; see also chapter 7). This process may also lead to a grass–shrub symbiosis when grass cover is low because shrubs create stable microenvironments for grass establishment and persistence. Understory grasses further reduce raindrop impact and promote local infiltration (Abrahams et al. 2003).

Though not fully understood, current concepts hold that major vegetation transitions at the Jornada are related to changes in soil water availability and its interaction with decomposition and nutrient availability (Gutierrez and Whitford 1987a), which differs from other sites (Lauenroth et al. 1978; Van Cleve et al. 1996; Shaver et al. 2001). Differences in C and N cycling patterns can be viewed as both consequences and drivers of vegetation change. Shrubland states may be sustained, for example, by increasingly localized production and nutrient cycling occurring in resource islands found beneath shrub patches (Schlesinger et al. 1990). Thus the interplay of rainfall patterns, soil degradation, and variable nutrient limitation in space and time may regulate the pace of vegetation change.

Resource island stability leads to the accumulation of material that is physically redistributed from interspaces between shrubs. In addition, biotic redistribution by animals attracted to resource islands, such as rodents, lagomorphs, birds, and ants, may be important (e.g., Dean et al. 1999). In contrast, the potential activity of termites, which are the major animal contributors to nutrient cycling, is little affected by changes associated with grass–shrub transitions (chapter 12). On many soils, termites appear to be ubiquitous and recruit rapidly to litter sources during favorable climatic conditions (Nash et al. 1999). Only extreme soil degradation associated with the formation or exposure of cemented soils in shrub interspaces seems likely to restrict termite activity. The patch-level consequences of biotic effects on nutrient cycles are as yet only partly understood.

Changes in vegetation physiognomy associated with shrub encroachment may favor populations of rodents and lagomorphs, leading to increased herbivore pressure on seedlings and the reproductive parts of adult grass and shrub plants (Nelson 1934; see also chapter 12). This effect may limit plant recruitment (figure 17-1). On the other hand, increased small mammal densities may increase rates of biopedturbation, improve rates of water infiltration in interspaces, and increase the likelihood of seed germination (Whitford and Kay 1999). For a given patch of mesquite shrubland, it remains unclear (1) whether biotic or abiotic limitations to grass recovery in shrub interspaces are most important and (2) whether particular taxa, such as small mammals, have a net positive, negative, or neutral effect on grass abundance at the patch scale.

Existing plot-level aboveground net primary productivity (ANPP) data in grasslands and shrublands suggest that reconfiguration of biological activity has not led to reductions in energy capture (assuming similar initial potential) at broader scales. A shift from grass to shrub dominance appears to involve changes in the identity of the producers, rather than a significant change in the overall ANPP production rates (Huenneke et al. 2002; see also chapters 5 and 11, this volume). Furthermore, because existing measurements do not consider belowground productivity, it is possible that total productivity (TNPP) and carbon sequestration are greater in shrublands than in grasslands (see House et al. 2003). Existing data, however, suggest that grasslands and shrublands have similar efficiencies with respect to the use of N and water, despite strong differences in how these nutrients are acquired (Reynolds et al. 1997; see also chapter 8). If borne out, this conclusion would support the view that TNPP is a constant property of the Jornada ecosystem (at least at the basin scale) that is constrained by energy and resource availability rather than species or functional group composition (Enquist and Niklas 2001; Brown 2004).

Variable Fluxes Drive Landscape Organization (and Reorganization)

The characteristics and dynamics of patches are clearly related to the movement of organisms and materials across the landscape. Although Jornada researchers have recognized this for some time with regard to particular processes (Schlesinger and Jones 1984; Wondzell et al. 1996; Gillette and Chen 2001), the means to measure and interpret interactions among these processes in a spatially explicit fashion have only recently become available (chapter 18). Nonetheless, the influence of the spatial organization of geology, soils, and plant communities on material redistribution at several spatial scales can be described in nonexplicit terms. Three vectors have been examined in detail at the Jornada: wind, water, and animals.

The influence of regional weather patterns is modified by the basin's internal spatial organization that in turn creates additional spatial patterning. For example, in exposed areas north of the Doña Ana Mountains, the southwesterly erosive winds have organized mesoscale spatial patterns of sand accumulation and erosional deflation (chapter 2) that influence soil texture, the depth of petrocalcic horizons, and thus plant community development and responses to disturbance (chapter 6). Within these zones at finer scales, erosive wind direction affects the spatial organization of mesquite shrubs and patterns and rates of grass mortality (Okin and Gillette 2001). In turn, the preponderance of honey mesquite on the extensive sandy basin floor results in a net flux of dust out of the basin (chapter 9). Similarly, the position of small mountain ranges interacts with moisture arriving to the basin at different times of year to create multiyear spatial patterning in rainfall amounts (chapter 3).

The precipitation arriving to different parts of the basin surface is redistributed across different distances and in different directions, depending on soil properties and slope. Localized water redistribution on the sandy areas of the basin floor may produce "spots" where petrocalcic horizons are absent ("playettes"; chapter 2) and aboveground plant production and grass cover are high relative to surrounding areas. Vegetation bands or "stripes" may be produced on the gentle slopes and loamy soils of lower piedmont positions (see Aguiar and Sala 1999). In upper piedmont positions, the distribution of surface flow alternates between narrow channels and broader beads that create yet another form of spatial regularity in plant community structure (chapter 7).

Although little moisture is available for groundwater recharge in the basin, rainfall events producing significant run-off lead to surface water transfers among landforms (Phillips et al. 1988; see also chapter 7). These transfers may be critical determinants of plant community patterns. Run-in water may be a significant factor in maintaining productive tobosa (*Pleuraphis mutica*) grasslands in lower piedmont and marginal basin floor positions (Herbel and Gibbens 1989; see also chapter 6). The soils of these areas may also accumulate unusually high amounts of organic carbon (chapter 4). Historical decreases in grass cover in upslope positions may have allowed increased surface water runoff, resulting in increasing cover downslope over the same period (Herbel et al. 1972). Although we currently have few spatially explicit data at suitable scales of space and time to examine, it is likely that basin-scale vegetation change can be understood as much by the redistribution of water as by local, within-patch changes emphasized in earlier work (Noy-Meir 1985; see also chapter 18).

Both wind and water fluxes interact with vegetation to drive changes in nutrient distributions and production (Breshears et al. 2003; see also chapters 5 and 11). For example, under historical grassland conditions on sandy soils, internal N cycling is generally much greater (50 kg N/ha/yr) than new inputs ($<$ 3 kg N/ha/yr). Mesquite dominance increases both the rate of symbiotic N fixation as well as redistribution within and outside of the basin due to wind flux. The change in cycling and redistribution of nutrients and water across the basin results in highly variable ANPP estimates such that basin-scale averages are difficult to assess. Consequently, current plot-scale measurements are ill equipped to detect the effects of basin-scale redistribution of key nutrients. It is possible that ANPP reductions at one scale are coupled with stability or even increases at other scales.

Like abiotic vectors, livestock have variable effects across the landscape. The distribution of fences, anthropogenic resources (livestock water tanks, mineral supplements), and patches dominated by different plant species influence livestock movements and their consequent effects on plants. Preferred dominant grasses, including dropseeds (*Sporobolus* spp.), black grama, and threeawns (*Aristida* spp.) are associated with sandy and gravelly soils, resulting in a tendency for livestock to aggregate in these areas. Heavy use of such areas, especially during drought, leads to rapid and persistent grass loss (chapter 13), further concentrating livestock movements to remaining preferred patches and eventually to less preferred species, depending on grass phenological state. These positive feedbacks and associated soil degradation can lead to nonlinear rates of grass loss and erosion across landscapes (van de Koppel et al. 2002). Soil degradation is exacerbated when preferred grass species are dominant and associated with erosion-susceptible landforms and soils. Fences and anthropogenic resources can be used to regulate the spatial distribution of livestock to minimize negative impacts, but short-term, mesoscale climatic variability imposes dynamic changes in the vulnerability of grass patches to extinction and has proven difficult to track effectively (and economically).

High Soil Heterogeneity Governs Basin-Level Variation in Key Processes

Geological and geomorphic processes create a template of soil differentiation and potential interconnections among soil units (Gile et al. 1981; McAuliffe 1994). Geomorphology strongly influences the nature of patch-level feedbacks and the rates and directions of fluxes within the Jornada Basin as well as fluxes into and out of the basin. There has been significant progress in recognizing that these linkages explain a wide range of ecological phenomena (Wondzell et al. 1996; see also chapters 2, 6, 9, and 14). These linkages will be explored more fully in the next phase of Jornada research (chapter 18).

The modern structure of the Jornada Basin has been traced from the depositional environments of Paleozoic basins and seas, Oligocene volcanism, block uplifts and basin subsidence from the Miocene to the present day, the extension of the Rio Grande Rift, the arrival of the river, and shifts in its location as well as the destination of the sediments it carried. In comparison to some other arid landscapes (Stafford Smith and Morton 1990), landforms and soils of the region are young, and their properties are determined by ongoing climatic cycles that drive shifting rates of erosion, deposition, soil formation, and soil destruction.

These processes have produced several strong gradients in soil properties at several scales and a high degree of spatial organization in plant communities (Wierenga et al. 1987). For example, the sedimentary bedrock alluvium on the eastern side of the basin and the alluvium derived from the ancestral Rio Grande in the central and western basin floor are primarily sands, and this pattern has a major influence on the patch dynamics and feedback behaviors described earlier. Within these areas, variation in the depth and development of calcium carbonate–rich soil horizons exerts a strong influence on soils and vegetation. In other areas, the presence of clay-rich horizons can have important positive effects on grass persistence (Gibbens and Beck 1987), and the development of these horizons may be reduced in the presence of high amounts of calcium carbonate in parent materials (Gile et al. 1981). The shift from rhyolite and monzonite to limestone-derived parent materials across the southern portion of the basin yields shifts in the availability of calcium carbonate as well as rates of weathering, and this affects the composition of plants, the identity of encroaching shrub species, and grass–shrub–animal interactions. Thus, the consequences of land use history are a function of both disturbance intensity and the inherent variation in geology and soils.

Biodiversity Exhibits Both Vulnerability and Resilience in a Dynamic Landscape

The processes structuring vegetation and soils in the Jornada Basin can be linked to some biodiversity patterns. The effects of grassland–shrubland transitions on ANPP, for example, are mirrored to some extent by their effects on biodiversity (Brown et al. 1997). Some elements of animal diversity appear not to respond to transitions, others decrease, and others increase in abundance. The net change in summary diversity measures may be low (e.g., Bestelmeyer and Wiens 2001b), but there is some turnover and loss of grassland obligate species, in some cases balanced by colonization of shrubland obligates (Naranjo and Raitt 1993; Pidgeon et al. 2001; see also chapter 12). Some grassland birds present at the time of European colonization may have already been driven regionally extinct and the fauna generally impoverished (Pidgeon et al. 2001). For taxa such as ants, however, the Jornada landscape may be more diverse than desert grasslands with few shrubs due to the abundance of native (and even rare) shrub-associated species (Bestelmeyer et al. 2005). Thus, shrub invasion may have enhanced certain aspects of animal species diversity.

The mosaic of vegetation and soil properties imparts a high degree of beta diversity for ants, lizards, and rodents across the Jornada Basin (chapter 12), but these patterns have been poorly examined for most taxa. Drought dynamics are superimposed on this mosaic to create strong spatiotemporal variability in resources used by animals. This leads to strong variation in population densities over time. Patterns of apparent local extinction and recolonization in some taxa (e.g., the hispid cotton rat, *Sigmodon hispidus*, occupying playa grasslands) suggest that meta-population dynamics may impart resilience to local species diversity in the Jornada ecosystem. As yet, however, there are not enough data on dispersal or patch occupancy for any species to evaluate the potential for habitat fragmentation to reduce species diversity. Consequently, it is not clear what kinds of habitat changes (e.g., exurban development, degrees of shrub encroachment) would produce habitat fragmentation for particular species.

Implications for Ecological Theory and a Long-Term Research Approach

Long-term ecological studies are inherently limited by the questions and concepts that framed them at the time studies are initiated. Emerging from a research tradition derived from the ARS and the International Biological Programme (IBP) of the 1970s, later research placed an emphasis on understanding whole ecosystems in terms of their component parts. These parts were typically trophic levels (represented by particular taxonomic groups) that exchanged and stored energy within a bounded ecosystem following the Lindeman ecosystem paradigm. It was hoped that data on biomass and energy flux through species populations could be assembled in systems models to predict dynamic behavior of the ecosystem and that these relationships could be generalized across "wide regions" (Golley 1993).

Jornada research initiated in the 1980s preserved the emphasis on production and trophic structure but recognized the need for more detail on the mechanisms underlying vegetation change. These studies, many of which are described in this volume, provide a detailed, multifaceted view of processes associated with desertification. Nonetheless, we recognize that this aggregate reductionist view continues to be constrained by (1) the small-scale, nonhierarchical, and spatially inexplicit nature of many observations and experiments; and (2) the opportunistic (and unfulfilled) integration of results across ecosystem components and individual investigators. These limitations were recognized by Eugene Odum at the inception of the IBP (Golley 1993), but the concepts and technologies to act on them are still maturing. Indeed, the need to address these limitations and develop holistic approaches spawned the subdiscipline of landscape ecology (Wiens 1999). This volume represents a first step in the development of an integrated, multiple-scale approach for the Jornada.

Nonetheless, with the benefit of recent conceptual advances and the new approaches being taken by the Jornada Basin research group, we can evaluate some long-held assumptions that have guided the practice and interpretation of some earlier research, including (1) patch-based correlations of physical and biological variables can be used to characterize ecosystem dynamics, (2) measurements of ongoing processes explain current patterns of ecosystem organization, and (3) site-specific conclusions can be generalized within and across biomes.

The Value of Patch-Based Correlations Is Limited

Different processes with different inherent scales of action influence patches at particular points within the Jornada. Thus, the value of correlations between local vegetation and local soil properties is limited. For example, satellite imagery and geomorphic studies reveal that the dynamics of northern basin floor positions are governed by eolian fluxes of soil from the border of the Rio Grande Valley, but these fluxes are buffered to the south by the Doña Ana Mountains such that dominant structuring processes become more localized. The importance of these distinctions is likely to vary with time and climatic conditions. Due to this spatiotemporal variation in processes, broad vegetation classifications such as those employed in the ANPP experiment (Huenneke et al. 2002) that are not spatially stratified and sufficiently replicated are bound to miss or obscure important differences in basin ecosystem properties.

Historical Events Exert a Powerful Influence on Ecosystem Processes

Much about the structure of the Jornada ecosystem cannot be explained by studies of contemporary processes. Historical events, including the effects of prehistoric Native American settlements and ranching enterprises of the eighteenth and nineteenth centuries, may have altered the long-term trajectory of localities by initiating desertification processes that continue to unfold. These effects are often unrecognized (and unrecognizable). The influence of sequential historical events on the spatial distribution of species confounds explanations based solely on the present-day species–environment relationships (Neilson 1986; Swetnam et al. 1999; Motzkin et al. 2002). Jornada Basin vegetation is clearly not in equilibrium at any scale, and its changing patterns are a product of both historical events and ongoing processes.

Generalizations about Arid Rangeland Behavior Are Inherently Limited

In addition to the processes just noted, much of the apparent uncertainty regarding the behavior of rangelands stems from a failure to account for regional differences in climate, spatially dominant soils, and the traits of plant species contained within broad functional groups (such as shrubs and grasses). A comparison of three patterns described by Walker (2002) for a generic rangeland with Jornada patterns serves to underscore this point.

Generic Pattern 1: Vegetation on Sandy Soils Is More Resilient than on Clayey Soils

The sandy loam and loamy sand soils of the Jornada were dominated by grasses in recent history and most have been converted to eroding shrubland, whereas clay loam soils often continue to be dominated by the original, dominant grasses (Gibbens and Beck 1988). One reason for this discrepancy is that Walker considered examples in which sandy soils were (apparently) originally dominated by shrubs. If we were to consider the behavior of postthreshold shrubland states at the Jornada, we would also consider them to be very resilient. From the point of view of historical composition (and the threats to remaining grasslands), we regard sandy soils as weakly resilient. The mechanisms underlying the Jornada's inconsistency with Walker's generalization are probably related to grazing history, differences in the palatability and life history of dominant grasses, and the relative landscape positions of sandy and clayey soils (Rietkerk et al. 1997).

Generic Pattern 2: Climatic Variation and Disturbance Reverse Grassland-Shrubland Transitions

In some rangelands, high interannual variation in rainfall constrains shrub dominance because drought causes woody plant mortality and recovery of shrubs is slow when compared to grasses. Even when woody plants establish under ideal conditions of wet years and fire absence to form even-aged stands, age-related senescence leads to grass reestablishment (Walker 2002). These mechanisms maintain a dynamic savanna structure over the long-term.

Our understanding of the Jornada situation does not conform to this pattern. Recruitment of mesquite on sandy soils may be episodic, but drought-induced or age-related mortality of adults is rarely observed (Goslee et al. 2003). Dominant shrub species (mesquite and creosotebush) are well equipped to survive drought (Reynolds et al. 1999a), and mesquite may live at least 60 years on the Jornada (Goslee et al. 2003) and up to 200 years elsewhere (McClaran 2003). Consequently, vegetation change has been directional and contagious with shrublands filling in grassland areas and not retreating over the Jornada's century-long record. Velvet mesquite (*Prosopis velutina*) shows a similar pattern in southeastern Arizona (McClaran 2003), even as other shrubs (burroweed, *Isocoma tenuisecta*) conform to Walker's pattern.

Generic Pattern 3: Over Sufficiently Long Time Scales There Is One Domain of Attraction

Grassland–shrubland transitions may appear to involve thresholds separating two domains of attraction over shorter time scales, but a single domain of attraction toward savanna may be revealed over sufficiently long periods, e.g., 40–50 years (Walker 2002; Valone et al. 2002). This is likely to be true in certain cases, even within some Chihuahuan Desert grasslands, but there is no evidence that this is universally true. Some grass–shrub transitions have lasted for at least a century and current processes indicate many will last much longer. Domains of attraction may shift due to the interplay of plant life history, soil degradation and loss, and climate change (Westoby 1980). Resilience times within a domain and the existence of alternative domains are highly variable across the Southwestern United States and within particular landscapes (Bestelmeyer et al. 2003a).

The Interface of Ecology and Rangeland Management

Although a primary focus of years of work in the Jornada Basin was to examine ecological processes and mechanisms underlying desertification, the dominant rangeland management theories of this period also contributed to the design, analysis, and interpretation of our experiments. Indeed, one of the primary goals was to determine the role of livestock grazing in the conversion of desert grasslands to shrublands (chapter 1). In turn, the assumption that an improved understanding of ecological processes would result in improved management serves as a basis for current rangeland research and applications. In this vein, we contrast the prevailing ideas that guided rangeland management (and its consequences) over the past century with what we now believe given hindsight and Jornada science.

Historical Ecological Assumptions

There are three cornerstone ideas that have underlain rangeland management decisions over the past century. First, it was implicitly assumed that Chihuahuan Desert grasslands possessed a level of resilience to grazing pressure similar to that of other grasslands in North America. This notion led to early management practices based on the notion that the decline of grass abundance following temporary overgrazing could be reversed during periods of increased rainfall. Although domestic livestock grazing had been present in the Jornada Basin since the late 1500s, the emergence of ranching as a commercial enterprise did not take hold until the latter part of the nineteenth century (chapter 13). The practitioners of this new culture immigrated to the Jornada Basin from the mesic prairies to the east and brought their concepts of grassland ecosystem behavior with them. The next century of grazing management practice, policy, and research were affected by those concepts.

Second, many believed that the primary challenge facing ranchers, researchers, and policy makers was to establish a grazing capacity (maximum stocking rate [ha/animal/year] possible, year after year, without reducing forage to vegetation ratios or other resources; Holechek et al. 1998a) (Jardine and Forsling 1922). A conservative stocking rate was viewed as an appropriate strategy for coping with spatial and temporal climatic variability because "attempts to adjust stocking rate to this highly variable basis of forage have had disastrous results. A breeding herd built up to use most of the forage crop in good or even average years cannot be maintained in dry years" (JER field-day report 1948 unpublished). Using a conservative strategy, adequate forage would be available in most (but not all) years. It was assumed that ungrazed forage produced during favorable years would be available to protect soil or be used as a forage reserve in drought years. It was also implicitly assumed that the infrequent periods of overuse would not have long-term consequences.

Third, many assumed that a more equitable spatial distribution of livestock grazing pressure would reduce instances of overuse of forage where animals had previously concentrated and underuse in areas that animals had avoided (Jardine and Forsling 1922). Thus, new parts of the landscape were made available for grazing, and provisions of nutrients and fencing have been used to distribute livestock more evenly across the forage resource, presumably reducing impacts on any given point. More recently, Herbel and Nelson (1969) advocated the opportunistic rotation of livestock among pastures to take advantage of spatially variable rainfall, plant production, and plant phenological stage, for example, flowering of soapweed (*Yucca elata*). Although these strategies accounted for spatial and temporal variation in the vulnerability of forage plants, they did not account for the fine scale of this variation. Most reasonably sized management units encompass significant spatial variability in soil properties, soil resource levels, and vegetation at the patch or patch-mosaic scale. Typical livestock grazing behavior (as currently managed) results in full utilization of palatable forage in patches before moving to the next forage patch (Bailey et al. 1996; Fuhlendorf and Smeins 1997). Therefore, the livestock use in any forage patch is largely inelastic to stocking rate and improved animal distribution.

Recognition of Heterogeneity, Thresholds, and Economic Constraints

We now know that management strategies based on the preceding assumptions have led to stocking rates that were too high at many places and in several periods, and this has led to the episodic loss of grass patches and, cumulatively, to desertification. Thus, it is important to ask what could have been done differently and, more important, how can the remaining grasslands be sustained? Foremost, it is clear that grazing capacity fluctuates greatly and stocking rates must be tightly controlled and adjusted rapidly from year to year given the high spatial and temporal variability in forage production. This idea is reflected in Herbel and Nelson's (1969) recommendations. This system sets objectives for pasture condition rather than animal production. Nonetheless, most grazing systems employed then, as today, have stocking rates based on a relatively fixed grazing capacity.

Due to the high spatiotemporal variability of NPP and threshold behavior, it has been argued that maintaining acceptable levels of forage harvest through even moderate droughts would have required an unrealistic level of economic flexibility on the part of individual ranchers. This suggests that ranchers could not have heeded Herbel and Nelson's advice in the late 1800s unless they changed their basic operations and their principle reliance on a cow-calf production system. Even today, creative management alternatives such as light stocking rates, fall calving season, and integration of complementary enterprises are strongly encouraged to sustain ranching in the Southwest (Ruyle et al. 2000).

Recent advances in technologies for tightly controlling animal distribution without fencing (Anderson 2001; Provenza 2003) offer hope that some economic constraints to sustainable grazing can be overcome. In addition, the identification and use of cattle breeds that minimize production costs and provide greater market flexibility during drought may facilitate opportunistic management by ranchers. Although technology offers improved tools for management, ecological solutions that are not tied to socioeconomic innovations are unlikely to stem the tide of grass loss.

This should not imply that all remaining grasslands can be preserved even if such approaches are successful. Current applications of chemical, mechanical, and management technologies to interdict degrading processes offer little chance of success when the mechanisms driving degradation derive from regional and landscape scales (chapters 14 and 18). Historical events may have catalyzed current degradation rates and be independent of subsequent livestock management (Jardine and Forsling 1922; Campbell 1929). Our understanding of soil–plant feedback processes and multiscale redistribution of soil resources (chapter 5) implies that simply treating one symptom (e.g., shrub increase) does little to mitigate grassland loss, especially in the short term, over which most economic analysis is performed.

A basic understanding of ecological processes is a prerequisite for reasonable decisions by land users. Ecological site descriptions offer a means of communicating how processes vary over time and among climate regions, landforms, and soils (USDA NRCS 1997). At the core of the descriptions, state-and-transition models summarize how particular processes discussed in this volume combine to produce reversible and difficult-to-reverse vegetation and soil changes (Bestelmeyer et al. 2003a) as well as indicators of these processes (Ludwig et al. 2000). By specifying such processes and their indicators, land users can evaluate management actions in light of the recognition of soil and vegetation heterogeneity at several scales, linkages with surrounding areas, and the likelihood of threshold behavior in vegetation dynamics.

Social Change and New Management Challenges

Two of the conclusions offered by John Wesley Powell in "Report on the Lands of the Arid Region" delivered to Congress in 1878 were that Western lands have distinct limits set by their aridity and cannot be appropriately managed if arbitrarily dissected into fractions by the political conventions of the day (de Buys 2001). In one sense, the history of research in the Jornada Basin has reaffirmed and refined these conclusions. We now understand variability in primary production, its low extremes, and the roles of scale in our use of this ecosystem. Our resource management institutions and principles recognize the necessity of working within biological limits, but social, economic, and administrative constraints often prevent actions based on this knowledge (Ruyle et al. 2000).

In a socioeconomic setting where property rights are paramount, management of Western lands by fractions has persisted since the nineteenth century, despite Powell's recommendation, with widely reported consequences for biodiversity and human welfare. Lately, however, creative alternatives have emerged around the Western United States that allow resource management to be coordinated over ecologically appropriate regional scales while accommodating ownership of fractions. Examples include grass banks and land management cooperatives. Our Jornada Basin program and other, regional long-term research (e.g., the Santa Rita Experimental Range in southeast Arizona) can contribute to these efforts by identifying the processes driving (or constraining) vegetation dynamics on particular soils, the role of linkages between areas, and thus the appropriate extents for management coordination.

As the biological and economic realities of traditional rangeland management have become clearer, the role of the land manager has changed. Public and private land managers that dealt exclusively with livestock-based agriculture are increasingly faced with urban–exurban populations seeking scenic amenities, including working agricultural landscapes. In New Mexico, as in much of the Intermountain West, there is a shift from traditional agriculture toward an economy based on services and professional industries (Rasker et al. 2003). For most communities in these regions, future growth will be tightly linked to environmental quality, an amenity often used by industry to attract employees. Nonetheless, livestock grazing continues to be a dominant land use in these regions. The increasing diversity of land uses imposes new values and criteria with respect to the acceptable structure and composition of ecosystems. New land uses also introduce novel processes to particular areas, such as the introduction of nonnative species and habitat fragmentation by roads and houses that increase the demand for information (e.g., the behavior of animal species; Maestas et al. 2003) that has not been a focus of past Jornada Basin research. It is critical that we adjust our scientific resources to track these changing needs.

Conclusions

The contributions of Jornada Basin research to our understanding of desertification processes, particularly at the patch scale, are profound. This work has been instrumental in directing desertification research across the globe. Perhaps even more important, the long-term multidisciplinary approach has described remarkable variability of desert ecosystem function, and its causes, across space and time. This perspective reinforces the need to develop scientific and management concepts and methods that account for the unexpected magnitude of ecological variability (Shrader-Frechette and McCoy 1993; O'Neill 2001; Archer and Bowman 2002; Simberloff 2004). We are implementing both long-term ecological research and proactive management strategies in the light of this realization (chapter 18). This will require (1) measurements that are not only spatially explicit and long-term but embedded in a process-based logic that makes use of spatial and temporal information, (2) observations of processes that can be linked across scales of space and time, (3) approaches that link mechanistic studies at long-term research sites to regional variations in pattern, and (4) monitoring and management strategies that can be adapted to the socioeconomic constraints and ecological processes regulating change in particular localities. These requirements, in turn, indicate that we must adopt truly interdisciplinary approaches in addition to multidisciplinary approaches that were formerly emphasized. We now have the tools to realize the convergence of ecosystem ecology, research, and landscape ecology foreseen by Eugene Odum 40 years ago. Cultivating the institutional structures to achieve this remains a significant challenge, but it is a challenge we are attempting to meet.

18

Future Directions in Jornada Research: Applying an Interactive Landscape Model to Solve Problems

Debra P. C. Peters
William H. Schlesinger
Jeffrey E. Herrick
Laura F. Huenneke
Kris M. Havstad

The long history of research at the Jornada Basin (through the Agricultural Research Service [ARS] since 1912, New Mexico State University in the late 1920s, and joined by the Long-Term Ecological Research [LTER] program in 1981) has provided a wealth of information on the dynamics of arid and semiarid ecosystems. However, gaps in our knowledge still remain. One of the most perplexing issues is the variation in ecosystem dynamics across landscapes. In this concluding chapter to this volume, we propose a new conceptual model of arid and semiarid landscapes that focuses explicitly on the processes and properties that generate spatial variation in ecosystem dynamics. We also describe how our framework leads to future research directions.

Many studies have documented variable rates and patterns of shrub invasion at the Jornada as well as at other semiarid and arid regions of the world, including the Western United States, northern Mexico, southern Africa, South America, New Zealand, Australia, and China (York and Dick-Peddie 1969; Grover and Musick 1990; McPherson 1997; Scholes and Archer 1997; see also chapter 10). In some cases, shrub invasion occurred very rapidly: At the Jornada, areas dominated by perennial grasses decreased from 25% to < 7% from 1915 to 1998 with most of this conversion occurring prior to 1950 (Gibbens et al. 2005; Yao et al. 2002a). In other cases, shrub invasion occurred slowly, and sites were very resistant to invasion; for example, perennial grasses still dominate on 12 out of 57 research quadrats originally established in black grama (*Bouteloua eropoda*) grasslands in the early twentieth century (Yao et al. 2002b). Soil texture, grazing history, and precipitation patterns are insufficient to account for this variation in grass persistence through time (Yao et al. 2002a). It is equally perplexing that although many attempts to remediate these shrublands back to perennial grasses have led to failure, some methods worked well, albeit with long (> 50 year) time lags (Rango et al. 2002; see also chapter 14).

Although variations in vegetation dynamics and shrub invasion are the most well known, other lesser known aspects of arid and semiarid systems have been found to be quite variable as well. In arid systems, aboveground net primary production (ANPP) can vary three- to fivefold, both between years at the same location and within the same year at different locations (Huenneke et al. 2001, 2002; see also chapter 11). Given the importance of arid and semiarid systems to current global issues, including carbon dynamics (Houghton et al. 1999; Pacala et al. 2001; Jackson et al. 2002), loss of biodiversity (Whitford et al. 2001), invasion of exotic species (DiTomaso 2000; Masters and Sheley 2001), wind and water erosion of soil and nutrients (Schlesinger et al. 1999; Wainwright et al. 2000), and emission of dust loads to the atmosphere (Gillette et al. 1992; Tegen et al. 1996; Gillette and Chen 2001), there is a critical need for a better understanding of ecosystem patterns and dynamics at multiple spatial and temporal scales. This improved understanding is critical to our ability to manage these landscapes as well as to make predictions about future dynamics with local, regional, and global impacts.

In this chapter, we describe and illustrate the utility of a conceptual framework that focuses on three interrelated aspects of landscapes: (1) feedbacks among plants, animals, and soils generated from interactions among biotic processes, a heterogeneous physical template, and the disturbance regime across a range of spatial and temporal scales; (2) neighborhood or contagious processes that generate fluxes and flows within and among spatial units; and (3) landscape characteristics, including the structure and spatial distribution of spatial units as well as the landscape context or the condition of the study area of interest relative to its surroundings that modify the transfers of materials.

A key feature of our conceptual model is that it facilitates an evaluation of the relative importance of each of these three aspects and recognizes that some or all may not be sufficiently important to be included for any given problem. Although arid and semiarid landscapes consist of a complex mosaic of vegetation, soil, and animal interactions, the errors in prediction that result from including poorly understood, noncritical spatial information do not warrant a complex model for all parts of a landscape. We first describe our spatially interactive model and compare it to previous models, and then we show how to identify the landscape locations where spatial processes and information (i.e., vegetation–animal–soil feedbacks, neighborhood processes, and landscape characteristics) are needed to understand and predict ecosystem dynamics. The local, regional, and global applications of our interactive landscape model to old and new problems are also discussed.

Current Models of Arid and Semiarid Systems

Complex interactions among a number of factors (effects of large and small animals, drought, fire, climate change, and soil properties) are often invoked to explain and predict heterogeneity in arid and semiarid landscapes. The issue of shrub invasion has been addressed by a number of authors (Humphrey 1958; Schlesinger et al. 1990; Allred 1996; Van Auken 2000; see also chapter 10), yet a general consensus does not exist regarding the key factors controlling different outcomes of shrub invasion under similar conditions. It is readily acknowledged that the problem is complicated by the presence of positive feedbacks and nonlinear thresholds (Archer 1989, 1994; Archer et al. 1995; Schlesinger et al. 1990; Reynolds et al. 1997; Rietkirk and van de Koppel 1997).

Inherently high spatial and temporal variation in the physical template (climate, soils) and disturbance regime (e.g., fire) plays an important role in ecosystem dynamics. Periodic droughts on 20–60-year cycles have variable effects that depend on the severity and length of the dry period as well as on confounding factors, such as livestock grazing (Herbel et al. 1972; Woodhouse and Overpeck 1998; Reynolds et al. 1999; Gibbens et al. 2005). Precipitation also has high variability between years; for example, 26 cm of rain fell in one year (1990) followed by 41 cm at the same location the next year (1991) in the Jornada Basin. Annual maximum daily rainfall at the Jornada also has high variability that is not related to mean average rainfall of the basin (chapter 3). Short-duration, spatially variable convective storms can also result in marked vegetation response across a landscape within a year. This fine-scale spatial variation in precipitation often translates into long-term patterns in ecosystem dynamics, such as highly variable spatial patterns in ANPP (Ludwig 1987; Huenneke et al. 2001, 2002).

Soil properties are also spatially and temporally heterogeneous. Arid landscapes are complex mosaics of soils of highly variable age formed from diverse parent materials and further modified by extensive redistribution by both wind and water (Gile et al. 1981; McAuliffe 1994; Wondzell and Ludwig 1995). Steep slopes exacerbate the effect of high-intensity convective storms (Wondzell et al. 1996). Even subtle differences in elevation ($< 0.6°$ slope) combined with plant–soil water feedbacks can generate "striped" patterns in vegetation (Montaña et al. 1990; Aguiar and Sala 1999). The disturbance regime is also highly variable; spatial variation in fires produces nonuniform effects on the vegetation. Lightning-ignited fires occur primarily in grasslands where fuel loads are sufficiently high and spatially continuous to carry a fire. Historically, fire may have played an important role in limiting shrub invasion, but the current dominance by shrubs throughout many arid landscapes limits the ability of fires to spread (Drewa et al. 2001).

The well-known resource-redistribution model of Schlesinger et al. (1990) incorporates both spatial variation in the physical environment and disturbance regime as well as positive plant–soil feedbacks in explaining desertification dynamics. This model focuses on small patches dominated either by a grass plant, a shrub, or bare area. As grasslands degrade, often associated with invasion by shrubs, patches of bare soil appear. Wind and water remove soil nutrients from these bare areas, and nutrients accumulate under individual shrubs to generate local islands of fertility. Grazing and drought act to move the system toward desertified shrublands, whereas the exclusion of grazers and extended wet periods favor grasslands. A basic assumption of this model is that all patches of one type (i.e., grass, shrub, and bare) function similarly. The scale of the most important heterogeneity is the contrast between plants and bare interspaces; smaller and larger spatial scales are assumed to be relatively homogeneous. Thus, heterogeneity within an individual plant is ignored, and landscape-scale patterns that result from plant-scale interactions are usually determined through linear extrapolation. No additional processes are assumed to be needed to extrapolate beyond the scale of an individual plant or area of bare soil. This model has been successfully applied to a number of ecosystem types and to a range of ecological problems in a variety of arid and semiarid systems at two spatial scales: the plant-interspace scale, where individual grasses and shrubs are compared, and the landscape scale, where large areas of relatively homogeneous grasslands and shrublands are compared (e.g., Augustine and Frank 2001; Hirobe et al. 2001; Maestre et al. 2001, 2002; MacGregor and O'Connor 2002; Maestre and Cortina 2002; Neave and Abrahams 2002).

Other models of arid landscapes include spatial and temporal heterogeneity in physical factors and disturbances at multiple scales, as well as redistribution of materials within and among spatial units. These landscape models typically aggregate plants into patches based on different characteristics (e.g., size, density, configuration, interpatch distance). Transfers of materials between patches can affect landscape function and response to disturbance (Ludwig et al. 1997; Reynolds et al. 1997; Reynolds and Wu 1999).

Interactive Model of Arid and Semiarid Landscapes

Although previous models have been successful in explaining certain aspects of arid and semiarid ecosystems, they are insufficient to account for the variation and complexity observed in dynamics across a range of spatial and temporal scales. We propose that a new model of arid and semiarid ecosystems is needed to understand these complex dynamics, predict future dynamics, and make effective management decisions. The new model builds on previous models, in particular that of Schlesinger et al. (1990), but explicitly includes (1) variation in biotic and abiotic processes and the disturbance regime as well as feedbacks among these system components, (2) landscape characteristics, and (3) contagious or neighborhood processes that connect different plants, patches, and landscape units (figure 18-1). Our focus is on understanding and predicting variations in ecosystem properties and dynamics and not just predicting average conditions through time or across a landscape.

In our conceptual model, landscapes are structured hierarchically and consist of increasingly larger spatial units, from individual plants to patches or patch mosaics to landscape units or small watersheds (figure 18-1). The smallest scale is a plant or its associated interspace, the scale at which much of the research has been conducted at the Jornada. The next scale of interest is the assemblage of similar plants and interspaces into patches. At the Jornada most patches are dominated by one of several species of shrubs (mesquite [*Prosopis glandulosa*], creosotebush [*Larrea tridentata*], and tarbush [*Flourensia cernua*]) or grasses (black

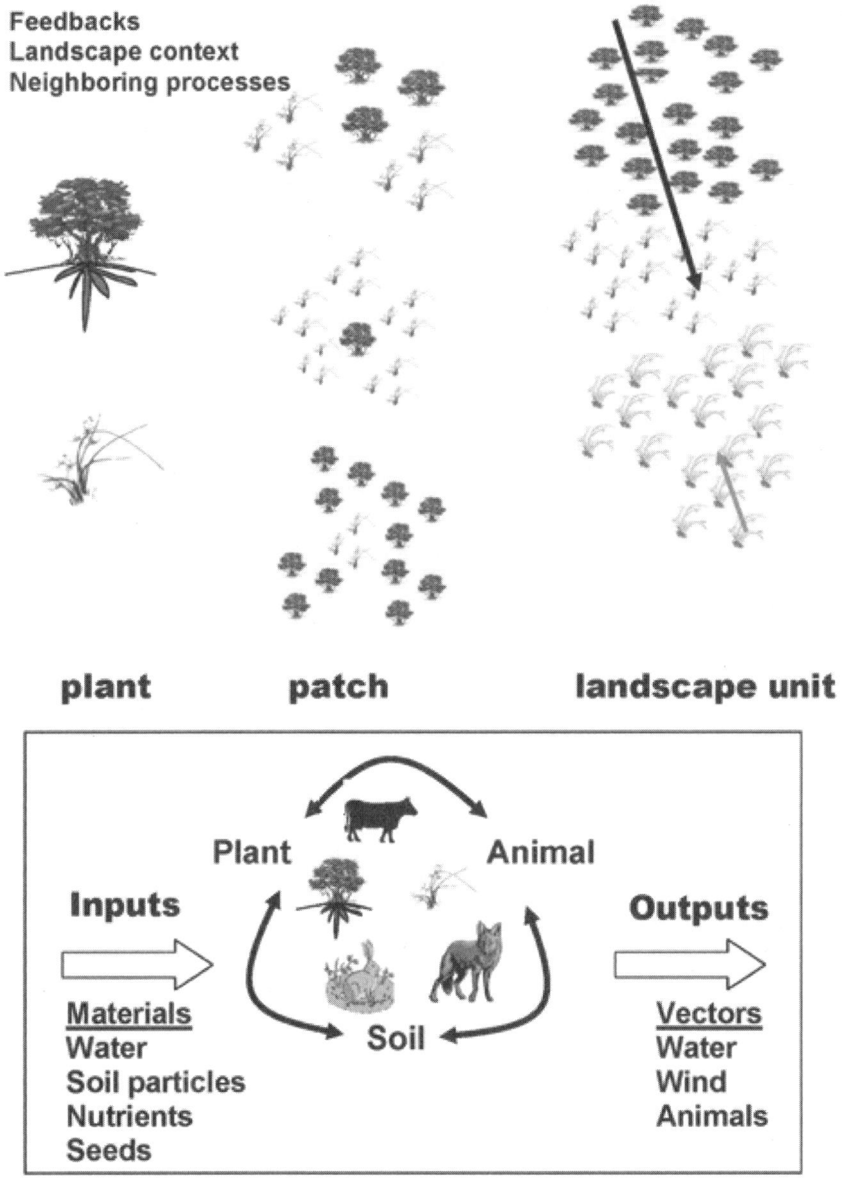

Figure 18-1. Interactive landscape model includes the importance of plant-animal-soil feedbacks, landscape context, and contagious or neighbor processes for a hierarchically structured landscape that consists of individual plants, patches, and landscape units. Positive and negative feedbacks result from complex interactions among plants, animals, and soil properties and processes across a range of spatial scales. Water, soil particles, nutrients, and seeds are redistributed among spatial units by water, wind, and animals. Landscape context is represented in the figure by the adjacency of different patch types (grasses, shrubs) where the patch response depends on the characteristics of neighboring patches. In the case of landscape units, soil particles can move easily by wind action through open canopy shrub patches (black line) with the resulting deposition of soil in adjacent grass patches. Movement of soil between different types of grass patches is very limited. Wind becomes relatively unimportant if the prevailing winds is from the opposite direction (red line), due to little movement of soil between patches.

grama and tobosa [*Pleuraphis mutica*]). These patches vary in size from several individual plants (< 5 m^2) to several hundred individuals (> 1,000 m^2). The next scale (landscape unit) consists of a number of interacting patches. Spatial variation within a landscape unit, such as small depressions, can collect water and nutrients, resulting in patchiness in soils and vegetation (Wainwright et al. 2002). The configuration of patches (i.e., size, number, and adjacency or between-plant distance) can have important effects on patch dynamics as well as on the function of the landscape unit. Important landscape units at the Jornada are bajadas, sandy basins, and playas. Landscape units are generally equivalent to ecological sites (USDA NRCS 1997) and are typically the unit of most interest for management. Thus, a landscape such as the Jornada Basin consists of a mosaic of interacting landscape units, each containing a mosaic of interacting patches that consist of a mosaic of interacting individual plants and bare areas.

The dynamics of each spatial unit are determined by local physical and biotic factors within the unit and transfers of materials among units (figures 18-2 and 18-3). Local physical factors that exert major controls on semiarid and arid ecosystem dynamics include geomorphology and soils, precipitation, temperature, and disturbance history. Geomorphic constraints include parent material, elevation, and slope aspect, length, and steepness that control the location of runoff and run-in areas (Monger 1999; see also chapters 4 and 7). Soil characteristics determine (1) the capture, retention, and supply of water; (2) the supply of nutrients through mineral weathering and organic matter mineralization; (3) erosion rates; and (4) the environment for root growth and soil biotic activity. Local precipitation inputs and temperature can vary among landscape units due to differences in elevation and location relative to areas with high topographic relief. Disturbance history is particularly important in areas such as the Jornada, where active management has occurred for decades.

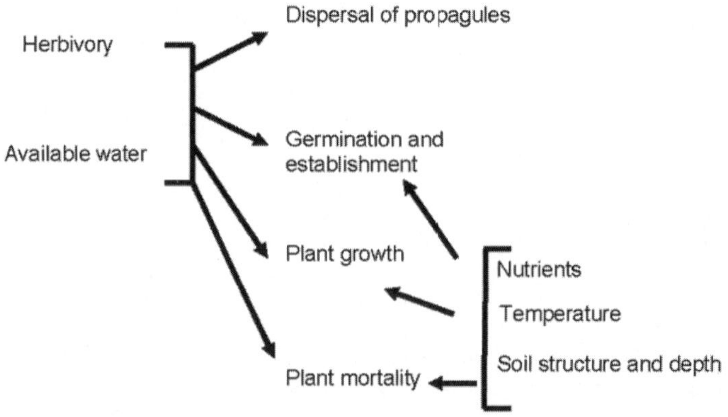

Figure 18-2. Conceptual model of within-spatial-unit controls on ecosystem structure and dynamics.

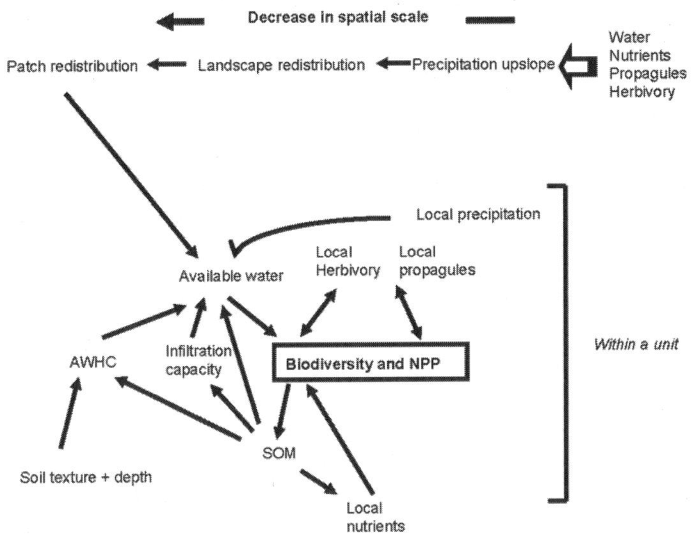

Figure 18-3. Conceptual model of flows of materials within and among spatial units. Water, nutrients, propagules, and herbivory are spatially distributed across a landscape and contribute to localized inputs within each landscape unit. Redistribution of water is shown as an example. Within a spatial unit, these materials exert controls on biological processes that influence patterns and dynamics in ecosystem properties, such as biodiversity and ANPP.

Biotic Processes, Interactions, and Feedbacks

Interactions and feedbacks among physical factors and biotic processes are critical to dynamics both within and among spatial units (figures 18-2 and 18-3). Here we focus on plant, animal, and soil processes that generate positive and negative feedbacks.

Plant–soil feedbacks are commonly associated with shrub invasion. As shrubs invade an area, resources become concentrated beneath shrubs and are less available to grasses, thus resulting in a positive feedback to shrub persistence (Schlesinger et al. 1990). Feedbacks between herbaceous species and soil properties can also result in positive feedbacks to grasses (Montaña 1992). For example, an increase in summer precipitation may promote the establishment of short-lived herbaceous species on some shrub-dominated sites (Peters and Herrick 1999b). This increase in plant biomass will result in an increase in soil organic matter and infiltration capacity of the soil with feedbacks to plant-available water and perennial grass establishment and recovery following shrub invasion. Plant–soil water feedbacks are also important in generating striped or spotted patterns in vegetation (Montaña 1992; Aguiar and Sala 1999).

Feedbacks among animals, plants, and soils are also important across a range of spatial scales (van de Koppel et al. 2002). Large herbivores respond to individual plants, patches, and landscape units that are often nonuniformly distributed across the landscape (Senft et al. 1987; Bailey et al. 1996). The response of large animals depends on the distribution of these spatial units of forage quality and quantity as well as abiotic features, such as topography and distance to water. Large animals can have major impacts on the vegetation and soils at multiple spatial scales. For example, the expansion of native shrubs into perennial grasslands occurs as a result of cattle (*Bos*) consuming seeds from species such as honey mesquite that have palatable pods. Viable seeds that pass through the animal's digestive tract can be redistributed at large distances from the source plant population. The dispersal of mesquite seeds followed by establishment and growth of adult plants in combination with herbivory that reduces the cover and competitive ability of grasses would lead to reduced forage availability with a negative feedback to cattle production and use of that area (Walker et al. 1981; Brown and Archer 1999).

Nonlinear interactions and feedbacks between small animals and the vegetation often result in complex ecosystem dynamics (Brown and Morgan Ernest 2002). For example, a high density of bannertail kangaroo rats (*Dipodomys spectabilis*) in grasslands versus a greater abundance of jackrabbits (*Lepus*) and rodents following shrub invasion (Moroka et al. 1982; Whitford 1997) has important effects on patterns and intensity of herbivory, granivory, and seed dispersal (chapter 12). Soil disturbances by small animals can alter soil wind and water erodibility through digging and tracking across the surface (Belnap and Gillette 1998; Neave and Abrahams 2001). These modifications in soil water availability can influence plant growth with feedbacks to animals.

Neighborhood Processes

Neighborhood processes affect the redistribution of materials across a landscape. Transfers of material can occur within as well as among spatial units (patch to patch) and between hierarchical levels (plant to patch and vice versa) (figure 18-3). Selective grazing on individual plants followed by fecal deposition at the patch scale can disperse materials from small to large areas and connect distant areas. By contrast, indiscriminate grazing at the patch scale followed by fecal deposition and localized seed germination within a bare area is a form of disaggregation from large to small areas. These transfers are particularly important in arid landscapes with high topographic and soil variation combined with intense, short-duration water and wind events that lead to redistribution of soil and water across a landscape (chapters 7 and 9). Furthermore, large and small animals are very effective at redistributing seeds and nutrients across spatial scales (chapter 12).

The major vectors of redistribution are water, wind, small and large animals, and disturbances such as fire and human activities (figure 18-3). Soil particles, water, nutrients, plant litter, and seeds are heterogeneously distributed across a landscape and contribute to localized inputs within each spatial unit. We discuss each of the vectors that generate neighborhood processes first, followed by the importance of landscape characteristics in modifying these rates of transfer.

Water

The horizontal transport of storm-water runoff across arid landscapes and its ultimate deposition—either as a contribution to soil water or to immediate evaporation to the atmosphere—is crucial to determining the availability of water for plant growth and reproduction. Water is also an important transport mechanism for nutrients, soil particles, and seeds. Increased nitrogen losses from desert hill slopes have been documented as grass cover declines and shrubs invade (Schlesinger et al. 1999, 2000). Erosion and deposition of soil particles occur across multiple spatial and temporal scales, from the redistribution of water between plants and bare interspaces to the movement of water in channels or rills within patches and in arroyos that connect landscape units (Schlesinger and Jones 1984; Wainwright et al. 2002; Parsons et al. 2003; see also chapter 7). Water is also an important dispersal agent for seeds, in particular for large-seeded species, such as creosotebush and sideoats grama (*Bouteloua curtipendula*) (Gutierrez-Luña 2000).

Wind

Redistribution of soil particles by wind has been well documented at the Jornada, in particular for sandy soils in mesquite shrublands (Marticorena et al. 1997; Okin and Gillette 2001). Other soils are also susceptible to wind erosion but are usually protected by vegetation or a strong physical or biological crust. Disturbance by cattle and humans is very important to note because it can remove the protective vegetation and disturbs the soil surface, exposing and creating more wind-erodible particles. Wind can also redistribute nutrients, both locally between plants and interspaces, and regionally or globally through atmospheric circulation patterns (Gillette et al. 1992; Okin and Gillette 2001; see also chapter 9). Furthermore, seeds of many herbaceous species have adaptations for wind dispersal (Howe and Smallwood 1982).

Small and Large Animals

Animals are effective agents of seed dispersal and redistribution of soil resources across spatial units. Cattle consume seeds of both shrubs and grasses and can deposit viable seeds in feces at large distances (several km) from the seed source (Janzen 1982; Chambers and MacMahon 1994). Small animals (rodents and lagomorphs in particular) can also have major impacts on ecosystem dynamics across multiple scales (Brown and Morgan Ernest 2002; Yates et al. 2002). Furthermore, small animals have spatial and temporal distributions that may or may not mirror large animal distribution patterns (chapter 12). Thus, complex interactions can occur within and across spatial scales.

Disturbance

Fire is a multiscale phenomenon that responds to vegetation fuel load and to environmental conditions of wind speed, temperature, and humidity that also vary spatially and temporally. In addition to direct effects on vegetation, fire influences transfers of materials by modifying animal behavior that affects spatial patterns in nutrient cycling dynamics (McNaughton 1985). By reducing plant cover, fire increases the redistribution of soil particles through wind and water erosion (Johansen et al. 2001; Whicker et al. 2002). Fire also releases particles into the atmosphere that can affect regional air quality (Wotawa and Trainer 2000).

Human activities also generate disturbance with potentially large impacts on transfers of materials. Traditionally managed as rangelands, aridlands are subject to increasing pressure for development and recreational use (Maestas et al. 2001). These human activities alter vegetative cover and biodiversity with increasing fragmentation of the landscape, resulting in increased wind and water erosion, as well as modifications to natural transport processes and movement of plants and animals. Dust loads in the atmosphere also increase during dry, windy periods with increasing urbanization and clearing of marginal land for industry and agriculture; this sequence of events occurred following the 1930s drought and has been cited as the primary cause of increases in suspended particulate matter in urban areas in the Southwestern United States.

Landscape Characteristics

The relative importance of transfers of materials within and among spatial scales is affected by (1) landscape characteristics, including structure (size, shape, and type of spatial units); (2) spatial distribution or configuration of spatial units; and (3) the context or location and characteristics of an area relative to its surroundings or nearby areas. These characteristics influence the connectivity of the landscape by modifying the ability of vectors of redistribution (water, wind, animals, and disturbance) to move materials horizontally. Highly connected landscapes consist of a mosaic of spatial units distributed in such a way as to promote the movement of materials via spatial processes, whereas landscapes with low connectivity may have barriers or spatial configurations of units that restrict horizontal movement of materials. Highly connected landscapes for one vector, such as water, may have low connectivity for another vector, such as animals.

Landscapes with high connectivity can amplify an environmental driver to result in spatially nonlinear responses and a cascading effect where the rate of movement across the landscape is faster than expected based on a linear extrapolation from small-scale responses. By contrast, disconnected landscapes can buffer environmental drivers, resulting in a slower rate of movement than expected. For example, a fire that occurs during periods of low humidity, strong winds, and high temperatures can spread quickly across a landscape with a highly connected fuel load. A landscape with a patchily distributed fuel load will limit the rate and spatial extent of the fire.

The relative importance of different contagious processes in generating these spatial nonlinearities and cascading effects can change through time. For example, at the Jornada, a 250-ha cattle exclosure was constructed in 1933 such that the northwest part of the exclosure was dominated by grassland and the southern part was dominated by honey mesquite (figure 18-4). Although cattle were excluded from the exclosure, mesquite continued to recruit and expand in the surrounding pasture that was grazed as well as in the exclosure as a result of the dispersal of mesquite seeds from small animals. Local dispersal from mesquite plants within the exclosure at the time of construction was also likely. From 1948 to 1987 the grazed pasture continued to fill in with new mesquite plants, and the established plants grew larger. At some point between 1987 and 1998, the density and spatial configuration of mesquite plants crossed a critical threshold such that wind erosion became prevalent and coppice dunes developed throughout both the grazed pasture and the nearby exclosure. These legacies, such as the historic location of shrub communities, can be particularly important in understanding current patterns in vegetation and soils.

Landscape context can directly affect ecosystem dynamics: A barrier may restrict animal movements from one vegetation type to another. A grass patch adjacent to a sandy mesquite patch is more likely to suffer negative impacts of sand deposition than a similar grass patch adjacent to a stable creosotebush-dominated area. Indirect effects of landscape context are also important. Few studies have included landscape context when examining factors influencing shrub invasion and grass persistence. Most studies have examined plot-level characteristics and have not accounted for the influence of the surrounding area. However, the very slow recovery (> 50 years) of black grama following shrub removal (Gibbens et al. 1993; Havstad et al. 1999) is likely related to the small amount of black grama initially within the plots and the lack of a significant seed source of this species within dispersal distance of the plots (Havstad et al. 2003). A recent analysis showed that distance to shrubland in 1915 was the best predictor of grass persistence in 1999; the farther a plot was from an existing shrubland, the greater the probability that grasses would persist through time (Yao et al. 2002a). Other explanatory factors commonly used in shrub invasion studies, such as soil texture, grazing intensity, and precipitation, were less important than distance to the nearest shrub-dominated areas. The importance of landscape context is also illustrated in figure 18-4: Placement of the exclosure in the western part of the grazed pasture would have delayed the expansion of dunelands in the exclosure, but merely as a result of distance from the advancing dunes and not protection from grazing.

Simplifying Complex Landscapes: How Do We Deal with Variation?

A particularly important part of our model is that spatial information is not needed for all parts of a landscape. Although arid and semiarid landscapes are complex mosaics of plants, soil, and animal interactions across spatial scales, it is not necessary and can even be detrimental to include spatial information for all locations on a landscape. The collection of spatial information on multiple ecosys-

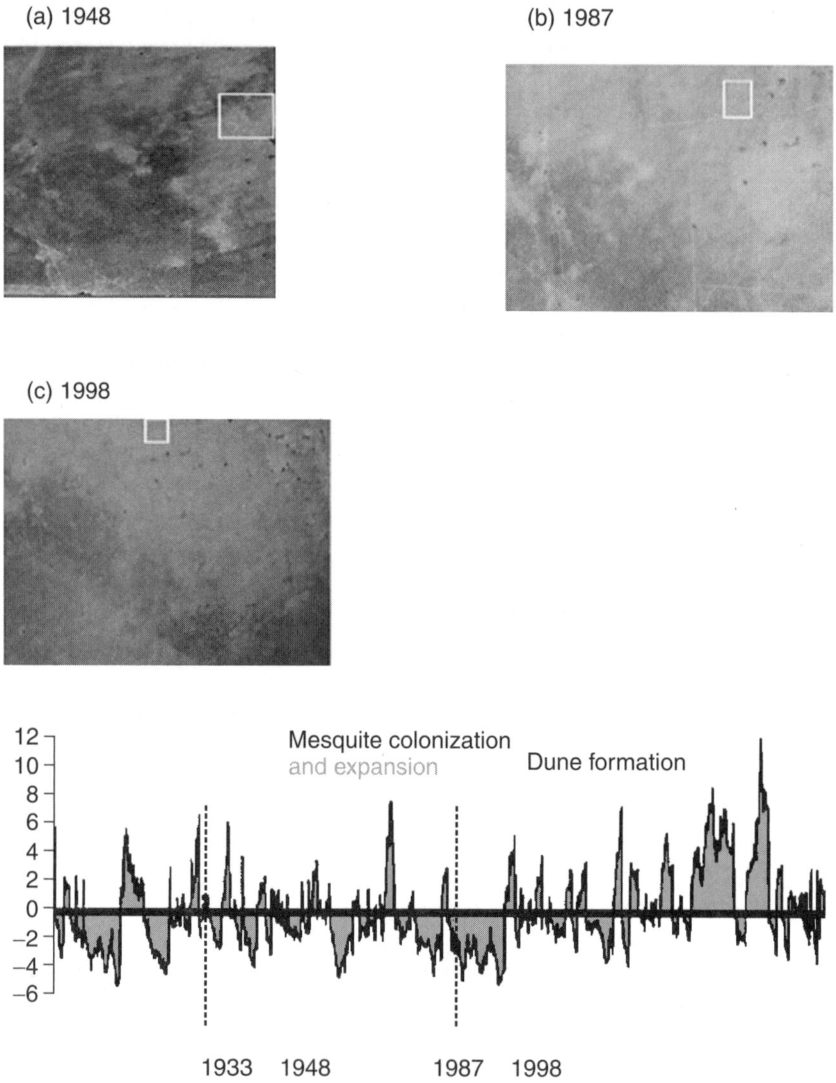

Figure 18-4. Images from natural revegetation exclosure through time showing nonlinear dynamics and importance of different spatial processes. A 1-mile by 1-mile exclosure was constructed in 1933 on the Jornada where the northwest part of the exclosure was dominated by grassland and the southern part was dominated by honey mesquite. Although cattle were excluded from the exclosure depicted by the white rectangle, mesquite continued to recruit and expand into the surrounding pasture that was grazed, primarily from east (right) to west (left). Almost 40 years were required for a large area to be dominated by mesquite dunelands shown in 1987 in the right side of the image. These dunelands then crossed a threshold and rapidly expanded; within 11 years the dunelands had dominated much of the pasture. Placement of the exclosure in the western part of the pasture would have delayed the expansion of dunelands into the exclosure, but not as a result of protection from grazing.

tem components (plants, animals, soils) across a range of spatial and temporal scales, including transfers of material among spatial units, is a time-consuming and costly venture. Furthermore, the inclusion of parameters, especially those that are poorly understood, will increase the errors associated with managing a large number of parameters and their estimation (O'Neill 1979; Gardner et al. 1980; Reynolds and Acock 1985). Finding the balance between errors of omitting key information and errors of including poorly known or measured parameters remains a critical challenge in ecology.

Our approach to simplifying complex landscapes is to determine the locations on a landscape where all three aspects of spatial information must be known for predictions to be accurate. We identify the key processes and their feedbacks that must be measured for each location. Management decisions and intensive sampling can then focus on those locations and key processes. Predictions for the remainder of the landscape can be obtained using estimates from representative sites that are extrapolated to similar areas using nonspatial and spatially implicit methods. Nonspatial methods using linear extrapolations are useful for locations where landscape characteristics and neighborhood processes, including feedbacks, are relatively unimportant (Lieth and Whittaker 1975; Schlesinger et al. 1990). Spatially implicit methods are needed when information on landscape characteristics is important, such as distance to the nearest shrubland, but not neighborhood processes and feedbacks. Spatially explicit methods are necessary when landscape characteristics, neighborhood processes, and feedbacks are important.

One approach to identifying critical locations and determining important spatial processes is to combine remotely sensed images with field data and spatial databases residing in a geographic information system. For example, remotely sensed images combined with field estimates of ANPP can be used both to determine the most appropriate vegetation index for each of the major ecosystem types and to identify the locations (hot spots) with extremely low and high ANPP values for each ecosystem type. Spatial databases can be used to determine the key processes operating to generate these extremely low or high values. Correlating indices from the remotely sensed images with the spatial databases can confirm the greater importance of spatial processes at these key locations compared with the rest of the landscape. This method has been used successfully to determine that different processes are important for different ecosystem types and for different time periods (wet versus dry years) (Peters et al. 2004). Identifying these locations and key processes is the first step in simplifying complex landscapes to prioritize management decisions and guide research questions and experimental designs.

Future Directions: Applications of Interactive Landscape Model to Old and New Problems

Our conceptual model of interactive landscapes provides new insight into several perplexing problems in arid and semiarid systems and promises to guide future research. Here we describe the use of this approach in addressing three pressing ecological issues: shrub invasion, remediation, and carbon storage and dynamics.

Shrub Invasion

Historically, landscapes of the Southwestern United States were a mosaic of grasslands and shrublands with a much greater proportional area dominated by grasses than occurs at present (Hastings and Turner 1965; Grover and Musick 1990; Gibbens et al. 2005). This mosaic was due to interactions among vegetation characteristics with topography and landform that influenced water availability (Gardner 1951; York and Dick-Peddie 1969; Stein and Ludwig 1979; Wondzell et al. 1996). Grasslands occurred primarily either on level uplands with sandy soils dominated by black grama or on playas or basin floors that received run-in water and were dominated by tobosa grass and burrograss. Shrublands occurred either on sites with shallow, calcareous soils susceptible to water erosion that were dominated by creosotebush (e.g., many upper alluvial fans) or on deep, sandy soils located along stream channels and arroyos that were dominated by honey mesquite.

Currently, most areas are dominated by shrubs with isolated patches of grasslands (Gibbens et al. 2005). The widespread expansion of native shrubs from specific locations into desert grasslands has occurred at unprecedented rates over the past century. This extensive expansion of shrubs has been attributed to a number of interacting factors (chapter 10). Disentangling the role of each factor has proven difficult. However, our landscape model provides new insights and testable hypotheses about the key processes involved and the patterns of expansion. We illustrate this approach for the two most common shrub species in these systems: honey mesquite and creosotebush.

Expansion of honey mesquite plants into grasslands likely occurred through the dispersal of seeds by cattle and small animals. Mesquite pods are palatable and can be dispersed long distances from the seed source (figure 18-5). Establishment and survival of these seedlings occurs frequently. Following this initial establishment, shrub density increases through time as plants fill in an area due to high seed availability (Goslee et al. 2003). Continued grazing by cattle on the few remaining grasses would further reduce their ability to compete against shrubs. Wind erosion that redistributes fine particles from bare interspaces to plants would lead to coppice dune formation. These islands of fertility would result in positive feedbacks to shrub persistence and growth through time. Thus the key processes involved in mesquite expansion include the widespread seed dispersal by animals, vegetation–soil–animal feedbacks that negatively affect grasses and promote shrubs, and wind erosion leading to dune formation and shrub persistence through time.

Expansion by creosotebush into grasslands over the past century was likely affected by a different set of spatial processes. Creosotebush seeds are relatively large and unpalatable, thus dispersal was most likely through sheet flow of water or in stream channels during floods. Constraints on seed germination and seedling establishment of creosotebush are poorly understood in the Chihuahuan Desert.

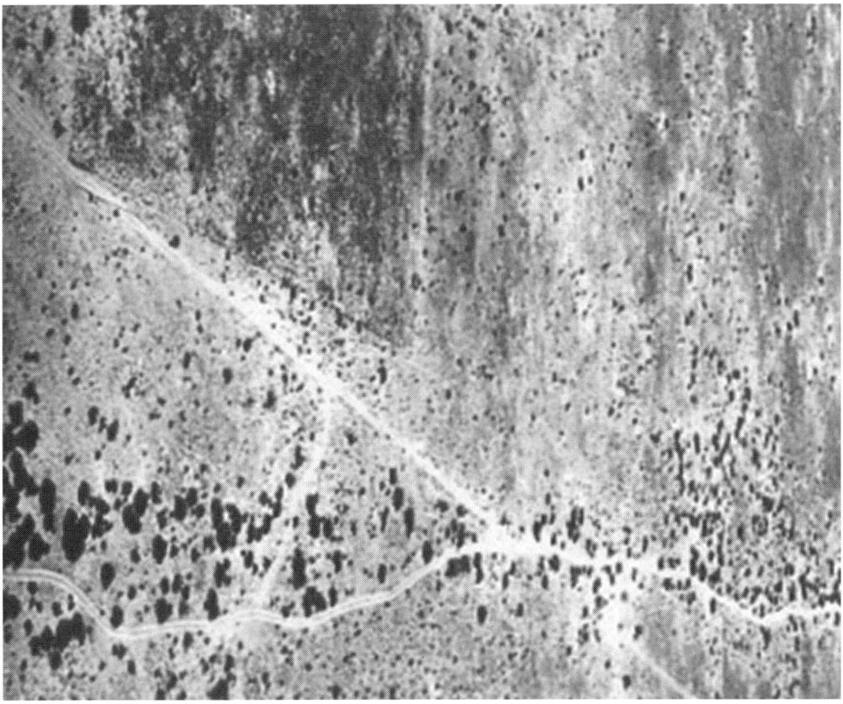

Figure 18-5. Aerial photo of Camino Real and spatial variation in mesquite plants. Most plants are currently found along the original Camino Real established by the Spanish colonizers in the seventeenth century with some dispersal away from the road.

However, observations suggest that recruitment events are episodic with large-scale events occurring following the droughts in the late 1800s and again in the 1950s. Recruitment at this time was likely promoted by the reduced grass cover due to heavy grazing by cattle over this time period (Fredrickson et al. 1998). Expansion of this species often occurs in areas where argillic horizons have been removed by erosion (McAuliffe 1994). Similar to mesquite-dominated areas, the remaining grasses would be preferentially grazed as shrub density increases, resulting in positive feedbacks to shrubs with reduced competition from grasses. Islands of fertility form in creosotebush stands due to water erosion of soil and nutrients from bare interspaces to plant canopies. Thus, the key processes involved in creosotebush expansion include localized seed dispersal by water, episodic recruitment events, and water redistribution at the plant-interspace scale that promotes continued dominance by shrubs.

Based on these processes, we can predict differences and similarities in the landscape-scale expansion dynamics of these two shrub species. Creosotebush likely expanded rapidly over large areas within short periods of time following a large-scale recruitment event. By contrast, mesquite likely expanded over a longer period of time due to multiple recruitment events. These statements are supported by the current structure of the Jornada landscape where mixed stands of honey mesquite and black grama still exist. Large, monodominant stands of creosotebush or mesquite can be found, but mixed stands of creosotebush and black grama rarely occur. The fastest expansion for both shrub species would have occurred on the landscape locations where positive shrub feedbacks were the strongest or the rates of transfer (seed dispersal, water or wind redistribution) favoring shrubs were the largest. Some data exist to support this statement; for example, sites closer to shrublands in the early 1900s converted to shrub dominance faster than sites farther away (Yao et al. 2002a).

Knowledge gained from studying native shrub expansion using a landscape perspective can be applied to the current invasion by exotic species or noxious weeds in many ecosystems worldwide. Most studies of invasive species have focused on species traits and environmental factors with less attention to biotic feedbacks and landscape properties (Rejmánek and Richardson 1996; Lonsdale 1999). Based on our work at the Jornada, identifying key processes, feedbacks, and sensitive parts of the landscape for different species is necessary for understanding and prediction of these complex dynamics.

Remediation

Numerous remediation approaches have been attempted since 1912 at the Jornada, either to limit shrub invasion or to return perennial grasses to dominance (chapter 14). Most attempts have low success rates and poor economic return. Although most remediation attempts have failed (Rango et al. 2002), there have been enough successes to indicate that the system can be manipulated (e.g., Cassady and Glendening 1940). In some cases, it has taken decades for positive effects to become apparent (Rango et al. 2002). Most remediation attempts in arid and semiarid ecosystems have had limited success for three major reasons: (1) the key processes limiting vegetation recovery at different spatial and temporal scales have not been identified, (2) nonlinear thresholds related to vegetation-soil-animal interactions and feedbacks have been crossed, and (3) the landscape context and importance of linkages among spatial units within a landscape have been ignored (Archer 1989; Peters and Betancourt 2001). Our landscape perspective has the potential to overcome these limitations and provide new insight and prediction into remediation efforts.

High spatial and temporal variability inherent in arid and semiarid ecosystems has led to the failure of many remediation attempts. However, this natural variability across a landscape and through time can be used to our advantage (Landres et al. 1999). Using our landscape perspective, the potentials and limits of different parts of a landscape, as well as the key processes affecting ecosystem dynamics can be identified. This approach also allows us to take advantage of extreme events, such as El Niño and drought, which interact with different parts of a landscape differently. In particular, we can focus on landscape locations where vegetation-soil-animal processes and landscape context are important in generating nonlinear dynamics (Illius and O'Connor 1999; Holmgren and Scheffer 2001), thus increasing the probability that an extreme event will produce a positive result.

Prioritizing efforts by concentrating on locations likely to change under certain conditions and focusing on key processes will provide guidelines and recommendations for future remediation efforts (Herrick et al. 1997).

Carbon and Nutrient Dynamics

Carbon sequestration and dynamics in arid and semiarid ecosystems have become an increasingly important issue relative to the global carbon budget. Recent estimates of carbon sinks in the coterminous United States from 1980 to 1990 indicate that grasslands and shrublands may account for similarly large amounts of carbon storage as in forests (Pacala et al. 2001). Shrub-dominated ecosystems are important contributors to carbon sinks as a result of their extensive area (44% of the total land area) and their large potential rates (Hibbard et al. 2001; Pacala et al. 2001). Increases in above- and belowground carbon storage as well as increases in emissions of NO_x and nonmethane hydrocarbons (e.g., terpenes, isoprene, and other aromatics) have resulted from the replacement of grasses by shrubs (Archer et al. 2001; Pacala et al. 2001; Jackson et al. 2002).

Although the total or average amount of carbon sequestered by grasslands and shrublands is of interest at the global scale, the spatial distribution and temporal dynamics of carbon across landscapes is of more interest and concern to land managers (Bird et al. 2002; Breshears and Allen 2002). Spatial patterns in carbon and other soil nutrients across landscapes may be complex due to processes such as wind and water erosion, disturbance, and animal redistribution of plant material and nutrients (Schlesinger and Pilmanis 1998). The redistribution of carbon and soil nutrients across a landscape may be as important to vegetation dynamics as local inputs, especially for nitrogen (chapter 6). Thus, incorporating these transfers of materials among spatial units is necessary to understand and predict carbon and nutrient dynamics for landscapes.

Most estimates for carbon sinks and losses have a high degree of uncertainty due to landscape-scale variation in edaphic and topographic factors (Schlesinger and Pilmanis 1998; Pacala et al. 2001; Hurtt et al. 2002). Providing estimates of carbon dynamics based on an average value for a landscape may be misleading. Some vegetation types will have extremely high standing biomass and production, whereas other types will have very low values (Huenneke et al. 2002). Weighting by the area covered by each type would improve the accuracy of the overall estimate, but spatial as well as temporal variation within each type still will not be accounted for. Thus, estimates based on relatively few samples that are extrapolated to the landscape likely misrepresent the values obtained by sampling locations with extreme values and accounting for transfers of materials among spatial units.

Conclusions

Decades of research at the Jornada through the ARS, New Mexico State University, and the LTER have provided a wealth of information on many aspects of arid and semiarid systems. However, a number of pressing problems and perplexing issues still remain. High spatial and temporal variation in ecosystem dynamics across multiple scales cannot be explained using current models of these systems (e.g., Schlesinger et al. 1990). We developed an interactive landscape model that incorporates three key properties of landscapes: (1) feedbacks among plants, animals, and soils; (2) contagious processes that generate fluxes and flows within and among spatial units; and (3) landscape characteristics, including structure and spatial configuration of spatial units as well as the characteristics of the study area or its context relative to its surroundings. This landscape scale perspective provides new insight into old problems (i.e., shrub invasion, remediation) and an operational approach to deal with new problems, such as carbon dynamics and global climate change. This perspective is central to our long-term efforts as the first century of research is completed in the Jornada Basin and we look toward the next decades of the twenty-first century.

References

Abernathy, G. H., and C. H. Herbel. 1973. Brush eradicating, basin pitting, and seeding machine for arid to semiarid rangeland. *Journal of Range Management* 26:189–192.

Abrahams, A. D., and A. J. Parsons. 1991. Relation between infiltration and stone cover on a semiarid hillslope, southern Arizona. *Journal of Hydrology* 122:49–59.

Abrahams, A. D., A. J. Parsons, and J. Wainwright. 1994. Resistance to overland flow on semiarid grassland and shrubland hillslopes, Walnut Gulch, southern Arizona. *Journal of Hydrology* 156:431–446.

Abrahams, A. D., A. J. Parsons, and J. Wainwright. 1995. Effects of vegetation change on interrill runoff and erosion, Walnut Gulch, southern Arizona. *Geomorphology* 13:37–48.

Abrahams, A. D., G. Li, and A. J. Parsons. 1996. Rill hydraulics on a semiarid hillslope, southern Arizona. *Earth Surface Processes and Landforms* 21:35–47.

Abrahams, A. D., A. J. Parsons, and J. Wainwright. 2003. Disposition of rainfall under creosotebush. *Hydrological Processes* 17:2555–2566.

Adams, S., B. R. Strain, and M. S. Adams. 1970. Water-repellent soils, fire, and annual plant cover in a desert scrub community of southeastern California. *Ecology* 51:696–700.

Aerts, R. 1996. Nutrient resorption from senescing leaves of perennials: Are there general patterns? *Journal of Ecology* 84:597–608.

Aguiar, M. R., and O. E. Sala. 1999. Patch structure, dynamics, and implications for the functioning of arid ecosystems. *Trends in Ecology and Evolution* 14:273–277.

Allred, K. W. 1990. Elmer Ottis Wooton and the botanizing of New Mexico. *Systematic Botany* 15:700–719.

Allred, K. W. 1996. Vegetative changes in New Mexico rangeland. *New Mexico Journal of Science* 36:168–229.

Allred, K. W. 2003. *A field guide to the flora of the Jornada Plain*, 4th edition. Range

Science Herbarium, Department of Animal and Range Sciences, New Mexico State University, Las Cruces, NM.

Anderson, D. M. 2001. Virtual fencing—A prescription range animal management tool for the 21st century. Pages 85–94 in *Proceedings of the "Tracking animals with GPS" conference*. Craigiebuckler, Aberdeen, Scotland.

Anderson, D. M., and J. L. Holechek. 1983. Diets obtained from esophageally fistulated heifers and steers simultaneously grazing semidesert tobosa rangeland. *Proceedings*, Western Section, American Society of Animal Science 34.

Anderson M. C., and F. R. Kay. 1999. Banner-tailed kangaroo rat burrow mounds and desert grassland habitats. *Journal of Arid Environments* 41:147–160.

Anderson, R. Y., A. Soutar, and T. C. Johnson. 1992. Long-term changes in El Niño/Southern Oscillation: Evidence from marine and lacustrine sediments. Pages 419–433 in H. F. Diaz, and V. Markgraf, editors, *El Niño. Historical and paleoclimatic aspects of the southern oscillation*. Cambridge University Press, Cambridge.

Andrade, E. R., and W. D. Sellers. 1988. El Niño and its effect on precipitation in Arizona and western New Mexico. *Journal of Climatology* 8:403–410.

Andreae, M. O. 1996. Raising dust in the greenhouse. *Nature* 380:389–390.

Andrén, O., and K. Paustian. 1987. Barley straw decomposition in the field: A comparison of models. *Ecology* 68:1190–1200.

Anthes, R. A. 1984. Enhancement of convective precipitation by mesoscale variations in vegetative covering in semiarid regions. *Journal of Climate and Applied Meteorology* 23:541–554.

Archer, S. 1989. Have southern Texas savannas been converted to woodlands in recent history? *American Naturalist* 134:545–561.

Archer, S., and A. Bowman. 2002. Understanding and managing rangeland plant communities. Pages 63–80 in A. C. Grice and K. C. Hodgkinson, editors, *Global rangelands: Progress and prospects*. CAB International, Wallingford, UK.

Archer, S., and F. E. Smeins. 1991. Ecosystem-level processes. Pages 109–139 in R. K. Heitschmidt and J. W. Stuth, editors, *Grazing management: An ecological perspective*. Timber Press, Portland, OR.

Archer, S., D. S. Schimel, and E. A. Holland. 1995. Mechanisms of shrubland expansion: Land use, climate or CO_2. *Climatic Change* 29:91–99.

Archer, S., T. W. Boutton, and K. A. Hibbard. 2001. Trees in grasslands: Biogeochemical consequences of woody plant expansion. Pages 115–138 in E. D. Schulze, S. P. Harrison, M. Heimann, E. A. Holland, J. Lloyd, I. C. Prentice, and D. S. Schimel, editors, *Global biogeochemical cycles in the climate system*. Academic Press, San Diego, CA.

Ares, F. N. 1974. *The Jornada Experimental Range: An epoch in the era of southwestern range management*. Range Monograph no. 1. Society for Range Management, Denver, CO.

Arkley, R. J. 1963. Calculation of carbonate and water movement in soil from climatic data. *Soil Science* 96:239–248.

Arnold, G. W., and M. L. Dudzinski. 1978. *Ethology of free-ranging domestic animals*. Elsevier, New York.

Arya, S. P. S. 1975. A drag partition theory for determining the large-scale roughness parameter and wind stress on Arctic pack ice. *Journal of Geophysical Research* 80: 3447–3454.

Atwood, T. L. 1987. *Influence of livestock grazing and protection from livestock grazing on vegetation characteristics of* Bouteloua eriopoda *rangelands*. Ph.D. dissertation. New Mexico State University, Las Cruces, NM.

Augustine, D. J., and D. A. Frank. 2001. Effects of migratory grazers on spatial heterogeneity of soil nitrogen properties in a grassland ecosystem. *Ecology* 82:3149–3162.

Austin, A. T., and O. E. Sala. 2002. Carbon and nitrogen dynamics across a natural precipitation gradient in Patagonia, Argentina. *Journal of Vegetation Science* 13:351–360.

Ayarbe, J. P., and T. L. Kieft. 2000. Mammal mounds stimulate microbial activity in a semiarid shrubland. *Ecology* 81:1150–1154.

Bach, L. B., P. J. Wierenga, and T. J. Ward. 1986. Estimation of the Philip infiltration parameters from rainfall simulation data. *Soil Science Society of America Journal* 50: 1319–1323.

Bachelet, D., S. M. Wondzell, and J. F. Reynolds. 1988. A simulation model using environmental cues to predict phenologies of winter and summer annuals in the northern Chihuahuan Desert. Pages 235–260 in A. Marani, editor, *Advances in environmental modelling*. Eslevier, Amsterdam.

Bagine, R. K. N. 1984. Soil translocation by termites of the genus *Odontotermes* (Holmgren) (*Isoptera:Macrotermitinae*) in an arid area of northern Kenya. *Oecologia* 64: 263–266.

Bahre, C. J. 1995. Human impacts on the grasslands of southeastern Arizona. Pages 230–264 in M. P. McClaran and T. R. VanDevender, editors, *The desert grassland*. University of Arizona Press, Tucson, AZ.

Bahre, C. J., and M. L. Shelton. 1993. Historic vegetation change, mesquite increases, and climate in southeastern Arizona. *Journal of Biogeography* 20:489–504.

Baier, W., and G. W. Robertson. 1966. A new versatile soil moisture budget. *Canadian Journal of Plant Science* 46:299–315.

Bailey, D. W., J. E. Gross, E. A. Laca, L. R. Rittenhouse, M. B. Coughenour, D. M. Swift, and P. L. Sims. 1996. Mechanisms that result in large herbivore grazing distribution patterns. *Journal of Range Management* 49:386–400.

Baldocchi, D., R. Valentini, S. Running, W. Oechel, and R. Dahlman. 1996. Strategies for measuring and modelling carbon dioxide and water vapor fluxes over terrestrial ecosystems. *Global Change Biology* 2:159–168.

Ball, M. C., V. S. Hodges, and G. P. Laughlin. 1991. Cold-induced photoinhibition limits regeneration of snow gum at tree line. *Functional Ecology* 5:663–668.

Ballín Cortés, J. R. 1987. *Estudio Preliminar de Desertificación en el Límite Sur del Desierto Chihuahuense*. San Luis Potosi, S. L. P., Mexico, Instituto de Investigacion de Zonas Deserticas, Universidad Autonoma de San Luis Potosi.

Barbault, R., C. Grenot, and Z. Uribe. 1978. Le partage des ressources alimentaires entre des espèces de lezards du desert de Mapimi (Mexique). *Terre et Vie* 52:135–150.

Barrow, J. R. 2003. Atypical morphology of dark septate fungal root endophytes of *Bouteloua* in southwestern USA rangelands. *Mycorrhiza* 13:239–247.

Barrow, J. R., and P. Osuna. 2002. Phosphorus solubilization and uptake by dark septate fungi in fourwing saltbush, *Atriplex canescens* (Pursh) Nutt. *Journal of Arid Environments* 51:449–459.

Barrow, J. R., K. M. Havstad, and B. D. McCaslin. 1997. Fungal root endophytes in fourwing saltbush, *Atriplex canescens*, on arid rangelands of southwestern USA. *Arid Soil Research and Rehabilitation* 11:177–185.

BassiriRad, H., D. C. Tremmel, R. A. Virginia, J. F. Reynolds, A. G. de Soyza, and M. H. Brunnell. 1999. Short-term patterns in water and nitrogen acquisition by two desert shrubs following a simulated summer rain. *Plant Ecology* 145:27–36.

Beatley, J. C. 1974. Phenological events and their environmental triggers in Mojave Desert ecosystems. *Ecology* 55:856–863.

Beck, R. F., and R. P. McNeely. 1997. A 28-year grazing study in southern New Mexico [abstract]. 50th Annual Meeting, Society for Range Management, Lakewood, CO, 50: 11.

Belnap, J. 2002. Nitrogen fixation in biological soil crusts from southeast Utah, USA. *Biology and Fertility of Soils* 35:128–135.

Belnap, J., and D. Gillette. 1997. Disturbance of biological soil crusts: Impacts on potential wind erodibility of sandy desert soils in southeastern Utah. *Land Degradation and Development* 8:355–362.

Belnap, J., and D. A. Gillette. 1998. Vulnerability of desert soil surfaces to wind erosion: The influences of crust development, soil texture, and disturbance. *Journal Arid Environments* 39:133–142.

Belsky, A. J., R. G. Amundson, J. M. Duxbury, S. J. Rhina, A. R. Ali, and S. M. Mwonga. 1989. The effects of trees on their physical, chemical, and biological environments in a semiarid savanna in Kenya. *Journal of Applied Ecology* 26:1005–1024.

Beltran-Przekurat, A., R. A. Pielke Sr., D. P. C. Peters, K. A. Snyder, and A. Rango. 2005. Feedbacks between vegetation change and near-surface climate in the Northern Chihuahuan Desert. *Ecosystems* (submitted).

Benemann, J. R. 1973. Nitrogen fixation in termites. *Science* 181:164–165.

Bertela, M. 1989. Inconsistent surface flux partitioning by the Bowen ratio method. *Boundary-Layer Meteorology* 49:149–167.

Bestelmeyer, B. T. 2005. Does desertification diminish biodiversity? Enhancement of ant diversity by shrub invasion in southwestern USA. *Diversity and Distributions* 11:45–55.

Bestelmeyer, B. T., and J. A. Wiens. 2001a. Ant biodiversity in semiarid landscape mosaics: The consequences of grazing vs. natural heterogeneity. *Ecological Applications* 11:1123–1140.

Bestelmeyer, B. T., and J. A. Wiens. 2001b. Local and regional-scale responses of ant diversity to a semiarid biome transition. *Ecography* 24:381–392.

Bestelmeyer, B.T., and J.A. Wiens. 2003. Scavenging ant foraging behaviour and variation in the scale of nutrient redistribution in semiarid grasslands. *Journal of Arid Environments* 53:373–386.

Bestelmeyer, B. T., J. R. Brown, K. M. Havstad, R. Alexander, G. Chavez, and J. E. Herrick. 2003a. Development and use of state-and-transition models for rangelands. *Journal of Range Management* 56:114–126.

Bestelmeyer, B.T., J.R. Miller, and J.A. Wiens. 2003b. Applying species diversity theory to land management. *Ecological Applications* 13: 1750–1761.

Bestelmeyer, B. T., D. A. Trujillo, A. J. Tugel, K. M. Havstad. 2005. A multi-scale classification of vegetation dynamics in arid lands: What is the right scale for models, monitoring, and restoration? *Journal of Arid Environments* 65:296–318.

Bethlenfalvay, G. J., S. Dakessian, and R. S. Pacovsky. 1984. Mycorrhizae in a southern California desert: ecological implications. *Canadian Journal of Botany* 62:519–524.

Bird, S. B., J. E. Herrick, M. M. Wander, and S. F. Wright. 2002. Spatial heterogeneity of aggregate stability and soil carbon in semiarid rangeland. *Environmental Pollution* 116:445–455.

Birkeland, P. W. 1999. *Soils and geomorphology*. Oxford University Press, New York.

Bisal, F., and J. Hsieh. 1966. Influence of moisture on erodibility of soil by wind. *Soil Science* 102:143–146.

Blagbrough, J. W. 1991. Late Pleistocene rock glaciers in the western part of the Capitan Mountains, Lincoln County, New Mexico: Description, age, and climatic significance. In J. M. Barker, B. S. Kues, G. S. Austin, and S. G. Lucas, editors, *42nd annual field conference guidebook*. New Mexico Geological Society, Socorro, NM.

Blair, T. C., and J. G. McPherson. 1994. Alluvial fan processes and forms. Pages 354–402 in A. D. Abrahams and A. J. Parsons, editors, *Geomorphology of desert environments*. Chapman and Hall, London.

Bock, C. E., and J. H. Bock. 1993. Cover of perennial grasses in southeastern Arizona in relation to livestock grazing. *Conservation Biology* 7:371–377.

Boeken, B., and B. Shachak. 1994. Desert plant communities in human-made patches—implications for management. *Ecological Applications* 4:702–716.

Bolton, S. M., T. J. Ward, and R. A. Cole. 1991. Sediment-related transport of nutrients from southwestern watersheds. *Journal of Irrigation and Drainage Engineering* 117: 736–747.

Bosch, D. D., and C. A. Onstad. 1988. Surface seal hydraulic conductivity as affected by rainfall. *Transactions of the American Society of Agricultural Engineers* 31:1120–1127.

Bovey, R. W. 2001. *Woody plants and woody plant management: Ecology, safety and environmental impact*. Marcel Dekker, New York.

Bowling, G. A. 1942. Introduction of cattle into colonial North America. *Journal of Dairy Science* 25:129–154.

Boyd, R. S., and G. D. Brum. 1983. Postdispersal reproductive biology of a Mojave Desert population of *Larrea tridentata* (Zygophyllaceae). *American Midland Naturalist* 110: 25–36.

Brackenridge, G. R. 1978. Evidence for a cold, dry, full-glacial climate in the American Southwest. *Quaternary Research* 9:22–40.

Brady, N. C., and R. R. Weil. 1996. *The nature and properties of soils*. Prentice Hall, Upper Saddle River, NJ.

Branscomb, B. L. 1958. Shrub invasion of a southern New Mexico desert grassland range. *Journal of Range Management* 11:129–133.

Bravar, L., and M. L. Kavvas. 1991. On the physics of droughts. I. A conceptual framework. *Journal of Hydrology* 129:281–297.

Breman, H., and C. T. de Wit. 1983. Rangeland productivity and exploitation in the Sahel. *Science* 221:1341–1347.

Breshears, D. D., and C. D. Allen. 2002. The importance of rapid, disturbance-induced losses in carbon management and sequestration. *Global Ecology and Biogeography* 11:1–5.

Breshears, D. D., P. M. Rich, F. J. Barnes, and K. Campbell. 1997. Overstory-imposed heterogeneity in solar radiation and soil moisture in a semiarid woodland. *Ecological Applications* 7:1201–1215.

Breshears, D. D., J. W. Nyhan, C. E. Heil, and B. P. Wilcox. 1998. Effects of woody plants on microclimate in a semiarid woodland: Soil temperature and evaporation in canopy and intercanopy patches. *International Journal of Plant Sciences* 159:1010–1017.

Breshears, D. D., J. J. Whicker, M. P. Johansen, and J. E. Pinder III. 2003. Wind and water erosion and transport in semiarid shrubland, grassland, and forest ecosystems: Quantifying the dominance of horizontal wind-driven transport. *Earth Surface Processes and Landforms* 28:1189–1209.

Bretz, J. H., and L. H. Horberg. 1949. Caliche in southeastern New Mexico. *Journal of Geology* 57:491–511.

Brisson, J., and J. F. Reynolds. 1994. The effect of neighbors on root distribution in a creosotebush (*Larrea tridentata*) population. *Ecology* 75:1693–1702.

Brisson, J., and J. F. Reynolds. 1997. Effects of compensatory growth on population processes: A simulation study. *Ecology* 78:2378–2384.

Brown, C. N. 1956. The origin of caliche on the northeastern Llano Estacado, Texas. Journal of Geology 64:1–15.

Brown, D. E. 1994. Chihuahuan Desert scrub. Pages 169–179 in D. E. Brown, editor, *Biotic communities: Southwestern United States and northwestern Mexico*. University of Utah Press, Salt Lake City, UT.

Brown, I. C., and M. Drosdoff. 1940. Chemical and physical properties of soils and their colloids developed from granitic materials in the Mojave Desert. *Journal of Agricultural Research* 61:338–344.

Brown, J. H. 2004. Toward a metabolic theory of ecology. *Ecology* 85:1771–1789.

Brown, J. H., and E. J. Heske. 1990. Control of a desert-grassland transition by a keystone rodent guild. *Science* 250:1705–1707.

Brown, J. H., and S. K. Morgan Ernest. 2002. Rain and rodents: Complex dynamics of desert consumers. *BioScience* 52:979–988.

Brown, J. H., T. J. Valone, and C. G. Curtin. 1997. Reorganization of an arid ecosystem in response to recent climate change. *Proceedings of the National Academy of Sciences* 94:9729–9733.

Brown, J. R., and S. Archer. 1989. Woody plant invasion of grasslands: Establishment of honey mesquite (*Prosopis glandulosa* var *glandulosa*) on sites differing in herbaceous biomass and grazing history. *Oecologia* 80:19–26.

Brown, J. R., and S. Archer. 1990. Water relations of a perennial grass and seedling vs adult woody plants in a subtropical savanna, Texas. *Oikos* 57:366–374.

Brown, J. R., and S. Archer. 1999. Shrub invasion of grassland: Recruitment is continuous and not regulated by herbaceous biomass or density. *Ecology* 80:2385–2396.

Brown, M. F. and W. G. Whitford. 2003. The effects of termites and straw mulch on soil N in a creosotebush (*Larrea tridentata*) dominated Chihuahuan Desert ecosystem. *Journal of Arid Environments* 53:15–20.

Brunel, J.-P., G. R. Walker, and A. K. Kennett-Smith. 1995. Field validation of isotopic procedures for determining sources of water used by plants in a semiarid environment. *Journal of Hydrology* 167:351–368.

Bryant, F. C., B. E. Dahl, R. D. Pettit, and C. M. Britton. 1989. Does short-duration grazing work in arid and semiarid regions? *Journal of Soil and Water Conservation* 44:290–296.

Bryant, J. P., F. D. Provenza, J. Pastor, P. B. Reichardt, T. P. Clarisen, and J. T. duTort. 1991. Interactions between woody plants and browsing mammals mediated by secondary metabolites. *Annual Review of Ecology and Systematics* 22:431–446.

Bryant, N. A., L. F. Johnson, A. J. Brazel, R. C. Balling, C. F. Hutchinson, and L. R. Beck. 1990. Measuring the effect of overgrazing in the Sonoran Desert. *Climatic Change* 17:243–264.

Buck, B. J., and G. H. Mack. 1995. Latest Cretaceous (Maastrichtian) aridity indicated by paleosols in the McRae Formation, south-central New Mexico. *Cretaceous Research* 16:559–572.

Buck, B. J., and H. C. Monger. 1999. Stable isotopes and soil-geomorphology as indicators of Holocene climate change, northern Chihuahuan Desert. *Journal of Arid Environments* 43:357–373.

Buffington, L. C., and C. H. Herbel. 1965. Vegetational changes on a semidesert grassland range from 1858 to 1963. *Ecological Monographs* 35:139–164.

Bull, W. B. 1977. The alluvial fan environment. *Progress in Physical Geography* 1:222–270.

Bull, W. B. 1979. Threshold of critical power in streams. *Bulletin of the Geological Society of America* 90:453–464.

Bull, W. B. 1991. *Geomorphic responses to climatic change.* Oxford University Press, New York.

Bull, W. B. 1997. Discontinuous ephemeral streams. *Geomorphology* 19:227–276.

Bulloch, H. E., and R. E. Neher. 1980. Soil survey of Doña Ana County area, New Mexico. U.S. Government Printing Office, Washington, DC.

Burgess, S. S. O., M. A. Adams, N. C. Turner, and C. K. Ong. 1998. The redistribution of soil water by tree root systems. *Oecologia* 115:306–311.

Burgess, S. S. O., J. S. Pate, M. A. Adams, and T. E. Dawson. 2000. Seasonal water acquisition and redistribution in the Australian woody phreatophyte, *Banksia prionotes.* Annals of Botany 85:215–224.

Burgess, T. L. 1995. Desert grassland, mixed shrub savanna, shrub steppe, or semidesert scrub? The dilemma of coexisting growth forms. Pages 31–67 in M. P. McClaran and T. R. VanDevender, editors, *The desert grassland.* University of Arizona Press, Tucson, AZ.

Burritt, E. A., and F. D. Provenza. 1989. Food aversion learning: Conditioning lambs to avoid a palatable shrub (*Cercocarpus montanus*). *Journal of Animal Science* 67:650–653.

Butler, J. H. A., and J. C. Buckerfield. 1979. Digestion of lignin by termites. *Soil Biology and Biochemistry* 11:507–513.

Cable, D. R. 1965. Damage to mesquite, Lehmann lovegrass, and black grama by a hot June fire. *Journal of Range Management* 18:326–329.

Cable, D. R. 1967. Fire effects on semidesert grasses and shrubs. *Journal of Range Management* 20:170–176.

Cable, J. M., and T. E. Huxman. 2004. Precipitation pulse size effects on Sonoran Desert soil microbial crusts. *Oecologia* 141:317–324.

Caldwell, M. M., T. E. Dawson, and J. H. Richards. 1998. Hydraulic lift: consequences of water efflux from the roots of plants. *Oecologia* 113:151–161.

Callahan, J. T. 1984. Long-term ecological research. *BioScience* 34:363–367.

Campbell, G. S. 1991. Simulation of water uptake by plant roots. Pages 273–285 in J. Hanks and J. T. Ritchie, editors, *Modeling plant and soil systems.* American Society of Agronomy, Madison, WI.

Campbell, G. S., J. D. Jungbauer Jr., S. Shiozawa, and R. D. Hungerford. 1993. A one-parameter equation for water sorption isotherm of soils. *Soil Science* 156:302–306.

Campbell, G. S., and J. M. Norman. 1998. *An introduction to environmental biophysics,* 2nd edition. Springer, New York.

Campbell, R. S. 1929. Vegetation succession in the Prosopis sand dunes of southern New Mexico. *Ecology* 10:392–398.

Campbell, R. S. 1931. Plant succession and grazing capacity on clay soils in southern New Mexico. *Journal of Agricultural Research* 43:1027–1051.

Campbell, R. S. 1936. Climatic fluctuations. Pages 135–150 in *The western range.* U.S. Senate Document no. 199, Washington, DC.

Campbell, R. S., and E. C. Crafts. 1938. Tentative range utilization standards; black grama (*Bouteloua eriopoda*). USDA, Forest Service, Southwest Forest and Range Experiment Station, Research Note 26.

Campbell, S. E. 1979. Soil stabilization by a prokaryotic desert crust: Implications for Precambrian land biota. *Origins of Life* 9:335–348.

Canfield, R. H. 1939. The effect of intensity and frequency of clipping on density and yield of black grama and tobosa grass. USDA Technical Bulletin mo. 681, Washington, DC.

Canfield, R. H. 1941. Application of the line intercept method in sampling range vegetation. *Journal of Forestry* 39:388–394.

Carpenter, D. E., M. G. Barbour, and C. J. Bahre. 1986. Old field succession in Mojave Desert scrub. *Madrono* 33:111–122.

Carson, M. A., and M. J. Kirkby. 1972. *Hillslope form and process*. Cambridge University Press, Cambridge.

Cassady, J. T., and G. E. Glendening. 1940. Revegetating semidesert rangelands in the Southwest. Civilian Conservation Corps, Federal Security Agency, Forestry Publication no. 8, Washington, DC.

Cassady, J. T., and K. A. Valentine. 1938. The yield and feed value of black grama grass range in southern New Mexico. Unpublished technical report.

Castillo, V. M., M. Martinez-Mena, and J. Albaladeja. 1997. Runoff and soil loss response to vegetation removal in a semarid environment. *Soil Science Society of America Journal* 61:1116–1121.

Cavazos, T., A. C. Comrie, and D. M. Liverman. 2002. Intraseasonal variability associated with wet monsoons in southeast Arizona. *Journal of Climate* 15:2477–2490.

Cerling, T. E. 1984. The stable isotopic composition of modern soil carbonate and its relationship to climate. *Earth Planetary Science Letters* 71:229–240.

Chambers, J. C. 2000. Using threshold and alternative state concepts to restore degraded or disturbed ecosystems. Pages 134–145 in *Proceedings of the High-Altitude Revegetation Workshop No. 14*. Colorado Water Resources Research Institute, Colorado State University, Fort Collins, CO.

Chambers, J. C., and J. A. MacMahon. 1994. A day in the life of a seed: Movements and fates of seeds and the implications for natural and managed systems. *Annual Review Ecology Systematics* 25:263–292.

Charley, J. L. 1972. The role of shrubs in nutrient cycling. Pages 182–203 in C. M. McKell, J. P. Blaisdell, and J. R. Goodin, editors, *Wildland shrubs—Their biology and utilization*. USDA Forest Service General Technical Report INT-1. U.S. Department of Agriculture, Ogden, UT.

Charley, J. L. 1975. Mineral cycling in rangeland ecosystems. Pages 215–256 in R. E. Sosebee, editor, *Rangeland plant physiology*. Society of Range Management, Denver, CO.

Charley, J. L., and N. E. West. 1975. Plant-induced soil chemical patterns in some shrub-dominated, semidesert ecosystems of Utah. *Journal of Ecology* 63:945–964.

Charley, J. L., and N. E. West. 1977. Micro-patterns of nitrogen mineralization activity in soils of some shrub-dominated, semidesert ecosystems of Utah. *Soil Biology and Biochemistry* 9:357–365.

Charney, J., P. H. Stone, and W. J. Quirk. 1976. Drought in the Sahara: Insufficient biogeophysical feedback? *Science* 191:100–102.

Charney, J. G. 1975. Dynamics of deserts and drought in the Sahel. *Quarterly Journal of the Royal Meteorological Society* 101:193–202.

Chen, C.L. 1983. Rainfall intensity-duration-frequency formulas. *Journal of Hydraulic Engineering* 109:1603–1621.

Chen, J. L., and J. F. Reynolds. 1997. GePSi: A generic plant simulator based on object-oriented principles. *Ecological Modelling* 94:53–66.

Chepil, W. S. 1956. Influence of moisture on erodibility of soil by wind. *Soil Science Society of America Proceedings* 20:288–292.

Chesson, P., and N. Huntly. 1997. The roles of harsh and fluctuating conditions in the dynamics of ecological communities. *American Naturalist* 150:519–553.

Chew, R. M. 1974. Consumers as regulators of ecosystems: an alternative to energetics. *Ohio Journal of Science* 74:359–370.

Chew, R. M., and A. E. Chew. 1965. The primary productivity of a desert-shrub (*Larrea tridentata*) community. *Ecological Monographs* 35:355–375.

Chew, R. M., and W. G. Whitford. 1992. A long-term positive effect of kangaroo rats (*Dipodomys spectabilis*) on creosotebushes (*Larrea tridentata*). *Journal of Arid Environments* 22:375–386.

Chin, E. 1977. Modeling daily precipitation occurrence process with Markov chain. *Water Resources Research* 13:949–956.

Clements, F. E. 1920. *Plant indicators.* Carnegie Institute of Washington, Publication 290. Washington, DC.

Clemons, R. E., and Seager, W. R. 1973. *Geology of Souse Springs quadrangle, New Mexico.* New Mexico Bureau of Mines and Mineral Resources, Bulletin 100. Socorro, NM.

Coffin Peters, D., and J. E. Herrick. 1999. Vegetation-soil feedbacks and sensitivity of Chihuahuan desert ecosystems to climate change. *Bulletin of the Ecological Society of America* 80:69.

Cole, D. R., and Monger, H. C. 1994. Influence of atmospheric CO_2 on the decline of C_4 plants during the last deglaciation. *Nature* 368:533–536.

Condie, K. C. 1982. Plate-tectonics model for Proterozoic continental accretion in the southwestern United States. *Geology* 10:37–42.

Condie, K. C., and A. J. Budding. 1979. *Geology and geochemistry of Precambrian rocks, central and south-central New Mexico.* New Mexico Bureau of Mines and Mineral Resources, Memoir 35. Socorro, NM.

Conley, W., M. R. Conley, and T. R. Karl. 1992. A computational study of episodic events and historical context in long-term ecological processes: Climate and grazing in the northern Chihuahuan Desert. *Coenoses* 7:55–60.

Connin, S. L., R. A. Virginia, and C. P. Chamberlain. 1997a. Carbon isotopes reveal soil organic matter dynamics following arid land shrub expansion. *Oecologia* 110:374–386.

Connin, S. L., R. A. Virginia, and C. P. Chamberlain. 1997b. Isotopic study of environmental change from disseminated carbonate in polygenetic soils. *Soil Science Society of America Journal* 61:1710–1722.

Cooke, R., A. Warren, and A. Goudie. 1993. *Desert geomorphology.* UCL Press, London.

Cooke, R. J., and R. W. Reeves. 1976. *Arroyos and environmental change in the American Southwest.* Clarendon Press, Oxford, U.K.

Coppinger, K. D., W. A. Reiners, I. C. Burke, and R. K. Olson. 1991. Net erosion on a sagebrush steppe landscape as determined by cesium-137 distribution. *Soil Science Society of America Journal* 55:254–258.

Cornelius, J. M., P. R. Kemp, J. A. Ludwig, and G. L. Cunningham. 1991. The distribution of vascular plant species and guilds in space and time along a desert gradient. *Journal of Vegetation Science* 2:59–72.

Cox, J. R., J. L. Morton, T. N. Johnson Jr., G. L. Jordan, S. C. Martin, and L. C. Fierro. 1984. Vegetation restoration in the Chihuahuan and Sonoran Deserts of North America. *Rangelands* 6:112–115.

Cox, J. R., A. De Alba-Avila, R. W. Rice, and J. N. Cox. 1993. Biological and physical factors influencing *Acacia constricta* and *Prosopis velutina* establishment in the Sonoran Desert. *Journal of Range Management* 46:43–48.

Crawford, C. S. 1988. Nutrition and habitat selection in desert detritivores. *Journal of Arid Environments* 14:111–121.

Crawford, D. S., and J. R. Gosz. 1982. Desert ecosystems: Their resources in space and time. *Environmental Conservation* 9:181–195.

Creusere, F. M., and W. G. Whitford. 1977. Ecological relationships in a desert anuran community. *Herpetologica* 32:7–18.

Cross, A. F., and W. H. Schlesinger. 1999. Plant regulation of soil nutrient distribution in the northern Chihuahuan Desert. *Plant Ecology* 145:11–25.

Cross, A. F., and W. H. Schlesinger. 2001. Biological and geochemical controls on phosphorus fractions in semiarid soils. *Biogeochemistry* 52:155–172.

Cunningham, G. L., and J. H. Burk. 1973. The effect of carbonate deposition layers ("caliche") on the water status of *Larrea divaricata*. *American Midland Naturalist* 90: 474–480.

Dabo, S. M. 1980. Botanical composition of blacktailed jackrabbit diets on semidesert rangeland. M.S. thesis. New Mexico State University, Las Cruces, NM.

Dabo, S. M., R. D. Pieper, R. F. Beck, and G. M. Southward. 1982. *Summer and fall diets of blacktailed jackrabbits on semidesert rangeland.* New Mexico State University Agricultural Experiment Station Research, Report 476, Las Cruces, NM.

Dahm, C. N., and D. I. Moore. 1994. The El Niño/Southern Oscillation phenomenon and the Sevilleta Long-Term Ecological Research site. Pages 12–20 in D. Greenland, editor, *El Niño and Long-Term Ecological Research (LTER) sites*. LTER Network Office, University of Washington, Seattle, WA.

Daniels, R. B., E. E. Gamble, and J. G. Cady. 1971. The relation between geomorphology and soil morphology and genesis. *Advances in Agronomy* 23:51–88.

Danielson, B. J. 1992. Habitat selection, interspecific interactions and landscape composition. *Evolutionary Ecology* 6:399–411.

Darrow, R. A. 1944. *Arizona range resources and their utilization—I. Cochise County.* University of Arizona, Agriculture Experiment Station, Bulletin 103. Tucson, AZ.

Davenport, D. W., D. D. Breshears, B. P. Wilcox, and C. D. Allen. 1998. Sustainability of piñon-juniper ecosystems—a unifying perspective of soil erosion thresholds. *Journal of Range Management* 51:231–240.

Davis, J. O., and Nials, F. 1988. Geomorphology of the study area. Pages 7–22 in J. C. Ravesloot, editor, *Archaeological resources of the Santa Teresa study area, south-central New Mexico*. BLM-NMGI-88-023-3120. Cultural Resources Management Division, Arizona State Museum, University of Arizona, Tucson, AZ.

Davis, S. D., J. S. Sperry, and U. G. Hacke. 1999. The relationship between xylem conduit diameter and cavitation caused by freezing. *American Journal of Botany* 86:1367–1372.

Davis, S. F. 1988. Simplified second-order Godunov-type methods. *SIAM Journal of Scientific and Statistical Computing* 9:445–473.

Day, T. A., and J. K. Detling. 1990. Grassland patch dynamics and herbivore grazing preference following urine deposition. *Ecology* 71:180–188.

Dean, W. R. J., S. J. Milton, and F. Jeltch. 1999. Large trees, fertile islands, and birds in an arid savanna. *Journal of Arid Environments* 41:61–78.

De Bruin, H. A. R., B. J. J. M. Van den Hurk, and L. J. M. Kroon. 1999. On the temperature-humidity correlation and similarity. *Boundary-Layer Meteorology* 93: 453–468.

de Buys, W. 2001. *Seeing things whole: The essential John Wesley Powell.* Island Press, Washington, DC.

Deng, Y., and J. B. Dixon. 2002. Soil organic matter and organic-mineral interactions. Pages 69–107 in J. B. Dixon and D. G. Schulze, editors, *Soil mineralogy with environmental applications*. SSSA Book Series no. 7. Madison, WI.

Detling, J. K. 1988. Grasslands and savannas: regulation of energy flow and nutrient cycling by herbivores. In J. J. Alberts and L. L. Pomeroy, editors, *Concepts of ecosystem ecology*. Springer-Verlag, New York.

Devine, D. L., M. K. Wood, and G. B. Donart. 1998. Runoff and erosion from a mosaic tobosa grass and burro grass community in the northern Chihuahuan Desert grassland. *Journal of Arid Environments* 39:11–19.

De Vries, A. C., J. C. Ritchie, M. Menenti, and W. P. Kustas. 1997. Aerodynamic roughness estimated from surface features for a coppice dune area using laser altimeter data. *12th Symposium on Boundary Layers and Turbulence* 12:289–290.

De Vries, A. C., J. C. Ritchie, W. Klaassen, M. Menenti, W. P. Kustas, A. Rango, and J. H. Prueger. 2003. Effective aerodynamic roughness estimated from airborne laser altimeter measurements of surface features. *International Journal of Remote Sensing* 24: 1545–1558.

Diaz, H. F., and G. N. Kiladis. 1992. Atmospheric teleconnections associated with the extreme phases of the Southern Oscillation. Pages 7–28 in H. F. Diaz and V. Markgraf, editors, *El Niño. Historical and paleoclimatic aspects of the Southern Oscillation*. Cambridge University Press, Cambridge.

Dick-Peddie, W. A. 1993. *New Mexico vegetation: past, present, and future*. University of New Mexico Press, Albuquerque, NM.

DiTomaso, J. M. 2000. Invasive weeds in rangelands: Species, impacts, and management. *Weed Science* 48:255–265.

Donahue, D. L. 1999. *The Western range revisited*. University of Oklahoma Press, Normal, OK.

Donovan, L. A., and J. R. Ehleringer. 1994. Water-stress and use of summer precipitation in a Great Basin shrub community. *Functional Ecology* 8:289–297.

Dregne, H. E. 1976. *Soils of arid regions*. Elsevier Scientific Publishers, Amsterdam.

Dregne, H., M. Kassas, and B. Rozanov. 1991. A new assessment of the world status of desertification. *Desertification Control Bulletin* 20:6–18.

Drewa, P. B., and K. M. Havstad. 2001. Effects of fire, grazing and the presence of shrubs on Chihuahuan Desert grasslands. *Journal of Arid Environments* 48:429–443.

Drewa, P. B., D. P. C. Peters, and K. M. Havstad. 2001. Fire, grazing, and shrub invasion in the Chihuahuan Desert. Pages 31–39 in T. P. Wilson and K. E. M. Galley, editors, *Proceedings of the Invasive Species Workshop: The role of fire in the control and spread of invasive species*. Fire Conference 2000: The National Congress on Fire Ecology, Prevention, and Management. Miscellaneous Publication no. 11, Tall Timbers Research Station, Tallahassee, FL.

Dugas, W. A. 1993. Micrometeorological and chamber measurement of CO_2 flux from bare soil. *Agricultural and Forest Meterology* 67:115–128.

Dugas, W. A., R. A. Hicks, and R. P. Gibbens. 1996. Structure and function of C_3 and C_4 Chihuahuan Desert plant communities. Energy balance components. *Journal of Arid Environments* 34:63–79.

Dugas, W. A., M. L. Heuer, and H. S. Mayeux. 1999. Seasonal carbon dioxide fluxes over coastal bermudagrass, native prairie, and sorghum. *Agricultural and Forest Meteorology* 83: 113–133.

Dunin, F. X., R. A. Nulsen, I. N. Baxter, and E. A. N. Greenwood. 1989. Evaporation from a lupin crop: A comparison of methods. *Agricultural and Forest Meteorology* 46:297–311.

Eberhart, L. L. 1978. Transect methods of population studies. *Journal of Wildlife Management* 42:1–31.

Ehleringer, J. R., and T. E. Dawson. 1992. Water-uptake by plants—perspectives from stable isotope composition. *Plant Cell and Environment* 15:1073–1082.

El-Ghonemy, A. A., A. Wallace, and E. M. Romney. 1978. Nutrient concentration in the natural vegetation of the Mojave Desert. *Soil Science* 126:219–229.

Elkins, N. Z., G. V. Sabol, T. J. Ward, and W. G. Whitford. 1986. The influence of subterranean termites on the hydrological characteristics of a Chihuahuan Desert ecosystem. *Oecologia* 68:521–528.

Elliot, E. T., H. W. Hunt, and D. E. Walter. 1988. Detrital food web interactions in North American grassland ecosystems. *Agriculture, Ecosystems and Environment* 24: 41–56.

Ellis, J. E., and D. M. Swift. 1988. Stability of African pastoral ecosystems: Alternative paradigms and implications for development. *Journal of Range Management* 41:450–459.

Elson, W. E., W. R. Seager, and R. E. Clemons. 1975. Emory cauldron, Black Range, New Mexico—source of the Kneeling Nun. Pages 283–292 in W. R. Seager, R. E. Clemons, and J. Callender, editors, *Guidebook of the Las Cruces country*. New Mexico Geological Society, Guidebook 26. New Mexico Geological Society, Socorro, NM.

Eltahir, E. A. B., and R. L. Bras. 1996. Precipitation recycling. *Reviews of Geophysics* 34: 367–378.

Encina-Rojas, A. 1995. *Detailed soil survey of the Jornada LTER (Long-Term Ecological Research) transect vicinity, southern New Mexico*. New Mexico State University, Las Cruces, NM.

Enquist, B. J., and K. J. Niklas. 2001. Invariant scaling relations across tree-dominated communities. *Nature* 410:655–660.

Estell, R. E., E. L. Fredrickson, D. M. Anderson, W. F. Mueller, and M. D. Remmenga. 1994. Relationship of tarbush leaf surface secondary chemistry to herbivory by livestock. *Journal of Range Management* 47:424–428.

Estell, R. E., E. L. Fredrickson, D. M. Anderson, K. M. Havstad, and M. D. Remmenga. 1998. Relationship of tarbush leaf surface terpene profile with livestock herbivory. *Journal of Chemical Ecology* 24:1–12.

Ethridge, D. E., R. D. Sherwood, R. E. Sosebee, and C. H. Herbel. 1997. Economic feasibility of rangeland seeding in the arid southwest. *Journal of Range Management* 50: 185–190.

Ettershank, G., J. A. Ettershank, M. Bryant, and W. G. Whitford. 1978. Effects of nitrogen fertilization on primary production in a Chihuahuan Desert ecosystem. *Journal of Arid Environments* 1:135–139.

Evans, R. D., and J. R. Johansen. 1999. Microbiotic crusts and ecosystem processes. *Critical Reviews in Plant Sciences* 18:183–225.

Everitt, J. H., D. E. Ecobar, I. Cavazos, J. R. Noriega, and M. R. Davis. 1995. A three-camera, multispectral, digital, video imaging system. *Remote Sensing of Environmen*t 54:333–337.

Everitt, J. H., M. A. Alaniz, M. R. Davis, D. E. Escobar, K. M. Havstad, and J. C. Ritchie. 1997. Light reflectance characteristics and video remote sensing of two range sites on the Jornada Experiment Range. Pages 485–495 in *Proceedings of the 16th Biennial Workshop on Videography and Color Photography in Resource Assessment*. American Society for Photogrammetry and Remote Sensing. Bethesda, MD

Fabry, J., and S. Pieper. n.d. A history of the New Mexico State University College Ranch. Unpublished manuscript.

Facelli, J. M., and D. J. Brock. 2000. Patch dynamics in arid lands: Localized effects of *Acacia papyrocarpa* on soils and vegetation of open woodlands of south Australia. *Ecography* 23:479–491.

Fatehi, M. 1986. *Comparative seasonal food habits of blacktailed jackrabbits and cattle on semidesert rangeland.* Ph.D. dissertation. New Mexico State University, Las Cruces, NM.

Fatehi, M., R. D. Pieper, and R. F. Beck. 1988. Seasonal food habits of blacktailed jackrabbits (*Lepus californicus*) in southern New Mexico. *Southwestern Naturalist* 33: 367–370.

Fischer, R. A., and N. C. Turner. 1978. Plant productivity in the arid and semiarid zones. *Annual Review of Plant Physiology* 29:277–317.

Fisher, F. M., and W. G. Whitford. 1995. Field simulation of wet and dry years in the Chihuahun Desert: Soil moisture, N mineralization and ion-exchange resin bags. *Biology and Fertility of Soils* 20:137–146.

Fisher, F. M., L. W. Parker, J. P. Anderson, and W. G. Whitford. 1987. Nitrogen mineralization in a desert soil: Interacting effects of soil moisture and nitrogen fertilizer. *Soil Society of America Journal* 51:1033–1041.

Fisher, F. M., J. C. Zak, G. L. Cunningham, and W. G. Whitford. 1988. Water and nitrogen effects on growth and allocation patterns of creosotebush in the northern Chihuahuan Desert. *Journal of Range Management* 41:387–391.

Floyd, T. 1996. Top-down impacts on creosotebush herbivores in a spatially and temporally complex environment. *Ecology* 77:1544–1555.

Forsling, C. L. 1919. *Chopped soapweed as emergency feed for cattle on southwestern ranges.* U.S. Department of Agriculture Bulletin 745.

Forsling, C. L. 1924. *Saving livestock from starvation on southwestern ranges.* U.S. Department of Agriculture Farmers' Bulletin 1428.

Fowler, H., and W. G. Whitford. 1980. Termites, microarthropods and the decomposition of senescent and fresh creosotebush (*Larrea tridentata*) leaf litter. *Journal of Arid Environments* 3:63–68.

Fowler, J. M. 1993. *Western livestock industry survey for 1991.* New Mexico Agricultural Experiment Station Research Report.

Fowler, J. M., and L. A. Torell. 1985. *The financial position of the New Mexico range livestock industry, 1940–1984.* New Mexico State University Agricultural Experiment Station Range Improvement Task Force Report no. 20.

Fowler, N. L. 1986. Microsite requirements for germination and establishment of three grass species. *American Midland Naturalist* 115:131–145.

Frank, D. A., and R. D. Evans. 1997. Effects of native grazers on grassland N cycling in Yellowstone National Park. *Ecology* 78:2238–2248.

Franklin, J., J. Duncan, and D. L. Turner. 1993. Reflectance of vegetation and soil in Chihuahuan Desert plant communities from ground radiometry using SPOT wavebands. *Remote Sensing of the Environment* 46:291–304.

Freckman, D. W., and R. Mankau. 1986. Abundance, distribution, biomass and eneretics of soil nematodes in a northern Mojave Desert ecosystem. *Pedobiologia* 29:129–142.

Freckman, D. W., W. G. Whitford, and Y. Steinberger. 1987. Effect of irrigation on nematode population dynamics and activity in desert soils. *Biology and Fertility of Soils* 3:3–10.

Fredrickson, E., K. M. Havstad, R. Estell, and P. Hyder. 1998. Perspectives on desertification: southwestern United States. *Journal Arid Environments* 39:191–207.

Fredrickson, E. L., J. P. Thilsted, R. E. Estell, and K. M Havstad. 1994. Effects of chronic ingestion of tarbush (*Flourensia cernua*) on ewe lambs. *Veterinary and Human Toxicology* 36:409–415.

Fredrickson, E. L., R. E. Estell, A. Laliberte, and D. M. Anderson. 2005. Mesquite recruitment in the Chihuahuan Desert: Historic and prehistoric patterns with long term impacts. *Journal of Arid Environments* 65:285–295.

Freeman, C. E. 1972. Pollen study of some Holocene alluvial deposits in Doña Ana County, southern New Mexico. *Texas Journal of Science* 24:203–220.

Fritschen, L. J., P. Qian, E. T. Kanemasu, D. Nie, E. A. Smith, J. B. Steward, S. B. Verma, and M. L. Wesely. 1992. Comparisons of surface flux measurement systems used in FIFE 1989. *Journal of Geophysical Research* 97:18697–18713.

Fuhlendorf, S. D., and F. E. Smeins. 1997. Long-term vegetation dynamics mediated by herbivores, weather and fire in a Juniperus-Quercus savanna. *Journal of Vegetation Science* 8:819–828.

Gadzia, J. S. G., and J. A. Ludwig. 1983. Mesquite age and size in relation to dunes and artifacts. *Southwestern Naturalist* 28:89–94.

Gallardo A., and W. H. Schlesinger. 1992. Carbon and nitrogen limitations of soil microbial biomass in desert ecosystems. *Biogeochernistry* 18:1–17.

Gallardo A., and W. H. Schlesinger. 1995. Factors determining soil microbial biomass and nutrient immobilization in desert soils. *Biogeochemistry* 28:55–68.

Gallardo A., J. J. Rodriguez-Saucedo, F. Covelo, and R. Fernandez-Ales. 2000. Soil nitrogen heterogeneity in a Dehesa ecosystem. *Plant and Soil* 222:71–82.

Gallegos, R. A. 1999. *Biogenic carbonate, desert vegetation, and stable carbon isotopes.* M.S. thesis. New Mexico State University, Las Cruces, NM.

Galloway, R. W. 1970. The full-glacial climate of the southwestern United States. *Annals of the American Association of Geographers* 60:245–256.

Galloway, R. W. 1983. Full-glacial southwestern United States: Mild and wet or cold and dry? *Quaternary Research* 19:236–248.

Gambill, M. D., M. K. Petersen, D. Hawkins, D. V. Dhuyvetter, M. Ward, I. Tovar, K. M. Havstad, and J. D. Wallace. 1994. Intake of cooked molasses blocks and subsequent performance of young postpartum cows grazing native range. *Proceedings of the Western Section of the American Society of Animal Science* 45:76–79.

Gamboa, G. J. 1975. Foraging and leaf-cutting of the desert gardening ant *Acromyrmex versicolor* (Pergande) (Hymenoptera: Formicidae). *Oecologia* 20:103–110.

Gao, Q., and J. F. Reynolds. 2003. Historical shrub-grass transitions in the northern Chihuahuan Desert: Modeling the effects of shifting rainfall seasonality and event size over a landscape gradient. *Global Change Biology* 9:1–19.

Garcia-Moya, E., and C. M. McKell. 1970. Contribution of shrubs to the nitrogen economy of a desert-wash plant community. *Ecology* 51:81–88.

Gardner, D. R. 1957. *The National Cooperative Soil Survey of the United States*. Ph.D. dissertation. Harvard University, Boston, MA. Issued October 1998 by USDA-NRCS.

Gardner, J. L. 1951. Vegetation of the creosotebush area of the Rio Grande Valley in New Mexico. *Ecological Monographs* 21:379–403.

Gardner, L. R. 1984. Carbon and oxygen isotope composition of pedogenic $CaCO_3$ from soil profiles in Nevada and New Mexico, USA. *Isotope Geoscience* 2:55–73.

Gardner, R. H., R. V. O'Neill, J. B. Mankin, and D. Kumar. 1980. Comparative error analysis of six predator-prey models. *Ecology* 61:323–332.

Gardner, W. R. 1958. Some steady-state solutions of unsaturated moisture flow equations with application to evaporation from a water table. *Soil Science* 85:228–232.

Garner, W., and Y. Steinberger. 1989. A proposed mechanism for the formation of "fertile islands" in the desert ecosystem. *Journal of Arid Environments* 16:257–262.

Garratt, J. R. 1993. Sensitivity of climate simulations to land-surface and atmospheric boundary-layer treatments—a review. *Journal of Climate* 6:419–449.

Gebauer, R. L. E., and J. R. Ehleringer. 2000. Water and nitrogen uptake patterns following moisture pulses in a cold desert community. *Ecology* 81:1415–1424.

Gerakis, P. A., and C. Z. Tsangarakis. 1970. The influence of *Acacia senegal* on the fertility of a sand sheet ('goz') soil in the central Sudan. *Plant and Soil* 33:81–86.

Gibbens, R., J. Tromble, J. Hennessy, and M. Cardenas. 1983. Soil movement in mesquite dunelands and former grasslands of southern New Mexico from 1933 to 1980. *Journal of Range Management* 36:145–148.

Gibbens, R. P., and R. F. Beck. 1987. Increase in number of dominant plants and dominance-classes on a grassland in the northern Chihuahuan Desert. *Journal of Range Management* 40:136–139.

Gibbens, R. P., and R. F. Beck. 1988. Changes in grass basal area for forb densities over a 64-year period on grassland types of the Jornada Experimental Range. *Journal of Range Management* 41:186–192.

Gibbens, R. P., and J. M. Lenz. 2001. Root systems of some Chihuahuan Desert plants. *Journal of Arid Environments* 49:221–263.

Gibbens, R. P., R. P. Beck, R. P. McNeely, and C. H. Herbel. 1992. Recent rates of mesquite establishment in the northern Chihuahuan Desert. *Journal of Range Management* 45:585--588.

Gibbens, R. P., K. M. Havstad, D. D. Billheimer, and C. H. Herbel. 1993. Creosotebush vegetation after 50 years of lagomorph exclusion. *Oecologia* 94:210–217.

Gibbens, R. P., R. P. McNeely, K. M. Havstad, R. F. Beck, and B. Nolen. 2005. Vegetation change in the Jornada Basin from 1858 to 1998. *Journal of Arid Environments* 61: 651–668.

Gile, L. H. 1961. A classification of ca horizons in the soils of the desert region, Doña Ana County, New Mexico. *Soil Science Society of American Proceedings* 25:52–61.

Gile, L. H. 1966a. Cambic and certain noncambic horizons in desert soils of southern New Mexico. *Soil Science Society of America Proceedings* 30:657–660.

Gile, L. H. 1966b. Coppice dunes and the Rotura soil. *Soil Science Society of America Proceedings 3*0:657–660.

Gile, L. H. 1967. Soils of an ancient basin floor near Las Cruces, New Mexico. *Soil Science* 103:265–276.

Gile, L. H. 1970. Soils of the Rio Grande Valley border in southern New Mexico. *Soil Science Society of America Proceedings* 34:465–472.

Gile, L. H. 1975a. Causes of soil boundaries in an arid region: I. Age and parent materials. *Soil Science Society of America Journal* 39:316–323.

Gile, L. H. 1975b. Causes of soil boundaries in an arid region: II. Dissection, moisture, and faunal activity. *Soil Science Society of America Proceedings* 39:324–330.

Gile, L. H. 1975c. Holocene soils and soil-geomorphic relations in an arid region of southern New Mexico. *Quaternary Research* 5:321–360.

Gile, L. H. 1986. Late Holocene displacement along the Organ Mountains fault in southern New Mexico—a summary. *New Mexico Geology* 8:1–4.

Gile, L. H. 1990. Chronology of lava and associated soils near San Miguel, New Mexico. *Quaternary Research* 33:37–50.

Gile, L. H. 1994. Soils, geomorphology, and multiple displacements along the Organ Mountains fault in southern New Mexico. New Mexico Bureau of Mines and Mineral Resources, Bulletin 133. Socorro, NM.

Gile, L. H. 1999. Eolian and associated pedogenic features of the Jornada Basin floor, southern New Mexico. *Soil Science Society of America Journal* 63:151–163.

Gile, L. H. 2002. Lake Jornada, an early-middle Pleistocene lake in the Jornada del Muerto Basin, southern New Mexico. *New Mexico Geology* 24:3–14.

Gile, L. H., and R. B. Grossman. 1968. Morphology of the argillic horizon in desert soils of southern New Mexico. *Soil Science* 106:6–15.

Gile, L. H., and R. B. Grossman. 1979. *The Desert Project soil monograph*. Soil Conservation Service, U.S. Department of Agriculture, National Soil Survey Center, Lincoln, NE.

Gile, L. H., and J. W. Hawley. 1966. Periodic sedimentation and soil formation on an alluvial-fan piedmont in southern New Mexico. *Soil Science Society of America Proceedings* 30:261–268.

Gile, L. H., and J. W. Hawley. 1968. Age and comparative development of desert soils at the Gardner Spring radiocarbon site, New Mexico. *Soil Science Society of America Proceedings* 32:709–719.

Gile, L. H., F. F. Peterson, and R. B. Grossman. 1965. The K horizon—a master soil horizon of carbonate accumulation. *Soil Science* 99:74–82.

Gile, L. H., F. F. Peterson, and R. B. Grossman. 1966. Morphological and genetic sequences of carbonate accumulation in desert soils. *Soil Science* 101:347–360.

Gile, L. H., R. B. Grossman, and J. W. Hawley. 1969. Effects of landscape dissection on soils near University Park, New Mexico. *Soil Science* 108:273–282.

Gile, L. H., J. W. Hawley, and R. B. Grossman. 1981. *Soils and geomorphology in the basin and range area of southern New Mexico—guidebook to the Desert Project*. New Mexico Bureau of Mines and Mineral Resources Memoir 39. Socorro, NM.

Gile, L. H., J. W. Hawley, R. B. Grossman, H. C. Monger, and C. E. Montoya. 1995. *Supplement to the Desert Project guidebook with emphasis on soil micromorphology*. New Mexico Bureau of Mines and Mineral Resources, Bulletin 142. Socorro, NM.

Gile, L. H., R. B. Grossman, J. W. Hawley, and H. C. Monger. 1996. Ancient soils of the Rincon Surface, northern Doña Ana County, in L. H. Gile and R. J. Ahrens, editors, *Studies of soil and landscape evolution in Southern New Mexico*. Soil Survey Investigations Report 44, Lincoln, NE.

Gile, L. H., R. P. Gibbens, and J. M. Lenz. 1997. The near-ubiquitous pedogenic world of mesquite roots in an arid basin floor. *Journal of Arid Environments* 35:39–58.

Gillette, D. 1978. Tests with a portable wind tunnel for determining wind erosion threshold velocities. *Atmospheric Environment* 12:2309–2313.

Gillette, D., and W. Chen. 1999. Size distributions of saltating grains: An important variable in the production of suspended particles. *Earth Surface Processes and Landforms* 24:449–462.

Gillette, D., and W. Chen. 2000. Particle production and aeolian transport from a "supply-limited" source area in the Chihuahuan Desert. *Journal of Geophysical Research* 106: 5267–5278.

Gillette, D., D. W. Fryrear, T. Gill, T. Ley, T. Cahill, and E. Gearhart. 1997a. Relation of vertical flux of PM_{10} to total aeolian horizontal mass flux at Owens Lake. *Journal of Geophysical Research* 102:26009–26015.

Gillette, D., E. Hardebeck, and J. Parker. 1997b. Large-scale variability of wind erosion mass flux rates at Owens Lake: The role of roughness change, particle limitation, change of threshold friction velocity, and the Owen effect. *Journal of Geophysical Research* 102:25989–25998.

Gillette, D. A., and W. Chen. 2001. Particle production and aeolian transport from a "supply-limited" source area in the Chihuahuan Desert, New Mexico, United States. *Journal of Geophysical Research* 106:5267–5278.

Gillette, D. A., and P. Stockton. 1989. The effect of nonerodible particles on wind erosion of erodible surfaces. *Journal of Geophysical Research* 94:12885–12893.

Gillette, D. A., I. H. Blifford Jr., and D. W. Fryrear. 1974. The influence of wind velocity on size distributions of soil wind erosion aerosols. *Journal of Geophysical Research* 79:4068–4075.

Gillette, D. A., R. N. Clayton, T. K. Mayeda, M. L. Jackson, and K. Sridhar. 1978. Tropospheric aerosols from some major dust storms of the southwestern United States. *Journal of Applied Meteorology* 17:832–845.

Gillette, D. A., G. Stensland, A. Williams, W. Barnard, D. Gatz, P. Sinclair, and T. Johnson. 1992. Emissions of alkaline elements calcium, magnesium, potassium, and sodium from open sources in the contiguous United States. *Global Biogeochemical Cycles* 6: 437–457.

Gillette, D. D., D. L. Wolberg, and A. P. Hunt. 1986. *Tyrannosaurus rex* from McRae Formation (Lancian, Upper Cretaceous), Elephant Butte Reservoir, Sierra County, New Mexico. Pages 235–238 in R. E. Clemons, W. E. King, G. H. Mack, and J. Zidek, editors, *Guidebook 37: Truth or Consequences region*. New Mexico Geological Society, Las Cruces, NM.

Gilmore, A. M., and M. C. Ball. 2000. Protection and storage of chlorophyll in overwintering evergreens. *Proceedings of the National Academy of Sciences of the United States of America* 97:11098–11101.

Glendening, G. E. 1952. Some quantitative data on the increase in mesquite and cactus on a desert grassland range in southern Arizona. Ecology 33:319–328.

Golitsyn, G., and D. A. Gillette. 1993. A joint Soviet-American experiment for the study of Asian desert dust and its impact on local meteorological conditions and climate. *Atmospheric Environment* 27a:2467–2470.

Golley, F. B. 1993. *A history of the ecosystem concept in ecology: More than the sum of the parts*. Yale University Press, New Haven, CT.

Goslee, S. C., K. M. Havstad, D. C. Peters, A. Rango, W. Schlesinger. 2003. High-resolution images reveal rate and pattern of shrub encroachment over six decades in New Mexico, USA. *Journal of Arid Environments* 54(4):755–767.

Gosz, J. R. 1993. Ecotone hierarchies. *Ecological Applications* 3:364–376.

Gosz, J. R., D. L. Moore, G. A. Shore, and H. D. Grover. 1995. Lightning estimates of precipitation location and quantity on the Sevilleta LTER, New Mexico. *Ecological Applications* 5:1141–1150.

Gosz, R. J., and J. R. Gosz. 1996. Species interactions on the biome transition zone in New Mexico: response of blue grama (*Bouteloua gracilis*) and black grama (*Bouteloua eriopoda*) to fire and herbivory. *Journal of Arid Environments* 34:101–114.

Goudie, A. 1973. *Duricursts in tropical and subtropical landscapes.* Oxford University Press, London.

Gough, L., C. W. Osenberg, K. L. Gross, and S. L. Collins. 2000. Fertilization effects on species density and primary productivity in herbaceous plant communities. *Oikos* 89: 428–439.

Greeley, R., and J. D. Iversen. 1985. *Wind as a geological process on Earth, Mars, Venus and Titan*. Cambridge University Press, Cambridge.

Grieu, R., J. M. Guehl, and G. Aussenac. 1988. The effects of soil and atmospheric drought on photosynthesis and stomatal control of gas exchange in three coniferous species. *Physiological Plant Pathology* 73:97–104.

Grossman, R. B., R. J. Ahrens, L. H. Gile, C. E. Montoya, and O. A. Chadwick. 1995. Areal evaluation of organic and carbonate carbon in a desert area of southern New Mexico. Pages 81–92 in R. Lal et al., editors, *Soils and global change*. CRC Press, Boca Raton, FL.

Grover, Herbert D., and H. Brad Musick. 1990. Shrubland encroachment in southern New Mexico, USA: An analysis of desertification processes in the American Southwest. *Climatic Change* 17:305–30.

Guilbault, M. R., and A. D. Matthias. 1998. Emission of N_2O from Sonoran Desert and effluent-irrigated grass ecosystems. *Journal of Arid Environments* 38:87–98.

Gustavson, T. C. 1991a. *Arid basin depositional systems and paleosols: Fort Hancock and Camp Rice formations (Pliocene-Pleistocene), Hueco Bolson, west Texas and adjacent Mexico.* Report of Investigations, Texas Bureau of Economic Geology, R101198.

Gustavson, T. C. 1991b. Buried vertisols in lacustrine facies of the Pliocene Fort Hancock formation, Hueco Bolson, west Texas and Chihuahua, Mexico. *Geological Society of America Bulletin* 103:448–460.

Gutierrez, J., and I. I. Hernandez. 1996. Runoff and interrill erosion as affected by grass cover in a semiarid rangeland of northern Mexico. *Journal of Arid Environments* 34: 287–295.

Gutierrez, J. R., and W. G. Whitford. 1987a. Chihuahuan Desert New Mexico USA annuals: Importance of water and nitrogen. *Ecology* 68:2032–2045.

Gutierrez, J. R., and W. G. Whitford. 1987b. Responses of Chihuahuan Desert herbaceous annuals to rainfall augmentation. *Journal of Arid Environments* 12:127–39.

Gutierrez, J. R., O. A. Da Silva, M. I. Pagani, D. Weems, and W. G. Whitford. 1988. Effects of different patterns of supplemental water and nitrogen fertilization on productivity and composition of Chihuahuan Desert annual plants. *American Midland Naturalist* 119:336–343.

Gutierrez, J. R., P. L. Meserve, L. C. Contreras, H. Vasquez, and F. M. Jaksie. 1993. Spatial distribution of soil nutrients and ephemeral plants underneath and outside the canopy of *Porlieria chilensis* shrubs (Zygophyllaceae) in arid coastal Chile. *Oecologica* 95: 347–352.

Gutierrez-Luna, R. 2000. *Low input remediation technique for seeding arid rangelands.* Ph.D. dissertation. New Mexico State University, Las Cruces, NM.

Gutschick, V. P. 1987. *A functional biology of crop plants.* Croom Helm, London/Timber Press, Beaverton, OR.

Hadley, N. F., and S. R. Szarek. 1981. Productivity of desert ecosystems. *BioScience* 31: 747–753.

Hagen, L., E. Skidmore, and A. Saleh. 1992. Wind erosion: Prediction of aggregate abrasion coefficients. *Transactions American Society of Agricultural Engineers* 35:1847–1850.

Hakkila, M. D. 1986. *Beef steer diets, forage intake, and fecal indices in south-central New Mexico.* Thesis. Animal and Range Sciences, New Mexico State University.

Hall, D. O., J. M. O. Scurlock, S. Ojima Dennis, and W. J. Parton. 2000. Grasslands and the global carbon cycle: modelling the effects of climate change. Pages 102–114 in T. M. L. Wigley and D. S. Schimel, editors, *The carbon cycle*. Cambridge University Press, Cambridge.

Hartley, A. E. 1997. *Environmental controls on nitrogen cycling in northern Chihuahuan Desert soils.* Ph.D. dissertation, Duke University, Durham, NC.

Hartley, A. E., and W. H. Schlesinger. 2000. Environmental controls on nitric oxide emission from northern Chihuahuan Desert soils. *Biogeochemistry* 50:279–300.

Hartley, A. E., and W. H. Schlesinger. 2002. Environmental controls on nitrogen fixation in northern Chihuahuan Desert soils. *Journal of Arid Environments* 52:293–304.

Hartmann, D. L. 1994. *Global physical climatology*. Academic Publishers, San Diego, CA.

Harvey, A. M. 1997. The role of alluvial fans in arid zone fluvial systems. Pages 231–259 in D. S. G. Thomas, editor, *Arid zone geomorphology*. Wiley, Chichester, U.K.

Hastings, J. R., and R. M. Turner. 1965. *The changing mile*. University of Arizona Press, Tucson, AZ.

Havstad, K. M. 1996. Legacy of Charles Travis Turney: The Jornada Experimental Range. *Archaeological Society of New Mexico Annual* 22:77–92.

Havstad, K. M., and W. H. Schlesinger. 1996. Reflections on a century of rangeland research in the Jornada Basin of New Mexico. Pages 23–25 in *Proceedings: Shrubland Ecosystem Dynamics in a Changing Climate.* May 1995, Las Cruces, New Mexico. USDA Forest Service, General Technical Report INT-GTR-338, Ogden, UT.

Havstad, K. M., R. P. Gibbens, C. A. Knorr, and L. W. Murray. 1999. Long-term influences of shrub removal and lagomorph exclusion on Chihuahuan Desert vegetation dynamics. *Journal of Arid Environments* 42:155–166.

Havstad, K. M., D. P. Peters, and L. W. Murray. 2003. Long-term dynamics of a degraded arid shrubland: delayed responses and the importance of spatial processes. Pages 353–355 in *Proceedings of the 7th International Rangeland Congress.* Durban, South Africa.

Hawke, S. D., and R. D. Farley. 1973. Ecology and behavior of the desert burrowing cockroach, *Arenivaga* sp. (Dictyoptera, Polyphagidae). *Oecologia* 11:263–279.

Hawley, J. W. 1972. Geologic-geomorphic mapping to serve soil resource development. Pages 24–30 in *Soil Conservation Society of America Proceedings*, 27th annual meeting.

Hawley, J. W. 1975a. Quaternary history of Doña Ana County region, south-central New Mexico. Pages 139–150 in *26th Annual Field Conference guidebook.* New Mexico Geological Society, Socorro, NM.

Hawley, J. W. 1975b. The Desert Soil-Geomorphology Project. Pages 183–185 in *26th Annual Field Conference guidebook.* New Mexico Geological Society, Socorro, NM.

Hawley, J. W. 1981. Pleistocene and Pliocene history of the international boundary area, southern New Mexico. Pages 26–32 in *El Paso Geological Society Field Trip, Geology of the Border, Southern New Mexico-Northern Chihuahua.*

Hawley, J. W. 1993. Geomorphic setting and late Quaternary history of pluvial-lake basins in the southern New Mexico region. New Mexico Bureau of Mines and Mineral Resources, Open File Report 387. Socorro, NM.

Hawley, J. W., and J. F. Kennedy. 2004. *Creation of a digital hydrogeologic framework model on the Mesilla Basin and southern Jornada del Muerto Basin.* New Mexico Water Resources Research Institute, New Mexico State University, Las Cruces, NM.

Hawley, J. W., and F. E. Kottlowski. 1969. Quaternary geology of the south-central New Mexico border region. Pages 89–115 in *Border Stratigraphy Symposium.* New Mexico Bureau of Mines and Mineral Resources, Circular 104. Socorro, NM.

Hawley, J. W., and R. P. Lozinsky. 1992. *Hydrogeologic framework of the Mesilla Basin in New Mexico and western Texas.* New Mexico Bureau of Mines and Mineral Resources, Open File Report 323. Socorro, NM.

Hawley, J. W., F. E. Kottlowski, W. S. Strain, W. R. Seager, W. E. King, and D. V. LeMone. 1969. The Santa Fe Group in the south-central New Mexico border region. Pages 52–76 in Border Stratigraphy Symposium. New Mexico Bureau of Mines and Mineral Resources, Circular 104. Socorro, New Mexico, USA.

Hawley, J. W., G. O. Bachman, and K. Manley. 1976. Quaternary stratigraphy in the basin and range and Great Plains provinces, New Mexico and western Texas. Pages 235–274 in W. C. Mahaney, editor, *Quaternary stratigraphy of North America.* Dowden, Hutchinson and Ross, Stroudsburg, PA.

Heitschmidt, R. K., and J. W. Stuth, editors. 1991. *Grazing management: An ecological perspective.* Timber Press, Portland, OR.

Hemmer, H. 1990. *Domestication: The decline of environmental appreciation.* Cambridge University Press, New York.

Hennessy, J. T., R. P. Gibbens, J. M. Tromble, and M Cardenas. 1983a. Vegetation changes from 1935 to 1980 in mesquite dunelands and former grasslands of southern New Mexico. *Journal of Range Management* 36:370–374.

Hennessy, J. T., R. P. Gibbens, J. M. Tromble, and M. Cardenas. 1983b. Water properties of caliche. *Journal of Range Management* 36:723–726.

Hennessy, J. T., R. P. Gibbens, J. M. Tromble, and M. Cardenas. 1985. Mesquite (*Prosopis glandulosa* Torr) dunes and interdunes in southern New Mexico—A study of soil properties and soil water relations. *Journal of Arid Environments* 9:27–38.

Hennessy, J. T., B. Kies, R. P. Gibbens, and J. M. Tromble. 1986. Soil sorting by forty-five years of wind erosion on a southern New Mexico range. *Soil Science Society of America Journal* 50:391–394.

Herbel, C. H. 1963. Fertilizing tobosa on flood plains in the semidesert grassland. *Journal of Range Management* 16:133–38.

Herbel, C. H. 1982. Grazing management on rangelands. *Journal of Soil and Water Conservation* 37:77–79.

Herbel, C. H., and R. P. Gibbens. 1985. Field water regimes of sandy loam soils on arid rangelands of southern New Mexico. Pages 514–516 in *Proceedings of the 15th International Grassland Conference*. Kyoto, Japan.

Herbel, C. H., and R. P. Gibbens. 1987. Soil water regimes of loamy sands and sandy loams on arid rangelands in southern New Mexico. *Journal of Soil and Water Conservation* 42:442–447.

Herbel, C. H., and R. P. Gibbens. 1989. Matric potential of clay-loam soils on arid rangelands in southern New Mexico. *Journal of Range Management* 42:386–392.

Herbel, C. H., and R. P. Gibbens. 1996. *Post-drought vegetation dynamics on arid rangelands of southern New Mexico.* Bulletin 776. New Mexico Agricultural Experimental Station, Las Cruces, NM.

Herbel, C. H., and L. H. Gile. 1973. Field moisture regimes and morphology of some arid land soils in New Mexico. Pages 119–152 in R. R. Bruce, K. W. Flach, and H. M. Taylor, editors, Field soil water regimes. Soil Science Society American Special Publication 5, Madison, WI.

Herbel, C. H., and W. L. Gould. 1995. *Management of mesquite, creosotebush and tarbush with herbicides in the northern Chihuahuan Desert.* Agricultural Experiment Station Bulletin. New Mexico State University, Las Cruces, NM.

Herbel, C. H., and A. B. Nelson. 1969. *Grazing management on semidesert ranges in southern New Mexico*. Jornada Experimental Range Report no. 1, 13 pages.

Herbel, C. H., F. Ares, and J. Bridges. 1958. Hand-grubbing mesquite in the semidesert grassland. *Journal of Range Management* 11:267–270.

Herbel, C. H., F. N. Ares, and R. A. Wright. 1972. Drought effects on a semidesert grassland range. *Ecology* 53:1084–1093.

Herbel, C. H., G. H. Abernathy, C. C. Yarbrough, and D. K. Gardner. 1973. Rootplowing and seeding arid rangelands in the Southwest. *Journal of Range Management* 26:193–197.

Herbel, C. H., L. H. Gile, E. L. Fredrickson, and R. P. Gibbens. 1994. Soil water and soils at soil water sites, Jornada Experimental Range. In L. H. Gile and R. J. Ahsens, editors, *Soil Survey Investigations Report mo. 44, Supplement to The Desert Project Soil Monograph, vol. 1.* U.S. Department of Agriculture, Lincoln, NE.

Herman, P. 2000. Biodiversity and evolution in mycorrizae of the desert. Pages 141–160 in C. W. Bacon and J. F. White Jr., editors, *Microbial endophytes*. Marcel Dekker, New York.

Herman, R. P., K. R. Provencio, J. Herrera-Matos, R. J. Torrez, and G. M. Seager. 1993.

Effect of water and nitrogen additions on free-living nitrogen fixer populations in desert grass root zones. *Applied and Environmental Microbiology* 59:3021–3026.

Herman, R. P., K. R. Provencio, J. Herrera-Matos, and R. J. Torrez. 1995. Resource islands predict the distribution of heterotrophic bacteria in Chihuahuan Desert soils. *Applied and Environmental Microbiology* 61:1816–1821.

Herrick, J. 1999. Soil organisms and rangeland soil hydrologic functions. Pages 91–100 in R. T. Meurisse, W. G. Ypsilantis, and C. A. Seybold, editors, *Soil organisms in Pacific Northwest forest and rangeland ecosystems—Population dynamics, functions and applications to management.* Proceedings of Symposium, March 17–19, 1998, Oregon State University, Corvallis.General Technical Report PNW-GTR-461. USDA Forest Service, Portland, OR.

Herrick, J. E., and M. M. Wander. 1998. Relationships between soil organic carbon and soil quality in cropped and rangeland soils: The importance of distribution, composition, and soil biological activity. Pages 405–425 in R. Lal, J. Kimble, R. Follett, and B. A. Stewart, editors, *Advances in soil science: Soil processes and the carbon cycle.* CRC Press, Boca Raton, F:.

Herrick, J. E., and W. G. Whitford. 1999. Integrating soil processes into management: From microaggregates to macrocatchments. Pages 91–95 in D. Eldridge and D. Freudenberger, editors, *Proceedings of the 6th International Rangeland Congress.* July 19–23, 1999, Townsville, Australia. Sixth International Rangeland Congress, Aitkenvale, Queensland, Australia.

Herrick, J. E., K. M. Havstad, and D. P. Coffin. 1997. Rethinking remediation technologies for desertified landscapes. *Journal Soil and Water Conservation* 52:220–225.

Herrick, J. E., B. T. Bestelmeyer, S. Archer, A. J. Tugel, and J. R. Brown. 2005. An integrated framework for science-based arid land management. *Journal of Arid Environments* 65:319–335.

Heske, E. J., J. H. Brown, and Q. Guo. 1993. Effects of kangaroo rat exclusion on vegetation structure and plant species diversity in the Chihuahuan Desert. *Oecologia* 95: 520–524.

Hibbard, K. A., S. Archer, D. S. Schimel, and D. W. Valentine. 2001. Biogeochemical changes accompanying woody plant encroachment in a subtropical savanna. *Ecology* 82:1999–2011.

Higgins, R. W., K. C. Mo, and Y. Yao. 1998. Interannual variability in the US summer precipitation regime with emphasis on the southwestern monsoon. *Journal of Climate* 11:2582–2606.

Hill, C. A., Y. V. Dublyansky, R. S. Harmon, and C. M. Schluter. 1995. Overview of calcite/opal deposits at or near the proposed high-level nuclear waste site, Yucca Mountain, Nevada, USA: Pedogenic, hypogene, or both. *Environmental Geology* 26:69–88.

Hipps, L. E., and W. P. Kustas. 2001. Spatial variations in evaporation. Pages 105–122 in R. Grayson and G. Bloschl, editors, *Spatial patterns in catchment hydrology: Observations and modeling.* Cambridge University Press, Cambridge.

Hipps, L. E., K. Ramalingam, W. P. Kustas, and J. H. Prueger. 1977. Surface fluxes and energy balance in an arid ecosystem. *AMS Proceedings of the 12th Symposium Boundary Layers Turbidity.*

Hipps, L. E., K. Ramalingam, W. P. Kustas, and J. H. Prueger. 1999. Difficulties in the determination of surface energy balance of arid landscapes. *American Meteorological Society 23rd Conference on Agricultural and Forest Meteorology* 392–394.

Hirobe, N., N. Karasawa, G. S. Zhang, L. H. Wang, and K. Yoshikawa. 2001. Plant species effects on the spatial patterns of soil properties in the Mu-us Desert ecosystem, Inner Mongolia, China. *Plant and Soil* 234:195–205.

Ho, M., R. E. Roisman, and R. A. Virginia. 1996. Using strontium and rubidium tracers to characterize nutrient uptake patterns in creosotebush and mesquite. *Southwestern Naturalist* 41:239–247.

Hobbs, N. T., D. S. Schimel, C. E. Owensby, and D. S. Ojima. 1991. Fire and grazing in the tallgrass prairie: Contingent effects on nitrogen budgets. *Ecology* 72:1374–1384.

Hobbs, R. J., and L. F. Huenneke. 1992. Disturbance, diversity and invasion: Implications for conservation. *Conservation Biology* 6:324–337.

Hochstrasser, T., G. Kroel-Dulay, D. P. C. Peters, and J. R. Gosz. 2002. Vegetation and climate characteristics of arid and semiarid grasslands in North America and their biome transition zone. *Journal of Arid Environments* 51:55–78.

Hodge, A. 2001. Foraging and the exploitation of soil nutrient patches: In defense of roots. *Functional Ecology* 15:416.

Holechek, J. L., and K. Hess Jr. 1994. Brush control considerations: A financial perspective. *Rangelands* 16:193–196.

Holechek, J. L., A. Tembo, A. Daniel, M. J. Fusco, and M. Cardenas. 1994. Long-term grazing influences on Chihuahuan Desert rangeland. *Southwest Naturalist* 39:342–349.

Holechek, J. L., R. D. Pieper, and C. H. Herbel. 1998a. *Range management: Principles and practices*. Prentice Hall, Upper Saddle River, NJ.

Holechek, J. L., H. deSouza-Gomes, F. Molinar, and D. Galt. 1998b. Grazing intensity: Critique and approach. *Rangelands* 20:15–18.

Holechek, J. L, M. Thomas, F. Molinar, and D. Galt. 1999. Stocking desert rangelands: What we've learned. *Rangelands* 21:8–12.

Holloway, J. W., W. T. Butts Jr., J. D. Beaty, J. T. Hopper, and N. S. Hall. 1979. Forage intake and performance of lactating beef cows grazing high or low quality pastures. *Journal of Animal Science* 48:692–700.

Holmgren, M., and M. Scheffer. 2001. El Niño as a window of opportunity for the restoration of degraded arid ecosystems. *Ecosystems* 4:151–159.

Hook, P. B., I. C. Burke, and W. K. Lauenroth. 1991. Heterogeneity of soil and plant N and C associated with individual plants and openings in North American shortgrass steppe. *Plant and Soil* 138:247–256.

Hooper, D. U., and L. Johnson. 1999. Nitrogen limitation in dryland ecosystems: responses to geographical and temporal variation in precipitation. *Biogeochemistry* 46:247–293.

Horst, T. W., and J. C. Weil. 1992. Footprint estimation for scalar flux measurements in the atmospheric surface layer. *Boundary-Layer Meteorology* 59:279–296.

Houghton, R. A., J. L. Hackler, and K. T. Lawrence. 1999. The US carbon budget: Contributions from land-use change. *Science* 285:574–578.

House, J. I., S. Archer, D. D. Breshears, and R. J. Scholes. 2003. Conundrums in mixed woody-herbaceous plant systems. *Journal of Biogeography* 30:1763–1777.

Howe, H. F., and J. Smallwood. 1982. Ecology of seed dispersal. *Annual Review Ecology and Systematics* 13:201–228.

Howes, D. A., and A. D. Abrahams. 2003. Modeling runoff and runon in a desert shrubland ecosystem, Jornada Basin, New Mexico. *Geomorphology* 53:45–73.

Huenneke, L. F., D. Clason, and E. Muldavin. 2001. Spatial heterogeneity in Chihuahuan Desert vegetation: Implications for sampling methods in semiarid ecosystems. *Journal of Arid Environments* 47:257–270.

Huenneke, L. F., J. P. Anderson, M. Remmenga, and W. H. Schlesinger. 2002. Desertification alters patterns of aboveground net primary production in Chihuahuan ecosystems. *Global Change Biology* 8:247–264.

Humes, K. S., W. P. Kustas, and M. S. Moran. 1994. Use of remote sensing and reference site measurements to estimate instantaneous surface energy balance components over a semiarid range watershed. *Water Resource Research* 30:1363–1373.

Humphrey, R. R. 1958. The desert grassland: A history of vegetational change and an analysis of causes. *Botanical Review* 24:193–252.

Humphrey, R. R. 1974. Fire in the deserts and desert grassland of North America. Pages 365–400 in T. T. Kozlowski and C. E. Ahlgren, editors, *Fire and ecosystems*. Academic Press, New York.

Hunter, R. 1991. Bromus invasions on the Nevada test site—present status of Brubens and Btectorum with notes on their relationship to disturbance and altitude. *Great Basin Naturalist* 51:176–182.

Huntley, N. 1991. Herbivores and the dynamics of communities and ecosystems. *Annual Review of Ecology and Systematics* 22:477–503.

Hurtt, G. C., S. W. Pacala, P. R. Moorcroft, J. Casperson, E. Shevliakova, R. A. Houghton, and B. Moore III. 2002. Projecting the future of the US carbon sink. *Proceedings of the National Academy of Sciences* 99:1389–1394.

Huxman, T. E., K. A. Snyder, D. Tissue, J. Leffler, K. Ogle, W. Pockman, D. Sandquist, and D. L. Potts. 2004. Precipitation pulses and carbon balance n semiarid and arid ecosystems. *Oecologia* 141:254–268.

Huxman, T. E., B. P. Wilcox, D. D. Breshears, R. L. Scott, K. A. Snyder, E. E. Small, K. Hultine, W. T. Pockman, and R. B. Jackson. 2005. Ecohydrological implications of woody plant encroachment. *Ecology* 86(2):308–319.

Illius, A. W., and T. G. O'Connor. 1999. On the relevance of nonequilibrium concepts to arid and semiarid grazing systems. *Ecological Applications* 9:798–813.

Iversen, J. D., and B. R. White. 1982. Saltation threshold on Earth, Mars, and Venus. *Sedimentology* 29:111–119.

Izett, G. A., and R. E Wilcox. 1982. Map showing localities and inferred distributions of the Huckleberry Ridge, Mesa Falls, and Lava Creek ash beds (Pearlette family ash beds) of Pliocene and Pleistocene age in the Western United States and southern Canada. Miscellaneous Investigation Series I-1325. U.S. Geological Survey, Denver, CO.

Izett, G. A., J. D. Obradovich, and H. H. Mehnert. 1988. The Bishop ash bed (Middle Pleistocene) and some older (Pliocene and Pleistocene) chemically and mineralogically similar ash beds in California, Nevada, and Utah. Pages 1–37 in U.S. Geological Survey Bulletin 1675. US Government Printing Office, Washington, DC.

Izett, G. A., K. L. Pierce, and N. D. Naeser. 1992. Isotopic dating of Lava Creek B tephra in terrace deposits along the Wind River, Wyoming—implications for post 0.6 MA uplift of the Yellowstone Hotspot. *Geological Society of America Abstracts with Programs*, A102.

Jackson, R. B., J. L. Banner, E. G. Jobbagy, W. T. Pockman, and D. H. Wall. 2002. Ecosystem carbon loss with woody plant invasion of grasslands. *Nature* 418:623–626.

Jackson, R. B., S. T. Berthrong, C. W. Cook, E. G. Jobbagy, R. L. McCulley. 2004. Comment on "A Reservoir of Nitrate Beneath Desert Soils." *Science* 304:51b.

Jackson, R. D., S. B. Idso, R. J. Reginato, and P. J. Pinter Jr. 1981. Canopy temperature as a crop water stress indicator. *Water Resource Research* 17:1133–1138.

James, L. F., D. B. Nielsen, and K. E. Panter. 1993. Impact of poisonous plants on the livestock industry. *Journal of Range Management* 45:3–8.

Janzen, D. H. 1982. Differential seed survival and passage rates in cows and horses, surrogate Pleistocene dispersal agents. *Oikos* 38:150–156.

Jardine, J. T., and C. L. Forsling. 1922. *Range and cattle management during drought*. U.S. Department of Agriculture Bulletin 1031. U.S. Government Printing Office, Washington, DC.

Jardine, J. T., and L. C. Hurtt. 1917. *Increased cattle production on southwestern ranges*. U.S. Department of Agriculture Bulletin 588. U.S. Government Printing Office, Washington, DC.

Jenkins, M. G., R. A. Virginia, and W. M. Jarrell. 1988. Depth distribution and seasonal populations of mesquite-nodulating rhizobia in warm desert ecosystems. *Soil Science Society of America Journal* 52:1644–1650.

Jenny, H. 1941. *Factors of soil formation*. McGraw-Hill, New York.

Jenny, H., and C. D. Leonard. 1934. Functional relationships between soil properties and rainfall. *Soil Science* 38:363–381.

Jobbágy, E. G., and R. B. Jackson. 2001. The distribution of soil nutrients with depth: Global patterns and the imprint of plants. *Biogeochemistry* 53:51–77.

Johansen, M. P., T. E. Hakonson, and D. D. Breshears. 2001. Post-fire runoff and erosion following rainfall simulation: Contrasting forests with shrublands and grasslands. Special Issue: Wildfire and Surficial Processes. *Hydrological Processes* 15:2953–2965.

Johnson, H. B., and H. S. Mayeux. 1990. *Prosopis glandulosa* and the nitrogen balance of rangelands: Extent and occurrence of nodulation. *Oecologia* 84:176–185.

Johnson, H. B., H. W. Polley, and H. S. Mayeux. 1993. Increasing CO_2 and plant-plant interactions: Effects on natural vegetation. *Vegetation* 104:157–170.

Johnson, K. A., and W. G. Whitford. 1975. Foraging ecology and relative importance of subterranean termites in Chihuahuan Desert ecosystems. *Environmental Entomology* 4:66–70.

Jones, H. G. 1992. *Plants and microclimate*, 2nd editon. Cambridge University Press, Cambridge.

Jones, T. A., and D. A. Johnson. 1998. Integrating genetic concepts into planning rangeland seedings. *Journal of Range Management* 51:594–606.

Jordan, G. L. 1981. *Range seeding and brush management on Arizona rangelands*. Cooperative Extension Service, Agricultural Experiment Station, University of Arizona, Tucson, AZ.

Jordan, P. W., and P. S. Nobel. 1984. Thermal and water relations of roots of desert succulents. *Annals of Botany* 54:705–717.

Jornada Experimental Range Staff. 1939. *Notes from Jornada Experimental Range*. Original copy in Jornada Experimental Range files, Las Cruces, NM.

Jornada Experimental Range Staff. 1958. *Annual report*. Las Cruces, NM. USDA-ARS Crops Research Division, Forage and Range Branch, Arid Pasture and Range Section.

Junge, C. E., and R. T. Werby. 1958. The concentration of chloride, sodium, potassium, calcium, and sulfate in rain water over the United States. *Journal of Meteorology* 15: 417–425.

Jurinak, J. J., L. M. Dudley, M. F. Allen, and W. G. Knight. 1986. The role of calcium oxalate in the availability of phosphorus in soils of semiarid regions. A thermodynamic study. *Soil Science* 142:255–261.

Kahya, E., and J. A. Dracup. 1993. U. S. streamflow patterns in relation to the El Niño/Southern Oscillation. *Water Resources Research* 29:2491–2503.

Katul, G. G., D. S. Ellsworth, and C. T. Lai. 2000. Modelling assimilation and intercellular CO_2 from measured conductance: a synthesis of approaches. *Plant, Cell and Environment* 23:1313–1328.

Kay, F. R., H. M. Sobhy, and W. G. Whitford. 1999. Soil microarthropods as indicators of exposure to environmental stress in Chihuahuan Desert rangelands. *Biology and Fertility of Soils* 28:121–128.

Kelly, E. F., O. A. Chadwick, and T. E. Hilinski. 1998. The effect of plants on mineral weathering. *Biogeochemistry* 42:21–53.

Kelt, D. A., and T. J. Valone. 1995. Effects of grazing on the abundance and diversity of annual plants in Chihuahuan Desert scrub habitat. *Oecologia* 103 :191–195.

Kemp, P. R. 1983. Phenological patterns of Chihuahuan Desert plants in relation to the timing of water availability. *Journal of Ecology* 71:427–436.

Kemp, P. R., and J. F. Reynolds. 2000. Variability in phenology and production of desert ephemerals: Implications for predicting resource availability for desert tortoises. Pages 34–39 in *Proceedings of the 24th Annual Meeting and Symposium of the Desert Tortoise Council*. St. George, UT.

Kemp, P. R., J. M. Cornelius, and J. F. Reynolds. 1992. A simple model for predicting soil temperatures in desert ecosystems. *Soil Science* 153:280–287.

Kemp, P. R., J. F. Reynolds, Y. Pachepsky, and J. L. Chen. 1997. A comparative modeling study of soil water dynamics in a desert ecosystem. *Water Resources Research* 33: 73–90.

Kemp, P. R., J. F. Reynolds, R. A. Virginia, and W. G. Whitford. 2003. Decomposition of leaf and root litter of Chihuahuan Desert shrubs: Effects of three years of summer drought. *Journal of Arid Environments* 53:21–39.

Kerley, G. I. H., and W. G. Whitford. 2000. Impact of grazing and desertification in the Chihuahuan Desert: Plant communities, granivores and granivory. *American Midland Naturalist* 144:78–91.

Kerley, G. I. H., W. G. Whitford, and F. R. R. Kay. 1997. Mechanisms for the keystone status of kangaroo rats: Gramivory rather than granivory? *Oecologia* 111:422–428.

Kerr, R. A. 1992. Another panel rejects Nevada disaster theory. *Science* 256:434–436.

Kidron, G. J., D. H. Yaalon, and A. Vonshak 1999. Two causes for runoff initiation on microbiotic crusts: Hydrophobicity and pore clogging. *Soil Science* 164:18–27.

Kieft, T. L. 1994. Grazing and plant-canopy effects on semiarid soil microbial biomass and respiration. *Biology and Fertility of Soils* 18:155–162.

Kieft, T. L., C. S. White, S. R. Loftin, R. Aguilar, J. A. Craig, and D. A. Skaar. 1998. Temporal dymanics in soil carbon and bitrogen resources at a grassland-shrubland ecotone. *Ecology* 79:671–683.

Kiladis, G. N., and H. F. Diaz. 1989. Global climatic anomalies associated with extremes in the Southern Oscillation. *Journal of Climate* 2:1069–1090.

Killingbeck, K. T., and W. G. Whitford. 1996. High foliar nitrogen in desert shrubs: An important ecosystem trait or defective desert doctrine? *Ecology* 77:1728–1737.

Killingbeck, K. T., and W. G. Whitford. 2001. Nutrient resorption in shrubs growing by design and by default in Chihuahuan Desert arroyos. *Oecologia* 128:351–359.

King, W. E. and J. W. Hawley. 1975. Geology and ground-water resources of the Las Cruces area. Pages 195–204 in *26th Annual Field Conference guidebook*. New Mexico Geological Society, Socorro, NM.

Kinnell, P. I. A. 1991. The effect of flow depth on sediment transport by raindrops impacting shallow flows. *Transactions of the American Society of Agricultural Engineers* 34:161–168.

Kirkpatrick, D. T, and M. S. Duran. 1998. Prehistoric peoples of the northern Chihuahuan Desert. Pages 41–45 in G. H. Mack, G. S. Austin, and J. M. Barker, editors, *Las Cruces Country II*. New Mexico Geological Society, Socorro, NM.

Kluth, C. F. 1986. Plate tectonics of the ancestral Rocky Mountains. Pages 353–369 in *Memoir 41*. American Association of Petroleum Geologists.

Kluth, C. F., and P. J. Coney. 1981. Plate tectonics of the ancestral Rocky Mountains. *Geology* 9:10–15.

Knapp, A. K., and M. D. Smith. 2001. Variation among biomes in temporal dynamics of aboveground primary production. *Science* 291:481–484.

Knapp, A. K., J. M. Briggs, D. C. Hartnett, and S. L. Collins. 1998. *Grassland dynamics: Long-term ecological research in tallgrass prairie*. Oxford University Press, New York.

Knipe, D., and C. H. Herbel. 1966. Germination and growth of some semidesert grassland species treated with aqueous extract from creosotebush. *Ecology* 47:775–781.

Korzdorfer, E. J. 1968. *Seeding, furrowing, brush removal, and rabbit exclusion effects on creosotebush-infested sites*. New Mexico State University, Las Cruces, NM.

Kotarba, A. 1980. Splash transport in the steppe zone of Mongolia. *Zeitzschrift fur Geomorphologie Supplementband* 35:92–102.

Kottlowski, F. E. 1953. Tertiary-Quaternary sediments of the Rio Grande Valley in southern New Mexico. Pages 144–148 in *4th Annual Field Conference guidebook*. New Mexico Geological Society, Socorro, NM.

Kottlowski, F. E. 1958. Geologic history of the Rio Grande near El Paso. Pages 46–54 in *Guidebook 1958 Field Trip*. West Texas Geological Society, Midland, TX.

Kottlowski, F. E. 1960a. *Reconnaissance geologic map of Las Cruces, 30-minute quadrangle*. New Mexico Bureau of Mines and Mineral Resources, Geologic Map 14. Socorro, NM.

Kottlowski, F. E. 1960b. *Summary of Pennsylvanian sections in southwestern New Mexico and southeastern Arizona*. New Mexico Bureau of Mines and Mineral Resources, Bulletin 66. Socorro, NM.

Kottlowski, F. E. 1965. Sedimentary basins of south-central and southwestern New Mexico. *American Association of Petroleum Geologists Bulletin* 49:2120–2139.

Kottlowski, F. E., R. H. Flower, M. L. Thompson, and R. W. Foster. 1956. *Stratigraphic studies of the San Andres Mountains, New Mexico*. New Mexico Bureau of Mines and Mineral Resources, Memoir 1. Socorro, NM.

Kraimer, R. A. 2003. *Mineralogical distinctions of carbonate in desert soils*. Ph.D. dissertation. New Mexico State University, Las Cruces, NM.

Kramer, P. J., and J. S. Boyer. 1995. *Water relations of plants and soils*. Academic Press, San Diego, CA.

Kuchler, A. W. 1964. *The potential natural vegetation of the coterminous United States*. American Geographical Society Special Publication 361.

Kurc, S. A., and E. E. Small. 2004. Dynamics of evapotranspiration in semiarid grassland and shrubland ecosystems during the summer monsoon season, central New Mexico. *Water Resources Research* 40.

Kustas, W. P., J. H. Blanford, D. I. Stannard, C. S. T. Daughtry, W. D. Nichols, and M. A. Weltz. 1994. Local energy flux estimates for unstable conditions using variance data in semiarid rangelands. *Water Resource Research* 30:1351–1361.

Kustas, W. P., J. H. Prueger, L. E. Hipps, J. L. Hatfield, and D. Meek. 1998. Inconsistencies in net radiation estimates from use of several models of instruments in a desert environment. *Agricultural and Forest Meteorology* 90:257–263.

Kustas, W. P., J. H. Prueger, J. L. Hatfield, K. Ramalingam, and L. E. Hipps. 2000. Variability in soil heat flux from a mesquite dune site. *Agricultural and Forest Meteorology* 103:249–264.

Lacey, J. R., and H. W. Van Poollen. 1981. Comparison of herbage production on moderately grazed and ungrazed western ranges. *Journal of Range Management* 34:210–212.

Lajtha, K. 1987. Nutrient reabsorption efficiency and the response to phosphorus fertilization in the desert shrub *Larrea tridentata* (DC.) Cov. *Biogeochemistry* 4:265–276.

Lajtha, K., and S. H. Bloomer. 1988. Factors affecting phosphate sorption and phosphate retention in a desert ecosystem. *Soil Science* 146:160–167.

Lajtha, K., and M. Klein. 1988. The effect of varying nitrogen and phosphorus availability on nutrient use by *Larrea tridentata*, a desert evergreen shrub. *Oecologia* 75:348–353.

Lajtha, K., and W. H. Schlesinger. 1986. Plant response to variations in nitrogen availability in a desert shrubland community. *Biogeochemistry* 2:29–37.

Lajtha, K., and W. H. Schlesinger. 1988. The effect of CaC03 on the uptake of phosphorus by two desert shrub species, *Larrea tridentata* (DC.) Cov. and *Parthenium incanum* H. B. K. *Botanical Gazette* 149:328–334.

Lajtha, K., and W. G. Whitford. 1989. The effect of water and nitrogen amendments on photosynthesis, leaf demography, and resource-use efficiency in *Larrea tridentata*, a desert evergreen shrub. *Oecologia* 80:341–348

Lambers, H., F. S. Chapin III, and T. L. Pons. 1998. *Plant physiological ecology*. Spinger-Verlag, New York.

Landres, P. B., P. Morgan, and F. J. Swanson. 1999. Overview of the use of natural variability concepts in managing ecological systems. *Ecological Applications* 9:1179–1188.

Lane, D. R., D. P. Coffin, and W. K. Lauenroth. 1998. Effects of soil texture and precipitation on above-ground net primary productivity and vegetation structure across the central grassland region of the United States. *Journal of Vegetation Science* 9:239–250.

Langford, R. P. 2000. Nabkha (coppice dunes) fields of south-central New Mexico, USA. *Journal of Arid Environments* 46:25–31.

Lau, N. C. 1992. Climate variability simulated in GCMs. Pages 617–642 in K. E. Trenberth, editor, *Climate system modeling*. Cambridge University Press, Cambridge.

Lauenroth, W. K., J. L. Dodd, and P. L. Sims. 1978. The effects of water-and nitrogen-induced stresses on plant community structure in a semiarid grassland. *Oecologia* 36: 211–222.

Lauenroth, W. K., and O. E. Sala. 1992. Long-term forage production of North American shortgrass steppe. *Ecological Applications* 2:397–403.

Le Houerou H. N. 1984. Rain use efficiency: A unifying concept in arid-land ecology. *Journal of Arid Environments* 7:213–247.

Le Houerou, H. N. 1996. Climate change, drought and desertification. *Journal of Arid Environments* 34:133–185.

Le Houerou, H. N., R. L. Bingham, and W. Skerbek. 1988. Relationship between the variability of primary production and the variability of annual precipitation in world arid lands. *Journal of Arid Environments* 15:1–18.

Leach, J. D., R. P. Mauldin, and H. C. Monger. 1998. The impact of eolian processes on archeological site size and characteristics in west Texas and southern New Mexico. *Bulletin of Texas Archeological Society* 69:92–108.

Leeder, M. R., G. H. Mack, J. Peakall, and S. L. Salyards. 1996. First quantitative test of alluvial stratigraphic models: Southern Rio Grande rift, New Mexico. *Geology* 24: 87–90.

Legrand, M., C. Pietras, G. Brogniez, M. Haeffelin, N. Abuhassen, and M. Sicard. 2000. A high-accuracy, multiwavelength radiometer for in situ measurements in the thermal infrared. Part I: Characterization of the instrument. *Journal of Atmospheric and Oceanic Technology* 71:1203–1214.

Leuning, R. 1988. Leaf temperatures during radiation frost. Part II. A steady state theory. *Agricultural and Forest Meteorology* 42:135–155.

Levin, S. A. 1992. The problem of pattern and scale in ecology. *Ecology* 73:1943–1967.

Li, B. L. 2000. Why is the holistic approach becoming so important in landscape ecology? *Landscape and Urban Planning* 50:27–41.

Lieth, H. 1975. Primary production of the major vegetation units of the world. Pages 203–215 in H. Lieth and R. H. Whittaker, editors, *Primary productivity of the biosphere*. Ecological Studies 14, Springer-Verlag, New York.

Lieth, H., and R. H. Whittaker, editors, 1975. *The primary productivity of the biosphere*. Springer, New York.

Lightfoot, D. C., and W. G. Whitford. 1987. Variation in insect densities on desert creosotebush: Is nitrogen a factor? *Ecology* 68:547–557.

Lightfoot, D. C., and W. G. Whitford. 1989. Interplant variation in creosotebush foliage characteristics and canopy arthropods. *Oecologia* 81:166–175.

Lightfoot, D. C., and W. G. Whitford. 1990. Phytophagous insects enhance nitrogen flux in a desert creosotebush community. *Oecologia* 82:18–25.

Lightfoot, D. C., and W. G. Whitford. 1991. Productivity of creosotebush foliage and associated canopy arthropods along a desert roadside. *American Midland Naturalist* 125: 310–322.

Linacre, E. T. 1973. A simpler empirical expression for actual evapotranspiration rates—A discussion. *Agricultural and Forest Meteorology* 11:451–452.

Link, S. O., H. Bolton Jr., M. E. Thiede, and W. H. Rickard. 1995. Responses of downy brome to nitrogen and water. *Journal of Range Management* 48:290–297.

Little, E. L. Jr. 1937. A study of poisonous drymaria on southern New Mexico ranges. *Ecology* 18:416–426.

Liu, X. 2002. *Calcium carbonate in subterranean termite foraging galleries in the northern Chihuahuan Desert.* Ph.D. dissertation, New Mexico State University, Las Cruces, NM.

Lockwood, J. W., and D. R. Lockwood. 1993. Catastrophe theory: A unified paradigm for rangeland ecosystem dynamics. *Journal of Range Management* 46:282–288.

Lodge, J. P., Jr., J. B. Pate, W. Basbergill, G. S. Swanson, K. C. Hill, E. Lorange, and A. L. Lazrus. 1968. *Chemistry of United States precipitation—Final report on the National Precipitation Sampling Network.* National Center for Atmospheric Research, Boulder, CO.

Loftis, S. G., and E. B. Kurtz. 1980. Field studies of inorganic nitrogen added to semiarid soils by rainfall and bluegreen algae. *Soil Science* 129:150–155.

Lohmiller, R. G. 1963. *Drought and its effect on condition and production of a desert grassland range*. M.S. thesis. New Mexico State University, Las Cruces, NM.

Loik, M. E., D. D. Breshears, W. K. Lauenroth, and J. Belnap. 2004. A multi-scale perspective of water pulses in dryland ecosystems: climatology and ecohydrology of the western USA. *Oecologia* 141(2):269–281.

Lonsdale, W. M. 1999. Global patterns of plant invasions and the concept of invisibility. *Ecology* 80:1522–1536.

Loreau, M., N. Mouquet, and R. D. Holt. 2003. Meta-ecosystems: A theoretical framework for a spatial ecosystem ecology. *Ecology Letters* 6:673–679.

Lovich, J. E., and D. Bainbridge. 1999. Anthropogenic degradation of the southern California desert ecosystem and prospects for natural recovery and restoration. *Environmental Management* 24:309–326.

Lozinsky, R. P., A. P. Hunt, and D. L. Wolberg. 1984. Late Cretaceous (Lancian) dinosaurs from the McRae Formation, Sierra County, New Mexico. *New Mexico Geology* 6:72–77.

Lucas, S. G., G. H. Mack, and J. W. Estep. 1998. The Ceratopsian dinosaur *Torosaurus* from the upper Cretaceous McRae Formation, Sierra County, New Mexico. Pages 223–227 in *New Mexico Geological Society, Guidebook 49*. Socorro, NM.

Ludwig, J., D. Tongway, D. Freudenberger, J. Noble, and K. Hodgkinson, editors. 1997. *Landscape ecology function and management: Principles from Australia's rangelands*. CSIRO, Australia.

Ludwig, J. A. 1977. Distributional adaptations of root systems in desert environments. Pages 85–90 in J. K. Marshall, editor, *The belowground ecosystem: A synthesis of plant-associated processes*. Range Science Series, no. 26. Colorado State University, Fort Collins, CO.

Ludwig, J. A. 1986. Primary production variability in desert ecosystems. Pages 5–17 in W. G. Whitford, editor, *Pattern and process in desert ecosystems*. University of New Mexico Press, Albuquerque, NM.

Ludwig, J. A. 1987. Primary productivity in arid lands: Myth and realities. *Journal of Arid Environments* 13:1–7.

Ludwig, J. A., and J. M. Cornelius. 1987. Locating discontinuities along ecological gradients. *Ecology* 68:448–450.

Ludwig, J. A., and W. G. Whitford. 1981. Short-term water and energy flow in arid ecosystems. Pages 271–299 in D. W. Goodall and R. A. Perry, editors, *Aridland ecosystems: Their structure, functioning and management*. Cambridge University Press, Cambridge.

Ludwig, J. A., J. A. Wiens, and D. J. Tongway. 2000. A scaling rule for landscape patches and how it applies to conserving soil resources in savannas. *Ecosystems* 3:84–97.

Lyford, F. P., and H. K. Qashu. 1969. Infiltration rates as affected by desert vegetation. *Water Resources Research* 5:1373–1376.

Lyons, T. J., P. Schwerdtfeger, J. M. Hacker, I. J. Foster, R. C. G. Smith, and H. Xinmei. 1993. Land-atmosphere interaction in a semiarid region: The bunny fence experiment. *Bulletin of the American Meteorology Society* 74:1327–1333.

Mabbutt, J. A. 1977. Desert Landforms. *An introduction to systematic geomorphology*. Volume 2. MIT Press, Cambridge, MA.

MacGregor, S. D., and T. G. O'Connor. 2002. Patch dieback of *Colophospermum mopane* in a dysfunctional semiarid African savanna. *Australian Ecology* 27:385–395.

Machette, M. N. 1985. Calcic soils of the southwestern United States. Pages 1–22 in D. L. Weide, editor, *Soils and quaternary geology of the Southwestern United States*. Geological Society of America Special Paper 203. Boulder, CO.

Machette, M. N. 1987. Preliminary assessment of paleoseismicity at White Sands Missile Range, southern New Mexico: Evidence for recency of faulting, fault segmentation, and repeat intervals for major earthquakes in the region. Pages 87–444 in U.S. Geological Survey, Open-File Report.

Mack, G. H. 1992. Paleosols as an indicator of climatic change at the early-late Cretaceous boundary, southwestern New Mexico. *Journal of Sedimentary Petrology* 62:483–494.

Mack, G. H., and W. C. James. 1986. Cyclic sedimentation in the mixed siliciclastic-carbonate Abo-Hueco transition zone (Lower Permian), southwestern New Mexico. *Journal of Sedimentary Petrology* 56:635–647.

Mack, G. H., and W. C. James. 1992. Calcic paleosols of the Plio-Pleistocene Camp Rice and Palomas Formations, southern Rio Grande rift. *Sedimentary Geology* 77:89–109.

Mack, G. H., and K. Suguio. 1991. Depositional environments of the Yeso Formation

(Lower Permian), southern Caballo Mountains, New Mexico. *New Mexico Geology* 13:45–49.

Mack, G. H., W. B. Kolins, and J. A. Galemore. 1986. Lower Cretaceous stratigraphy, depositional environments and sediment dispersal in southwestern New Mexico. *American Journal of Science* 286:309–331.

Mack, G. H., F. E. Kottlowski, and W. R. Seager. 1998a. The stratigraphy of south-central New Mexico. Pages 135–154 in *49th Annual Field Conference guidebook.* New Mexico Geological Society, Socorro, NM.

Mack, G. H., G. S. Austin, and J. M. Barker, editors. 1998b. *Las Cruces Country II. 49th Annual Field Conference guidebook.* New Mexico Geological Society, Socorro, NM.

Mack, G. H., S. L. Salyards, and W. C. James. 1993. Magnetostratigraphy of the Plio-Pleistocene Camp Rice and Palomas Formations in the Rio Grande Rift of southern New Mexico. *American Journal of Science* 293:49–77.

Mack, G. H., W. R. Seager, and J. Kieling. 1994a. Late Oligocene and Miocene faulting and sedimentation and evolution of the southern Rio Grande Rift, New Mexico, USA. *Sedimentary Geology* 92:79–96.

Mack, G. H., W. C. James, and S. L. Salyards. 1994b. Late Pliocene and Early Pleistocene sedimentation as influenced by intrabasinal faulting, southern Rio Grande Rift. Pages 257–264 in Geological Society of America Special Paper 291.

Mack, G. H., D. R. Cole, W. C. James, T. H. Giordano, and S. L. Salyards. 1994c. Stable oxygen and carbon isotopes of pedogenic carbonate as indicators of Plio-Pleistocene paleoclimate in the southern Rio Grande Rift, south-central New Mexico. *American Journal of Science* 294:621–640.

Mack, G. H., T. F. Lawton, and C. R. Sherry. 1995. Fluvial and estuarine depositional environments of the Abo Formation (Early Permian), Caballo Mountains, south-central New Mexico. Pages 181–187 in New Mexico Museum of Natural History and Science Bulletin 6.

Mack, G. H., W. C. McIntosh, M. R. Leeder, and H. C. Monger. 1996. Plio-Pleistocene pumice floods in the ancestral Rio Grande, southern Rio Grande Rift, USA. *Sedimentary Geology* 103:1–8.

Mack, G. H., D. W. Love, and W. R. Seager. 1997. Spillover models for axial rivers in regions of continental extension: The Rio Mimbres and Rio Grande in southern Rio Grande Rift, USA. *Sedimentology* 44:637–652.

Mack, R. N., and J. N. Thompson. 1982. Evolution in steppe with few large, hooved mammals. *American Naturalist* 119:757–773.

MacKay, W. and E. MacKay. 2002. *The ants of New Mexico.* Edwin Mellen Press, New York.

MacKay, W. P. 1991. The role of ants and termites in desert communities. Pages 113–150 in G. A. Polis, editor, *The ecology of desert communities.* University of Arizona Press, Tucson, AZ.

MacKay, W. P., F. M. Fisher, S. Silva, and W. G. Whitford. 1987a. The effects of nitrogen, water and sulfur amendments on surface litter decomposition in the Chihuahuan Desert. *Journal of Arid Environments* 12:223–232.

MacKay, W. P., S., Silva, S. J. Loring, and W. G. Whitford. 1987b. The role of subterranean termites in decomposition of above-ground creosotebush litter. Sociobiology 13:235–239.

MacKay, W. P., J. C. Zak, and W. G. Whitford. 1989. The natural history and role of subterranean termites in the northern Chihuahuan Desert. Pages 53–77 in J. Schmidt, editor, *Special biotic relationships in the arid Southwest.* University of New Mexico Press, Albuquerque, NM.

MacKay, W. P., S. J. Loring, J. C. Zak, S. I. Silva, F. M. Fisher, and W. G. Whitford. 1994. Factors affecting loss in mass of creosotebush leaf-litter on the soil surface in the northern Chihuahuan Desert. *Southwestern Naturalist* 39:78–82.

MacMahon, J. A., and D. J. Schimpf. 1981. Water as a factor in the biology of North American desert plants. Pages 114–171 in D. D. Evans and J. L. Thames, editors, *Water in desert ecosystems*. Dowden, Hutchinson, and Ross, Stroudsburg, PA.

MacMahon, J. A., and F. H. Wagner. 1985. The Mojave, Sonoran, and Chihuahuan Deserts of North America. Pages 105–202 in M. Evenari, I. Noy-Meir, and D. W. Goodall, editors, *Hot deserts and arid shrublands*, Vol 12A: *Ecosystems of the world*. Elsevier, Amsterdam.

MacNeish, R. S., and J. G. Libby, editors. 2004. *Pendejo Cave*. University of New Mexico Press, Albuquerque, NM.

Maestas, J. D., R. L. Knight, and W. C. Gilbert. 2001. Biodiversity and land-use change in the American Mountain West. *Geographical Review* 91:509–524.

Maestas, J. D., R. L. Knight, and W. C. Gilgert. 2003. Biodiversity across a rural land-use gradient. *Conservation Biology* 17:1425–1434.

Maestre, F. T., and J. Cortina. 2002. Spatial patterns of surface soil properties and vegetation in a Mediterranean semiarid steppe. *Plant and Soil* 24:279–291.

Maestre, F. T., S. Bautista, J. Cortina, and J. Bellot. 2001. Potential for using facilitation by grasses to establish shrubs on a semiarid degraded steppe. *Ecological Applications* 11:1641–1655.

Maestre, F. T., M. Huesca, E. Zaady, S. Bautista, and J. Cortina. 2002. Infiltration, penetration resistance and microphytic crust composition in contrasted microsites within a Mediterranean semiarid steppe. *Soil Biology and Biochemistry* 34:895–898.

Marion, G. M., and K. L. Babcock. 1977. The solubilities of carbonates and phosphates in calcareous soil suspensions. *Soil Science Society of America Journal* 41:724–728.

Marion, G. M., W. H. Schlesinger, and P. J. Fonteyn. 1985. CALDEP: A regional model for soil $CaCO_3$ (caliche) deposition in southwestern deserts. *Soil Science* 139:468–481.

Marion, G. M., and W. H. Schlesinger. 1994. Quantitative modeling of soil forming processes in deserts: The CALDEP and CALGYP Models. Pages 129–145 in R. B. Bryant and R. W. Arnold, editors, *Quantitative modeling of soil-forming processes*. Soil Science Society of America Special Publication 39, Madison, WI.

Marshall, J. K. 1971. Drag measurements in roughness arrays of varying density and distribution. *Agricultural Meteorolgy* 8:269–292.

Marticorena, B., and G. Bergametti. 1995. Modeling the atmospheric dust cycle: 1. Design of a soil-derived dust emission scheme. *Journal of Geophysical Research* 100:16415–16430.

Marticorena, B., G. Bergametti, D. Gillette, and J. Belnap. 1997. Factors controlling threshold friction velocity in semiarid and arid areas of the United States. *Journal of Geophysical Research* 102:23277–23287.

Martin, S. C. 1975. *Ecology and management of southwestern semidesert grass-shrub ranges: The status of our knowledge*. USDA Forest Service, Research Paper RM-156.

Martin, S. C., and K. E. Severson. 1988. Vegetation response to the Santa Rita grazing system. *Journal of Range Management* 41:291–296.

Martinez-Meza, E., and W. G. Whitford. 1996. Stemflow, throughfall and channelization of stemflow by roots in three Chihuahuan Desert shrubs. *Journal of Arid Environments* 32:271–287.

Masters, R. A., and R. L. Sheley. 2001. Principles and practices for managing rangeland invasive plants. *Journal of Range Management* 54:502–517.

Matson, P. A., and M. D. Hunter. 1992. The relative contributions of top-down and bottom-up forces in population and community ecology. *Ecology* 73:723.

Mauchamp, A., and J. L. Janeau. 1993. Water funnelling by the crown of *Flourensia cernua*, a Chihuahuan Desert shrub. *Journal of Arid Environments* 25:299–306.

Mauchamp, A., C. Montaña, J. Lepart, and S. Rambal. 1993. Ecotone-dependent recruitment of a desert shrub, Flourensia cernua, in vegetation stripes. Oikos 68:107–116.

Mayeux, H. S., H. B. Johnson, and H. W. Polley. 1991. Global change and vegetation dynamics. Pages 62–74 in L. F. James, J. O. Evans, M. H. Ralphs, and B. J. Sigler, editors, *Noxious range weeds*. Westview Press, Boulder, CO.

Mazzarino, M. J., L. Oliva, A. Abril, and M. Acosta. 1991. Factors affecting nitrogen dynamics in a semiarid woodland (Dry Chaco, Argentina). *Plant and Soil* 138:85–98.

Mazzarino, M. J., M. B. Bertiller, C. Sain, Satti and F. Coronato. 1998. Soil nitrogen dynamics in northeastern Patagonia steppe under different precipitation regimes. *Plant and Soil* 202:125–131.

McAuliffe, J. 1988. Markovian dynamics of simple and complex desert communities. *American Naturalist* 131:459–490.

McAuliffe, J. R. 1994. Landscape evolution, soil formation, and ecological patterns and processes in Sonoran Desert bajadas. *Ecological Monographs* 64:111–143.

McAuliffe, J. R. 2003. The interface between precipitation and vegetation: the importance of soils in arid and semiarid environments. Pages 9–27 in J. F. Weltzin and G. R. McPherson, editors, *Changing precipitation regimes and terrestrial ecosystems*. University of Arizona Press, Tucson, AZ.

McClaran, M. P. 1995. Desert grassland and grasses. Pages 1–30 in M. P. McClaran and T. R. VanDevender, editors, *The desert grassland*. University of Arizona Press, Tucson, AZ.

McClaran, M. P. 2003. A century of vegetation change on the Santa Rita Experimental Range. Pages 16–33 in *RMRS-P-30 Proceedings, Santa Rita Experimental Range: 100 years (1903–2003) of accomplishments and contributions*. 30 October–1 November 2003; Tucson, AZ. USDA Forest Service, Rocky Mountain Research Station, Ogden, UT.

McClaran, M. P., and T. R. VanDevender, editors. 1995. *The desert grassland*. University of Arizona Press, Tucson, AZ.

McDonald, J. N. 1981. *North American bison, their classification and evolution*. University of California Press, Berkeley, CA.

McFadden, L. D., and J. C. Tinsley. 1985. Rate and depth of pedogenic carbonate accumulation in soils: Formation and testing of a compartment model. Pages 23–41 in D. L. Weide, editor, *Soils and quaternary geology of the Southwestern United States*. Geological Society of America Special Paper 203. Boulder, CO.

McFadden, L., S. G. Wells, and M. J. Jercinovinch. 1987. Influences of eolian and pedogenic processes on the origin and evolution of desert pavements. *Geology* 15:504–508.

McIntosh, W. C., L. L. Kedzie, and J. F. Sutter. 1991. *Paleomagnetism and* $^{40}Ar/^{39}Ar$ *ages of ignimbrites, Mogollon-Datil volcanic field, southwestern New Mexico*. New Mexico Bureau of Mines and Mineral Resources, Bulletin 135. Socorro, NM.

McIntyre, D. S. 1958. Permeability measurements of soil crusts formed by raindrop impact. *Soil Science* 85:185–189.

McKenna-Neuman, C., and W. G. Nickling. 1994. Momentum extraction with saltation: Implications for experimental evaluation of wind profile parameters. *Boundary-Layer Meteorology* 68:35–50.

McMillan, N. J. 1998. Temporal and spatial magmatic evolution of the Rio Grande Rift. Pages 107–116 in *49th Annual Field Conference guidebook*. New Mexico Geological Society, Socorro, NM.

McNaughton, S. J. 1979. Grazing as an optimization process: grass-ungulate relationships in the Serengeti. *American Naturalist* 113:691–703.

McNaughton, S. J. 1985. Ecology of a grazing ecosystem: The Serengeti. *Ecological Monographs* 55:259–294.

McNaughton, S. J. 1991. Evolutionary ecology of large tropical herbivores. Pages 509–522 in P. W. Price, T. M. Lewinsjohn, G. W. Fernandes, and W. W. Benson, editors, *Plant-animal interactions: Evolutionary ecology in tropical and temperate regions*. Wiley, New York.

McNaughton, S. J. 1993. Grasses and grazers, science and management. *Ecological Applications* 3:17–20.

McPherson, G. R. 1995. The role of fire in desert grasslands. Pages 130–151 in M. P. McClaran and T. R. VanDevender, editors, *The desert grassland*. University of Arizona Press, Tucson, AZ.

McPherson, G. R. 1997. *Ecology and management of North American savannas*. University of Arizona Press, Tucson, AZ.

Meentemeyer, V. 1978. Macroclimate and lignin control of litter decomposition. *Ecology* 59:465–472.

Melillo, J. M., J. D. Aber, and J. F. Muratore. 1982. Nitrogen and lignin control of hardwood leaf litter decomposition dynamics. *Ecology* 63:621–626.

Melzer, R. 2000. *Coming of age in the Great Depression: The Civilian Conservation Corps experience in New Mexico, 1933–1942*. Yucca Tree Press, Las Cruces, NM.

Menenti, M., and J. C. Ritchie. 1994. Estimation of effective aerodynamic roughness of Walnut Gulch watershed with laser altimeter measurements. *Water Resources Research* 30:1329–1337.

Menenti, M., J. C. Ritchie, K. S. Humes, R. Parry, Y. Pachepsky, D. Gimenez, and S. Leguizamon. 1996. Estimation of aerodynamic roughness at various scales. Pages 39–58 in J. B. Steward, E. T. Engman, R. A. Feddes and Y. Kerr, editors, *Scaling up to hydrology using remote sensing*. Wiley, London.

Mielnick, P., W. A. Dugas, K. Mitchell, and K. Havstad. 2005. Long-term measurements of CO_2 flux and evapotranspiration in Chihuahuan desert grassland. *Journal of Arid Environments* 60:423–436.

Milchunas, D. G., O. E. Sala, and W. K. Lauenroth. 1988. A generalized model of the effects of grazing by large herbivores on grassland community structure. *American Naturalist* 132:87–106.

Miller, B., R. Reading, J. Hoogland, T. Clark, G. Ceballos, R. List, S. Forrest, L. Hanebury, P. Manzano, J. Pacheco, and D. Uresk. 2000. The role of prairie dogs as a keystone species: Response to Stapp. *Conservation Biology* 14:318–321.

Miller, J. F., R. H. Frederick, and R. J. Tracey. 1973. *Precipitation-frequency atlas of the coterminous Western United States (by states)*, vol. 4: *New Mexico*. National Oceanic and Atmospheric Administration Atlas 2. National Weather Service, Silver Spring, MD.

Miller, R. E., and L. F. Huenneke. 2000. The relationship between density and demographic variation within a population of Larrea tridentate. *Southwestern Naturalist* 45(3):313–321.

Miller, R. F., and G. B. Donart. 1979. Response of Bouteloua eriopoda (Torr.) Torr. and Sporobolus flexuosus (Thurb.) Rybd. to season of defoliation. *Journal of Range Management* 32:63–67.

Milton, S. J., W. R. J. Dean, G. I. H. Kerley, M. T. Hoffman, and W. G. Whitford. 1998. Dispersal of seeds as nest material by the cactus wren. *Southwestern Naturalist* 43: 449–452.

Minnick, T. J., and D. P. Coffin. 1999. Geographic patterns of simulated establishment of two Bouteloua species: Implications for distributions of dominants and ecotones. *Journal of Vegetation Science* 10:343–356.

Molles, M. C., and C. N. Dahm. 1990. A perspective on El Niño and La Niña: Global implications for stream ecology. *Journal of the North American Benthological Society* 9:68–76.

Monger, H. C. 1990. *Mineralogical transformations in a southern New Mexico Aridison.* Ph.D. dissertation. New Mexico State University, Las Cruces, NM.

Monger, H. C. 1993. *Soil-geomorphic and paleoclimatic characteristics of the Fort Bliss maneuver areas, southern New Mexico and western Texas.* Historic and Natural Resources Report 10. Fort Bliss, TX.

Monger, H. C. 1995. Pedology in arid lands archaeological research: an example from southern New Mexico-western Texas in pedological perspectives. In M. Collins, editor, *Archeological Research.* Soil Science Society of America Special Publication 44: 35–50.

Monger, H. C. 1999. Natural cycles of desertification in the Chihuahuan Desert, North America. Pages 209–233 in *Proceedings of the Fifth International Conference on Desert Development.* Texas Technological University Press, Lubbock, TX.

Monger, H. C. 2003. Millennial-scale climate variability and ecosystem response at the Jornada LTER site. In D. Greenland, editor, *Climate variability and ecosystem response at the LTER sites.* Oxford University Press, London.

Monger, H. C., and H. P. Adams. 1996. Micromorphology of calcite-silica deposits, Yucca Mountain, Nevada. *Soil Science Society of America Journal* 60:519–530.

Monger, H. C., and L. A. Daugherty. 1991a. Neoformation of palygorskite in a southern New Mexico aridisol. *Soil Society of America Journal* 55:1646–1650.

Monger, H. C., and L. A. Daugherty. 1991b. Pressure solution: possible mechanism for silicate grain dissolution in a petrocalcic horizon. *Soil Science Society of America Journal* 55:1625–1629.

Monger, H. C., and R. A. Gallegos. 2000. Biotic and abiotic processes and rates of pedogenic carbonate accumulation in the southwestern United States—relationship to atmospheric CO_2 sequestration. Pages 273–289 in R. Lal et al., editors, *Global climate change and pedogenic carbonates.* CRC Press, Boca Raton, FL.

Monger, H. C., and W. C. Lynn. 1996. Clay mineralogy of the Desert Project and Rincon Soils. In L. H. Gile and R. J. Ahrens, editors, *Studies of soil and landscape evolution in southern New Mexico: Supplement to the Desert Project soil monograph*, vol. 2. Soil Survey Investigations Report 44, Lincoln, NE.

Monger, H. C., L. A. Daugherty, W. C. Lindemann, and C. M. Liddell. 1991a. Microbial precipitation of pedogenic calcite. *Geology* 19:997–1000.

Monger, H. C., Daugherty, L. A., and Gile, L. H. 1991b. A microscopic examination of pedogenic calcite in an aridisol of southern New Mexico. Pages 37–60 in W. D. Nettleton, editor, *Occurrence, characteristics, and genesis of carbonate, gypsum, and silica accumulations in soils.* Soil Science Society of America Special Publication 26. Madison, WI.

Monger, H. C., D. R. Cole, J. W. Gish, and T. H. Giordano. 1998. Stable carbon and oxygen isotopes in Quaternary soil carbonates as indicators of ecogeomorphic changes in the northern Chihuahuan Desert, USA. *Geoderma* 82:137–172.

Montaña, C. 1992. The colonization of bare areas in two-phase mosaics of an arid ecosystem. *Journal of Ecology* 80:315–327.

Montaña, C., E. Ezcurra, A. Carrillo, and J. P. Delhoume. 1988. The decomposition of litter in grasslands of northern Mexico: A comparison between arid and nonarid environments. *Journal of Arid Environments* 14:55–60.

Montaña, C., J. Lopez-Portillo, and A. Mauchamp. 1990. The response of two woody species to the conditions created by a shifting ecotone in an arid ecosystem. *Journal of Ecology* 78:789–798.

Mooney, H. A., B. B. Simpson, and O. T. Solbrig. 1977. Phenology, morphology, physiology. Pages 26–45 in B. B. Simpson, editor, *Mesquite: Its biology in two desert ecosystems.* Dowden, Hutchinson, and Ross, Stroudsberg, PA.

Moore, D. C., and M. J. Singer. 1990. Crust formation effects on soil erosion processes. *Soil Science Society of America Journal* 54:1117–1123.

Moore, I. D. 1981. Effect of surface sealing on infiltration. *Transactions of the American Society of Agricultural Engineers* 24:1546–1552.

Moorhead, D. L., and J. F. Reynolds. 1989a. The contribution of abiotic processes to buried litter decomposition in the northern Chihuahuan Desert. *Oecologia* 79:133–135.

Moorhead, D. L., and J. F. Reynolds. 1989b. Mechanisms of surface litter mass loss in the northern Chihuahuan Desert: A reinterpretation. *Journal of Arid Environments* 16: 157–163

Moorhead, D. L., F. M. Fisher, and W. G. Whitford. 1988. Cover of spring annuals on nitrogen-rich kangaroo rat mounds in a Chihuahuan Desert grassland. *American Midland Naturalist* 120:443–447.

Moorhead, D. L., J. F. Reynolds, and P. J. Fonteyn. 1989. Patterns of stratified soil water loss in a Chihuahuan Desert community. *Soil Science* 148:244–249.

Moroka, N., R. F. Beck, and R. D. Piper. 1982. Impact of borrowing activity of the bannertail kangaroo rat on southern New Mexico desert rangeland. *Journal of Range Management* 35:707–710.

Morrisey, R. J. 1951. The northward expansion of cattle ranching in New Spain, 1550–1600. *Agricultural History* 25:115–121.

Morton, H. L., and A. Melgoza. 1991. Vegetation changes following brush control in creosotebush communities. *Journal of Range Management* 44:133–139.

Motzkin, G., R. Eberhardt, B. Hall, D. Foster, J. Harrod, and D. MacDonald. 2002. Vegetation variation across Cape Cod, Massachusetts: Environmental and historical determinants. *Journal of Biogeography* 29:1439–1454.

Muldavin, E. H., P. Neville, and G. Harper. 2001. Indices of grassland biodiversity in the Chihuahuan Desert ecoregion derived from remote sensing. *Conservation Biology* 15: 844–855.

Mun, H. T., and W. G. Whitford. 1989. Effects of nitrogen amendment on annual plants in the Chihuahuan Desert. *Plant and Soil* 120:225–231.

Mun, H. T., and W. G. Whitford. 1990. Factors affecting annual plants assemblages on banner-tailed kangaroo rat mounds. *Journal of Arid Environments* 18:165–173.

Mun, H. T., and W. G. Whitford. 1998. Change in mass and chemistry of plant roots during long-term decomposition on a Chihuahuan Desert watershed. *Biology and Fertility of Soils* 26:16–22.

Murphy, K. L., J. M. Klopatek, and C. C. Klopatek. 1998. The effects of litter quality and climate on decomposition along an elevational gradient. *Ecological Applications* 8: 1061–1071.

Myneni, R. B., S. O. Los, C. J. Tucker. 1996. Satellite-based identification of linked vegetation index and sea surface temperature anomaly areas from 1982–1990 for Africa, Australia, and South America. *Geophysical Research Letters* 23:729–732.

Naranjo, L. G., and R. J. Raitt. 1993. Breeding bird distribution in Chihuahuan Desert habitats. *Southwestern Naturalist* 38:43–51.

Nash, M. H. 1985. *Numerical classification, spatial dependence and vertical kriging of soil sites in southern New Mexico.* M.S. thesis. New Mexico State University, Las Cruces, NM.

Nash, M. H., and W. G. Whitford. 1995. Subterranean termites: Regulators of soil organic matter in the Chihuahuan Desert. *Biology and Fertility of Soils* 19:15–18.

Nash, M. S., P. J. Wierenga, and A. Gutjahr. 1991. Time-series analysis of soil moisture and rainfall along a line transect in arid rangeland. *Soil Science* 152:189–198.

Nash, M. S., A. Toorman, P. J. Wierenga, A. Gutjahr, and G. L. Cunningham. 1992. Estimation of vegetative cover in an arid rangeland based on soil moisture using cokriging. *Soil Science* 154:25–36.

Nash, M. S., J. P. Anderson, and W. G. Whitford. 1999. Spatial and temporal variability in relative abundance and foraging behavior of subterranean termites in desertified and relatively intact Chihuahuan Desert ecosystems. *Applied Soil Ecology* 12:149–157.

National Science Foundation. 1979. *Long-term ecological research concept statement and measurement needs.* Summary of a Workshop, Institute of Ecology, June 25–27, 2979, Indianapolis, Indiana. Available: http://intranet.lternet.edu/archives/documents/foundations/79Workshop.html

Navar, J., and R. Bryan 1990. Interception loss and rainfall redistribution by three semiarid growing shrubs in northeastern Mexico. *Journal of Hydrology* 115:51–63.

Neave, M., and A. D. Abrahams. 2001. Impact of small mammal disturbances on sediment yield from grassland and shrubland ecosystems in the Chihuahuan Desert. *Catena* 44: 285–303.

Neave, M., and A. D. Abrahams. 2002. Vegetation influences on water yields from grassland and shrubland ecosystems in the Chihuahuan Desert. *Earth Surface Processes and Landforms* 27:1011–1020.

Neher, R. E., and O. R. Bailey. 1976. *Soil survey of White Sands Missile Range, New Mexico.* U.S. Government Printing Office, Washington, DC.

Neilson, R. P. 1986. High-resolution climatic analysis and Southwest biogeography. *Science* 232:27–34.

Neilson, R. P. 1993. Transient ecotone response to climatic change: some conceptual and modeling approaches. *Ecological Applications* 3:385–395.

Nelson, E. W. 1934. *The influence of precipitation and grazing upon black grama grass range.* USDA Technical Bulletin no. 409, Washington, DC.

Neuenschwander, L. H., S. H. Sharrow, and H. A. Wright. 1975. Review of tobosa grass (Hilaria mutica). *Southwestern Naturalist* 2:255–263.

New Mexico Agricultural Experiment Station. 1949. *Sixtieth Annual Report (1947–48).* New Mexico College of Agriculture and Mechanical Arts, State College, NM.

New Mexico Agricultural Experiment Station. 1950. *Sixtieth Annual Report (1949–50).* New Mexico College of Agriculture and Mechanical Arts, State College, NM.

Nicholls, N. 1988. El Niño-Southern Oscillation and rainfall variability. *Journal of Climate* 1:418–421.

Nikiforoff, C. C. 1937. General trends of the desert type of soil formation. *Soil Science* 43:105–131.

Nishita, H., and R. M. Haug. 1973. Distribution of different forms of nitrogen in some desert soils. *Soil Science* 116:51–58.

Nobel, P. S. 1978. Surface temperatures of cacti—influences of environmental and morphological factors. *Ecology* 59:986–996.

Norris, J. J. 1950. *The effect of rodents, rabbits, and cattle on two vegetation types in semidesert rangeland.* New Mexico Agricultural Experiment Station Bulletin 353, Las Cruces, NM.

Noy-Meir, I. 1973. Desert ecosystems: Environment and producers. *Annual Review of Ecology and Systematics* 4:25–51.

Noy-Meir, I. 1974. Desert ecosystems: Higher trophic levels. *Annual Review of Ecology and Systematics* 5:195–214.

Noy-Meir, I. 1979/80. Structure and function of desert ecosystems. *Israel Journal of Botany* 28:1–19.

Noy-Meir, I. 1985. Desert ecosystem structure and function. Pages 93–103 in M. Evenari, I. Noy-Meir, and D. W. Goodall, editors, *Hot deserts and arid shrublands.* Elsevier Science, Amsterdam.

Oakes, C. 2000. *History and consequence of keystone mammal eradication in the desert grasslands: The Arizona blacktailed prairie dog* (Cynomys ludovicianus arizonensis). Ph.D. dissertation. University of Texas, Austin, TX.

Oberlander, T. M. 1994. Global deserts: a geomorphic comparison. Pages 13–35 in A. D. Abrahams and A. J. Parsons, editors, *Geomorphology of desert environments.* Chapman and Hall, London.

Oesterheld, M., O. E. Sala, and S. J. McNaughton. 1992. Effects of animal husbandry on herbivore carrying capacity at a regional scale. *Nature* 356:234–236.

Oesterheld, M., J. Loreti, M. Semmartin, and O. E. Sala. 2001. Interannual variation in primary production of a semiarid grassland related to previous-year production. *Journal of Vegetation Science* 12:137–142.

Ogle, K. 2003. *A Bayesian inverse modeling approach to reconstructing plant root area and water uptake profiles.* M.S. thesis. Duke University, Durham, NC.

Ogle, K., and J. F. Reynolds. 2002. Desert dogma revisited: Coupling of stomatal conductance and photosynthesis in the desert shrub, *Larrea tridentata. Plant Cell and Environment* 25:909–922.

Ogle, K., and J. F. Reynolds. 2004. Plant responses to precipitation in desert ecosystems: integrating functional types, pulses, thresholds, and delays. *Oecologia* DOI: 10.1007/s00442-004-1507-5:1432–1939 (online).

Ogle, K., R. L. Wolpert, and J. F. Reynolds. 2004. Reconstructing plant root area and water uptake profiles. *Ecology* 85:1967–1978.

Okin, G. S. 2002. Toward a unified view of biophysical land degradation processes in arid and semiarid lands. Pages 95–109 in J. F. Reynolds and D. M. Stafford Smith, editors, *Global desertification: Do humans cause deserts*? Dahlem University Press, Berlin.

Okin, G. S., and D. A. Gillette. 2001. Distribution of vegetation in wind-dominated landscapes: Implications for wind erosion modeling and landscape processes. *Journal Geophysical Research* 106:9673–9684.

O'Neill, R. V. 1979. Natural variability as a source of error in model predictions. Pages 23–32 in G. S. Innis and R. V. O'Neill, editors, *Systems analysis of ecosystems.* International Cooperative, Fairland, MD.

O'Neill, R. V. 2001. Is it time to bury the ecosystem concept? (With full military honors of course!). *Ecology* 82:3275–3284.

O'Neill, R. V., A. R. Johnson, and A. W. King. 1989. A hierarchical framework for the analysis of scale. *Landscape Ecology* 3:193–205.

Oren, R., J. S. Sperry, G. G. Katul, D. E. Pataki, B. E. Ewers, N. Phillips, and K. V. R.

Schäfer. 1999. Survey and synthesis of intra-and interspecific variation in stomatal sensitivity to vapour pressure deficit. *Plant, Cell and Environment* 22:1515–1526.

Osmond, C. B., O. Björkman, and D. J. Anderson. 1980. *Physiological processes in plant ecology*. Springer-Verlag, New York.

Ostfeld, R. S., and F. Keesing. 2000. Pulsed resources and community dynamics of consumers in terrestrial ecosystems. *Trends in Ecology and Evolution* 15:232–237.

Otterman, J. 1974. Baring high-albedo soils by overgrazing: a hypothesized desertification mechanism. *Science* 186:531–533.

Otterman, J. 1975. Surface albedo and desertification. *Science* 189:1012–1015.

Owen, P. R. 1964. Saltation of uniform sand grains in air. *Journal Fluid Mechanics* 20: 225–242.

Pacala, S. W., G. C. Hurtt, D. Baker, P. Peylin, R. A. Houghton, R. A. Birdsey, L. Heath, E. T. Sundquist, R. F. Stallard, P. Ciais, P. Moorcroft, J. P. Caspersen, E. Shevliakova, B. Moore, G. Kohlmaier, E. Holland, M. Gloor, M. E. Harmon, S.-M. Fan, J. L. Sarmiento, C. L. Goodale, D. Schimel, and C. B. Field. 2001. Consistent land-and atmosphere-based US carbon sink estimates. *Science* 292:2316–2320.

Pachepsky, Y. A., and J. C. Ritchie. 1998. Seasonal changes in fractal landscape surface roughness estimated from airborne laser altimetry data. *International Journal of Remote Sensing* 19:2509–2516.

Pachepsky, Y., D. Timlin, B. Acock, H. Lemmon, and A. Trent. 1993. *2DSOIL-a new modular simulator of soil and root processes.* Release 02, U.S. Department of Agriculture, Beltsville, MD.

Pachepsky, Y. A., J. C. Ritchie, and D. Gimenez. 1997. Fractal modeling of airborne laser altimetry data. *Remote Sensing of Environment* 61:150–161.

Parker, K. W. 1938. *Effect of jackrabbits on the rate of recovery of deteriorated rangeland.* New Mexico Agricultural Experiment Station Press Bulletin 839. Albuquerque, NM.

Parker, L. W., D. W. Freckman, Y. Steinberger, L. Driggers, and W. G. Whitford. 1984a. Effects of simulated rainfall and litter quantities on desert soil biota: Soil respiration, microflora, and protozoa. *Pedobiologia* 27:185–195

Parker, L. W., P. F. Santos, J. Phillips, and W. G. Whitford. 1984b. Carbon and nitrogen dynamics during the decomposition of litter and roots of a Chihuahuan Desert annual. *Ecological Monographs* 54:339–360.

Parsons, A. J., A. D. Abrahams, and J. R. Simanton. 1992. Microtopography and soil-surface materials on semiarid piedmont hillslopes, southern Arizona. *Journal of Arid Environments* 22:107–115.

Parsons, A. J., J. Wainwright, A. D. Abrahams, and J. R. Simanton. 1997. Distributed dynamic modeling of interrill overland flow. *Hydrological Processes* 11:1833–1859.

Parsons, A. J., J. Wainwright, P. M. Stone, and A. D. Abrahams. 1999. Transmission losses in rills on dryland hillslopes. *Hydrological Processes* 13:2897–2905.

Parsons, A. J., J. Wainwright, W. H. Schlesinger, and A. D. Abrahams. 2003. The role of overland flow in sediment and nitrogen budgets of mesquite dunefields, southern New Mexico. *Journal of Arid Environments* 53:61–71.

Parton, W. J., D. S. Schimel, C. V. Cole, and D. S. Ojima. 1987. Analysis of factors controlling soil organic matter levels in Great Plains grasslands. *Soil Science Society of America Journal* 51:1173–1179.

Paruelo, J. M., W. K. Lauenroth, I. C. Burke, and O. E. Sala. 1999. Grassland precipitation-use efficiency varies across a resource gradient. *Ecosystems* 2:64–68.

Patterson, E., and D. Gillette. 1977. Commonalities in measured size distributions for aerosols having a soil-derived component. *Journal of Geophysical Research* 82:2074–2082.

Paulsen, H. A. Jr., and F. N. Ares. 1961. Trends in carrying capacity and vegetation on an arid southwestern range. *Journal of Range Management* 14:78–83.

Paulsen, H. A. Jr., and F. N. Ares. 1962. *Grazing values and management of black grama and tobosa grasslands and associated shrub ranges of the Southwest.* USDA Forest Service Technical Bulletin no. 1270, Washington DC.

Peplow, E. H. Jr. 1958. *History of Arizona*, vol. 2. Lewis Historical Publishing, New York.

Peterjohn, W. T., and W. H. Schlesinger. 1991. Factors controlling denitrification in a Chihuahuan Desert ecosystem. *Soil Science Society of America Journal* 55:1694–1701.

Peters, A. J., B. C. Reed, M. D. Eve, and K. M. Havstad. 1993. Satellite assessment of drought impact on native plant communities of southeastern New Mexico, USA. *Journal of Arid Environments* 24:305–319.

Peters, D., and J. E. Herrick. 2001. Modelling vegetation change and land degradation in semiarid and arid ecosystems: An integrated hierarchical approach. *Advances in Environmental Monitoring and Modelling.* Web page at www.kcl.ac.uk/kis/schools/hums/geog/advemm/vol1no2.html.

Peters, D. P. 2003. Spatially explicit simulation models: the relevance for management of invasives [abstract]. 7th International Conference on the Ecology and Management of Alien Plant Invasions. November 3–7, 2003, Fort Lauderdale, FL, p. 68.

Peters, D. P. C. 2000. Climatic variation and simulated patterns in seedling establishment of two dominant grasses at a semiarid-arid grassland ecotone. *Journal of Vegetation Science* 11:493–504.

Peters, D. P. C. 2002a. Plant species dominance at a grassland-shrubland ecotone: An individual-based gap dynamics model of herbaceous and woody species. *Ecological Modelling* 152:5–32.

Peters, D. P. C. 2002b. Recruitment potential of two perennial grasses with different growth forms at a semiarid-arid transition zone. *American Journal of Botany* 89:1616–1623.

Peters, D. P. C., and J. Betancourt. 2001. Climatic triggers for nonlinear and threshold responses of rangelands. *Bulletin Ecological Society America* 82:31.

Peters, D. P. C. and K. M. Havstad. 2005. Nonlinear dynamics in Chihuahuan Desert ecosystems: Interactions among drivers and processes across scales. *Journal of Arid Environments* 65:196–206.

Peters, D. P., and J. E. Herrick. 1999a. Landscape-scale processes and sensitivity of Chihuahuan Desert ecosystems to climate change [abstract]. 5th World Congress, International Association for Landscape Ecology, Snowmass, CO. II(L–Z):122.

Peters, D. P. C., and J. E. Herrick. 1999b. Vegetation-soil feedbacks and sensitivity of Chihuahuan Desert ecosystems to climate change. *Bulletin of the Ecological Society of America* 80:69.

Peters, D. P. C., R. A. Pielke Sr., B. T. Bestelmeyer, C. D. Allen, S. Munson-McGee, and K. M. Havstad. 2004. Cross-scale interactions, nonlinearities, and forecasting catastrophic events. *Proceedings of the National Academy of Sciences* 101:15130–15135.

Peterson, F. F. 1981. *Landforms of the basin and range province.* Nevada Agricultural Experiment Station, Technical Bulletin 28. University of Nevada, Reno, NV.

Phillips, F. M. 1994. Environmental tracers for water movement in desert soils of the American Southwest. *Soil Science Society of America Journal* 58:15–24.

Phillips, F. M., L. A. Peters, and M. K. Tansey. 1986. Paleoclimatic inferences from an isotopic investigation of ground water in the central San Juan Basin, New Mexico. *Quaternary Research* 26:179–193.

Phillips, F. M., J. L. Mattick, T. A. Duval, D. Elmore, and P. W. Kubik. 1988. Chlorine 36 and tritium from nuclear weapons fallout as tracers for long-term liquid and vapor movement in desert soils. *Water Resources Research* 24:1877–1891.

Phillips, W. S. 1963. Depth of roots in soil. *Ecology* 44(2):424.

Pianka, E. R. 1967. On lizard species diversity: North American flatland deserts. *Ecology* 48:333–351.

Pidgeon, A. M., N. E. Mathews, R. Benoit, and E. V. Nordheim. 2001. Response of avian communities to historic habitat change in the northern Chihuahuan Desert. *Conservation Biology* 15:1772–1788.

Pielke, R. A., R. Avissar, M. Raupach, A. J. Dolman, X. Zeng, and A. S. Denning. 1998. Interactions between the atmosphere and terrestrial ecosystems: Influence on weather and climate. *Global Change Biology* 4:461–475.

Pieper, R. 1994. Ecological implications of livestock grazing. Pages 177–211 in M. Vavra, W. Laycock, and R. Pieper, editors, *Ecological implications of livestock herbivory in the West.* Society for Range Management, Denver, CO.

Pieper, R. D., J. C. Anway, M. A. Ellstrom, C. H. Herbel, R. L. Packard, S. L. Pimm, R. J. Raitt, E. E. Staffeldt, and J. G. Watts. 1983. *Structure and function of North American desert grassland ecosystems.* New Mexico State University Agricultural Experiment Station Special Report 39.

Pilz, W. R. 1983. *Nesting ecology and diet of Swainson's hawk in the Chihuahuan Desert, south-central New Mexico.* M.S. thesis. New Mexico State University, Las Cruces, NM.

Pinnick, R., G. Fernandez, B. Hinds, C. Bruce, R. Schaefer, and J. Pendleton. 1985. Dust generated by vehicular traffic on unpaved roadways: Sizes and infrared extinction characteristics. *Aerosol Science and Technology* 4:99–121.

Pinnick, R., G. Fernandez, E. Martinez-Andazola, B. Hinds, A. Hansen, and K. Fuller. 1993. Aerosols in the arid southwestern United States: Measurements of mass loading, volatility, size distribution, absorption characteristics, black carbon content, and vertical structure to 7 km above sea level. *Journal of Geophysical Research* 98:2651–2666.

Plamboeck, A. H., H. Grip, and U. Nygren. 1999. A hydrological tracer study of water uptake depth in a Scots pine forest under two different water regimes. *Oecologia* 119: 452–460.

Pockman, W. T., J. S. Sperry, and J. W. O'Leary. 1995. Sustained and significant negative water pressure in xylem. *Nature* 378:715–716.

Poesen, J. W. A. 1992. Mechanisms of overland-flow generation and sediment production on loamy and sandy soils with and without rock fragments. Pages 275–305 in A. J. Parsons and A. D. Abrahams, editors, *Overland flow: Hydraulics and erosion Mechanics.* UCL Press, London.

Polis, G. A. 1994. Food webs, trophic cascades and community structure. *Australian Journal of Ecology* 19:121–136.

Polley, H. W., H. B. Johnson, and H. S. Mayeux. 1992. Carbon dioxide and water fluxes of C_3 annuals and C_3 and C_4 perennials at subambient CO_2 concentrations. *Functional Ecology* 6:696–703.

Polley, H. W., H. B. Johnson, H. S. Mayeux, and C. R. Tischler, C. R. 1996. Are some of the recent changes in grassland communities a response to rising CO_2 concentrations? Pages 177–195 in C. Korner and F. A. Bazzaz, editors, *Carbon dioxide, populations, and communities.* Academic Press. San Diego, CA.

Potts, D. L., and D. G. Williams. 2004. Response of tree ring holocellulose d13C to moisture availability in Populus fremontii at perennial and intermittent stream reaches. *Western North American Naturalist* 64:27–37.

Prince, S. D., E. B. de Colstoun, and L. L. Kravitz. 1998. Evidence from rain-use efficiencies does not indicate extensive Sahelian desertification. *Global Change Biology* 4: 359–374.

Privette, J. L., G. P. Asner, J. Conel, K. F. Huemmrich, R. Olson, A. Rango, A. F. Rahman, K. Thome, and E. A. Walter-Shea. 2000. The EOS Prototype Validation Exercise (PROVE) at Jornada: Overview and lessons learned. *Remote Sensing of Environment* 74:1–12.

Provenza, F. D., J. J. Villalba, C. D. Cheney, and S. J. Werner. 1998. Self-organization of foraging behavior: From simplicity to complexity without goals. *Nutrition Research Reviews* 11:199–222.

Provenza, F. D. 2003. *Foraging behavior: Managing to survive in a world of change. Behavioral principals for human, animal, vegetation, and ecosystem management.* Agricultural Experimental Station, Utah State University, Logan, UT.

Pykala, J. 2000. Mitigating human effects on European biodiversity through traditional animal husbandry. *Conservation Biology* 14:705–712.

Raitt, R. J., and R. L. Maze. 1968. Densities and species composition of breeding birds of a creosotebush community in southern New Mexico. *Condor* 70:193–205.

Raitt, R. J., and S. L. Pimm. 1978. Dynamics of bird communities in the Chihuahuan Desert. *Condor* 78:427–442.

Ramalingam, K. 1999. *Determination of surface fluxes of heat and water vapor in an arid ecosystem.* M.S. thesis, Utah State University, Logan, UT.

Rango, A., J. C. Ritchie, W. P. Kustas, T. J. Schmugge, K. L. Brubaker, K. M. Havstad, J. H. Prueger, and K. S. Humes. 1996. JORNEX: A remote sensing campaign to quantify rangeland vegetation change and plant community-atmospheric interactions. Pages 445–446 in *Proceedings of the Second International Conference on GEWEX.* Washington, DC.

Rango, A., J. Chopping, J. Ritchie, K. Havstad, W. Kustas, and T. Schmugge. 2000. Morphological characteristics of shrub coppice dunes in desert grasslands of southern New Mexico derived from scanning LIDAR. *Remote Sensing of Environment* 74:27–44.

Rango, A., S. Goslee, J. Herrick, J. Chopping, K. Havstad, L. Huenneke, R. Gibbens, R. Beck, and R. McNeely. 2002. Remote sensing documentation of historic rangeland remediation treatments in southern New Mexico. *Journal of Arid Environments* 50: 549–572.

Rasker, R., B. Alexander, and P. Holmes. 2003. *The changing economy of the west: Employment and personal income trends, by region, state, and industry, 1970 to 2000.* Sonoran Institute, Bozeman, MT.

Rasmussen, E. M. 1985. El Niño and variations in climate. *American Scientist* 73:168–177.

Rasmussen, E. M., and J. M. Wallace. 1983. Meteorological aspects of the El Niño/Southern Oscillation. *Science* 222:1195–1202.

Rastetter, E. B., J. D. Aber, D. P. C. Peters, D. S. Ojima, and I. C. Burke. 2003. Using mechanistic models to scale ecological processes across space and time. *BioScience* 53:1–9.

Rawls, W. J., D. L. Brakensiek, J. R. Simanton, and K. D. Kohl. 1990. Development of a crust factor for a Green Ampt model. *Transactions of the American Society of Agricultural Engineers* 33:1224–1228.

Redmond, K. T., and R. W. Koch. 1991. Surface climate and streamflow variability in the western United States and their relationship to large-scale circulation indices. *Water Resources Research* 27:2381–2399.

Rejmánek, M., and D. M. Richardson. 1996. What attributes make some plant species more invasive? *Ecology* 77:1655–1661.

Renard, K. G. 1970. *The hydrology of semiarid rangeland watersheds*. ARS 41-62, USDA-ARS, Washington, DC.

Reynolds, H. G., and G.E. Glendening. 1949. Merriam kangaroo rate a factor in mesquite propagation on southern Arizona rangelands. *Journal of Range Management* 2:193–197.

Reynolds, J. F. 1986. Adaptive strategies of desert shrubs with special reference to the creosotebush (*Larrea tridentata* [DC] Cov.). Pages 19–49 in W. G. Whitford, editor, *Pattern and process in desert ecosystems*. University of New Mexico Press, Albuquerque, NM.

Reynolds, J. F., and B. Acock. 1985. Predicting the response of plants to increasing carbon dioxide: A critique of plant growth models. *Ecological Modelling* 29:107–129.

Reynolds, J. F., and B. Acock. 1997. Modularity and genericness in plant and ecosystem models. *Ecological Modelling* 94:7–16.

Reynolds, J. F., and G. L. Cunningham. 1981. Validation of a primary production model of the desert shrub *Larrea tridentata* using soil-moisture augmentation experiments. *Oecologia* 51:357–363.

Reynolds, J. F., and P.W. Leadley. 1992. Modeling the response of arctic plants to changing climate. Pages 413–438 in F. S. Chapin III, R. Jefferies, J. F. Reynolds, G. Shaver, and J. Svoboda, editors, *Arctic ecosystems in a changing climate*. Academic Press, San Diego, CA.

Reynolds, J. F., and J. Wu. 1999. Do landscape structural and functional units exist? Pages 275–298 in J. D. Tenhunen and P. Kabat, editors, *Integrating hydrology, ecosystem dynamics, and biogeochemistry in complex landscapes*. Wiley, New York.

Reynolds, J. F., D. W. Hilbert, J.-L. Chen, P. C. Harley, P. R. Kemp, and P. W. Leadley. 1992. *Modeling the response of plants and ecosystems to elevated CO_2 and climate change*. DOE/ER-60490T-H1, National Technical Information Service, U.S. Department of Commerce, Springfield, VA.

Reynolds, J. F., D. W. Hilbert, and P. R. Kemp. 1993. Scaling ecophysiology from the plant to the ecosystem: A conceptual framework. Pages 127–140 in J. Ehleringer and C. Field, editors, *Scaling processes between leaf and the globe*. Academic Press, New York.

Reynolds, J. F., P. R. Kemp, B. Acock, J.-L. Chen, and D. L. Moorhead. 1996a. Progress, limitations, and challenges in modeling the effects of elevated CO_2 on plants and ecosystems. Pages 347–380 in G. Koch and H. A. Mooney, editors, *Carbon dioxide and terrestrial ecosystems*. Academic Press, San Diego, CA.

Reynolds, J. F., J. D. Tenhunen, P. W. Leadley, H. Li, D. L. Moorhead, B. Ostendorf, and F. S. Chapin. 1996b. Patch and landscape models of arctic tundra: potentials and limitations. Pages 293–324 in J. F. Reynolds and J. D. Tenhunen, editors, *Landscape function and disturbance in arctic tundra*. Ecological Studies Series, vol. 120, Springer-Verlag, Berlin.

Reynolds, J. F., R. A. Virginia, and W. H. Schlesinger. 1997. Defining functional types for models of desertification. Pages 194–214 in T. M. Smith, H. H. Shugart, and F. I. Woodward, editors, *Plant functional types: Their relevance to ecosystem properties and global ghange*. Cambridge University Press, Cambridge.

Reynolds, J. F., R. J. Fernández, and P. R. Kemp. 1999a. Drylands and global change: The effect of rainfall variability on sustainable rangeland production. Pages 73–86 in K. N. Watanabe and A. Komanine, editors, *Proceedings of the Twelfth Toyota Conference: Challenge of Plant and Agricultural Sciences to the Crisis of Biosphere on the Earth in the 21st Century*. November 25–28, Mikkabi, Shizuoka, Japan. Landes Biosciences, Austin, TX.

Reynolds, J. F., R. A. Virginia, P. R. Kemp, A. G. de Soyza, and D. C. Tremmel. 1999b. Impact of drought on desert shrubs: Effects of seasonality and degree of resource island development. *Ecological Monographs* 69:69–106.

Reynolds, J. R., P. R. Kemp, and J. D. Tenhunen. 2000. Effects of long-term rainfall variability on evaporation and soil water distribution in the Chihuahuan Desert: A modeling analysis. *Plant Ecology* 150:145–159.

Reynolds, J. F., H. Bugmann, and L. Pitelka. 2001. How much physiology is needed in forest gap models for simulating long-term vegetation response to global change? Limitations, potentials, and recommendations. *Climatic Change* 51:541–557.

Reynolds, J. F., and D. M. Stafford Smith, editors. 2002. *Global desertification: Do humans cause deserts*? Dahlem Workshop Report 88. Dahlem University Press, Dahlem, Germany.

Reynolds, J. F., D. M. Stafford Smith, and E. Lamben. 2003. Do humans cause deserts? An old problem through the lens of a new framework: The Dahlem desertification paradigm. Pages 2042–2048 in N. Allsop, A. R. Palmer, S. J. Miedon, K. P. Kukman, G. Kerley, C. Hurt, and C. Brown, editors, *Proceedings of the VIIth International Rangeland Congress*. July 26–August 1, 2003, Durban, South Africa.

Reynolds, J. F., P. R. Kemp, K. Ogle, and R. J. Fernández. 2004. Modifying the "pulse-reserve" paradigm for deserts of North America: Precipitation pulses, soil water and plant responses. *Oecologia* 141(2):194–210.

Richards, J. H., and M. M. Caldwell. 1987. Hydraulic lift: Substantial nocturnal water transport between soil layers by *Artemisia tridentata* roots. *Oecologia* 73:486–489.

Richardson, A. J. 1982. Relating Landsat digital count values to ground reflectance for optically thin atmospheric conditions. *Applied Optics* 21:1457–1464.

Richmond, G. M. 1986. Stratigraphy and correlation of glacial deposits of Rocky Mountains, the Colorado Plateau, and the ranges of the Great Basin. Pages 99–127 in V. Sibrava, D. Q. Bowen, and G. M. Richmond, editors, *Quaternary glaciation in the Northern Hemisphere*. Pergamon Press, New York.

Rietkirk, M., and J. van de Koppel. 1997. Alternative stable states and threshold effects in semiarid grazing systems. *Oikos* 79:69–76.

Rietkerk, M., F. van den Bosch, and J. van de Koppel. 1997. Site-specific properties and irreversible vegetation changes in semiarid grazing systems. *Oikos* 80:241–252.

Ritchie, J. C. 1996. Remote sensing applications to hydrology: Airborne laser altimeters. *Hydrological Sciences Journal* 41:625–636.

Ritchie, J. C., A. Rango, W. P. Kustas, T. J. Schmugge, K. Brubaker, K. M. Havstad, B. Nolen, J. H. Prueger, J. H. Everitt, M. R. Davis, F. R. Schiebe, J. D. Ross, K. S. Humes, L. E. Hipps, J. Menenti, W. G. M. Bastiaanssen, and H. Pelgrum. 1996. JORNEX: An airborne campaign to quantify rangeland vegetation change and plant community-atmospheric interactions. Pages II-55–II-66 in *Proceedings of the Second International Airborne Remote Sensing Conference and Exhibition*.

Ritchie, J. C., J. E. Herrick, and C. A. Ritchie. 2003. Variability in soil redistribution in the northern Chihuahuan Desert based on ^{137}cesium measurements. *Journal of Arid Environments* 55:737–746.

Rockström, J., and M. Falkenmark. 2000. Semiarid crop production from a hydrological perspective: Gap between potential and actual yields. *Critical Reviews in Plant Sciences* 19:319–346.

Roels, J. M. 1984. Surface runoff and sediment yield in the Ardeche rangelands. *Earth Surface Processes and Landforms* 9:371–381.

Ropelewski, C. F., and M. S. Halpert. 1986. North American precipitation and temperature patterns associated with the El Niño/Southern Oscillation (ENSO). *Monthly Weather Review* 114:2352–2362.

Ropelewski, C. F., and M. S. Halpert. 1987. Global-and regional-scale precipitation patterns associated with the El Niño/Southern Oscillation. *Monthly Weather Review* 115:1606–1626.

Rosenfeld, J. S. 2002. Functional redundancy in ecology and conservation. *Oikos* 98:156–162.

Rosiere, R. E., R. F. Beck, and J. D. Wallace. 1975. Cattle diets on semidesert grassland: Botanical composition. *Journal of Range Management* 28(2):89–93.

Rostagno, C. M. 1989. Infiltration and sediment production as affected by soil surface conditions in a shrubland of Patagonia, Argentina. *Journal of Range Management* 42: 382–385.

Rostagno. C. M., H. F. del Valle, and L. Videla. 1991. The influence of shrubs on some chemical and physical properties of an aridic soil in northeastern Patagonia, Argentina. *Journal of Arid Environments* 20:179–188.

Roundy, B. A., and S. H. Biedenbender. 1995. Revegetation in the desert grassland. Pages 265–303 in M. P. McClaran and T. R. VanDevender, editors, *The desert grassland.* University of Arizona Press, Tucson, AZ.

Rouse, J. F. 1977. *The Criollo: Spanish cattle in the Americas.* University of Oklahoma Press, Norman, OK.

Royer, D. L. 1999. Depth to pedogenic carbonate horizon as a paleoprecipitation indicator? *Geology* 27:1123–1126.

Rubio, H. O., M. K. Wood, M. Cardenas, and B. A. Buchanan. 1992. The effect of polyacrylamide on grass emergence in south-central New Mexico. *Journal of Range Management* 45:296–300.

Rubio Arias, H. O. 1988. *The effects of polyacrylamide on grass emergence in south-central New Mexico.* Ph.D. dissertation. New Mexico State University, Las Cruces, NM.

Ruhe, R. V. 1962. Age of the Rio Grande Valley in southern New Mexico. *Journal of Geology* 70:151–167.

Ruhe, R. V. 1964. Landscape morphology and alluvial deposits in southern New Mexico. *Annals of the Association of American Geographers* 54:147–159.

Ruhe, R. V. 1967. *Geomorphic surfaces and surficial deposits in southern New Mexico.* New Mexico Bureau of Mines and Mineral Resources, Memoir 18. Socorro, NM.

Ruhe, R. V. 1975. *Geomorphology.* Houghton Mifflin, Boston, MA.

Rundel, P. W., and A. C. Gibson. 1996. *Ecological communities and processes in a Mojave Desert ecosystem: Rock Valley, Nevada.* Cambridge University Press, Cambridge.

Rundel, P. W., E. T. Nilsen, M. R. Sharifi, R. A. Virginia, W. M. Jarrell, D. H. Kohl, and G. B. Shearer. 1982. Seasonal dynamics of nitrogen cycling for a *Prosopis* woodland in the Sonoran Desert. *Plant and Soil* 67:343–353.

Ruyle, G.B., R. Tronstad, D.W. Hadley, P. Herlman, and D.A. King. 2000. Commercial livestock operations in Arizona. Pages 379–418 in R. Jemison and C. Raish, editors, *Livestock management in the American Southwest: Ecology, society, and economics.* Elsevier Press, St. Louis, MO.

Ryel, R. J., M. M. Caldwell, C. K. Yoder, D. Or, and A. J. Leffler. 2002. Hydraulic redistribution in a stand of *Artemisia tridentata*: Evaluation of benefits to transpiration assessed with a simulation model. *Oecologia* 130:173–184.

Sala, O. E., R. A. Golluscio, W. K. Lauenroth, and A. Soriano. 1989. Resource partitioning between shrubs and grasses in the Patagonian Steppe. *Oecologia* 81:501–505.

Sala, O. E., and W. K. Lauenroth. 1982. Small rainfall events: An ecological role in semi-arid regions. *Oecologia* 53:301–304.

Sala, O. E., W. J. Parton, L. A. Joyce, and W. K. Lauenroth. 1988. Primary production of the central grassland region of the United States. *Ecology* 69:40–45.

Sala, O. E., W. K. Lauenroth, and W. J. Parton. 1992. Long-term soil-water dynamics in the shortgrass steppe. *Ecology* 73:1175–1181.

Saleh, A., and D. W. Fryrear. 1995. Threshold wind velocities of wet soils as affected by windblown sand. *Soil Science* 160:304–309.

Sandor, J. A., P. L. Gersper, and J. W. Hawley. 1990. Prehistoric agricultural terraces and soils in the Mimbres area, New Mexico. *Soils and Early Agriculture: World Archaeology* 22:70–86.

Santos, P. F., and W. G. Whitford. 1981. The effects of microarthropods on litter decomposition in a Chihuahuan Desert ecosystem. *Ecology* 62:654–663.

Santos, P. F., E. DePree, and W. G. Whitford. 1978. Spatial distribution of litter and microarthropods in a Chihuahuan Desert ecosystem. *Journal of Arid Environments* 1: 41–49.

Santos, P. F., J. Phillips, and W. G. Whitford. 1981. The role of mites and nematodes in early stages of buried litter decomposition in a desert. *Ecology* 62:664–669.

Santos, P. F., N. Z. Elkins, Y. Steinberger, and W. G. Whitford. 1984. A comparison of surface and buried *Larrea tridentata* leaf litter decomposition in North American hot deserts. *Ecology* 65:278–284.

Sarna-Wojcicki, A. M., and M. S. Pringle Jr. 1992. Laser-fusion $^{40}Ar/^{39}Ar$ ages of the Tuff of Taylor Canyon and Bishops Tuff, E. California–W. Nevada [abstract]. Geological Society of America, Abstracts with Programs, p. 633.

Sarr, D. A. 2002. Riparian livestock exclosure research in the western United States: A critique and some recommendations. *Environmental Management* 30:516–526.

Scanlon, B. R., and R. S. Goldsmith. 1997. Field study of spatial variability in unsaturated flow beneath and adjacent to playas. *Water Resources Research* 33:2239–2252.

Scanlon, B. R., R. P. Langford, and R. S. Goldsmith. 1999. Relationship between geomorphic settings and unsaturated flow in an arid setting. *Water Resource Research* 35: 983–999.

Scanlon, B. R., D. G. Levitt, R. C. Reedy, K. E. Keese, and M. J. Sully. 2005. Ecological controls on water cycle response to climate variability in deserts. *Proceedings of the National Academy of Sciences* 102:6033–6038.

Schaefer, D., Y. Steinberger, and W. G. Whitford. 1985. The failure of nitrogen and lignin control of decomposition in a North American desert. *Oecologia* 65:382–386.

Schaefer, D. A., and W. G. Whitford. 1981. Nutrient cycling by the subterranean termite, *Gnathamitermes tubiformans*, in a Chihuahuan Desert ecosystem. *Oecologia* 48:277–283.

Schaeffer, S. M., D. G. Williams, and D. C. Goodrich. 2000. Transportation of cottonwood/willow forest estimated from sap flux. *Agricultural and Forest Meteorology* 105:257–270.

Schenk, H. J., and R. B. Jackson. 2002. Rooting depths, lateral root spreads and below-ground/above-ground allometries of plants in water-limited ecosystems. *Journal of Ecology* 90:480–494.

Schiebe, F. R., D. A. Waits, J. H. Everitt, and J. Duncan. 2001. Evaluation of the SST crop reflectance management system. *Proceedings of the 18th Biennial Workshop on Color Photography and Videography in Resource Management.* Society of Photogrammetry and Remote Sensing, Amherst, MA.

Schlesinger, W. H. 1985. The formation of caliche in soils of the Mojave Desert, California. *Geochimica et Cosmochimica Acta* 49:7–66.

Schlesinger, W. H., and C. S. Jones. 1984. The comparative importance of overland runoff and mean annual rainfall to shrub communities of the Mojave Desert. *Botanical Gazette* 145:116–124.

Schlesinger, W. H., and W. T. Peterjohn. 1991. Processes controlling ammonia volatilization from Chihuahuan Desert soils. *Soil Biology and Biochemistry* 23:637–642.

Schlesinger, W. H., and A. M. Pilmanis. 1998. Plant-soil interactions in deserts. *Biogeochemistry* 42:169–187.

Schlesinger, W. H., P. J. Fonteyn, and W. A. Reiners. 1989. Effects of overland flow on plant water relations, erosion, and soil water percolation on a Mojave Desert landscape. *Soil Science Society of America Journal* 53:1567–1572.

Schlesinger, W. H., J. F. Reynolds, G. L. Cunningham, L. Huenneke, W. M. Jarrell, R. A. Virginia, and W. G. Whitford. 1990. Biological feedbacks in global desertification. *Science* 247:1043–1048.

Schlesinger, W. H., J. A. Raikes, A. E. Hartley, and A. F. Cross. 1996. On the spatial pattern of soil nutrients in desert ecosystems. *Ecology* 77:364–374.

Schlesinger, W. H., A. D. Abrahams, A. J. Parsons, and J. Wainwright. 1999. Nutrient losses in runoff from grassland and shrubland habitats in southern New Mexico: I. Rainfall simulation experiments. *Biogeochemistry* 45:21–34.

Schlesinger, W. H., T. J. Ward, and J. Anderson. 2000. Nutrient losses in runoff from grassland and shrubland habitats in southern New Mexico: II. Field plots. *Biogeochemistry* 49:69–86.

Schmid, H. P. and T. R. Oke. 1990. A model to estimate the source area contributing to turbulent exchange in the surface layer over patchy terrain. *Quarterly Journal of the Royal Meteorological Society* 116:965–988.

Schmidt, R. H. Jr. 1979. A climatic delineation of the "real" Chihuahuan Desert. *Journal of Arid Environments* 2:243–250.

Schmugge, T., A. French, J. C. Ritchie, A. Rango, and H. Pelgrum. 2001. Temperature and emissivity separation from multispectral thermal infrared observations. *Remote Sensing of Environment* 78:1–10.

Schmutz, E. M., E. L. Smith, P. R. Ogden, M. L. Cox, J. O. Klemmedson, J. J. Norris, and L. C. Fierro. 1992. Desert grasslands. Pages 337–362 in R. T. Coupland, editor, *Ecosystems of the World 8A*. Elsevier Science, Amsterdam.

Scholes, R. J., and S. R. Archer. 1997. Tree-grass interactions in savannas. *Annual Review of Ecology and Systematics* 28:517–544.

Schowalter, T. D. 1996. Arthropod associates and herbivory on tarbush in southern New Mexico. *Southwestern Naturalist* 41:140–144.

Schowalter, T. D., D. C. Lightfoot, and W. G. Whitford. 1999. Diversity of arthropod responses to host-plant water stress in a desert ecosystem in southern New Mexico. *American Midland Naturalist* 142:281–290.

Schuepp, P. H., M. Y. Leclerc, J. I. MacPherson, and R. L. Desjardins. 1990. Footprint prediction of scalar fluxes from analytical solutions of the diffusion equation. *Boundary-Layer Meteorology* 50:355–373.

Schulze, E. D. 1986. Carbon dioxide and water vapor exchange in response to drought in the atmosphere and in the soil. *Annual Review of Plant Physiology* 37:247–274.

Schulze, E. D., M. M. Caldwell, J. Canadell, H. A. Mooney, R. B. Jackson, D. Parson, R. Scholes, O. E. Sala, and P. Trimborn. 1998. Downward flux of water through roots (i.e., inverse hydraulic lift) in dry Kalahari sands. *Oecologia* 115:460–462.

Schumacher, A., and W. G. Whitford. 1975. The foraging ecology of two species of Chihuahuan Desert ants: *Formica perpilosa* and *Trachymyrmex smithii neomexicanus* (Hymenoptera: Formicidae). *Insectes Sociaux* 21:317–330.

Schwinning, S., and J. R. Ehleringer. 2001. Water use trade-offs and optimal adaptations to pulse-driven arid ecosystems. *Journal of Ecology* 89:464–480.

Schwinning, S., and O. E. Sala. 2004. Hierarchy of responses to resource pulses in arid and semiarid ecosystems. *Oecologia* 141(2):211–220.

Scoging, H. M. 1992. Modelling overland-flow hydrology for dynamic hydraulics. Pages 89–103 in A. J. Parsons and A. D. Abrahams, editors, *Overland flow: Hydraulics and erosion mechanics*. UCL Press, London.

SCS. 1963. Standard soil survey of Jornada Experimental Range, New Mexico. Unpublished report by the Soil Conservation Service. Available online at www.usda-ars.nmsu.edu.

Scurlock, D. 1998. *From the Rio to the Sierra: An environmental history of the middle Rio Grande Basin*. USDA Forest Service General Technical Report RMRS-GTR-5.

Seager, W. R. 1975. Cenozoic tectonic evolution of the Las Cruces area, New Mexico. Pages 241–250 in *26th Annual Field Conference guidebook*. New Mexico Geological Society, Socorro, NM.

Seager, W. R. 1981. *Geology of Organ Mountains and southern San Andres Mountains, New Mexico*. New Mexico Bureau of Mines and Mineral Resources, Memoir 36. Socorro, NM.

Seager, W. R. 1995. *Geology of southwest quarter of Las Cruces and northwest El Paso 1° × 2° sheets, New Mexico*. New Mexico Bureau of Mines and Mineral Resources, Geologic Map 60. Socorro, NM.

Seager, W. R. and J. W. Hawley. 1973. *Geology of Rincon Quadrangle, New Mexico*. New Mexico Bureau of Mines and Mineral Resources, Bulletin 101. Socorro, NM.

Seager, W. R. and G. H. Mack. 1986. *Laramide paleotectonics of southern New Mexico*. Pages 660–685 in American Association of Petroleum Geologists, Memoir 41.

Seager, W. R., and G. H. Mack. 1995. Jornada Draw fault: A major Pliocene-Pleistocene normal fault in southern Jornada del Muerto. *New Mexico Geology* 17:37–43.

Seager, W. R., and A. B. Mayer, A.B. 1988. Uplift, erosion and burial of Laramide fault blocks, Salado mountains, Sierra County, New Mexico. *New Mexico Geology* 10:49–53, 60.

Seager, W. R., R. E. Clemons, and J. F. Callendar, editors. 1975a. *26th Annual Field Conference guidebook*. New Mexico Geological Society, Socorro, NM.

Seager, W. R., R. E. Clemons, and J. W. Hawley. 1975b. *Geology of Sierra Alta Quadrangle Doña Ana County, New Mexico*. New Mexico Bureau of Mines and Mineral Resources, Bulletin 102. Socorro, NM.

Seager, W. R., F. E. Kottlowski, and J. W. Hawley. 1976. *Geology of Doña Ana Mountains, New Mexico*. New Mexico Bureau of Mines and Mineral Resources Circular 147. Socorro, NM.

Seager, W. R., M. Shafiqullah, J. W. Hawley, and R. F. Marvin. 1984. New K-Ar dates from basalt and the evolution of the southern Rio Grande rift. *Geological Society of America Bulletin* 95:87–99.

Seager, W. R., G. H. Mack, M. S. Raimonde, and R. G. Ryan. 1986. Laramide basement-cored uplift and basins in south-central New Mexico. Pages 123–130 in *37th Annual Field Conference guidebook*. New Mexico Geological Society, Socorro, NM.

Seager, W. R., J. W. Hawley, F. E. Kottlowski, and S. A. Kelley. 1987. *Geology of east half of Las Cruces and northeast El Paso 1° × 2° sheets, New Mexico*. New Mexico Bureau of Mines and Mineral Resources, Geologic Map 57. Socorro, NM.

Seager, W. R., G. H. Mack, and T. F. Lawton. 1997. Structural kinematics and depositional history of a Laramide uplift-basin pair in southern New Mexico: Implications for development of intraforeland basins. *Geological Society of America Bulletin* 109: 1389–1401.

Senft, R. L., M. B. Coughenour, D. W. Bailey, L. R. Rittenhouse, O. E. Sala, and D. M. Swift. 1987. Large animal foraging and ecological hierarchies. *BioScience* 37:789–795.

Senock, R. S., D. M. Anderson, L. W. Murray and G. B. Donart. 1993. Tobosa tiller defoliation patterns under rotational and continuous stocking. *Journal of Range Management* 46:500–505.

Seybold, C. A., J. E. Herrick, and J. J. Brejda. 1999. Soil resilience: A fundamental component of soil quality. *Soil Science* 164:224–234.

Seyfried, M. S., S. Schwinning, M. A. Walvoord, W. T. Pockman, B. D. Newman, R. B. Jackson and F. M. Phillips. 2005. Ecohydrological control of deep drainage in arid and semiarid regions. *Ecology* 86: 277–287.

Shachak, M., and G. M. Lovett. 1998. Atmospheric deposition to a desert ecosystem and its implications for management. *Ecological Applications* 8:455–463.

Shaffer, D. T., and W. G. Whitford. 1981. Behavioral responses of a predator, the round-tailed horned lizard, *Phrynosoma modestum*, and its prey, honey pot ants, *Myrmecocystus* spp. *American Midland Naturalist* 105:209–216.

Shantz, H. L., and R. Zon. 1924. *Natural vegetation. Atlas of American agriculture*. Part 1, Section E (Map). U.S. Department of Agriculture, Washington, DC.

Shao, Y., M. R. Raupach, and P. A Findlater. 1993. The effect of saltation bombardment on the entrainment of dust by wind. *Journal of Geophysical Research* 98D:12719–12726.

Sharifi, M. R., F. C. Meinzer, E. T. Nilsen, P. W. Rundel, R. A. Virginia, W. M. Jarrell, D. J. Herman, and P. C. Clark. 1988. Effect of manipulation of water and nitrogen supplies on the quantitative phenology of *Larrea tridentata* (creosotebush) in the Sonoran Desert of California. *American Journal of Botany* 75:1163–1174.

Shaver, G. R., S. M. Bret-Harte, M. H. Jones, J. Johnstone, L. Gough, J. Laundre, and F. S. Chapin. 2001. Species composition interacts with fertilizer to control long-term change in tundra productivity. *Ecology* 82:3163–3181.

Shen, W., J. Wu, P. R. Kemp, J. F. Reynolds, and N. B. Grimm. 2004. Simulating the dynamics of primary productivity of a Sonoran ecosystem: Model parameterization and validation. *Ecological Modelling* (submitted).

Sheppard, P. R., M. K. Hughes, A. C. Comrie, G. D. Packin, and K. Angersbach. 2002. The climate of the US Southwest. *Climate Research* 21:219–238.

Shrader-Frechette, K., and E. D. McCoy. 1993. *Method in ecology: Strategies for conservation*. Cambridge University Press, Cambridge.

Shreve, F. 1917. A map of the vegetation of the United States. *Geological Review* 3:119–125.

Shuttleworth, W. J., and R. J. Gurney. 1990. The theoretical relationship between foliage temperature and canopy resistance in sparse crops. *Quarterly Journal of the Royal Meteorology Society* 116:497–519.

Silva, S. I., W. P. MacKay, and W. G. Whitford. 1985. The relative contributions of termites and microarthropods to fluff grass litter disappearance in the Chihuahuan Desert. *Oecologia* 67:31–34.

Silva, S. I., W. P. MacKay, and W. G. Whitford. 1989a. Temporal patterns of microarthropod population densities in fluff grass (*Erioneuron pulchellum*) litter; relationship to subterranean termites. *Pedobiologia* 33:333–338.

Silva, S., W. G. Whitford, W. M. Jarrell, and R. A. Virginia. 1989b. The microarthropod fauna associated with a deep-rooted legume, *Prosopis glandulosa*, in the Chihuahuan Desert. *Biology and Fertility of Soils* 7:330–335.

Silvertown, J., and J. B. Wilson. 1994. Community structure in a desert perennial community. *Ecology* 75:409–417.

Simberloff, D. 2004. Community ecology: is it time to move on? *American Naturalist* 163:787–799.

Simonson, R. W. 1959. Outline of a generalized theory of soil genesis. *Soil Science Society of America Proceedings* 23:152–156.

Sims, P. L., and J. S. Singh. 1978. The structure and function of ten western North American grasslands. II. Intra-seasonal dynamics in primary producer compartments. *Journal of Ecology* 66:547–572.

Singer, A. 1989. Illite in the hot aridic soil environment. *Soil Science* 147:126–133.

Sisson, W. B., and G. O. Throneberry. 1986. Seasonal nitrate reductase activity of three genotypes of *Atriplex canescens* in the northern Chihuahuan Desert. *Journal of Ecology* 74:579–589.

Skujins, J. 1981. Nitrogen cycling in desert ecosystems. Pages 477–491 in F. E. Clark and T. Rosswall, editors, *Terrestrial nitrogen cycles*. Ecological Bulletin no. 33. Swedish Natural Resources Council, Stockholm.

Small, E. 2005. Climatic controls on diffuse groundwater recharge in semiarid environments on the southwestern United States. *Water Resources Research* 41.

Smith, D. M., N. A. Jackson, J. M. Roberts, and C. K. Ong. 1999. Reverse flow of sap in tree roots and downward siphoning of water by *Grevillea robusta*. *Functional Ecology* 13:256–264.

Smith, J. G. 1899. *Grazing problems in the Southwest and how to meet them*. U.S. Department of Agriculture Bulletin no. 16. US Government Printing Office, Washington, DC.

Smith, J. L., J. J. Halvorson, and H. Bolton. 1994. Spatial relationships of soil microbial biomass and C and N mineralization in a semiarid, shrub-steppe ecosystem. *Soil Biology and Biochemistry* 26:1151–1159.

Smith, J. N. 1993. *Diurnal diet composition and behavior of cattle and sheep on Chihuahuan Desert Range*. Dissertation, New Mexico State University.

Smith, R. E., and J. Y. Parlange. 1978. A parameter-efficient hydrologic infiltration model. *Water Resources Research* 14:533–538.

Smith, R. E., D. L. Chery Jr., K. G. Renard, and W. R. Gwinn. 1981. *Supercritical flow flumes for measuring sediment-laden flow*. U.S. Department of Agriculture Technical Bulletin 1655. U.S. Government Printing Office, Washington, DC.

Smith, S. D., and R. S. Nowak. 1990. Ecophysiology of plants in the intermountain lowlands. Pages 179–241 in C. B. Osmond, L. F. Pitelka, and G. M. Hidy, editors, *Plant bology of the basin and range*. Springer-Verlag, New York.

Smith, T. M., H. H. Shugart, and F. I. Woodward, editors. 1997. *Plant functional types*. Cambridge University Press, Cambridge.

Snow, J. T., and T. M. McClelland. 1990. Dust devils at White Sands Missile Range, New Mexico. 1. Temporal and spatial distributions. *Journal of Geophysical Research* 95: 13707–13721.

Snyder, K. A., and S. L. Tartowski. In Press. Multi-scale temporal variation in water availability: implications for vegetation dynamics in arid and semiarid ecosystems. *Journal of Arid Environments*.

Snyder, K. A., and D. G. Williams. 2000. Water sources used by riparian trees varies among stream types on the San Pedro River, Arizona. *Agricultural and Forest Meteorology* 105:227–240.

Sobecki, T. M., and L. P. Wilding. 1983. Formation of calcic and argillic horizons in selected soils of the Texas coast prairie. *Soil Science Society of America Journal* 47: 707–715.

Society for Ecological Restoration Science and Policy Working Group. 2002. *The SER primer on ecological restoration.* Available online at www.ser.org.

Soil Survey Staff. 1951. *Soil survey manual.* U.S. Department of Agriculture, Handbook 18. US Government Printing Office, Washington, DC.

Soil Survey Staff. 1975. *Soil taxonomy—A basic system of soil classification for making and interpreting soil surveys.* U.S. Department of Agriculture, Soil Conservation Service, Handbook 436, U.S. Government Printing Office.

Soil Survey Staff. 1999. *Soil taxonomy—A basic system of soil classification for making and interpreting soil surveys*, 2nd edition. U.S. Department of Agriculture, Soil Conservation Service, Handbook 436, U.S. Government Printing Office.

Sokolik, I., and G. Golitsyn. 1993. Investigation of optical and radiative properties of atmospheric dust aerosols. *Atmospheric Environment* 27a:2509–2517.

Sokolik, I. N., and O. B. Toon. 1996. Direct radiative forcing by anthropogenic airborne mineral aerosols. *Nature* 381:681–683.

Sokolik, I., A. Andronova, and T. C. Johnson. 1993. Complex refractive index of atmospheric dust aerosols. *Atmospheric Environment* 27a:2495–2502.

Springer, M. E. 1958. Desert pavement and vesicular layer of some soils of the Lahontan Basin, Nevada. *Soil Science Society of America Journal* 22:63–66.

Stafford Smith, D. M., and S. R. Morton. 1990. A framework for the ecology of arid Australia. *Journal of Arid Environments* 18:255–278.

Stannard, D. I., J. H. Blanford, W. P. Kustas W. D. Nichols, S. A. Amer, T. J. Schmugge, and M. A. Weltz. 1994. Interpretation of the surface flux measurements in heterogeneous terrain during the monsoon '90 experiment. *Water Resources Research* 30: 1227–1239.

Stein, R. A., and J. A. Ludwig. 1979. Vegetation and soil patterns on a Chihuahuan Desert bajada. *American Midland Naturalist* 101:28–37.

Steinberger, Y., and W. G. Whitford. 1983a. The contribution of rodents to decomposition processes in a desert ecosystem. *Journal of Arid Environments* 6:177–181.

Steinberger, Y., and W. G. Whitford. 1983b. The contribution of shrub pruning by jackrabbits to litter input in a Chihuahuan Desert ecosystem. *Journal of Range Management* 28:453–457.

Steinberger, Y., and W. G. Whitford. 1984. Spatial and temporal relationships of soil microarthropods on a desert watershed. *Pedobiologia* 26:275–284.

Steinberger, Y., and W. G. Whitford. 1985. Microarthropods of a desert tobosa grass (*Hilaria mutica*) swale. *American Midland Naturalist* 114:225–234.

Stockton, P. H., and D. A. Gillette. 1990. Field measurement of the sheltering effect of vegetation on erodible land surfaces. *Land Degradation and Rehabilitation* 2:77–85.

Strahler, A. N., and A. H. Strahler. 1987. *Modern physical gography.* Wiley, New York.

Strain, W. S. 1966. *Blancan mammalian fauna and Pleistocene formations, Hudspeth County, Texas.* University of Texas-Austin, Texas Memorial Museum, Bulletin 10.

Strojan, C. L., D. C. Randall, and F. B. Turner. 1987. Relationship of leaf litter decomposition rates to rainfall in the Mojave Desert. *Ecology* 68:741–744.

Strong, D. R. 1988. Insect host range. *Ecology* 69:885.

Sud, Y. C., W. C. Chao, and G. K. Walker. 1993. Dependence of rainfall on vegetation:

Theoretical considerations, simulation experiments, observations, and inferences from simulated atmospheric soundings. *Journal of Arid Environments* 25:5–18.

Sviridenkov, V., D. A. Gillette, A. A. Isakov, I. N. Sokolik, V. V. Smirnov, B. D. Belan, M. V. Panchenko, A. V. Andronova, S. M. Kolomiets, V. H. Zhukov, and D. A. Zhukovsky. 1993. Microphysical characteristics of dust aerosol measured during the Soviet-American experiment in Tadzhikistan, 1989. *Atmospheric Environment* 27a: 2481–2486.

Swetnam, T. W., and J. L. Betancourt. 1990. Fire–southern oscillation relations in the southwestern United States. *Science* 249:1017–1020.

Swetnam, T. W., and J. L. Betancourt. 1992. Temporal patterns of El Niño/Southern Oscillation—wildfire teleconnections in the southwestern United States. Pages 259–270 in H. F. Diaz and V. Markgraf, editors, *El Niño: Historical and paleoclimatic aspects of the Southern Oscillation*. Cambridge University Press, Cambridge.

Swetnam, T. W., C. D. Allen, and J. L. Betancourt. 1999. Applied historical ecology: Using the past to manage for the future. *Ecological Applications* 9:1189–1206.

Szarek, S. R. 1979. Primary production in four North American deserts: Indices of efficiency. *Journal of Arid Environments* 2:187–209.

Tadingar, T. 1982. *Utilization of standing crop by cattle under seasonal, short-duration grazing and seasonal, continuous grazing on a semidesert tobosa rangeland.* M.S. thesis. New Mexico State University, Las Cruces, NM.

Tanner, B. D. 1988. Use requirements for Bowen ratio and eddy correlation determination of evapotranspiration. Pages 605–616 in H. DeLynn, editor, *Planning now for irrigation and drainage in the 21st century*. Irrigation and Drainage Division of the American Society of Civil Engineers, Reston, VA.

Tanner, B. D., M. S. Tanner, W. A. Dugas, E. C. Campbell, and B. L. Bland. 1985. Evaluation of an operational eddy correlation system for evapotranspiration measurement. Pages 87–99 in *Advances in Evapotranspiration. Proceedings of the National Conference on Advances in Evaportranspiration,* American Society of Agricultural Engineers, St. Joseph, MI.

Teague, R., R. Borchardt, J. Ansley, B. Pinchak, J. Cox, J. K. Foy, and J. McGrann. 1997. Sustainable management strategies for mesquite rangeland: The Waggoner Kite project. *Rangelands* 195:4–8.

Teaschner, T. B. 2001. *Influence of soil depth and texture on mesquite (*Prosopis glandulosa*) density and canopy cover in the northern Chihuahuan Desert, New Mexico.* M.S. thesis. New Mexico State University, Las Cruces, NM.

Tedford, R. H. 1981. Mammalian biochronology of the late Cenozoic basins of New Mexico. *Geological Society of American Bulletin* 92:1008–1022.

Tegen, I., and I. Fung. 1995. Contribution to the atmospheric mineral aerosol load from land surface modification. *Journal of Geophysical Research* 100:18707–18726.

Tegen, J., A. A. Lacis, and I. Fung. 1996. The influence of climate forcing of mineral aerosols from disturbed soil. *Nature* 380:380–384.

Thomas, H. 1986. Water use characteristics of *Dactylis glomerata* L., *Lolium perenne* L., and L. *multiflorum* L American plants. *Annals of Botany* 57:211–223.

Thompson, M. M., and H. Gruner. 1980. Foundations in photogrammetry. Pages 1–36 in C. C. Slama, C. Theurer, and S. W. Henriksen, editors, *Manual of photogrammetry*. American Society of Photogrammetry, Falls Church, VA.

Tiedemann, A. R., and J. O. Klemmedson. 1973. Effect of mesquite on physical and chemical properties of the soil. *Journal of Range Management* 26:27–29

Tiedemann, A. R., and J. O. Klemmedson. 1986. Long-term effects of mesquite removal on soil characteristics: 1. Nutrients and bulk density. *Soil Science Society of America Journal* 50:472–475.

Tongway, D. J., and J. A. Ludwig. 1994. Small-scale resource heterogeneity in semiarid landscapes. *Pacific Conservation Biology* 1:201–208.

Tongway, D. J., and J. A. Ludwig. 1997. The conservation of water and nutrients within landscapes. In J. Ludwig, D. Tongway, D. Freudenberger, J. Noble, and K. Hodgkinson, editors, *Landscape ecology, function and management: Principles from Australia's rangelands.* CSIRO, Collingwood, Australia.

Torell, L. A., J. M. Fowler, M. E. Kincaid, and J. M. Hawkes. 1992. *The importance of public lands to livestock production in the US.* New Mexico Agricultural Experiment Station, Range Improvement Task Force Report no. 32. New Mexico State University, Las Cruces, NM.

Trenberth, K. E. 1976. Spatial and temporal variations of the Southern Oscillation. *Quarterly Journal of the Royal Meteorological Society* 102:639–653.

Trenberth, K. E., J. W. Hurrell. 1995. Decadal coupled atmosphere: Ocean variations in the North Pacific Ocean. *Canadian Special Publication of Fisheries and Aquatic Sciences* 121:15–24.

Tromble, J. M. 1983. Interception of rainfall by creosotebush (*Larrea tridentata*). Pages 373–375 in *Proceedings of the 14th International Grassland Congress.* Westview Press, Boulder, CO.

Tschirley, F. H., and S. C. Martin. 1960. Germination and longevity of velvet mesquite seed in soil. *Journal of Range Management* 13:94–97.

Tucker, C. J., W. W. Newcomb, and H. E. Dregne. 1994. AVHRR data sets for determination of desert spatial extent. *International Journal of Remote Sensing* 15:3547–3565.

Turner, F. B., and R. M. Chew. 1981. Production by desert animals. Pages 199–260 in D. W. Goodall and R. A. Perry, editors, *Arid-land eosystems: Structure, functioning, and management* (vol. 2). Cambridge University Press, Cambridge.

Turner, M. G. 1989. Landscape ecology: The effect of pattern on process. *Annual Review of Ecology and Systematics* 20:171–197.

U.S. Census Bureau. 2005. *United States Census 2000.* Retrieved March 15, 2005 from www.census.gov/main/www/cen2000.html.

Upchurch, G. R. Jr., and G. H. Mack. 1998. Latest Cretaceous leaf megafloras from the Jose Creek Member, McRae Formation of New Mexico. Pages 209–222 in *49th Annual Field Conference guidebook.* New Mexico Geological Society, Socorro, NM.

USDA. 2001. *New Mexico agricultural statistics—1990.* USDA, New Mexico Agricultural Statistics Service. Las Cruces, NM. Available online at www.usda.gov/nass/nm.

USDA Natural Resources Conservation Service. 1997. *National range and pasture handbook.* U.S. Department of Agriculture, Grazinglands Institute. 190-vi-NRPN. Washington, DC.

Valentine, K. A. 1941. *Some observations of the deterioration of the mesquite sandhills range type and the factors responsible for the deterioration.* Ranch Day Bulletin. The Southwestern Forest and Range Experiment Station, New Mexico Agricultural Experiment Station and New Mexico Extension Service Ranch. Las Cruces, NM.

Valentine, K. A. 1942. *Mesquite sandhills artificial revegetation experiments, 1934–1936.* USDA-ARS Jornada Experimental Range. Las Cruces, NM. Original copy in Jornada Experimental Range files, Las Cruces, NM.

Valentine, K. A. 1947. *Effect of water-retaining and water-spreading structures in revegetating semidesert rangeland.* New Mexico Agricultural Experiment Station. Las Cruces, NM. Original copy in Jornada Experimental Range files, Las Cruces, NM.

Valentine, K. A. 1970. *Influence of grazing intensity on improvement of deteriorated black grama range.* New Mexico State University Agriculural Experiment Station Bulletin 553. Las Cruces, NM.

Valone, T. J., M. Meyer, J. H. Brown, and R. M. Chew. 2002. Timescale of perennial grass recovery in desertified arid grasslands following livestock removal. *Conservation Biology* 16:995–1002.

Van Auken, O. W. 2000. Shrub invasions of North American semiarid grasslands. *Annual Review Ecology Systematics* 31:197–215.

Van Cleve, K., L. A. Viereck, and C. T. Dyrness. 1996. State factor control of soils and forest succession along the Tanana River in interior Alaska, USA. *Arctic and Alpine Research* 28:388–400.

van de Koppel, J., M. Reitkerk, F. van Langevelde, L. Kumar, C. A. Klausmeier, J. M. Fryxell, J. W. Hearne, J. van Andel, N. de Ridder, A. Skidmore, L. Stroosnijder, and H. H. T. Prins. 2002. Spatial heterogeneity and irreversible vegetation change in semi-arid grazing systems. *American Naturalist* 159:209–218.

Vanden Heuvel, R. C., 1966, The occurrence of sepiolite and attapulgite in the calcareous zone of a soil near Las Cruces, New Mexico. Pages 193–207 in *Clays and clay minerals: Proceedings of the 13th National Conference on Clays and Clay Minerals.* Pergamon Press, New York.

Vanderhill, J. B. 1986. *Lithostratigraphy, vertebrate paleontology, and magnetostratigraphy of Plio-Pleistocene sediments in the Mesilla Basin, New Mexico.* Ph.D. dissertation. University of Texas-Austin, Austin, TX.

VanDevender, T. R. 1990. Late Quaternary vegetation and climate of the Chihuahuan Desert, United States and Mexico. Pages 104–133 in J. L. Betancourt, T. R. VanDevender, and P. S. Martin, editors, *Packrat middens. The last 40,000 years of biotic change.* University of Arizona Press, Tucson, AZ.

VanDevender, T. R. 1995. Desert grassland history: Changing climates, evolution, biogeography, and community dynamics. Pages 68–99 in M. P. McClaren and T. R. VanDevender, editors, *The desert grassland.* University of Arizona Press, Tucson, AZ.

VanDevender, T. R, and W. G. Spaulding. 1979. Development of vegetation and climate in the southwestern United States. *Science* 204:701–710.

van Genuchten, M. Th. 1980. A closed-form equation for predicting the hydraulic conductivity of unsaturated soils. *Soil Science Society of America Journal* 44:892–898.

Van Zee J. W., W.G. Whitford, and W. E. Smith. 1997. Mutual exclusion by dolichoderine ants on a rich food source. *Southwestern Naturalist* 42:229–231.

Veatch, J. O. 1918. The soils of the Jornada Range Reserve, New Mexico. Unpublished report by U.S. Bureau of Soils. Available online at www.usda-ars.nmsu.edu.

Virginia, R. A., and W. M. Jarrell. 1983. Soil properties in a mesquite-dominated Sonoran Desert ecosystem. *Soil Science Society of America Journal* 47:138–144.

Virginia, R. A., W. M. Jarrell, and E. Franco-Vizcaino. 1982. Direct measurement of denitrification in a *Prosopis* (mesquite)-dominated Sonoran Desert ecosystem. *Oecologia* 53:120–122.

Vorhies, C. T., and W. P. Taylor. 1933. *Life history and ecology of jackrabbits, Lepus alleni and Lepus californicus, ssp., in relation to grazing in Arizona.* Arizona Agricultural Experiment Station Technical Bulletin 49.

Wagner, D., M. J. E. Brown, and D. M. Gordon. 1997. Harvester ant nests, soil biota, and soil chemistry. *Oecologia* 112:232–236.

Wagner, M. W., K. M. Havstad, D. E. Doornbos, and E. L. Ayers. 1986. Forage intake of rangeland beef cows with varying degrees of crossbred influence. *Journal of Animal Science* 63:1484–1490.

Wainwright, J. 2005. Climate and climatological variations in the Jornada Experimental Range and neighbouring areas of the US Southwest. *Advances in Environmental Monitoring and Modelling* 2(1).

Wainwright, J., A. J. Parsons, and A. D. Abrahams. 1999a. Rainfall energy under creosotebush. *Journal of Arid Environments* 43:111–120.

Wainwright, J., M. Mulligan, and J. B. Thornes. 1999b. Plants and water in drylands. Pages 78–126 in A. J. Baird and R. L. Wilby, editors, *Ecohydrology*. Routledge, London.

Wainwright, J., A. J. Parsons, and A. D. Abrahams. 2000. Plot-scale studies of vegetation, overland flow and erosion interactions: Case studies from Arizona and New Mexico. *Hydrological Processes* 14:2921–2943.

Wainwright, J., A. J. Parsons, W. H. Schlesinger, and A. D. Abrahams. 2002. Hydrology-vegetation interactions in areas of discontinuous flow on a semiarid bajada, southern New Mexico. *Journal of Arid Environments* 51:219–358.

Walker, B. H. 2002. Ecological resilience in grazed rangeland: A generic case study. Pages 183–194 in L. H. Gunderson and L. Pritchard Jr., editors, *Resilience and the behavior of large-scale systems*. Island Press, Washington, DC.

Walker, B. H., D. Ludwig, C. S. Holling, and R. M. Peterman. 1981. Stability of semiarid savanna grazing systems. *Journal of Ecology* 69:473–498.

Walker, D. A., and M. D. Walker. 1991. History and pattern of disturbance in Alaskan arctic terrestrial ecosystems: A hierarchical approach to analysing landscape change. *Journal of Applied Ecology* 28:244–276.

Wallace, A., E. M. Romney, and R. B. Hunter. 1980. The challenge of a desert: Revegetation of disturbed desert lands. *Great Basin Naturalist Memoirs* 4:216–225.

Wallwork, J. A. 1982. *Desert soil fauna*. Praeger, New York.

Wallwork, J. A., and D. C. Weems. 1984. *Jornadia larreae* n. gen. n. sp., a new genus of oribatid mite (Acari: Cryptostigmata) from the Chihuahuan Desert. *Acarologia* 25: 77–80.

Wallwork, J. A., B. W. Kamill, and W. G. Whitford. 1985. Distribution and diversity patterns of soil mites and other microarthropods in a Chihuahuan Desert site. *Journal of Arid Environments* 9:215–231.

Wallwork, J. A., M. MacQuitty, S. Silva, and W. G. Whitford. 1986. Seasonality of some Chihuahuan Desert soil oribatid mites (Acari: Cryptostigmata). *Journal of Zoology, London* (A) 208:403–416.

Walter, D. E. 1988. Nematophagy by soil arthropods from the shortgrass steppe, Chihuahuan Desert, and Rocky Mountains of the central United States. *Agriculture, Ecosystems, and Environment* 24:307–316.

Walter, H. 1971. Natural savannahs as a transition to the arid zone. Pages 238–265 in J. H. Burnett, editor, *Ecology of tropical and subtropical vegetation*. Van Nostrand Reinhold, New York.

Walter, H. 1979. *Vegetation of the earth and ecological systems of the geosphere*. Springer, New York.

Walton, M., J. E. Herrick, R. P. Gibbens, and M. Remmenga. 2001. Persistence of biosolids in a Chihuahuan Desert rangeland 18 years after application. *Arid Land Research and Management* 15:223–232.

Wansi, T., R. D. Pieper, R. F. Beck, and L. W. Murray. 1992. Botanical content of black-tailed jackrabbit diets on semidesert rangeland. *Great Basin Naturalist* 52:300–308.

Warren, S. D. 2001. Synopsis: Influence of biological soil crusts on arid land hydrology and soil stability. In J. Benlap and O. L. Lange, editors, *Biological soil crusts: Structure, function and management*. Springer-Verlag, Berlin.

Webb, R. H., J. W. Steiger, and R. M. Turner. 1987. Dynamics of Mojave Desert shrub assemblages in the Panamint Mountains, California. *Ecology* 68:478–490.

Wellman, P. I. 1954. *Glory, God and gold: A narrative history*. Doubleday, Garden City, NY.

Wells, P. V. 1979. An equable glaciopluvial in the West—pleniglacial evidence of increased precipitation on a gradient from the Great Basin to the Sonoran and Chihuahuan Deserts. *Quaternary Research* 12:311–325.

Weltz, M. A., M. R. Kidwell, and H. D. Fox. 1998. Influence of biotic and abiotic factors in measuring and modeling soil erosion on rangelands: State of knowledge. *Journal of Range Management* 51:482–495.

Weltzin, J. F., and G. R. McPherson. 1997. Spatial and temporal soil moisture resource partitioning by trees and grasses in a temperate savanna, Arizona, USA. *Oecologia* 112:156–164.

Weltzin, J. F., and G. R. McPherson, editors. 2003. *Changing precipitation regimes and terrestrial ecosystems. A North American perspective*. University of Arizona Press, Tucson, AZ.

Weltzin, J. F., S. Archer, and R. K. Heitschmidt. 1997. Small-mammal regulation of vegetation structure in a temperate savanna. *Ecology* 78:751–763.

Weltzin, J. F., M. E. Loik, S. Schwinning, D. G. Williams, P. Fay, B. Haddad, J. Harte, T. E. Huxman, A. K. Knapp, G. Lin, W. T. Pockman, M. R. Swaw, E. E. Small, M. D. Smith, S. D. Smith, D. T. Tissue, and J. C. Zak. 2003. Assessing the response of terrestrial ecosystems to potential changes in precipitation. *Bioscience* 53:941–952.

West, N. E. 1990. Structure and function of microphytic soil crusts in wildland ecosystems of arid to semiarid regions. *Advances in Ecological Research* 20:179–223.

West, N. E., and J. O. Klemmedson. 1978. Structural distribution of nitrogen in desert ecosystems. Pages 1–16 in N. E. West and J. Skujins, editors, *Nitrogen in desert ecosystems*. Dowden, Hutchinson and Ross, Stroudsburg, PA.

West, N. E., and J. Skujms. 1977. The nitrogen cycle in North American cold-winter semi-desert ecosystems. *Oecologia Plantarum* 12:45–53.

West, N. E., K. H. Rea, and R. O. Harniss. 1979. Plant demographic studies in sagebrush-grass communities of southeastern Idaho. *Ecology* 60:376–388.

Westerman, R. L., and T. C. Tucker. 1979. In situ transformations of nitrogen-15 labeled materials in Sonoran Desert soils. *Soil Science Society of America Journal* 43:95–100.

Westoby, M. 1980. Elements of a theory of vegetation dynamics in arid rangelands. *Israeli Journal of Botany* 28:169–194.

Wetselaar, R. 1968. Soil organic nitrogen mineralization as affected by low soil water potentials. *Plant and Soil* 29:9–17.

Wezel, A., J. L. Rajor, and C. Herbrig. 2000. Influence of shrubs on soil characteristics and their function in Sahelian agro-ecosystems in semiarid Niger. *Journal of Arid Environments* 44:383–398.

Whicker, J. J., D. D. Breshears, P. J. Wasiolek, T. B. Kirchner, R. A. Tavani, D. A. Schoep, and J. C. Rodgers. 2002. Temporal and spatial variation of episodic wind erosion in unburned and burned semiarid shrubland. *Journal of Environmental Quality* 31:599–612.

Whisenant, S. G. 1996. Initiating autogenic restoration on degraded arid lands. *Proceedings of the Fifth International Rangeland Congress*. Denver, CO.

Whisenant, S. G. 1999. *Repairing damaged wildlands: A process-oriented, landscape-sale approach*. Cambridge University Press, Cambridge.

White, P. S., and S. T. A. Pickett. 1985. Natural disturbance and patch dynamics: an introduction. Pages 3–9 in S. T. A. Pickett and P. S. White, editors, *The ecology of natural disturbance and patch dynamics*. Academic Press, San Diego, CA.

Whitford, W. G. 1974. *Jornada validation site report*. US/IBP Desert Biome Research Memorandum 74-4. Ecology Center, Utah State University, Logan, UT.

Whitford, W. G. 1976. Temporal fluctuations in density and diversity of desert rodent populations. *Journal of Mammalogy* 57:351–369.

Whitford, W. G. 1978. Structure and seasonal activity of Chihuahuan Desert ant communities. *Insectes Sociaux* 25:79–88.

Whitford, W. G. 1979. Foraging in seed harvesting ants, *Pogonomyrmex* spp. *Ecology* 59: 185–189.

Whitford, W. G. 1989. Abiotic controls on the functional structure of soil food webs. *Biology and Fertility of Soils* 8:1–6.

Whitford, W. G. 1991. Subterranean termites and long-term productivity of desert rangelands. *Sociobiology* 19:235–243.

Whitford, W. G. 1993. Animal feedbacks in desertification: An overview. *Revista Chilena de Historia Natural* 66:243–251.

Whitford, W. G. 1997. Desertification and animal biodiversity in the desert grasslands of North America. *Journal of Arid Environments* 37:709–720.

Whitford, W. G. 2000. Keystone arthropods as webmasters in desert ecosystems. Pages 25–41 in D. C. Coleman and P. F. Hendrix, *Invertebrates as webmasters in ecosystems*. CABI, Wallingford, Oxon, U.K.

Whitford, W. G. 2002. *Ecology of desert systems*. Academic Press, San Diego, CA.

Whitford, W. G., and M. Bryant. 1979. Behavior of a predator and its prey: The horned lizard (*Phrynosoma cornutum*) and harvester ants (*Pogonomyrmex* spp.). *Ecology* 60: 686–694.

Whitford, W. G., and F. M. Creusere. 1977. Seasonal and yearly fluctuations in Chihuahuan Desert lizard communities. *Herpetologica* 33:54–65.

Whitford, W. G., and R. DiMarco. 1995. Variability in soils and vegetation associated with harvester ant (*Pogonomyrmex rugosus*) nests on a Chihuahuan Desert watershed. *Biology and Fertility of Soils* 20:169–173.

Whitford, W. G., and G. Ettershank. 1975. Factors affecting foraging activity in Chihuahuan Desert harvester ants. *Environmental Entomology* 4:689–696.

Whitford, W. G., and F. R. Kay. 1999. Biopedturbation by mammals in deserts: A review. *Journal of Arid Environments* 41:203–230.

Whitford, W. G., and K. H. Meltzer. 1976. Changes in O_2 consumption, body water and lipid in burrowed desert juvenile anurans. *Herpetologica* 32:23–25.

Whitford, W. G., and L. W. Parker. 1989. Contributions of soil fauna to decomposition and mineralization processes in semiarid and arid ecosystems. *Arid Soil Research and Rehabilitation* 3:199–215.

Whitford, W. G., P. Johnson, and J. Ramirez. 1976. Comparative ecology of the harvester ants *Pogonomyrmex barbatus* (F. Smith) and *Pogonomyrmex rugosus* (Emery). *Insectes Sociaux* 23:117–132.

Whitford, W. G., D. J. Depree, and R. K. Johnson Jr. 1978a. The effects of twig girdlers (*Cerambycidae*) and node borers (*Bostrichidae*) on primary production in mesquite (*Prospis glandulosa*). *Journal of Arid Environments* 1:345–350.

Whitford, W. G., S. Dick-Peddie, D. Walters, and J. A. Ludwig. 1978b. Effects of shrub defoliation on grass cover and rodent species in a Chihuahuan Desert ecosystem. *Journal of Arid Environments* 1:237–242.

Whitford, W. G., E. DePree, and P. Johnson. 1980a. Foraging ecology of two Chihuahuan

Desert ant species: *Novomessor cockerelli* and *Novomessor albisetosus*. *Insectes Sociaux* 27:148–156.

Whitford, W. G., M. Bryant, G. Ettershank, J. Ettershank, and P. F. Santos. 1980b. Surface litter breakdown in a Chihuahuan Desert ecosystem. *Pedobiologia* 20:243–245.

Whitford, W. G., D. J. DePree, P. Hamilton, and G. Ettershank. 1981a. Foraging ecology of seed-harvesting ants, *Pheidole* spp., in a Chihuahuan Desert ecosystem. *American Midland Naturalist* 105:159–167.

Whitford, W. G., R. Repass, L. W. Parker, and N. Z. Elkins. 1981b. Effects of initial litter accumulation and climate on litter disappearance in a desert ecosystem. *American Midland Naturalist* 108:105–110.

Whitford, W. G., Y. Steinberger, and G. Ettershank. 1982. Contributions of subterranean termites to the economy of Chihuahuan Desert ecosystems. *Oecologia* 55:298–302.

Whitford, W. G., Y. Steinberger, W. P. MacKay, L. W. Parker, D. Freckman, J. A. Wallwork, and D. Weems. 1986. Rainfall and decomposition in the Chihuahuan Desert. *Oecologia* 68:512–515.

Whitford, W. G., K. Stinnett, and J. Anderson. 1988a. Decomposition of roots in a Chihuahuan Desert ecosystem. *Oecologia* 75:8–11.

Whitford, W. G., K. Stinnett, and Y. Steinberger. 1988b. Effects of rainfall supplementation on microarthropods on decomposing roots in the Chihuahuan Desert. *Pedobiologia* 31:147–155.

Whitford, W. G., G. S. Forbes, and G.I.H. Kerley. 1995. Diversity and spatial variability and functional roles of invertebrates in desert grassland ecosystems. Pages 152–195 in M. P. McClaran and T. R. VanDevender, editors, *The desert grassland.* University of Arizona Press, Tucson, AZ.

Whitford, W. G., J. Anderson, and P. M. Rice. 1997. Stemflow contributions to the "fertile island" effect in creosotebush, *Larrea tridentata. Journal of Arid Environments* 35: 451–457.

Whitford, W. G., J. Van Zee, M. H. Nash, W. E. Smith, and J. E. Herrick. 1999. Ants as indicators of exposure to environmental stressors in North American desert grasslands. *Environmental Monitoring and Assessment* 54:143–171.

Whitford, W. G., R. Neilson, and A. de Soyza. 2001. Establishment and effects of establishment of creosotebush, *Larrea tridentata*, on a Chihuahuan Desert watershed. *Journal of Arid Environments* 47:1–10.

Whittaker, R. H. 1975. *Communities and ecosystems*, 2nd edition. MacMillan, New York.

Wiegand, T., H. A. Snyman, K. Kellner, and J. M. Paruelo. 2004. Do grasslands have a memory: Modeling phytomass production of a semiarid South African grassland. *Ecosystems* 7:243–258.

Wiens, J. A. 1999. The science and practice of landscape ecology. Pages 371–383 in J. M. Klopatek and R. H. Gardner, editors, *Landscape ecological analysis: Issues and applications.* Springer-Verlag, New York.

Wierenga, P. J., J. M. H. Hendrickx, M. H. Nash, J. Ludwig, and L. A. Daugherty. 1987. Variation of soil and vegetation with distance along a transect in the Chihuahuan Desert. *Journal of Arid Environments* 13:53–63.

Wierenga, P. J., R. G. Hills, and D. B. Hudson. 1991. The Las Cruces trench site: Characteristics, experimental results, and one-dimensional flow predictions. *Water Resources Research* 27:2695–2705.

Williams, J., and M. Bonnell. 1988. Influence of scale of measurement on the spatial and temporal variability of the Philip infiltration parameters: An experimental study in an Australian savannah woodland. *Journal of Hydrology* 104:35–51.

Williams, J., J. Dobrowolski, N. West, and D. Gillette. 1995. Microphytic crust influence on wind erosion. *Transactions of American Society of Agricultural Engineers* 38:131–137.

Windberg, L. A., F. F. Knowlton, S. M. Ebbert and B. T. Kelly. 1997. Aspects of coyote predation on Angora goats. *Journal of Range Management* 50:226-230.

Wisdom, W. A., and W. G. Whitford. 1981. Effects of vegetation change on ant communities of arid rangelands. *Environmental Entomology* 10:893–897.

Wolberg, D. L., R. P. Lozinsky, and A. P. Hunt. 1986. Late Cretaceous (Maastrichtian-Lancian) vertebrate paleontology of the McRae Formation, Elephant Butte area, Sierra County, New Mexico. Pages 227–234 in *37th Annual Field Conference guidebook.* New Mexico Geological Society, Socorro, NM.

Wondzell, S. M., and J. A. Ludwig. 1995. Community dynamics of desert grasslands: Influences of climate, landforms, and soils. *Journal of Vegetation Science* 6:377–390.

Wondzell, S. M., G. L. Cunningham, and D. Bachelet. 1987. A hierarchical classification of landforms: Some implications for understanding local and regional vegetation dynamics. Pages 15–23 in E. F. Aldon, C. Gonzales Vincent, and W. N. Mori, editors, *Strategies for Classification and management of Native Vegetation for Food Production in Arid Zones.* U.S. Forest Service Rocky Mountain Forest and Range Experiment Station, General Technical Report RM-150.

Wondzell, S. M., G. L. Cunningham, and D. Bachelet. 1996. Relationships between landforms, geomorphic processes, and plant communities on a watershed in the northern Chihuahuan Desert. *Landscape Ecology* 11:351–362.

Wong, S. C., I. R. Cowan, and G. D. Farquhar. 1979. Stomatal conductance correlates with photosynthetic capacity. *Nature* 282:424–426.

Wood, J. C., M. K. Wood, and J. M. Tromble. 1987. Important factors influencing water infiltration and sediment production on and lands in New Mexico. *Journal of Arid Environments* 12:111–118.

Wood, J. E. 1969. *Rodent populations and their impact on desert rangelands.* Pages 1–37 in New Mexico State University, Agricultural Experiment Station Bulletin 555.

Wood, M. K., E. L. Garcia, and J. M. Tromble. 1991. Runoff and erosion following mechanical and chemical control of creosotebush (*Larrea tridentata*). *Weed Technology* 5:48–53.

Woodhouse, C. A., and J. T. Overpeck. 1998. 2000 years of drought variability in the Central United States. *American Meteorological Society Bulletin* 79:2693–2714.

Wooding, R., E. Bradley, and J. Marshall. 1973. A low-speed wind tunnel for model studies in micrometeorology. *Boundary-Layer Meteorology* 5:285–308.

Woodmansee, R. G. 1988. Ecosystem processes and global change. Pages 11–27 in T. Rosswall, R. G. Woodmansee, and P. G. Risser, editors, *Spatial and temporal variability of biospheric and geospheric processes.* Wiley, New York.

Wooton, E. O. 1908. *The range problem in New Mexico.* New Mexico Agricultural Experiment Station Bulletin 66.

Wooton, E. O. 1915. *Factors affecting range management in New Mexico.* U.S. Department of Agriculture Bulletin 211.

Wotawa, G., and M. Trainer. 2000. The influence of Canadian forest fires on pollutant concentrations in the United States. *Science* 288:324–328.

Wright, H. A. 1980. *The role and use of fire in the semidesert grass-shrub type.* USDA Forest Service General Technical Report INT-85.

Wright, N., and L. J. Streetman. 1958. Past performance and future potential of black grama for southwestern ranges. *Journal of Range Management* 11:207–214.

Wright, R. A. 1982. Aspects of desertification in *Prosopis* dunelands of southern New Mexico, USA. *Journal of Arid Environments* 5:277–284.

Wright, R. A., and J. H. Honea. 1986. Aspects of desertification in southern New Mexico, USA: Soil properties of a mesquite duneland and a former grassland. *Journal of Arid Environments* 11:139–145.

Wright, R. A., and G. M. Van Dyne. 1976. Environmental factors influencing semidesert grassland perennial grass demography. *Southwestern Naturalist* 21:259–274.

Yaalon, D. H. 1965. Downward movement and distribution of anions in soil profiles with limited wetting. Pages 157–164 in E. D. Hallsworth and D. V. Crawford, editors, *Experimental Pedology*. Butterworth, London.

Yamaguchi, Y., A. B. Kahle, H. Tsu, T. Kawakami, and M. Pniel. 1998. Overview of advanced spaceborne thermal emission and reflection radiometer (ASTER), *IEEE Transactions on Geoscience and Remote Sensing 36:1062-1071.*

Yang, C., J. H. Everitt, C. Mao, and M. R. Davis. 2001. An airborne hyperspectral imaging system for agricultural and natural resources applications. *Proceedings of the 18th Biennial Workshop on Color Photography and Videography in Resource Assessment*, Society for Photogrammetry and Remote Sensing, Amherst, MA.

Yao, J., D. P. C. Peters, R. P. Gibbens, and K. M. Havstad. 2002a. Response of perennial grasses to precipitation in the northern Chihuahuan Desert: Implications for grassland restoration. *Bulletin of the Ecological Society of America* 83:302.

Yao, J., D. P. C. Peters, K. M. Havstad, R. P. Gibbens, and J. E. Herrick. 2002b. *Spatial variation in shrub invasion and loss of perennial grasses in the Chihuahuan Desert: A multi-scale approach.* International Association for Landscape Ecology, U.S. Regional Association, 17th Annual Meeting, Lincoln, NE.

Yates, T., J. N. Mills, C. A. Parmenter, T. G. Ksiazek, R. R. Parmenter, J. R. Vande Castle, C. H. Calisher, S. T. Nichol, K. D. Abbott, J. C. Young, M. L. Morrison, B. J. Beaty, J. L. Dunnum, R. J. Baker, J. Salazar-Bravo, and C. J. Peters. 2002. The ecology and evolutionary history of an emergent disease: Hantavirus pulmonary syndrome. *BioScience* 52:989-998.

York, J. C., and W. A. Dick-Peddie. 1969. Vegetation changes in southern New Mexico during the past hundred years. Pages 157–166 in W. G. McGinnies and B. J. Goldman, editors, *Arid lands in perspective*. University of Arizona Press, Tucson, AZ.

Yoshie, F. 1986. Intercellular CO_2 concentration and water-use efficiency of temperate plants with different life-forms and from different microhabitats. *Oecologia* 68:370–374.

Young, J. A., R. A. Evans, and R. E. Eckert Jr. 1981. Environmental quality and the use of herbicides on *Artemesia*/grasslands of the US intermountain area. *Agriculture and the Environment* 6:53–61.

Zaady, E., P. M. Groffman, and M. Shachak. 1996. Litter as a regulator of N and C dynamics in macrophytic patches in Negev Desert soils. *Soil Biology and Biochemistry* 28:39–46.

Zimmer, K. J. 1993. *Spatial and temporal variation in the breeding and foraging ecology of black-throated sparrows*. Ph.D. dissertation. New Mexico State University, Las Cruces, NM.

Zobisch, M. A. 1993. Erosion susceptibility and soil loss on grazing lands in some semiarid and subhumid locations of eastern Kenya. *Journal of Soil and Water Conservation* 48:445–448.

Index

Page numbers in italics are tables or figures; those in bold are photographs or maps.